MICROPROGRAMMED STATE MACHINE DESIGN

Michel A. Lynch, Ph.D.

Department of Electrical Engineering
University of Florida
Gainesville, Florida

CRC Press
Taylor & Francis Group
Boca Raton   London   New York

CRC Press is an imprint of the
Taylor & Francis Group, an **informa** business

CRC Press
Taylor & Francis Group
6000 Broken Sound Parkway NW, Suite 300
Boca Raton, FL 33487-2742

© 1993 by Taylor & Francis Group, LLC
CRC Press is an imprint of Taylor & Francis Group, an Informa business

No claim to original U.S. Government works

ISBN 13: 978-0-8493-4464-0 (hbk)

Visit the Taylor & Francis Web site at
http://www.taylorandfrancis.com

and the CRC Press Web site at
http://www.crcpress.com

Library of Congress Cataloging-in-Publication Data

Lynch, Michel A.
    Microprogrammed state machine design / Michel A. Lynch
       p.   cm.
    Developed for use in the undergraduate Digital computer
architecture course at the University of Florida.
    Includes bibliographical references and index.
    ISBN 0-8493-4464-6
    1. Microprogramming. 2. Electronic digital computers—Design and
construction. 3. Machine design. I. Title.
QA76.635.L95 1993
005.6—dc20

                                              92-35287
                                            CIP

Library of Congress Card Number 92-32641

PREFACE

This book is for students of computer architecture and those who wish to design their own solutions to data processing problems both for commercial and research purposes. It is intended to follow an introductory course in digital design although the elements of such a course are outlined briefly in the early chapters.

The first six chapters form an introduction to the tools and elements of design that are used in microprogrammed state machines. Chapter 1 establishes the context of the book in the overall area of producing machine-assisted solutions to engineering problems. The strengths and weaknesses of a solution are inherent in the choice of whether quantity and/or time are made discrete. Quantity is forced to be discrete in a digital solution by the fixed length of the storage element that contains it. The application areas of the digital solution is currently expanding at a rapid pace due to rapid growth in the technology of silicon processing. The analog realm still has areas of application today including interfacing digital systems to the "real" world. The student must be aware of progress in various associated areas that may again push an analog, or other, solution to dominate areas that are currently commanded by digital solutions. He must be able to transfer his knowledge when needed without having to "start over again".

Chapter 2 reviews number systems with the intent of showing that the choice of such systems can have a beneficial effect on hardware design. The best hardware in a solution is that which does not need to be designed or fabricated. It is easy to test and doesn't fail in the field. It is in this area that number systems and algorithms have their greatest effect.

Chapters 3 and 4 introduce the hardware building blocks that will be used in the designs and examples that follow. An effort has been made to show that the use intended for a component controls the placement of the elements that make it up. One frequently overlooks the most important element in the dazzle of transistors. This element is the wire or data path. It is the interconnection of transistors that creates the building blocks. It is these building blocks with which we begin our study of computational elements in Chapter 3. We finish the chapter with some major architectural elements that enable the modern designer to seriously consider "dedicated" solutions to engineering problems in preference to general purpose solutions, i.e., computers shared by many users. The microprogrammed controllers discussed in Chapter 4 are the other half of such a solution; i.e., the skills needed to produce a dedicated solution are similar to those that are used in the general purpose solution.

Chapter 5 introduces assemblers for use in preparing the control program. The question of software development is important in the context of this subject since a solution must be reached with as little waste of time and effort as possible. The field of computer programming has advanced a reasonable distance over the years since many people have been active in producing many solutions. The method of attack using structured programming principles, editors, high-level languages, and simulators has, in many cases, resulted in reasonably efficient solutions produced at reasonable costs. Hardware design has acquired similar tools in the form of computer-based schematic capture, logic description, and simulation programs. We introduce assemblers in this book since they form the most hardware-intimate and simplest-to-modify method to deal with our solution. The purpose of introducing three assemblers is to acquaint the student, using a standard microprogramming meta-assembler, with the needs of the problem of providing software support at this level. The second assembler, typical of a fully developed tool for a host computer, is intended to show how a commonly available tool is adapted to serve as a first- or second-level assembler for a machine that is under design. The third assembler, being generally available in its source-code form, allows the user to further customize it for his purposes. The author has not introduced a high-level language as a tool for the target system because of the space available. A "shareware" C compiler called Small C by J. Hendrix is available. The author's students have customized it in various ways to adapt it for use with systems under design. One version of adaptation involved converting the code generation section to a "table driven" form so that the compiler can be used to support any solution for which a table can be generated.

The last chapter in the section on tools and components, Chapter 6, deals with simulation. The author has taken the view that the student should understand how a simulator works by writing his own. While there are many good simulation tools that the student will encounter in his professional life, many of these are very complex and require a long "learning curve" before they are useful. If the student learns what to expect from a simulator by writing simple ones, then he will be better prepared to tackle the professional tools; i.e., he will know something about what to expect in terms of capabilities and be less prone to expecting the unattainable.

The core of the book is found in Chapters 7 and 8 where several examples of first- and second-level microprogrammed state machines are given. The author has tried to present several algorithms and methods for their simplification before realization in hardware. Many designers like to pick the components and then fit the problem to them. While some of this is important, in this day and age one can frequently design custom parts for a particular application and implement them as programmable logic devices, gate arrays, or custom logic. A top-down approach is encouraged that examines the algorithm mathematically to exploit simplifications resulting from the proper choice of representation and application of algebraic manipulation. This step is followed by a cycle of design and simulation where the number of hardware components is reduced, hopefully, and the number of steps in the execution sequence increases to the limit allowed by the problem specification. A cycle of hardware design and simulation should follow, but is beyond the scope of this book as is the actual fabrication stage. If the design sequence has been followed thoroughly, then the number of "surprises" encountered at each succeeding stage should be relatively few and also "fixable".

Chapter 9 serves to wrap up several subjects introduced in earlier chapters. The cycle of design and simulation steps is illustrated with an example encountered in implementing the simple computer during simulation. It is intended to bring home in an intimate way through experience the types and importance of things that can be learned through simulation. The student is discouraged from producing only one simple-minded solution to an assignment. The hardware test and verification department does not appreciate such efforts when they must somehow make such crude designs work as they come off the production line. The time devoted to such "heroic" rescue efforts should be employed in the design stage. If the student realizes that his first solution is just that, the beginning of the design process, then an important goal of this book will be realized.

Chapters 8 and 9 contain most of the "computer architecture" material discussed in the book. By showing a simplified microprogrammed ASM implementation of a Complex Instruction Set Computer (CISC) in Chapter 8, the opportunity is created to discuss the important sequences that *make* a computer. The speed penalty incurred by the fetch-execute cycle in the second-level machine is considered, and methods to overcome it are presented. These continue in Chapter 9 where instruction and data pipelines and caches are introduced. A contrasting design philosophy, the Reduced Instruction Set Computer (RISC), is also shown. Two specialized application areas, digital signal processing and video display processing, are introduced with examples in Chapter 7. Further discussion and some integrated solutions are given in Chapter 9.

A Bibliography is presented that contains a list of books and papers that the author has used in the course of preparing this book. It is given both as a reference for certain points in the text and as a list of books that the reader can use for further study.

Another goal of the book is to encourage the design of specific solutions to problems that would currently, or in the near future, be addressed using general purpose computers; in particular, supercomputers. A practical, dedicated solution is normally built as an adjunct to a general purpose machine where the computer manipulates data and handles other housekeeping tasks. In this approach, design time is spent on creating hardware data paths as well as software. Current programming solutions use the designer's own custom software running on someone else's hardware paths, e.g., on a supercomputer. In the past, the programming solution was used since the building blocks were so large and costly. Today, with a single floating point unit available in an integrated circuit for an affordable price, the author hopes that many will decide that they can adopt a combination hardware/software solution to their problems.

The book is organized to be read in the order of its chapters, i.e., starting at the beginning. If it is used in course work, the demands of running a laboratory may require that some reordering occur. Chapter 7, with its progression of examples, can form the basis of a laboratory sequence. This sequence can then be followed by the second-level machine example in Chapter 8. This means that if one wishes to begin the laboratory exercises with the ROM/Latch example in Section 7.2 then there is some groundwork that must be laid quickly. Unfortunately, the order of the chapters cannot be followed. To perform any of the ROM/Latch problems, Chapters 1 and 5 will be needed immediately. Chapter 1 should be read for perspective on the whole field of problems and then a review of the algorithmic state machine (ASM) and the ASM diagram undertaken in Section 1.3. The microprogram assembler fundamentals should be studied in Sections 5.1 through 5.3. If a real microprogram assembler is available, then Sections 5.4 through 5.6 will serve as an introduction to the assembler-specific literature. Most of the modern microprogram assemblers can deal directly with the syntax shown in Sections 5.4 through 5.6 since the Microtec assembler was one of the first generally available. The modern assemblers add macro features and other syntax flexibility that make them easier to use.

If a copy of A68K by Charlie Gibbs can be obtained, then the student has the advantage of using a copy on his own machine instead of sharing limited licenses with other students in the class. The absolute assembly nature of the program makes entering data into target machines and EPROM programmers much easier. Study of Section 5.15 (which includes examples for the ROM/Latch problems) and Appendix B in Chapter 5 should be all that is necessary to begin work. If neither assembler is available, then the student may use any macroassembler available on local machines, including the DEC VAX, IBM mainframe, IBM PC, and others. The methods used in Sections 5.7 through 5.14 along with the hints given in Section 5.15 will allow the student to adapt any macroassembler for use at the first- or second-level of his machine. Complications arise because none of these assemblers can directly produce absolute output so the step of recovering this information through a program like VAXFMT (discussed in Section 5.14) is necessary.

Once the laboratory program starts, the instructor can use class time to introduce the complex logic elements in Chapter 3 and the microprogrammed controllers in Chapter 4.

The author has found that the introduction of simulation in the form of program statements that describe the elements in conjunction with the logic diagram allows the student to work into simulation concepts gradually. Therefore, each time a structure is introduced in Chapter 3, a short program that describes it can be appropriated from Chapter 6. A series of homework assignments are possible, while studying the elements in Chapters 3 and 4, that encourage the student to write simulators for progressively more complex machines using the principles in Chapter 6. This is also one way to show the construction and use of a hardware description language.

By the time the first five chapters have been covered, the laboratory is ready to include more complex elements. Since the first labs concentrated on the controller, i.e., the ASM contained very little architecture, the logical progression is to replace the ROM/Latch controller with a modern one. At this point the lab equipment can change in one of two ways. The expensive path is to purchase evaluation boards from the vendors of the complex parts. Texas Instruments, Inc. (TI) and Advanced Micro Devices (AMD) make such boards for educational use. Each company has produced several boards as their parts have evolved over the years. To keep the lab experience from concentrating solely on programming, it is desirable to acquire boards that allow the attachment of user-designed modules. The second path that lab equipment design can take is to use locally produced evaluation boards. The machines shown in Figures 7-3, 7-5, and 7-6 can each be placed on a prototyping board that uses push-in component matrices. The number of wires involved will probably remain connected during the course of the lab. The student can perform the fabrication in one lab period. More complex structures may be fabricated using wrapped-wire techniques or printed circuit (PC) boards. Construction will generally take longer than the time available in a typical lab period, so some assignment method of longer duration must be employed. The second method is probably the most rewarding since the student has access to all of the lines in the device without having to work around the expediencies of another designer.

Due to its wiring complexity, the pedagogical machine in Figure 7-7 can not be fabricated on prototyping boards but *must* be made using wrapped-wire or PC boards. A very flexible version may be made that includes the Register Arithmetic/Logic Unit (RALU) and memory additions as well as the integer Multiplier/Accumulator (MAC) outlined in Figure 7-13. Such designs can support the examples shown as well as the addition of Analog-to-Digital (A/D) and Digital-to-Analog (D/A) converters for interfacing to the real world. Several digital signal processing algorithms can be studied with these additions.

It is probably not practical to build boards in this course that include the Floating Point Units (FPUs) since they are such large integrated circuits. Their use is possible by obtaining the manufacturers' evaluation boards or building boards in the context of some other work, e.g., masters or senior design projects. If they can be obtained and used, then the FPU's capabilities will be easier to demonstrate.

The author has configured his laboratory workstations to contain the following equipment:

1. IBM PC clone
2. Digital voltmeter
3. Power supply
4. Dual channel, dual-delayed trigger oscilloscope
5. Logic state analyzer (32 channels)
6. Prototyping board

Any of the suggested methods can be accommodated in this workstation. The PC serves both to control any associated boards and perform editing and assembling of source programs. Another important use for it is to communicate with the student by allowing him to bring work from other computers (including his own) into the laboratory. An EPROM programmer is shared by all in the laboratory. This choice was made since some of the more exotic devices (e.g., the complete microprogrammed controllers from Altera along with other PLDs, like PALs) require programmers that are too expensive to place at each workstation.

The work in Chapter 8 has been supported in the author's lab by simulators. These can be used in the laboratory or at the student's own work place, depending on the simulators used. The author and some students wrote an integrated circuit level simulator that allows the student to perform first- and second-level programming just as he would on a hardware version. Extensions have been made using the material in Chapter 6 to allow the student to build new structures using the "integrated circuits" included in the computer example. Extensions suggested in Chapter 9 may be studied in this environment. While the locally written behavioral simulation modules have the major advantage that the student can use them outside of the lab, the professional environment that he will see will be based on a workstation running advanced description and simulation software. The lines suggested in both Chapters 7 and 8 may be followed on such systems with a realism that includes everything except solder. The cost in time of producing such a "board" is also very high; thus, these designs must be made outside of class or by the instructor.

The author would like to extend his appreciation to many people that have helped this book come into being. The text material grew out of notes developed for use in the undergraduate "Digital Computer Architecture" course, EEL4713, at the University of Florida. I wish to thank the students in this course for the last several semesters for their patience and feedback, which has allowed evolution toward a better book to occur.

Several individuals have helped produce parts of this work and I wish to thank them for their support at this time. Messrs. Bruce Kleinman and Brian Miller wrote one of many outstanding simulations of the computer example in Chapter 8. The software has become the basis for several lab and homework assignments where later generations of students may do some "real" microprogramming. The clarity of their code and their adherence to high programming standards has allowed it to be run on all machines encountered by the author since it was written. It has served as a model for illustration in this book as well as the basis of the evolution to reusable modules discussed in Chapter 6.

The example discussed in Section 7.5 began as a project for a group of graduate students in our department in association with Messrs. Ron Drafz and Ansel Goldgar at Texas Instruments, Inc. The author performed the basic design following some studies on efficient addressing of data used in several calculation algorithms, including convolution, done by Dr. Eric Dowling. Other contributing members of the group were Drs. Michael Griffen and Michael Sousa. Fabrication of the boards and their testing and demonstration was led by Mr. Ahmad Ansari, as part of his Master's research, along with the help of several undergraduates. He is responsible for drawing Figures 7-20 through 7-27. The TI group was instrumental in obtaining the critical parts for us.

I would like to acknowledge the significant contributions of several people that, through the twisted paths of learning, led to the creation of this book. The trait that these people share is that of being a *teacher*, their professions notwithstanding. They are Messrs. Louis C. Foree, Dubert Dennis, and Robert Gorrell, and Drs. Colin A. Plint, C. C. Lin, Thomas D. Carr, and Alex G. Smith. I hope, someday, to be as good a teacher as they are. I wish to acknowledge the support of my wife and son in this endeavor. They provide the reasons for continuing such a project.

AUTHOR

Dr. Michel A. Lynch received his Ph.D. from the University of Florida in 1972. Because of his interests in physics and radio astronomy and in the acquisition and processing of data acquired from the natural world, he entered the emerging field of microprocessor-based computers. As a Senior Development Engineer with Biomega Corporation, a subsidiary of Physio-Control Corporation, he was active in the development of noninvasive blood pressure measurement algorithms and systems. The systems were optimized for use in the emergency medical field to support patient transport in helicopters and ambulances. He holds a patent for a successful system. The author is now a Lecturer in the Department of Electrical Engineering at the University of Florida in Gainesville, Florida. His interests include digital system design and the application of such systems in the fields of scientific computation and computer music.

TO MARY and DAVID

TABLE OF CONTENTS

TABLE OF FIGURES

Chapter 8

A MICROPROGRAMMED COMPUTER

Chapter 9

IMPROVEMENTS, VARIATIONS, AND CONCLUSION

MICROPROGRAMMED STATE MACHINE DESIGN

Chapter 1

INTRODUCTION

1.1. GOALS

The author has three goals for *Microprogrammed State Machine Design*. The first is to show methods by which microprogrammed Algorithmic State Machines (ASMs) can be designed and built to solve problems encountered in the sciences and engineering. The application of large-scale dedicated state machines has been made possible by the recent appearance of very complex combinatorial networks such as the floating point multiplier/accumulator. It is now possible to design very fast, dedicated solutions to problems that would have ordinarily been handled using standard programming techniques in conjunction with supercomputers. The cost in time and components incurred by this method is decreasing to and now approaches that of the conventional method, i.e., programming a supercomputer.

The second goal is associated with understanding the steps to implementing a solution to a problem so that the solution can be reached in an efficient manner. Tools must be created and used to support this goal. The important tools are built in software and run on a computer as part of a development system. Chapter 6 is dedicated to the discussion of simulation methods by which the designer can understand his problem before any effort is committed to fabrication. In addition to the various simulation programs, the designer also needs language processors for the system under development. Chapter 5 shows the development and use of Assembler language processors.

The third goal comes from the recognition that in the last generation we have experienced rapid growth in the application of computers to many of our daily problems. These machines are also ASMs; however, they have a second level of control. The current state of application is such that these machines are ubiquitous. Chapter 8 describes how one important group, those that are microprogrammed, works by introducing a very simple example in order not to obscure the subject with detail. The example is evolved in Chapter 9 to increase its speed by a judicious introduction of additional hardware. Comparisons are made with other important implementations including the Reduced Instruction Set Computer (RISC) and the Digital Signal Processor (DSP) to contrast the various methods by which designers have optimized their solutions for target applications.

1.2. THE PROBLEM

Figure 1-1 gives the appearance of solutions to problems that can be solved by the methods used in the sciences and engineering. Inputs $(X1,...,Xj)$ are applied to a "black box" containing data transformations. Outputs $(Z1,...,Zk)$, produced by the "box", represent the desired results. The outputs may also be reintroduced to the data transformation, as illustrated in the Z_i equation, in which case the system is said to have "feedback".

A simple example of a data transformation without feedback derived from the academic world involves an instructor's preparation of a student's average grade on an exam from the scores on individual questions. The data transformation Z1 is just the sum of all of the scores divided by the number of scores. This is an example of a type of digital filter having Finite Impulse Response (FIR) and equal (unity) coefficients. If one considers the system and its related data transformations to contain not only the student's single exam but all of

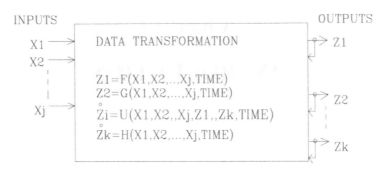

Figure 1-1: Solution to an analytic problem.

the exams he takes in a course, then one is able to identify a system with feedback where the data transformations are not as easily represented. It may be possible to write some data transformation that describes the student's reaction to the grade from the first exam and the additional study that it induces to produce a prediction of a score for the second exam.

The image given in Figure 1-1 does not imply how the inputs are generated, the representation of the data transformations, or the means by which the results are produced. We cannot deal with a problem given in such nebulous terms; thus the remainder of this chapter will proceed a step at a time to the point at which we can handle the problem using methods that will be set forth in the rest of this book.

1.3. THE SOLUTION

There are several related methods by which a problem may be represented. The first that presents itself to a mathematically inclined person is that of a system of equations. An equation is a sentence describing, in terms of mathematical operations, the combinations of the inputs that will be made to produce a desired result. A system of equations is needed if many outputs are to be produced. This is a very natural form for the representation of a solution to a problem because of the long association of mathematics with the sciences and other fields. While the equations are different, the "black box" appears similar to the user whether the equations represent the evolution of a star or a "spread sheet" to describe the operation of a business. The new outputs reflect the current inputs and any previous outputs that were included in the data transformation equations.

Another way to represent the solution of a problem is both older than equations and younger. This method involves the algorithm, a finite sequence of elementary steps that produces the desired result. If one were to actually perform the calculations indicated in a equation, then one would be following an algorithm. The equation with its operations and rules of combination contains the information necessary to produce the algorithm (sequence of steps) to solve it. There is information in the equation that is not in the algorithm, however. The form of the equation may give insight into the physical operation of the thing described. If two entities, say a pendulum and an oscillating electron, have equations of motion that have a similar form, then one is tempted to ascribe similar behavior to the forces causing the motion. This has been exploited in all areas of science and engineering such that great leaps in understanding in one area can be made by using analogies from other, better explored areas. The "form" of an algorithm does not contribute the same type of information.

It is first necessary to have a means to write a description of an algorithm. In this book we will present two methods, the ASM diagram and the Hardware Description Language (HDL). The ASM diagram does contain form and we exploit this form in implementing solutions in hardware but we will not make the grand analogies that have resulted in the progress that the sciences have made since the Renaissance.

1.3.1. Solution – Continuous or Discrete?

Now that we have decided to represent the problems we want to solve as equations or algorithms, we must make some more choices before we are ready to produce results. Most of our academic life we have used numbers to represent quantities. Even though the "nouns" of our equations were variables, we really meant that the result could be produced by doing the indicated operations on the variables replaced by a meaningful set of values as represented in some number system. How were the values obtained? Are they presented to us as a list of digits on a piece of paper?

At this point, we need to discriminate between numbers represented by a discrete representation and numbers given in a continuous representation. The numbers given by a combination of digits on a sheet of paper are inherently discrete values. The counting numbers are infinite in number and they may be mapped onto the complete range of values that a variable may have. However, when we prepare the sheet of paper with the sets of digits we will in one way or another limit the number of digits in the representation of a quantity. This is done for very good reason. If we are using a disciplined representation, we impart information about the accuracy to which a value is actually known by writing the appropriate number of digits. There are other representations of numbers that have similar limits, but we tend to ignore them when we use them.

Two representations will be considered here. They are an electrical current and a current of water. Both are ultimately discrete at the atomic level; however, they appear continuous at the level normally encountered by humans. Our point here is that we must discriminate between representing numbers by continuous or discrete quantities because the effects of the representation will be far reaching.

1.3.2. Solution – Continuous

Let us consider an example in which all inputs and outputs are represented by continuous quantities as is implied by our capability to represent the transformations as equations. Most of our calculus is based on the behavior of continuous functions and even a few discontinuities in the function can provide plenty of excitement.

Analog computation is the automatic (machine-assisted) calculation of a data transformation using variables represented by continuous quantities. This method of computation includes many mechanical devices (including the slide rule) as well as the electrical analog computer that represents the most advanced expression of this method. Certain problems, especially those expressible as differential equations, can be very simply and rapidly solved using electrical analog computation.

The method *as implemented* has two basic shortcomings. These involve absolute calibration and temporal drift of results. Within the limitations produced by these problems, many electrical analog computers are still active and perform their "data transformations" better than the corresponding digital implementation.

Two examples will suffice here. Consider the "linear regulated" power supply found in many computers as well as other equipment. The analog computer in these devices performs a feedback calculation that sets the output voltage at a constant value regardless of load current or input voltage variation. The absolute calibration and temporal drift are usually maintained within limits of 0.1% once initial calibration has been accomplished. This tolerance is appropriate for many applications.

Another example is the audio amplifier that performs a transformation related to scaling. If the typical preamplifier with its "bass" and "treble" frequency controls is included, then we see a very complex transformation, output = f(V_{in}, frequency), being performed using very few parts and with less cost and distortion than a corresponding digital unit. The problems of absolute calibration and temporal drift are handled by noting that sound does not contain a Direct Current (DC) component; therefore, if only the oscillatory part of the signal is passed to the next stage, the absolute calibration and drift will not be a problem. Note: the ear is not especially sensitive to gain drifts on the order of those easily attainable.

Portions of both of these examples are slowly yielding to the digital onslaught. In order to make the digital versions acceptable from a cost point of view, an order of magnitude increase in the functionality of the new version must be provided. The digital "switching" power supply has managed to squeeze out most of the analog circuitry by replacing all but the voltage comparison circuitry with switches (inherently digital devices). The maturity of this device has resulted in prices that are comparable to linear power supplies as well as a major improvement in efficiency. The digital version of the audio amplifier has not progressed so rapidly. The preamplifier has appeared with digital elements assisting the frequency compensation circuitry, but the cost is so high that other functions such as room ambience (reverberation) management must be included.

Analog computation is one method by which to implement the solution of a problem. It is currently in a slight eclipse but should not be overlooked by any designer since its areas of application are important and may change in its favor with continuing technological improvements. Remember that the digital domain must be reached by sampling the real world with analog-to-digital conversion devices; that is, a digital designer looks at the world through analog "eyes".

1.3.3. Solution – Discrete

Our interest in this book is to present methods by which problems can be solved using digital techniques. Let us consider the numbers given on a sheet of paper to be used as inputs to our problem and take seriously that they represent discrete samples of the real world. What are the consequences of a discrete (or digital) representation?

1.3.3.a. Discrete Data

The data itself is composed of numbers, each of which is composed of a string of digits. We are all familiar with the decimal number system in which the digits available are those from 0 to 9. Numbers are written using one or more digits in such a way that the position of a digit within the number imparts information on the size (weight) of that digit with respect to the whole number. The decimal number system is built such that each position has 10 times the weight of the position to the right of it. A marker, the decimal point, is used to indicate the position that has unit weight. Positions to the right of the decimal point *also* have a weight of 10 times the weight of the position to their right or, put another way, they have a weight of 0.1 times the weight of the position to their left. The weight of a position is a scale factor used to multiply the value of the digit in that position. The value of the number is then the sum of the scaled digits for all of the digits in the number as follows:

Value of 834.52

$$\text{Value} = (8 * 10^2) + (3 * 10^1) + (4 * 10^0) + (5 * 10^{-1}) + (2 * 10^{-2})$$

The first point to notice in using a discrete notation is that one has only so many positions to use in writing the value of each number. This problem was not obvious when we prepared the list of numbers above, but was present there nonetheless. The problem is caused by the way we plan to implement the solution. We must provide places to save numbers. These places will play the part of the variables in the equations shown previously but they will not generally have the flexibility with respect to digit count in a number that we had in the algebraic representation. We felt in that case that the number of digits in an input represented the accuracy with which we had measured its value. The number of digits retained in an intermediate calculation in the course of evaluating the equation was as many as "necessary", sometimes disregarding the meaning of significance. At no time was the number of digits fixed by means other than our own laziness or understanding.

From this point on, the number of digits in a number will be fixed by our implementation of the solution in hardware. We will need to make decisions that take into account the range of values available on inputs as well as their precision. Furthermore, the range and precision of intermediate calculations will have to be anticipated since a shortcoming here can completely dominate the results, especially in a system with feedback.

1.3.3.b. Systems Having Continuous Time

Analog computation of data transformations involves time in continuous form. One knows the state of the system at all times no matter how small the increment of time is made. One version of digital data transformation can be performed with the same view of time even though the data are represented discretely. If the data transformations can be represented by Boolean algebraic equations with no overall sampling implied (say, by an external clock), then the outputs will be available after some time delay and will follow the inputs and feedback. The time delay is caused by the same physical reason as that experienced in a corresponding analog implementation. If the computation is implemented electrically, then the time delay is the propagation time for an electrical input through the appropriate circuit path. Using either analog or digital circuitry in this way to implement the equations of transformation results in the outputs following changes in the inputs and feedback terms.

The time delays through a network of this type vary depending on the "path length" that a change in input (or output feedback term) must "travel" to reach the output in question. As the equations of transformation become more complex, the paths become longer and even more importantly vary in length depending on the input being examined. The output may actually assume an unplanned state between the instant an input changes and the effect of the change as it travels to the output. This is called a "glitch" in some quarters and can limit the application of this implementation method.

This point will be considered in more detail later, but it should be noted here that at some point the digital circuitry reaches a complexity where this and other problems make it unusable. The point at which this occurs is in some cases able to be pushed back by technological advances and very high-quality digital circuit design, but it usually lies far short of the complexity implied by the equations of transformation. A way around this problem is shown in Figure 1-2. The figure shows a group of Boolean function networks that implement a small part of one of the equations of transformation shown in Figure 1-1. Each network is of a size that can be readily implemented at the present time in our technological development. The output of each network is collected into a memory which is also a digital structure. At this point, we will not describe the memory any further than to say that it can faithfully remember the data given to it once it is told to do so by a signal applied from outside the system. This signal is called a clock (here SYSCLK). The clock signals a *point* in time when the inputs to the memory are *sampled* or sensed and remembered. The outputs of the memory may serve as inputs to more Boolean function networks which in turn send out-

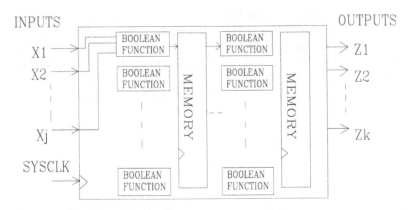

INPUTS   OUTPUTS

Figure 1-2: An implementable solution.

puts to another stage of memory. This structure is repeated until the entire data transformation equation has been evaluated.

There is a strong parallel between this digital implementation and the corresponding analog version with the exception of the introduction of sampling. Sampling was introduced to allow all of the outputs of the Boolean networks to settle to their steady-state values. The clock was then used to signal the memories to remember their inputs. The maximum clock rate is determined by the longest time required for the Boolean networks to settle. This seemingly innocuous solution to a strictly circuit-related problem will open a large box of problems as we shall see next.

1.3.3.c. Systems Having Discrete Time

Time becomes a discrete variable. Not only do we imply that time, like any other variable, is represented by a number of limited precision, but that the inputs will be sampled at discrete times. No information is available about the real world except at those points in time that we choose to sample. This has a major effect on the appearance of the results and their meaning.

For particularly simple circumstances, a rule exists that allows one to plan how and when to sample a system. The rule, called the Sampling or Nyquist Theorem, relates the sample rate to the behavior of the system being sampled. Basically, the system must be sampled more than twice as fast as the frequency of a sine wave that represents the highest frequency component of a spectrum of the variable being sampled. It is important to notice that the Sampling Theorem assumes that the sample rate is *constant*.

If a pendulum is oscillating sinusoidally about its equilibrium position with a frequency of 10 Hz, then to know the position of the pendulum bob within the meaning of the Sampling Theorem one must sample its position at a rate of *at least* 20 samples per second. If the pendulum were to slow down (because someone lengthened it slightly), one would still be able to detect the change and eventually calculate from the data the value of the new period. However, notice the effect of speeding up the pendulum by the same amount. The data would look the same as before and one would calculate a value that would show that the pendulum had changed speed but its period would be indeterminate.

This experience is an example of the stroboscopic effect that you have seen at the movies. It is inherent in all sampled systems no matter how the sampling was accomplished. Therefore, the best that the Sampling Theorem can tell us is that if we sample, we must do it at a constant rate and at a rate more than twice the maximum frequency that the variable we are measuring will *ever* change. This applies to buttons as well as sound, radar data, etc.

1.3.3.d. Repetition and Iteration

The digital implementation of a solution to our data transformation shown in Figure 1-2 contains a very large number of parts for any problem that we might actually want to solve. Furthermore the transformation equations are hardwired and therefore are not amenable to easy change. The structure shown in Figure 1-2 suggests a modification that will ultimately lead to a very flexible, simple design.

The memory blocks indicate the changes needed. If the Boolean network's outputs have indeed settled at the time they were clocked into the memories, then the time until the clock is applied to the next stage has no maximum. The Sampling Theorem places an effective maximum on it but the implementation does not. Furthermore as shown in the figure, the Boolean networks are each wired to solve one part of a particular equation only and are not inherently flexible. However, our equation is ultimately composed of a small set of operations and a collection of variables. We suspect from our math background that the more complex operations used in the equations can be performed using some of the basic operations repetitively. A modification to Figure 1-2 might proceed as follows.

1. Make the Boolean networks perform the basic mathematical operations

2. Let the first memory block supply its outputs back to the Boolean networks on its inputs such that each network receives appropriate incoming data as well as feedback terms

3. Discard the remaining networks and memory structures

We have produced a structure that we will call an *architecture*. An architecture contains combinatorial elements, memory, and data paths. We have made the Boolean networks more general purpose in nature. In order to perform the operations suggested by the original equations, we must provide a means to route data to the appropriate network and to select the appropriate basic operation for it to perform. The networks that are retained in our design can be identified in terms of the basic mathematical operations with names like "adder", "subtracter", and "multiplier". In many cases, what appeared as a fundamental operation in our equation (e.g., an integral) must be reduced to a sequence of steps (an algorithm) in terms of the operations that we are capable of implementing as Boolean networks.

The dividing line between "basic" and "complex" operations is established by circuit design skills and technology at any given point in time. This dividing line has been slowly moving toward the complex direction such that, at the present time, multiplication of either integer or floating point numbers can be performed by a combinatorial network in one integrated circuit. The integral is not implemented as a combinatorial network in the digital domain yet, but has been one in the analog domain from the beginning. We will explore the structure of these combinatorial networks and the architectures in which they are embedded in the following chapters.

Inasmuch as the Boolean networks in Figure 1-2 were evolved into networks implementing the basic mathematical operations, we must introduce a means to select the operation needed for the equation being evaluated. The equation could not be evaluated in a "flow-through" sense as was implied by the analog or combinatorial network implementation, but must be handled in a piece-wise manner using repeated applications of the same network on the intermediate results stored in memory. The complex operations must be decomposed into steps involving fundamental operations that can be implemented using Boolean networks. All of this decomposition implies that there is some sort of "fine-grained" control that must be applied to the system. The sample time shown earlier must therefore be further subdivided into fine increments such that all of the steps required can be performed before the next sample from the world is introduced. The same requirement concerning the settling time of the combinatorial networks is in force so that the maximum real world sampling rate is constrained by the settling time of the networks times the number of iterations required to evaluate the equation. We have traded a possibility for relatively high speed by using dedicated hardware for a small amount of general purpose hardware. This trade-off must always be evaluated for every project such that the resulting implementation performs as specified and is buildable.

1.4. THE ALGORITHMIC STATE MACHINE

A new element is required in our implementation by the introduction of "fine-grained" control, i.e., a control mechanism. Figure 1-3 shows a diagram of an ASM. The elements of our implementation from Figure 1-2 have been placed in the right-hand box of the ASM, the *architecture*. The structure on the left is the source of the "fine-grained" control signals and is called a *controller*. The basic building blocks of an ASM are combinatorial elements, memory, and data paths.

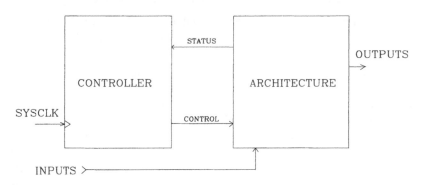

Figure 1-3: Structure of an algorithmic state machine, ASM.

An actual implementation of a controller is also an architecture since it contains the same elements. However, we single it out for further study since it is the source of the control signals that must present the right command at the right time to cause the desired equation to be evaluated in the general purpose architecture we have placed in the box on the right. The controller is also clocked, but generally at a rate much higher than the sample rate required by the Sampling Theorem for the problem at hand. This accommodates the many steps that must be done for each sample acquired from outside.

We have identified the complete structure in the figure as an ASM. It is a machine on which our transformation equations can be evaluated by using a finite sequence of steps (an algorithm). This structure will be discussed more thoroughly in the following chapters.

Associated with the ASM is a design method that is in a graphical form similar to a flowchart. It is illustrated in Figures 1-4 and 1-5. Since the machine is "resting" between each clock pulse, we imagine it to be *in a state*, represented by the rectangular boxes in the figure. The numbers on the upper right corners of each box are used to denote the state of the machine. They are derived from values of the state variables stored in the memory in the controller. The symbols Z1, Z2, etc. are the outputs of the ASM and correspond to those shown in Figure 1-1. The machines in the lower part of Figure 1-4 are called Moore machines since their outputs are functions only of the values of the state variables.

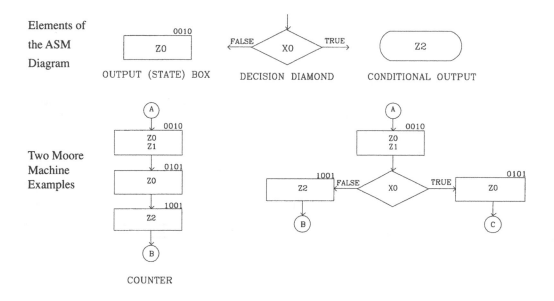

Figure 1-4: The ASM diagram elements and examples.

The symbols X0, X1, etc. in the ASM diagram in Figures 1-4 and 1-5 are inputs to the machines and correspond to the inputs in Figure 1-1. The diamond in the ASM diagram indicates that a question is being asked about the inputs. The arrows that leave the diamond in the right-hand part of Figure 1-5 are labeled with values of X2 and X1 (in that order). The system will move from the state 0100 to 0000 if both X2 and X1 are 1 but will move from state 0100 to 1001 if X2 is 0 and X1 is 1. This diagram implies that an ASM can make decisions on inputs.

The box with the rounded ends containing the outputs Z2 and Z3 in the right-hand part of Figure 1-5 indicates a *conditional output*. If an output is in a box representing a state (e.g., Z1 in state 0100 in the right-hand ASM diagram) then any time the system is in that state the output will be *On* or *True*. The conditional output is a method to show that an output is to be turned on only if a particular input combination is true, as well as that the system is in a particular state. In this example, Z2 and Z3 will be *On* if the system is in state 0100, *and* X2 is 1, *and* X1 is 0. The conditional output "happens" during the time occupied when the system is in the state "above" it in the ASM diagram. A system having outputs as a function of the state and the inputs is called a Mealy machine.

In summary the ASM and its associated ASM diagram will serve as a "thinking" structure. We will fit the logical structures that we design in the remainder of this book into such a form. The ASM diagram will be evolved into a form that is more appropriate to a "programmed" system, but the information it contains will be preserved.

The examples of the ASM given in the figures show very simple data transformations. The architecture being manipulated by the controller consists at most of a few NAND gates. Our goal is to control much more complex architectures for the purpose of evaluating equations of essentially unlimited complexity. In order to do this, we will increase the capabilities of the controller and examine various ways that the controller can be implemented. We will then settle on the microprogrammed controller since this allows us to easily implement very complex, flexible control sequences.

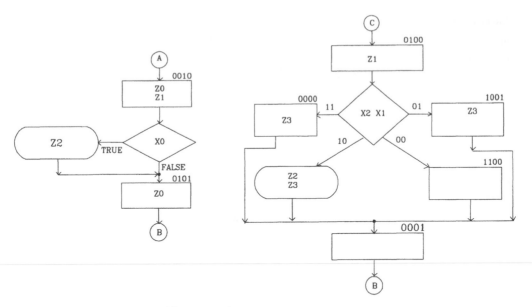

Figure 1-5: Two Mealy machine examples.

1.5. THE INSTRUCTION SET PROCESSOR

In the previous section we introduced the concept of the division of a complex operation into a sequence of basic operations. We also discussed a machine to implement such a structure using a controller and an architecture, the ASM. It has been noted that our goal is to handle systems of unlimited complexity and that this would have some effect on the means that we used to describe our intentions. Let us consider some of the means to cope with this level of complexity.

Our transformation equations are written in terms of operations being applied to variables. Complex operations can be performed by dividing them into sequences of basic operations. The basic operations that we can implement directly as combinatorial networks are the fundamental arithmetic and logical operations on numbers represented in one of several formats. We have given the controller the capability to generate the lengthy sequences of control signals required for the evaluation of our equations. The implication is that, by some method, the values of the control signals needed in the sequence are "built in", to the controller. If the sequence is truly "built in," then the system is no more flexible than the structure arrived at after evolving Figure 1-2; however, it is buildable.

While there are methods to place the sequences in the controller such that they can be changed, let us examine the problem first while we are still standing back and seeing the whole picture. Can our concept of the state machine itself contribute to the flexibility of the implementation?

The answer is in the affirmative if we introduce another layer of control. First assume that we have placed permanently in our controller the sequences necessary to make our architecture, with its general purpose but controllable networks, perform all of the basic operations that are needed to evaluate the transformation equations. Not only would we build in the sequences needed for the basic operations, but we could also place sequences for division, integration, or whatever seemed appropriate for the possibly limited hardware resources available in the machine.

For the purpose of our discussion, let us name each of these sequences. We could call the sequence that adds two numbers ADD, the one that subtracts two numbers SUB, etc. There would be a name for each sequence. If we connected the controller to a keyboard with switches labelled ADD, SUB, etc., then the controller could determine which operation we wanted to perform by sensing the keys as inputs. If we also provided keys for the input of data, we would have "invented" the calculator. Sitting outside the calculator (and the ASM that implements it), we don't see the fine-grained structure of each operation but we see only the operation itself. If our transformation equation involves only addition and subtraction or even multiplication and division, we see these as indivisible entities and not as the complex sequences that underlie them.

Our problem is now expressible in symbols (or steps) that correspond one-to-one to the symbols in our transformation equations. We have made a fundamental improvement in our capability to evaluate equations since we don't have to work with the fine-grained "detail" every time we evaluate an equation. If we left an integration operation out of our set of supported operations then we are faced with the detail sequence required for it.

One important step has happened here though; that involves the fact that we can't "see inside" the calculator so we must implement our integral using only the keys or operations visible to us. This makes our job simpler, usually but probably makes the actual evaluation as seen from inside the hardware much less efficient. We have traded efficiency (and execution time) for ease of use.

Now let us assume that we are handed a class roll containing all of the grades that a class has made on all quizzes, homework, and lab assignments for the semester. The data transformation equation required to determine the total scores for all of the students in the class might look something like this:

SCORE = 30% of Midterm Exam + 40% of Final Exam

+ 18% of Total Lab Grade + 9% of Total Homework Grade

We must punch this into our calculator once for each student for all students in the section. The important point to realize is that the algorithm for producing the score for each student is the same for *all* students. Can we get the state machine to help us by not requiring us to type in the algorithm for each student?

Let us assign a numeric code to each of the keys on the calculator. The value is obvious for the data entry keys, but we will also assign a code to the operation keys as well. The philosophy of this move is that if we can come up with a way to store the values of the data entry keys in memory, then we should be able to store similar quantities representing the operation keys in a different but similarly constructed memory. We will call the newly added memory "program memory". It will contain codes for each of the operation keys that we pushed in the course of evaluating the above equation for one student. We will need to add some functions to help the controller get data from the keyboard and to evaluate results.

Have you noticed the "=" key on your calculator? Think what it means in the context of this discussion. The controller will be given some new sequences also. One sequence is needed to tell it to record the next several keys pressed in its program memory (how about "PROGRAM"?). We need a key to tell the controller that the program is complete (how about "END"?). Another sequence is needed to tell the controller to execute the sequence of operations from its program memory (how about "GO"?).

Once the concept is grasped, we find that we want to add a key and then an "instruction" to REPEAT the program in program memory. This repetition can be made conditional on our holding down a key or maybe on the result of a calculation. We are well on our way to inventing a computer.

The device we have constructed in our mind is a major step above the calculator with which we began. The controller has been given a sequence with which it starts and ends all other sequences; i.e., it fetches the instruction codes. A memory is added that holds these sequences of codes. The sequences of codes are called a "program" and the program memory is logically part of the memory structure of the controller. Programs in the program memory are composed of sequences of instruction codes.

We have added a new level of control to our machine, an instruction set level. Our machine has become an Instruction Set Processor (ISP). We, as preparers of instruction sequences (programs), do not see the fine-grained sequences needed to implement the instructions that we issue. The level of graininess has increased in size.

Our thoughts can be composed of bigger words; hence, we can think bigger thoughts. The sophistication of our world view, relativity, the Big Bang, etc. is based on the conceptual language, mathematics, having a compact and powerful vocabulary. Imagine Maxwell's equations written in other than a differential or integral form. The problems we expect to evaluate with a computer will be solved more easily when the granules of evaluation (basic operations) are applicable to the problem and are as large as possible.

Figure 1-6 shows the overall structure of the most common version of an ISP, i.e., the form of computer attributed to von Neumann. The second-level instructions are stored in as-

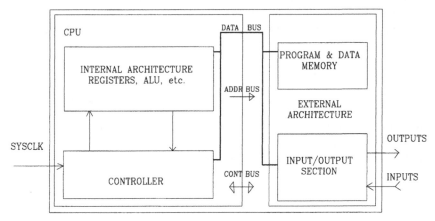

Figure 1-6: The von Neumann computer.

sociation with the data in the external architecture. Additional elements are added in this section to assist with data transfers between the outside world and the computer. The section called the Central Processing Unit (CPU) implements sequences that fetch instructions from the external architecture and interpret them. Both data and instructions travel over the same path (the data bus) to the CPU. Algorithms in the controller of the CPU separate the instruction and data stream after it enters the CPU. Notice that the CPU is itself an ASM.

A contrasting structure, called the Harvard architecture, maintains separate program and data memories. Each have separate physical paths into the CPU. This allows data to be transferred between the CPU and data memory at the same time that instructions are being fetched, resulting in a faster machine. This architecture was used infrequently in the past because of the cost of the extra memory and data path. It is now frequently employed within microprocessors since it yields such a significant speed increase.

In summary, consider Figure 1-7. Each level depicts the ways to represent a solution to an engineering problem that was discussed above. When a choice of implementation

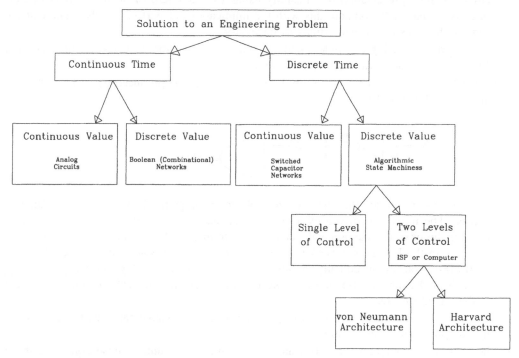

Figure 1-7: A hierarchy of choices.

method is made, constraints are imposed on the solution. For example, continuous time/data implies an analog solution with its attendant absolute accuracy problems. However, if discrete time is chosen, then the solution is limited by the Sampling Theorem. The subject of using a single level of control is discussed in Chapter 7, while the subject of the computer will be considered in much greater depth in Chapter 8.

EXERCISES

1. Speculate on the how the basic logic functions (e.g., OR) are implemented using the flow of water in a pipe as the working medium instead of the electrical analog used in this book. Show a way to implement a mechanical flip-flop. Both structures exist and have been described in the literature.
2. Show the "form" of a decision-making step and a counted loop using the ASM diagram elements. How does "form" help the designer understand and state the solution to his problem?
3. Draw and describe a "network" that performs integration using the continuous electric current analog. Do the same thing for an analog based on a continuous flow of water.
4. Draw and describe operational amplifier-based circuits that can perform integration and differentiation. How are the variables entered into the circuits? How are the circuits ini-

tialized or can they run forever? Draw a schematic of a circuit that will integrate the equation of motion for a point mass falling in a gravitational field. Specify how and where the values of the gravitational constant, g, and the initial velocity, v_0, are entered into the circuit.

5. Why does the representation of a quantity using a limited number of digits imply that the quantity is discrete? Compare this with the implications of performing the same calculations by hand using as many digits as necessary.

6. Demonstrate, by using a simple combinatorial digital network that has no memory, why time is a continuous variable for the circuit.

7. Define "glitches" or "hazards". Show how a glitch may be produced using the following simple combinatorial network.

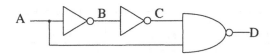

Draw a timing diagram that displays the voltage at the points B, C, and D as a function of time. The input A is driven by a logic signal that is low for the first 50 nanoseconds, high for the next 100 nanoseconds, and low again after that. Assume the components are drawn from the LSTTL databook (e.g., the inverters are 74LS04s and the NAND gate is a 74LS00). You are specifically interested in the propagation delay times t_{pLH} and t_{pHL} for each of the devices. Be sure to make the timing diagram to scale. Identify the glitch on the signal line D.

8. Why does the introduction of a memory driven by a clock signal cause us to be aware of the Sampling Theorem constraints?

9. Show the ASM diagram for a 10-state, 7-bit, Up/Down counter whose codes are those used by one digit of a 7-segment display. The counter must count directly in the codes that will light the proper segments of a 7-segment display to show the digits 0 through 9. Each state variable will be used to drive one of the display segments.

10. Prepare an ASM diagram for a machine having eight state variables, Q0 through Q7 and a clock input called CLK. The values of the state variables should follow the timing diagram given as follows.

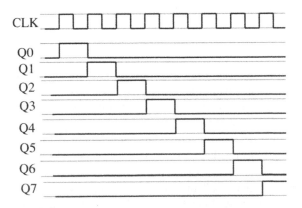

After the value of Q7 is high, the pattern should repeat endlessly. A network such as this can be used as an 8-phase clock.

11. The conditional output as shown in Figure 1-5 is dangerous to use as a control signal in an architecture. Assume the network controlled by Z2 in that figure has a relatively long propagation time before registers to receive its output can get stable inputs. Discuss, in this case, why the use of the conditional control signal causes problems.

12. Most modern *modem*s, devices used to attach a computer to a telephone line, have a set of commands that allow a program running on the computer to dial the phone, monitor call progression and perform other operations related to the task of computer-to-computer communication. Obtain a databook for one such "intelligent" modem and list its commands. Discuss the appropriateness of this *instruction set* for the task at hand. Compare this instruction set with the instruction set of a typical microprocessor. Which set is better for the task as seen by the programmer? Define "better" in this context. How well does the modem's instruction set perform computation on floating point numbers? How is the modem's instruction set made to do its job? Is the implementation the same as that for the microprocessor's instruction set?

13. Extend Figure 1-7 to include other methods to implement solutions to the engineering problem discussed in this chapter.

Chapter 2

REPRESENTATION OF NUMBERS

2.1. INTRODUCTION

We must now consider the form taken by the variables in our solutions. The representation chosen will be affected by the capabilities of the hardware that we use. The digital domain is chosen since it represents a cost-effective environment. The logic cell of choice in this domain stores data having one of two values. A two-state variable leads to the choice of the binary number system for representation.

Our circuit elements are derived from the analog circuits that were used to implement the analog computers discussed earlier. They have been optimized to throw away or ignore the continuum of possible input and output values used there in favor of the two values appropriate to the digital domain. It has sometimes been suggested that a three-state (or 10-state) cell could be built. The current experience, however, is that many binary cells (enough to hold the 10 suggested states) can be built more cheaply and compactly than a single 10-state device. It appears that this single point will keep us in a binary digital domain for the foreseeable future.

2.2. REPRESENTATION OF INTEGERS

Since we have settled on a two-state combinatorial cell, we should consider how we plan to represent the large number of possible values that a variable in our equations can assume. The next several paragraphs should be a review of the representations normally encountered in the digital world. A Binary digIT is called the "bit" and a multidigit binary number is interpreted similar to a multidigit decimal number where each digit position relative to the units position marker has a weight. The weight in the binary number system is a power of two and the unit position marker is called the "binary point".

The simplest number encountered is the "logical" representation. A single binary digit can be used to represent the two states that a logical number can assume: TRUE or FALSE. In most computers, a multidigit binary number is used for the convenience of implementation.

2.2.1. Unsigned Binary Integers

The first arithmetic number representation to be discussed is the unsigned integer. This multibit representation is still the most commonly used at all stages of computation and is also used to contain logical variables in most computer language implementations.

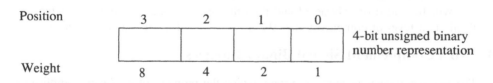

The box shown above illustrates a 4-digit binary number. The numbers above the box label the bit position relative to the units position at bit 0. The numbers below the box represent the weight of each position relative to the units position. Since the notation contains 4 bits, it can enumerate 16 entities (has 16 states) but cannot directly represent a negative number.

If we were hand-calculating using this notation, then we could freely extend the number of bits in the notation if needed as a result of our calculation. However, in a digital solution we must anticipate the largest number that we will ever encounter in all of the variables that we will use and supply enough bits to hold it. This is more easily done when the transformation equations do not contain feedback; however, the task is nontrivial in most cases. It is a very important part of system design. The notation is said to *overflow* when the result does not equal that expected. This can happen in an unsigned notation when two numbers are added and the result would occupy one more bit than the notation allows. The rules of unsigned subtraction require that the larger number be "subtracted from" since a negative result has no meaning and in this sense cannot be represented. Under these conditions, overflow can not occur upon subtraction. At this point we suspect the overflow condition can be signaled by the appearance of a Carry Out from the most significant bit of the sum. We will wait until we look more closely at the implementation of arithmetic later in the chapter to make this a certain identification.

2.2.2. Signed Binary Integers

In order to represent a signed number, we will use one of the bit positions to represent the sign, which can assume only one of two values — positive or negative. Frequently, the left-most bit position in the number is used for this task since it more or less matches the notation we have used in our written system.

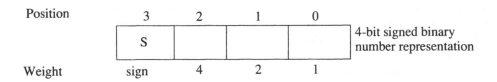

If the remainder of the representation is interpreted as an unsigned number (here bits 2 through 0), then the representation is called *"sign"* and *"magnitude"*. The rules of arithmetic are different for the bit position associated with the sign and those associated with the magnitude of the number. In order to add two numbers, the same rules of arithmetic apply to the magnitude portion as are used for unsigned arithmetic. The sign bits are combined using a set of four rules that relate to whether the signs are the same (or different). If subtraction is performed, the larger magnitude must be determined since the remaining magnitude portion can only represent size. From a hardware viewpoint, this requires the use of two separate units to do arithmetic on signed numbers, one for the sign and the other for the magnitude portion. The sign arithmetic is started first because it determines whether the magnitudes will be added or subtracted and then the resulting sign is tidied up at the end. All in all this is a level of complexity that should be avoided at such a simple stage.

2.2.3. Two's Complement Signed Binary Integers

The two's complement number system avoids the problem of hardware complication described for the sign/magnitude notation. If we make a list of all of the unsigned numbers representable by a 4-bit notation and then "map" onto it 16 signed numbers centered about zero we can create the two's complement notation as below.

Unsigned number	Binary	Signed number
0	0000	0
1	0001	1
2	0010	2
3	0011	3
4	0100	4
5	0101	5
6	0110	6
7	0111	7
8	1000	-8
9	1001	-7
10	1010	-6
11	1011	-5
12	1100	-4
13	1101	-3
14	1110	-2
15	1111	-1

Now take the table and rearrange it noting that the most significant bit of the binary number correlates with the sign of the signed number (i.e., a positive signed number has a 0 in the bit-3 position, while a negative signed number has a 1 in that bit position).

Rearranged Table

Unsigned Number	Binary	Signed Number
8	1000	-8
9	1001	-7
10	1010	-6
11	1011	-5
12	1100	-4
13	1101	-3
14	1110	-2
15	1111	-1
0	0000	0
1	0001	1
2	0010	2
3	0011	3
4	0100	4
5	0101	5
6	0110	6
7	0111	7

Looking at the signed number column shows that we have produced a list of numbers that progresses from the most negative to the most positive. There is one signed number for each bit combination. We have therefore created a signed representation that has the following format given at the top of the next page.

It is similar to our sign/magnitude notation but it can be shown to have the property that all bits *including* the sign are combined using the same rules of arithmetic.

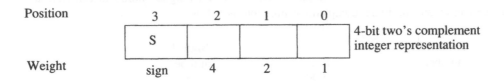

Position		3	2	1	0	
		S				4-bit two's complement integer representation
Weight		sign	4	2	1	

2.3. ADDITION OF SIGNED AND UNSIGNED BINARY NUMBERS

Let's now consider the actual operations necessary to add binary numbers. We will start by studying the addition of two 1-bit binary numbers, A and B, and will add to them the 1-bit Carry In. The outputs will be the 1-bit Sum and 1-bit Carry Out.

Truth Table for 1-Bit Binary Addition.

Carry In	A	B	Sum	Carry Out
0	0	0	0	0
0	0	1	1	0
0	1	0	1	0
0	1	1	0	1
1	0	0	1	0
1	0	1	0	1
1	1	0	0	1
1	1	1	1	1

The truth table is given in canonical form since we expect to be able to generate a Boolean algebraic equation for both Sum and Carry Out in order to implement our basic addition operation. The examples given will illustrate the use of this truth table in the calculation of multibit binary sums.

Consider the problem in which +5 is added to +2:

	Decimal	4 bit binary	operation
Carry In	0	0	No Carry In
A	+5	0101	From Table above
B	+2	0010	From Table above
	+7	0111	

The 4-bit sum is determined by adding each bit position beginning with the right most (least significant) just as in multidigit decimal addition. The truth table above is used to determine the sum bit given the three inputs to each column (Carry In, A, and B). The beginning Carry In into the least significant bit position is always 0. The sum for this column is 1 and the Carry Out (into the next column) is 0, as determined from line 3 of the truth table. This operation proceeds until all bits *including* the sign have been added *using the same truth table*. The result after consulting the rearranged table above is the code for a +7 which is consistent with the expected result.

Now consider the addition of two negative numbers, -3 and -2:

	Decimal	4-bit binary operation	
Carry In	0	0	No Carry In
A	-3	1101	From table above
B	-2	1110	From table above
Sum	-5	1 1011	

Using the same addition rules as above, the 4-bit result is 1011 which translates to a -5 using the rearranged table above. However, a Carry Out of 1 was generated. If we try all possible combinations of numbers in our 4-bit representation, we will find that this algorithm always works for addition if we ignore the Carry Out of the most significant bit position.

An example involving adding a positive and a negative number is appropriate at this point.

	Decimal	4-bit binary operation	
Carry In	0	0	No Carry In
A	-3	1101	From table above
B	+5	0101	From table above
Sum	+2	1 0010	

Again, by applying the binary addition truth table 1-bit at a time we find that the 4-bit result is indeed a +2. We ignore the Carry Out of the most significant bit as before.

2.3.1. Overflow

Now we need to consider the overflow situation since we expected it to occur in the unsigned case when we added two "large" numbers. Let's see what happens when we add +5 and +6, both numbers representable in our 4-bit notation:

	Decimal	4-bit binary operation	
Carry In	0	0	No Carry In
A	+5	0101	From table above
B	+6	0110	From table above
Sum	+11	0	1011

The binary result translates to -5 upon consulting our table above. The answer is incorrect and we will use the term *overflow* to name this kind of action. We could have detected overflow after the fact by examining the signs of A and B and comparing the binary result with the expected value of the sign. If A and B are positive, then the sign of the result is positive with a similar rule for A and B both negative. If A and B have opposite signs, then no overflow is possible since the result will be closer to zero upon addition than the larger of the absolute magnitudes of A or B. If we had to carry out the sign and magnitude comparison steps implied here every time we performed an operation, then we would not have improved our situation over the sign and magnitude representation. Let's go back and look at the results of our calculations to see if we have thrown away information.

Overflow?		No		No		Yes		Yes
Carry In		0		0		0		0
A	+2	0010	-2	1110	+2	0010	-2	1110
B	+5	0101	-5	1011	+7	0111	-7	1001
Sum	+7	0111	-7	1001	+9	1001	-9	0111
Carry Out of MS bit		0		1		0		1
Carry In to MS bit		0		1		1		0

We have shown four possible sums, two that overflowed (result has incorrect sign and magnitude) and two that did not. Additional information has been shown for each problem that can be tested for all possible combinations in the representation and correlated with the occurrence of overflow. In the cases shown here, notice that the Carry In and the Carry Out of the *most significant bit* (MS bit) position are the same if overflow does not occur and are different if it does. These Carries are generated in the normal course of performing addition using the 1-bit truth table above. They occur at virtually the end of the algorithm as one proceeds from right to left a bit at a time. An equation that indicates overflow can be easily generated:

$$\text{Overflow} = \text{Carry Out}_{\text{of MS Bit}} \ \textbf{XOR} \ \text{Carry In}_{\text{to MS Bit}}$$

This equation can be implemented combinatorially (no memory or sequence of steps) and its result will be available at the same time the sum is produced. This and the fact that the sign bit is operated on by the same rules (hence circuits) as the remaining part of the number assures that the two's complement number system will be used to represent signed integers for a long time to come.

In summary, unsigned numbers may be added using the 1-bit truth table shown above. If a Carry Out of 1 is produced from the most significant bit position in an unsigned number, then the result is said to have overflowed the notation. All bits of a two's complement signed number representation may be summed using the same rules as for unsigned binary addition. An additional piece of information is calculated using the Carry In and Carry Out from the MS bit position in the calculation. If these two bits are the same, then no overflow occurred in the two's complement notation; however, if the bits are different, then overflow occurred. Another point to notice is that the same hardware can perform both two's complement (signed) addition and unsigned binary addition. The two's complement overflow detection is done in "parallel" to the addition of the most significant bit. The bits are added using the same hardware whether the MS bit is interpreted as a sign or just the most significant bit of an unsigned number. Interpretation of the result as to whether overflow occurred, or even whether it is a signed number, is left until after the addition operation. One test (a single step) can then be used to determine the type of overflow. If the carry out of the MS bit is tested, then the result has been interpreted as an unsigned number. If the combinatorially compared Carry In and Carry Out of the MS bit position is tested, then the result has been interpreted as a two's complement (signed) number. This obviously makes a little hardware go a long way.

2.4. TWO'S COMPLEMENT AND SIGN/MAGNITUDE CONVERSION

For human readability, we must convert the negative half of the two's complement number set to a sign/magnitude notation. The trade-off in employing the two's complement number system for arithmetic has been the loss of readability of half of the representation. If a means to convert this portion to a sign/magnitude representation can be found, then the loss will be relatively small.

Let's copy part of the rearranged signed number table down and add the one's complement of the binary number in a column to the right.

Unsigned number	Binary	Signed number	One's complement
8	1000	-8	0111
9	1001	-7	0110
10	1010	-6	0101
11	1011	-5	0100
12	1100	-4	0011
13	1101	-3	0010
14	1110	-2	0001
15	1111	-1	0000

The one's complement of the binary number is formed by *complementing* (substituting 1 for 0 and 0 for 1) each bit in the number. This operation can be performed in parallel on all bits by using one logic inverter for each bit. Notice that the binary pattern that results is one less than the magnitude of the signed number. Our conversion rule will then take the following form:

Sign/magnitude notation:

Positive number = one's complement of the negative number + 1

If we wish to determine the binary pattern for a negative number we do the same thing including the addition of one. The operation of two's complementation can be summarized as:

Two's complement(N) = One's complement(N) + 1

or

$-N = /N + 1$, where the slash "/" means logical NOT

2.5. SUBTRACTION OF SIGNED AND UNSIGNED BINARY INTEGERS

The operation of subtraction could be implemented by writing a truth table for the operation similar to the one provided for addition. This truth table could be implemented as a multibit binary network that provided the desired combinatorial result when data was placed on its inputs. An arithmetic unit could be configured that consisted of one addition network and one subtraction network along with data steering networks. The controller would then supply the signals necessary to steer data to the proper network at the proper time.

Instead of doing this let us back up and ask ourselves if there is a way of avoiding the subtraction network altogether. Consider the following equation:

$$C = A - B \qquad \text{where A, B, and C are signed numbers}$$
$$C = A + (-B) \qquad \text{always true for signed numbers}$$

If A, B, and C are represented by two's complement numbers, then

$$C = A + [(\text{One's Complement of B}) + 1] \qquad \text{from } -B = /B + 1$$

The original subtraction has been converted to two additions and a one's complement operation. If the complement can be performed combinatorially, then we can implement subtraction by using the addition network and routing the number to be subtracted through the one's complementer on the way. The second addition operation consists of adding a 1 to the sum of A and the one's complement of B. This can be performed by setting the initial Carry In to the operation to 1 instead of the 0 used for the addition operations above. We will look in much greater detail at the *arithmetic unit* in a later chapter, but we should expect to implement signed and unsigned binary subtraction using this method.

2.6. ENCODED BINARY NOTATIONS

In some circumstances, it is convenient to avoid the conversion of a binary notation to a displayable, human readable form. One notation, Binary Coded Decimal (BCD), provides this convenience. In this case, the binary digits are collected in groups of four and used to represent a decimal digit. This means that a 16-bit binary notation can represent a 4-digit decimal number. There is some inefficiency in the storage use of this notation since 6 of the possible 16 different states that can be represented by a 4-bit binary notation have been discarded. Addition and subtraction can be implemented using methods similar to the above except that the truth table for addition will be written 4 bits at a time. The table has 9 inputs (4 for A, 4 for B and 1 for Carry In) and 5 outputs (4 for SUM and 1 for Carry Out). Notice that many of the lines in the truth table have invalid BCD data for one or both of the inputs. The outputs are meaningless, which should prompt us to provide another 1-bit output in addition to Carry Out that signals the presence of one of the invalid combinations. BCD arithmetic capabilities can be added to a normal 4-bit binary adder by noting that the BCD Carry Out must be generated at a point in the truth table different from the one for binary addition. This can be handled by a network in parallel to the binary Carry Out network. In the event of a BCD Carry Out in BCD addition, the Sum must be "corrected" by adding 6 to the value determined. This may be implemented with an additional network or can be handled by the controller generating an "Add 6 to Sum" operation if BCD Carry Out is True. The signed BCD notation is usually used in the sign/magnitude form. Generally, a 4-bit field is used to contain the sign. Many of the BCD notations place this sign field at the right end of the number reflecting the serial arithmetic procedures frequently used. Many business applications use this notation and the length of the number is usually made a part of the number description so that some of the flexibility of paper and pencil can be retained.

There are many other encoded notations. Each of them can support arithmetic operations of some kind. The notations fall into two classes. The first have properties that make them easy to record, communicate with, or control equipment. Examples of this group include the Group Code Redundant (GCR) and Grey Code. The second have properties that increase the speed of arithmetic operations when they are used. The first example of this is the two's complement number system which was described at length above. A second example includes the Logarithmic Number System (LNS) and the Residue Number System (RNS). Both systems improve multiplication operations by reducing them to addition, and the RNS example results in simpler hardware for the operation. All encoded representations suffer from the need to convert them to/from a human-readable form.

2.7. FIXED POINT REPRESENTATION

In normal mathematical operations, one needs to be able to represent values that are between 0 and 1. If the binary integer notation is reexamined, then we see that it can be adapted to represent fractional values.

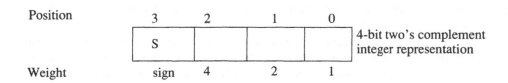

Consider the weights assigned to the bit positions as shown in the lowest line above. Bit position 0 has a weight of 1, while bit position 1 has a weight of 2. The weight is just two to the power of the bit position. If this were a decimal number, the weight of a digit position would be 10 and the decimal point would necessarily be to the right of digit position 0. The same applies in the binary case whether the number is in unsigned or two's complement (signed) form. Let's try a longer notation, one whose binary point is *not* at the right-hand end but in the middle.

In this example the 8-bit signed binary number has the binary point placed between bits 4 and 3. The weights are shown below the figure.

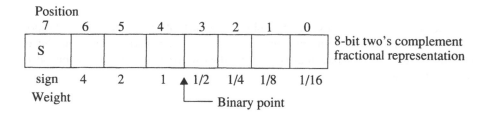

The following example shows how to interpret the binary number: 0110.0101 Separate the integer part and the fractional part.

$$(0110) + (.0101)$$

Interpret the integer part as before: $(0110) = +6$

The fractional part = (0 * 1/2) + (1 * 1/4) + (0 * 1/8) + (1 * 1/16)

The fractional part is most easily added decimally:

    0.0000   0 * 1/2
    0.2500   1 * 1/4
    0.0000   0 * 1/8
    0.0625   1 * 1/16
    _____

    0.3125 Decimal

The binary number is equivalent to +6.3125 in decimal.

This notation is called *fixed point* since, while it allows the representation of fractional values, it does not contain any information that allows the binary point to be manipulated by the hardware. The binary point *does not appear* explicitly in the notation. Its position is assigned by the designer and must be kept track of by him during the course of the implementation of the solution. In many cases, this is quite adequate; however, it becomes another item in the design.

How do we implement arithmetic operations on fixed point numbers? If we notice that the binary point does not appear explicitly in the notation then from our example above, 0110.0101 looks like 01100101 which is the 8-bit two's complement notation for 101 decimal. If we keep the binary points aligned during an addition or subtraction, then we can use the same networks developed for two's complement numbers above. In many circumstances, this is relatively easy to do and it results in a very simple solution to a wide range of problems.

There is an important shortcoming to this and all other fractional notations (including the floating point notation that will be introduced later). An integer notation represents the value of an integer *exactly*. There is a one-to-one correspondence between each binary pattern and the integer it is to represent. Unfortunately, the same cannot be said for the fractional representations. Consider the simple decimal number: +5.4

Let's convert it to the 8-bit fixed point format shown above. The +5 is easily done using the table shown for 4-bit two's complement numbers above.

    +5 → 0101

The fractional part (0.4) is converted by subtracting each of the weights of the bit positions to the right of the binary point from the number remaining after the previous bit is determined. If the result of the subtraction is positive then a one is placed in that bit position, otherwise a 0 is used.

        0.4
    -   0.5   (1/2, the weight of the first position)
        ____

       -0.1

The value of the first bit position to the right of the binary point must be 0 since the fractional part of the number is less than 1/2. Now we try the next bit position's weight:

```
   0.4      this did not change above
-  0.25     (1/4, the weight of the second bit position)
   ────
   0.15
```

The value of the second bit position to the right of the binary point must be 1 since the result was positive. Now we try the third bit position's weight:

```
   0.15       the amount left from the previous step
-  0.125      (1/8, the weight of the third bit position)
   ────
   0.025
```

The value of the third bit position is one also. Now we try the fourth and last bit position in our 8-bit notation.

```
   0.025      the amount left from the previous step
-  0.0625     (1/16, the weight of the fourth bit position)
   ─────
 - 0.0375
```

The value of the fourth bit position is 0 since the result was negative. Notice however that if we had a fifth position to the right of the binary point we would try it using the 0.025 remaining above. Our fixed point representation is:

$$+5.4 \rightarrow 0101.0110$$

We are a little short (by 0.025) of an exact representation. The number 0101.0110 is exactly +5.375. This error can be decreased by including more bits to the right of the binary point, but it can never be eliminated.

The error increases in certain types of arithmetic operations. Consider the subtraction of two positive numbers that are nearly equal. For example,

```
          +5.4         0101.0110
minus     +5.375       0101.0110
          ─────        ─────────
Result    +0.025       0000.0000
```

This effect can be critical and it has led to many efforts to minimize the effect. The floating point notation shown next will minimize but *not* eliminate it. This is a point that must be included in the design of any hardware or program that manipulates fractional quantities.

Now let us look at a negative number in this representation. Consider the number -3.75. Finding the representation is similar to that explained above. First, find the representation for the positive number, i.e., +3.75. The integer part in the 8-bit fixed point format in our example is 0011. The fractional part is 1100 since there is one 2^{-1}, one 2^{-2}, and no smaller terms. The entire positive number is represented by the following:

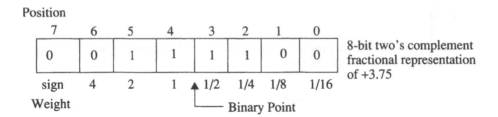

Position

7	6	5	4	3	2	1	0
0	0	1	1	1	1	0	0

sign 4   2   1 ▲ 1/2 1/4 1/8 1/16

Weight   └── Binary Point

8-bit two's complement fractional representation of +3.75

The negative representation is created by forming the two's complement of the 8-bit binary pattern. First, the one's complement of the pattern is 11000011. Adding 000000001 looks like the following:

```
Carry In              0000  0110
One's complement      1100  0011
One                   0000  0001
                      ─────────────
-3.75                 1100  0100
```

Position

7	6	5	4	3	2	1	0
1	1	0	0	0	1	0	0

sign 4   2   1 ▲ 1/2 1/4 1/8 1/16

Weight   └── Binary Point

8-bit two's complement fractional representation of -3.75

Notice that adding this number to the representation for +3.75 results in 0.

2.8. THE FLOATING POINT REPRESENTATION

A floating point notation is intended to increase the dynamic range of a fixed-length representation. The dynamic range is the distance between the largest and smallest number representable by a notation. The 16-bit two's complement notation can represent 65535 entities having a value from -32768 to +32767. The smallest increment representable is 1. The 8-bit fixed point notation shown in the previous section has a range from -8.0 to +7.9375 with the precision (i.e., the smallest increment) being 0.0625.

The floating point notation borrows the concept of scientific notation to increase its dynamic range. The scientific notation appends a power of 10 (the scale factor) to a number that represents the value to the precision desired. The exponent may be positive or negative with the values less than 0 being used to represent numbers between 1 and 0. The fractional part of the scientific notation (the mantissa) is also signed to reflect the sign of the entire number.

A floating point representation has similar fields as follows:

Sign of exp.	Exponent	Sign of fraction	Fractional part	Floating point representation

The sign fields require only 1 bit to represent them. The number of bits in the exponent determines the dynamic range of the representation and the number of bits in the fraction determines the precision. The floating point representation described below comes from the IEEE 754 Standard (see Bibliography) and is implemented in many devices available to the designer of architectures. Other similar representations have been defined by IBM and DEC for use on their computers.

2.8.1. The IEEE 754 Floating Point Representation

The single precision IEEE format is 32 bits long and has the following format:

Bit Position	Function
31	Sign of fractional part
30-23	Biased Exponent
22-0	Fraction

The bit positions are numbered from the right (least significant) end of the 32-bit field. The biased exponent means that the value of the exponent (i.e., power to which two is raised) is calculated by subtracting 127 from the number formed from the 8-bit exponent field (bit positions 23 through 30). This format is another example of an offset notation that is commonly used to carry sign information in the exponent field of floating point numbers.

The fraction uses an unsigned binary fixed point notation. It is normalized, i.e., it has been shifted left until the left-most bit is 1 and the exponent has been adjusted accordingly. The normalization operation preserves some of the precision that would be lost in an operation *if* the designer provides a wider arithmetic unit to process the bits to the right of the end of the notation. One more shift is needed to reach the normalized state. Since a 1 will always be next to the binary point (between bits 22 and 23), why store it? It is therefore shifted out and only the remaining 23 bits are retained. The number has 24 bits of precision since the 1 must still be made available for computation.

The fraction is given in sign/magnitude form in the single-precision IEEE format as follows.

+1.bbbbbbbbbbbbbbbbbbbbbbb

Each "b" represents a bit position in the fractional part. The exponent is an offset binary number with the following relationship between its value and its representatior:

For 0 < Biased Exp < 255, the Exponent value = Biased Exp - 127

Exp Value	Biased Exp	Representation
-127	0	00000000
-126	1	00000001
•	•	•
-1	126	01111110

```
        0                    127                    01111111
        1                    128                    10000000
        •                     •                         •
       127                   254                    11111110
```

If both the Biased Exp and Fraction are 0 then the number is defined to be 0. The sign is significant. If the Biased Exp is 255 decimal (11111111 binary) and the fraction is 0 then the number represents plus or minus infinity based on the sign bit. Denormalized numbers have a zero biased exponent and a non-zero fractional part. These are frequently used to retain small values that would become zero upon normalization. They will also be encountered in the course of calculations such as aligning the binary point before adding.

The number of bits available in a 32-bit single, or 64-bit double precision format is greater than the number of valid combinations of floating point numbers. The remaining combinations are flagged as NANs (Not A Number). The serious designer must be sure that his network or algorithm can cope with such situations since they may be encountered inadvertently during use. A major part of the IEEE 754 specification is devoted to rules for handling these situations.

We will now examine some examples of the single precision format. The first number is +5.4 which was shown in an 8-bit fractional format earlier. The hexadecimal form followed by the same number in binary with spaces separating the digits is:

40ACCCCD = 0100 0000 1010 1100 1100 1100 1100 1101

The binary representation with spaces between the various fields in the IEEE single precision notation is:

```
    31         30    23    22                      0
    0       10000001      01011001100110011001101
   Sign     Biased Exp       Fraction
                          1010110011001100110011001101
                          Fraction with the leading 1 restored
        The complete fractional part with sign and binary point is:
              + 1.01011001100110011001101
     Biased Exponent: 10000001 = 129
     Exponent Value = 129 - 127 = 2
     Complete number: + 101.01100110011001101₁ or +5.4_d
```

Complete number: + $101.011001100110011001101_b$ or $+5.4_d$

The second number is -3.75 which was also shown in an 8-bit fractional format earlier:

C0700000 = 1100 0000 0111 0000 0000 0000 0000 0000

The binary representation with spaces between the various fields in the IEEE single precision notation is shown below.

A *double* precision format is also defined as follows:

Bit Position	Function	No.
63	Sign of fractional part	1
62–52	Biased Exponent	11
51–0	Fraction	52(+1)

```
 31          30    23      22                            0
 1          10000000       1110000000000000000000000
 Sign       Biased Exp      Fraction
                           11110000000000000000000000
                           Fraction with the leading 1 restored
 The complete fractional part with sign and binary point is:
        -  1.11100000000000000000000
 Biased Exponent: 10000000 = 128
 Exponent Value = 128 - 127 = 1
 Complete Number: - 11.110000000000000000000000_b  or  -3.75_d
```

The precision of the fraction is greatly increased and the dynamic range has been slightly increased over the single precision format. An example of the IEEE double precision format is now shown using the number +3.375:

$$400B000000000000 = 0100\ 0000\ 0000\ 1011\ 0000\ 0000\ ....$$

```
 63          62    52      51                     ...0
 0          10000000000     1011000000000000000000000..  .
 Sign       Biased Exp      Fraction
                           11011000000000000000000000.....
                           Fraction with the leading 1 restored
 The complete fractional part with sign and binary point is:
        +  1.1011000000000000000000000.......
 Biased Exponent: 100 0000 0000 = 2048
 Exponent Value = 2048 - 2047 = 1
 Complete Number: + 11.10110000000000000000000_b...  or  +3.375_d
```

In each of the above examples, remember to compare the resulting complete number with that calculated using a sign/magnitude fixed point format. The integer part of the each example can be read directly as a positive binary number, i.e., treat the most significant bit as a number and not a sign.

An *extended* precision format is also defined, but not to the bit level. Many implementers of architectures use a total length for the representation of 80 to 96 bits, with most of the length being added in the fraction. To preserve precision, the combinatorial data paths are made very wide for addition and subtraction since these operations can cause precision to be lost.

2.8.2. Floating Point Addition and Subtraction

Addition and subtraction of floating point numbers, diagrammed in Figure 3-24, begins by partitioning each number into its exponent and fractional parts and treating the parts separately. The binary point must be located in the same place in both numbers, i.e., the exponents must be equal. To accomplish this, the fraction of the smaller number is shifted to the right and its exponent adjusted until it is the same as the exponent of the larger number. This requires that the exponents of the two numbers be compared. In the IEEE notation, this comparison is done as if the exponent was represented as an unsigned number. The comparison is performed by doing integer subtraction as described earlier and interpreting the meaning of the Carry Out as a borrow. Once the smaller number is identified, its fraction is shifted to the right (or *denormalized*) while preserving the number's value by incrementing its exponent for each bit position shifted. Notice that a one must be shifted in for the first step since the most significant bit is assumed but not stored in this representation. Zeros are shifted in for the remaining steps since the fractions are themselves unsigned. Denormalization must be performed before each addition or subtraction. The reverse process, *normalization*, is performed after the result is computed in order to preserve as much precision as possible. These processes lend themselves to being implemented as algorithms using either the underlying integer arithmetic hardware or separate combinatorial structures. We will examine the specialized floating point structures in more detail later since they greatly affect the speed of floating point operations.

Once the binary points are aligned on the two input numbers, the fractional parts are added or subtracted as if they were integers, i.e., using the integer addition and subtraction networks. Since the representation of the sign is explicit, the fractional part is treated as an unsigned number and the decision to add or subtract is made before the operation begins. As indicated earlier, the results are normalized to preserve precision. This is possible only if the data path width through the summation network is wider than the fractional part of the representation. Many implementers of such networks provide as many as 80 bits of fractional precision in this part of the data path for this purpose.

2.8.3. Floating Point Multiplication

Multiplication, as diagrammed in Figure 3-23, is performed by first multiplying the fractional parts of the input numbers; then the exponents are added. In this notation, 127 must be subtracted from the resulting exponent. The fractional part of the result must be normalized by shifting it to the right by at most one place. The multiplication of the fractional parts is performed just as it would be for two unsigned integers. If this is implemented using a combinatorial network, then the multiplication operation is the fastest of all arithmetic operations on floating point numbers.

Division is usually performed algorithmically using approximation techniques such as the Newton-Raphson Algorithm when a fast floating point multiplier is available; otherwise, it can be performed using algorithms similar to those for decimal numbers.

2.8.4. Floating Point Overflow

There is no overflow possible for the fractional part of the number since the data path is maintained wider than the fraction zone. Overflow in floating point numbers occurs when the exponent becomes larger than the space allowed for it. It is easy to see how this can be caused by multiplication since the exponents are added during the operation. This can also occur on addition of very large positive numbers or subtraction of large negative numbers as well. The overflow is by a very small amount and occurs during normalization of the result.

Another situation occurs in which the exponent is expected to be smaller (more negative) than that allowed. In the case of the IEEE notation, this occurs when the exponent has a value less than -127 or a representation less than 1. The situation is called *underflow*.

In the cases where floating point numbers are applied, overflow (or underflow) is very devastating since the results are incorrect by a very large amount. Circuitry or software tests must always be used to detect the occurrence of the out of range condition and then a "fix-up" consistent with the solution of the problem should be taken. The design of the hardware should assist in this detection in a manner similar to that used for integer overflow.

2.8.5. Summary of the Floating Point Representation

The floating point representation is important in problems where the dynamic range of a variable is greater than can be practically represented in one of the integer formats. Since it is a fractional notation, it will inherently accumulate error though not at a rate as fast as the fixed point notation. If the algorithm and data are well understood, then this error can be contained at a harmless level. With the advent of integrated floating point units (FPU), one can employ this notation in very high-speed operations. The integer representations can still run at a faster rate for the same precision; however, their dynamic range is not as great.

With the advent of inexpensive fast memory, the original arguments against long (hence wide dynamic range) integer formats is not as strong as it once was. The integer formats are the only solution if wide dynamic range *and* high precision must be maintained simultaneously. While the floating point notation seems to "take care of the decimal" for you in a more or less automatic sense, you should not consider that it is a "use and forget about" notation. The fact that every floating point number is an approximation to a decimal value and that error accumulates during many types of operations means that its use can have insidious side effects.

EXERCISES

Hint: The most instructive way to proceed on the following exercises is to analyze the procedures required to convert the assigned number between representations and then to write a program in your favorite language that performs the task. The program's input and output should normally come from the "console" or "terminal" of the computer on which the exercise is performed. The programs may be kept short and to the point by making them handle only a single type of conversion. Do not use the data formatting instructions in the computer language — their task is what you are trying to study in these exercises. Input and output data are normally in character format while the binary, fixed point and exponential and fractional parts of floating point numbers can be manipulated using the integer data type.

1. a. Convert +85 to an 8-bit unsigned binary notation.
   b. Convert +85 to an 8-bit two's complement notation.
   c. Find the representation of -85 in eight bit two's complement notation.

2. Determine the value of each of the following IEEE single precision floating point numbers:
   a. 42FE0000
   b. 00000000
   c. C3000000

3. a. Convert +9.25 to an 8-bit sign/magnitude fixed point notation. Place the binary point between bit positions 2 and 3 and the sign in bit position 7 ($0 \rightarrow +$, $1 \rightarrow -$).
   b. Convert -9.25 to an 8-bit two's complement fixed point notation with the binary point as in part a.
   c. Convert -9.25 to a single precision IEEE floating point notation.

4. The number +5.4 was given in the text in an example of a single precision floating point number. The floating point value was given as 40ACCCCD.
    a. Is this an exact representation of +5.4? Your answer should include the value of the next bit to the right of the D in the floating point number.
    b. What number does 40ACCCCD represent exactly?

5. Write the value of each of the following IEEE single precision floating point numbers in an appropriate sign/magnitude fixed point format by extracting suitable parts from the numbers given.
    a. 43A3D000
    b. BAA3D70A

6. Perform these subtraction problems by using a 4-bit two's complement binary representation for the numbers. Take advantage of the relationship, $F = A - B = A + /B + 1$, that holds for this number system. Be sure to check for two's complement overflow.

	3		3		-6		6
minus	4	minus	-4	minus	-2	minus	-2
	___		____		____		____

7. Some programming languages, including C and Fortran, allow you to print the contents of a variable using a hexadecimal format, i.e., you can see the actual storage format that your compiler (or machine — as appropriate) uses for a variable. Write a short program that prints the storage formatted number, in hexadecimal or binary, on your screen (or printer). Hint, the formatting code in C is %x while it is Z in many versions of Fortran. Some compilers will not allow you to print out a floating point format with the hexadecimal formatting code. In these cases, you will need to perform your own conversions. Try the floating point numbers 0, 0.5, +3.3, 100, and -256.

Chapter 3

ARCHITECTURAL COMPONENTS

3.1. INTRODUCTION

The focus of this chapter is on the structures in which data can be combined and stored. A quick review of the basic logic elements will be conducted first. Then the architectural components, i.e., combinatorial elements, memory, and data paths, will be discussed individually. They will then be combined to form structures, or architectures, that can be controlled. In many cases, the intended use of an architecture has influenced its structure. This is pointed out in the cases of the Q register in the Register Arithmetic/Logic Unit (RALU) and the arrangement of elements in the Multiplier/Accumulator. Other structures are optimized for a particular data representation. This is shown in the discussion of signed and unsigned arithmetic using the Arithmetic/Logic Unit (ALU) and in the implementation of Floating Point Units (FPUs).

The architectures described in this chapter are intended to be used in the "architecture" section of an algorithmic state machine, Figure 1-3. The discussion in this chapter will be limited to the data combinations and transfers that can be accomplished during a single clock cycle. Multicycle operations will be covered more thoroughly in Chapters 7 and 8.

3.2. REVIEW OF BASIC DIGITAL ELEMENTS

This section is a summary of the elements from which the architectures discussed in this and the next chapter will be constructed. The fundamental logical operations, AND, OR, and NOT are shown in Figure 3-1. For each logical operation, a hardware representation has been constructed in several different forms. While we are interested in electrical implementations, we must not forget that there are other analogs (e.g., water) to physical variables that can be used besides an electrical current or voltage difference. Circuit families have been implemented in the course of the past 30 years that simulate all of the logic functions and many other structures as well. One family of importance is the Advanced Schottky (AS) and its low-power version, ALS, as implemented in Transistor-Transistor Logic (TTL). An important Complementary Metal Oxide Semiconductor (CMOS) transistor-based family has a speed approaching the AS family and a power dissipation that is much less. Many very complex architectures are being implemented in it for these reasons. The current versions of these families have corresponding elements that have the same package pinout structure as well as the same electrical interface characteristics. Comparison can be made between the older LS (low power Schottky TTL) logic family and the HCT (high speed CMOS with TTL thresholds) logic family. A similar trend is developing for devices in the ALS and AS families and the ACT (advanced CMOS with TTL thresholds) family. In summary, each family contains integrated circuits that perform the fundamental logic operations of AND, OR, and NOT. There is a strong similarity in package style, pinout, and input/output (I/O) characteristics between elements in the families built using the Schottky TTL and CMOS processes. It should be noted that the basic gate in all of the TTL families is the NAND (NOT AND) gate. From this gate can be constructed the OR and the NOT functions by the judicious use of DeMorgan's law. This implies that all digital structures can be constructed using this one element. A similar structure exists for the CMOS process as well. It is the NOR (NOT OR) gate. The same statements can be made for its utility as for the TTL NAND gate. Over the years, both the OR and the NOT (as well as the NAND and NOR) have been added as separate devices to both families so that the designer has a complete set of elements at his disposal. The Exclusive OR function has also been added since it occurs in so many operations, including addition.

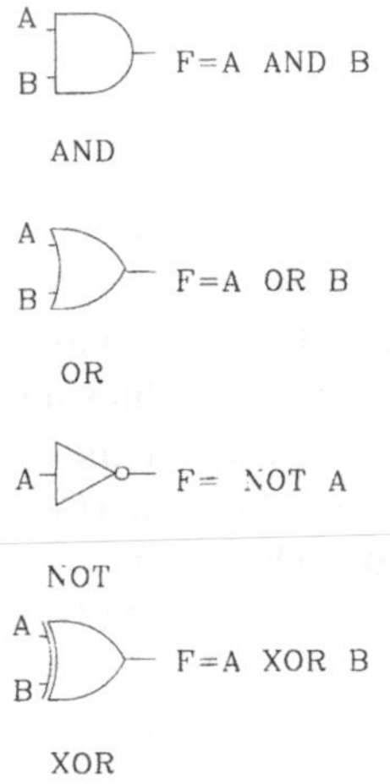

Figure 3-1: Symbols for the fundamental logic operations.

The third symbol from the top in Figure 3-1 is actually a composite of a buffer symbol, the triangle, and an invert "bubble". The inversion bubble indicates the assertion level is a low voltage at that point. Notice the absence of the bubble on the other terminals of the devices shown. At each point where the bubble is missing the assertion level is a high voltage. The assertion level is the voltage at a point when the logical value is True. The third symbol is interpreted as a voltage inverter. It is also a logic inverter when using a positive logic interpretation. In mixed logic, the presence of a bubble at the opposite end of the output connection determines whether a logical inversion has occurred. If the bubble is absent or present at both ends of the wire then no logical inversion occurs. If the opposite ends of the wire have differing symbols then a logic inversion will be present. The truth tables that describe the functions for two input variables in Figure 3-1 are shown next.

AND				OR				NOT			XOR		
A	B	F		A	B	F		A	F		A	B	F
0	0	0		0	0	0		0	1		0	0	0
0	1	0		0	1	1		1	0		0	1	1
1	0	0		1	0	1					1	0	1
1	1	1		1	1	1					1	1	0

The Boolean math operators for the basic functions are "*" for AND, "+" for OR and "/" for NOT. The asterisk, "*", will also be attached to a symbol, e.g., U*, to indicate that the symbol is True when a Low voltage is present.

3.3. SIMPLE COMBINATORIAL STRUCTURES

The next level of complexity is represented by a pair of structures that can be used to steer data between several sources and destinations. These structures are presented here in their 1-bit form. Multibit numbers may be steered if a multiplexer (MUX) is provided for each bit in the number representation.

Figure 3-2: Two-input multiplexer.

The first example is the *multiplexer*. Look at the symbol labeled MUX in Figure 3-2. The variables I0 and I1 are 1-bit binary numbers that are placed on the two inputs of the multiplexer. The logical variable S is a control input so we place it in a special place in the diagram. The output of the multiplexer can be described by the following truth table:

MUX

S	I1	I0	Output
0	0	0	0
0	0	1	1
0	1	0	0
0	1	1	1
1	0	0	0
1	0	1	0
1	1	0	1
1	1	1	1

The data steering behavior of this network can be emphasized by using "don't cares" (X) in the truth table as follows.

MUX

S	I1	I0	Output
0	X	0	0
0	X	1	1
1	0	X	0
1	1	X	1

or

MUX

S	I1	I0	Output
0	X	I0	I0
1	I1	X	I1

The right-hand form replaces the two possible values for I0 (i.e., 0 and 1), with the symbol for I0. This form of the table may also be called a Function Table and is useful for

the initial statement of the problem. It will have frequent use as the networks we are discussing become more complex.

The logic equation for Output can be written in its Sum-of-Products (SOP) form by
ORing the products (AND terms) from the first truth table that make the Output equal to 1.

Output   = /S * /I1 * I0 + /S * I1 * I0
                + S * I1 */I0 + S * I1 * I0

The equation can be minimized to a simpler form by using I1 + /I1 = 1 in the first
line and /I0 + I0 = 1 in the second line.

Output   = /S * I0 + S * I1

This commonly recognized form of the equation for a multiplexer can be extended to
many inputs by making S contain more bits, enough so that each of the inputs can be selected by each of the combinations of the bits in S. A four-input multiplexer then has the
following equation.

Output   = /S1 * /S0 * I0 + /S1 * S0 * I1
                + S1 * /S0 * I2 + S1 * S0 * I3

The control variable S requires two bits to be able to point at all of the inputs.

The reverse operation to the MUX is called the *demultiplexer*. A function table for a
two-output demultiplexer is given below.

DEMUX

S	I	O0	O1
0	I	I	F
1	I	F	I

This function table can be expanded to a truth table from which can be derived the
SOP form of the equations for the outputs O0 and O1 as follows.

O0 = /S * I
O1 = S * I

A demultiplexer with more outputs will require that the control variable S be represented by a multibit number as above so that a four-output demultiplexer has the following
equations.

O0 = /S1 * /S0 * I
O1 = /S1 * S0 * I
O2 = S1 * /S0 * I
O3 = S1 * S0 * I

It is seen that the two bits S1 and S0 of the control variable S take on the binary values of the output number selected.

A re-interpretation of the meaning of the control and input signals for a demultiplexer produces a very important structure called a *decoder*. Let's interpret the input signal I as an enable signal EN and interpret S as a multibit number A that we desire to "decode". Rewriting the equations for the four-output demultiplexer above we get:

O0 = /A1 * /A0 * EN	A = 0
O1 = /A1 * A0 * EN	A = 1
O2 = A1 * /A0 * EN	A = 2
O3 = A1 * A0 * EN	A = 3

As long as enable (EN) is true, the output that corresponds to the binary encoded number A will be true. If the value of A changes, then the selected output will change while all others are false. This decoder converts a binary encoded variable A to a "one-of-four" output, thus "decoding" it. There are many applications for decoders and we will find them in many of the structures that we describe in this and following chapters.

Note that decoders can be built that combinatorially convert many different coded numbers (BCD, Excess 3, Grey Code, etc.) to outputs of the one-of-N form. Another useful form drives display devices such as 7-segment LED displays. In this case, seven decoders use the same inputs to produce output signals, one for each segment of the display.

3.4. THE SIMPLEST ARITHMETIC STRUCTURE — THE ADDER

The HALF ADDER obeys the following Truth Table:

A	B	HSUM	HCout
0	0	0	0
0	1	1	0
1	0	1	0
1	1	0	1

This network does not use the Carry In as shown in the truth table for 1-bit binary addition in Chapter 1. An inspection of the portion of the truth table devoted to the HSUM shows the following equation:

$$HSUM = A \ XOR \ B$$

To see this, compare the HSUM column in the truth table with that for the Exclusive OR function in Section 3.2. The Carry Out of this network has a similarly simple form by comparing the output HCout with the truth table for the AND function above:

HCout = A AND B

The truth table for the half adder is seen to consist of that portion of the binary addition truth table in Chapter 1 in which the Carry In is set to 0. We can introduce a second half adder into our network to produce the sum of the three inputs A, B, and Carry In by doing the following operation:

SUM = HSUM XOR Carry In

Substitute for HSUM the equation produced above and we have:

SUM = A XOR B XOR Carry In

This is one of the common forms for the 1-bit binary summation network. The implementation of this equation and the one for Carry Out is called the *full adder*. A symbol for this network is shown in Figure 3-3.

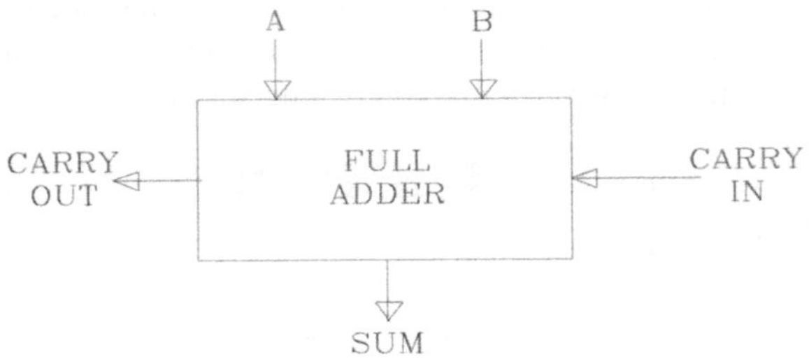

```
                SUM  =  A  XOR  B  XOR  Cin
Cout  =  (A  AND  B)  OR  (B  AND  Cin)  OR  (A  AND  Cin)
```

Figure 3-3: The 1-bit full adder.

The equation for Carry Out is most easily derived directly from the truth table for three input binary addition shown in Chapter 1. The SOP form will reduce to the following after some manipulation.

Carry Out = A*B + B*Cin + A*Cin

3.5. COMBINATIONS OF SIMPLE NETWORKS

This section examines networks whose forms are arrays of the simpler networks discussed earlier. In some cases, in order to minimize the signal propagation delay through the network, extensive modifications will be made to the simple array concept.

3.5.1. The Multibit Full Adder

Figure 3-4 shows a 4-bit full adder network formed by placing four 1-bit full adders side by side. This is a conceptual operation intended to emphasize the operation implied by a multibit full adder (ADDER). While there are two 4-bit wide inputs and one 4-bit wide output shown in the figure, it should be noted that there is only a single-bit carry input and a single-bit carry output since all other carry inputs and outputs are connected in a daisy chain *inside* the structure.

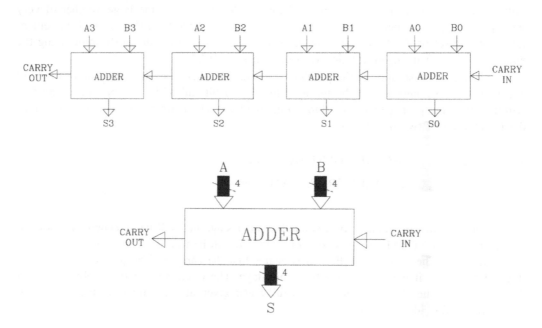

Figure 3-4: A 4-bit full adder with ripple carry.

If one examines the equation for Carry Out for a 1-bit full adder shown in the previous section, one can identify the number of physical gates that a signal change must pass through to produce the final Carry Out from the 4-bit network.

$$\text{Carry Out} = A_3{*}B_3 + (A_3 + B_3){*}Cin_3$$

$$= A_3{*}B_3 + (A_3 + B_3){*}Cout_2$$

$$= A_3{*}B_3 + (A_3 + B_3){*}[A_2{*}B_2+(A_2+B_2){*}Cin_2]$$

This continues until the final form containing only the four bits of A and B and the Carry In to the least significant bit is produced.

$$\text{Carry Out} = A_3{*}B_3 + (B_3+A_3) * \{A_2{*}B_2 + (B_2+A_2) * [A_1{*}B_1 + (B_1+A_1) * (A_0{*}B_0 + (B_0+A_0) * \text{Carry In})]\}$$

An implementation of this equation using the basic gates will result in a very long propagation path (the signal traverses several gates) from A0, B0, or Carry In to the Carry Out. This illustrates the fundamental limitation to speed in an arithmetic system, especially to one composed of a series of 1-bit full adders.

An alternative solution for a relatively narrow (4-bit) full adder involves writing a truth table for the Sum and Carry Out. This truth table has nine input terms (four for A, four for B, and one for Carry In) and five output terms (four for Sum and one for Carry Out). The 512 lines of the truth table can be used to generate a SOP equation for each of the five output bits. The immediate shortcoming of this implementation is the large number of very wide product terms, especially for the least significant bit of the Sum (S0). Actual implementation of this set of SOP equations directly can be performed by judiciously minimizing the equations. One such implementation is present in the 74LS283.

The present solution to the problem is produced either from minimization of the SOP equations or rearrangement of the terms in the four 1-bit full adder solution in a way that will minimize the propagation time through the network. The rearrangement method involves the recognition of two new terms.

$G_i = A_i * B_i$ the CARRY GENERATE term

$P_i = A_i + B_i$ the CARRY PROPAGATE term

These terms are defined for a single-bit adder stage but will be extended over several bits later. The GENERATE term is so named since it will be true if a Carry Out is generated in the ith term. The PROPAGATE term indicates that the carry will propagate into the next stage of the adder if there is a Carry In to this stage. These terms are actually obtained from the equation for the Carry Out of the 1-bit full adder given above and shown here in a form to emphasize this identification.

$$\text{Carry Out}_{i+1} = A_i * B_i + (B_i + A_i) * Cin_i$$
$$= G_i + P_i * Cin_i$$

The Carry Out from the first stage is:

$$C_1 = G_0 + P_0 * Cin_0$$

The Carry Out from the second stage is:

$$C_2 = G_1 + P_1 * (G_0 + P_0 * Cin_0)$$

This can be rewritten algebraically as:

$$C_2 = G_1 + P_1 * G_0 + P_1 * P_0 * Cin_0$$

If one built a 2-bit full adder, then the propagation time across this equation would be the same as the SOP implementation or about two gate delays. This network is still simple enough to implement. The Carry Out from the third stage is:

$$C_3 = G_2 + P_2 * Cin_2$$

which becomes:

$$C_3 = G_2 + P_2 * (G_1 + P_1 * G_0 + P_1 * P_0 * Cin_0)$$

Algebraic simplification results in:

$$C_3 = G_2 + P_2 * G_1 + P_2 * P_1 * G_0 + P_2 * P_1 * P_0 * Cin_0$$

Again, it should be noted that the number of gate delays has not increased over the previous stage and the complexity of the equation is still implementable in hardware.

The Carry Out from the 4-bit full adder shown in these equations is of the form:

$$\text{Carry Out} = G_3 + P_3 {*} G_2 + P_3 {*} P_2 {*} G_1 + P_3 {*} P_2 {*} P_1 {*} G_0 + P_3 {*} P_2 {*} P_1 {*} P_0 * \text{Carry In}$$

The algebra to simplify the form is not shown here but is similar to that above. The number of gate delays across the 4-bit full adder has not increased over the "narrower" full adders shown above.

The equation for the Sum of two 4-bit numbers can be similarly represented. The Sum for the *i*th term is:

$$S_i = A_i \text{ XOR } B_i \text{ XOR } Cin_i$$

The equations for each bit of the 4-bit Sum are now shown in terms of the P_i and G_i terms for the Carry In to each stage:

$$S_0 = A_0 \text{ XOR } B_0 \text{ XOR Cin}$$
$$S_1 = A_1 \text{ XOR } B_1 \text{ XOR } (G_0 + P_0 * \text{Cin})$$
$$S_2 = A_2 \text{ XOR } B_2 \text{ XOR } (G_1 + P_1 * G_0 + P_1 * P_0 * \text{Cin})$$
$$S_3 = A_3 \text{ XOR } B_3 \text{ XOR } (G_2 + P_2 * G_1 + P_2 * P_1 * G_0 + P_2 * P_1 * P_0 * \text{Cin})$$

These equations should emphasize the reduction in the time delay across the summation network as compared to the original non-minimized 4-bit full adder composed of four 1-bit full adders.

At some point, the number of terms in the most significant bit of the Sum and the Carry Out reaches a number that cannot be implemented by the current technology. One can then build an N-bit full adder by connecting the carry inputs and carry outputs of a set of M- (e.g., 4-) bit blocks in a daisy chain as was done to create the original 4-bit full adder from 1-bit full adders shown in Figure 3-2. If the sum bits and carry output in each 4-bit block are implemented using the G and P terms, then the time delay across the N-bit structure is a minimum and is N/M times the delay across each M-bit structure.

A further minimization of the time delay may be accomplished by defining a G and P output term for the M-bit structure with the intent of using these as a means to speed carry calculations across the entire N-bit structure. If we use the Carry Out equation for a 4-bit full adder expressed in terms of the G_i and P_i terms, then we can see the form of the newly defined G and P terms for the 4-bit block.

$$\text{Carry Out} = G_3 + P_3*G_2 + P_3*P_2*G_1 + P_3*P_2*P_1*G_0 + P_3*P_2*P_1*P_0 * \text{Carry In}$$

The following equations define the block G and P terms:

$$G = G_3 + P_3*G_2 + P_3*P_2*G_1 + P_3*P_2*P_1*G_0$$
$$P = P_3*P_2*P_1*P_0$$

In terms of them the Carry Out of the 4-bit block is just:

$$\text{Carry Out} = G + P * \text{Carry In}$$

This equation has the same form as the original G_i and P_i expressed Carry Out from the 1-bit full adder. It suggests that a mechanism can be implemented whereby the Carry Out from an N-bit network composed of blocks of M-bit wide networks is computed without paying the time delay penalties suggested above.

Assume that a 16-bit wide network is to be made using four 4-bit wide networks each implemented as above. The Carry In to the least significant block is Carry In. Instead of using the Carry Out of the least significant block (block 0) as the Carry In to the next more significant block (block 1), we will use the block G and P terms to generate the Carry In to block 1.

$$\text{Cin}(1) = G(0) + P(0)*\text{Carry In}$$

Do not confuse the block subscripts (i) with the bit subscripts "i" used above. The Carry In to the next more significant block (block 2) has the form:

$$\text{Cin}(2) = G(1) + P(1)*P(0)*\text{Carry In}$$

This exploits the similarity in form to that seen in developing the interbit carries earlier. Note that as the complexity of the Carry In equation to a block increases, the time delay across the Carry In evaluation network does not.

At some point the technology no longer provides a means of implementing the Cin, G, and P networks. This requires another application of the identification sequence shown for this level of block structure. In this case, imagine that a 64-bit full adder is implemented using four of the 16-bit full adder structures that were built from 4-bit full adder blocks designed above. Another level of "superblock" G and P terms could be implemented from the same equations shown for the block G, P, and Carry Ins shown above. This structure could be continued as long as necessary and introduce far less additional propagation time than an N-bit structure using 1-bit full adders in a ripple-carry arrangement.

Just as the 4-bit full adder including the Carry Out and block G and P outputs can be thought of as a logical structure implementable as a single integrated circuit, the block G, P, and Carry In networks can be integrated together as a single device. This device is called a Lookahead Carry Generator. At any given time, the block size will be set by the capabilities of the implementation technology. Therefore the G and P terms were commonly brought out of the 4- and 8-bit wide adders that were popular in the recent past. The advent of 32-bit wide structures has made expansion less likely and therefore G and P outputs are usually not present even though they may be used internally in the implementation.

3.5.2. The Shifter

Figure 3-5 shows a structure composed of several three input multiplexers. The structure is called a *shifter* because a multibit number applied to its input (Din) can be "shifted" to the left or right by 1 bit with respect to the corresponding bit positions in the output (Dout). The symbol for the shifter, shown in the lower part of the figure emphasizes the broad, main data path Din and Dout, but also identifies two narrower paths SIO0 and SIO3. SIO0 contains two signals SHIFT LEFT IN and SHIFT RIGHT OUT that can be seen in the upper drawing. These paths extend the width of the main data path (Din and Dout) so that

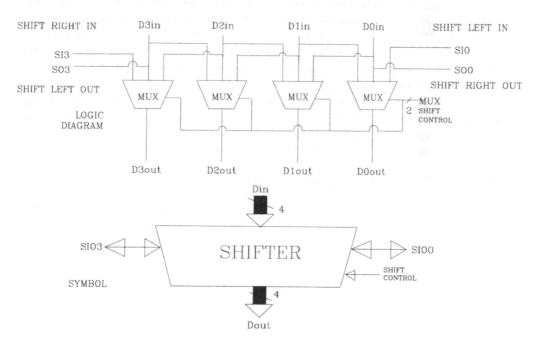

Figure 3-5: 4-bit shifter.

the least significant bit (lsb) shifted out during a right shift can be captured in external memory or a bit can be supplied to the lsb during a left shift.

When the SHIFT RIGHT IN and SHIFT LEFT OUT are similarly combined, the MUX SHIFT CONTROL must be decoded to enable the SO0 and SO3 paths as well as provide the proper input select code to the three input multiplexers in the parallel data path Din to Dout. This allows a complete 4-bit wide data path with the proper fill bits for 1-bit left and right shifts of data on Din.

The width of the main data path, Din to Dout, can be increased by adding more three input multiplexers, one for each bit in the wider main data path. The select control for each added MUX is connected to the same MUX SHIFT CONTROL signal applied to the other MUXes.

A network that can shift data by more than 1 bit position can be implemented by reconnecting the MUX inputs to the desired bit position in the incoming bus, Din. If it is desired to shift the incoming data to the left by two bit positions, then D0in should be connected to the input of the D2out MUX instead of that of the D1out MUX. D1in should run to the input of D3out MUX, etc. A 2-bit SHIFT LEFT IN path will be needed. SI0 would go to the input of the D0out MUX and SI1 would go to the input of the D1out MUX. A similar arrangement is required for the shift out path at the left end of the network. SO3 comes from D2in and SO4 comes from D3in. All reconnections are made to the MUX inputs with the same select code.

A *barrel shifter* is a network that can shift an incoming path by one of a large number of bit positions, usually the maximum number meaningful for the incoming data path width. In this example using a 4-bit incoming data path, the maximum left shift would most likely be 4 bit-positions while the maximum right shift would also be 4 bit-positions. As an example, consider the network that would provide all of the bit-position shifts for a 4-bit data path Din from two bit-positions left to two bit-positions right. This requires the three input MUXes in Figure 3-3 to be replaced by five input MUXes. The shift in and shift out data paths SI3, SO3, SI0, and SO0 would each be 2 bits wide, as in the previous paragraph. The MUX SHIFT CONTROL which points the shift MUXes now requires a 3-bit code.

Several arithmetic and logical operations require shifting by an arbitrary (controllable) number of bit-positions. If the shift operation can be performed in a single pass through the shift network, then the operation can be much faster than with a single-bit shift network and the control algorithm will be simpler. Some of the operations that incorporate shifting include bit isolation, binary point alignment, and normalization.

3.5.3. The Arithmetic/Logic Unit (ALU)

Figure 3-6 shows the symbol for a structure of combinatorial elements that performs the major arithmetic and logic operations that are needed in our implementation of the data translation equations. Consider the function table for a hypothetical Arithmetic/Logic Unit (ALU) below:

ALU Function Table

F SEL	Function
0	Bitwise One's Complement the bits in R
1	Bitwise AND the bits in R with those in S
2	Bitwise OR the bits in R with those in S
3	Bitwise Exclusive OR R with S
4	Two's Complement ADD R to S with Cin
5	Two's Complement SUBTRACT S from R with Cin
6	Two's Complement SUBTRACT R from S with Cin

In principle, the network can be thought of as one copy of each of the components needed to perform the operation required for each bit in the outgoing data path F. The incoming data A and B are routed to the proper combinatorial network where the combination is performed and the result is conducted to output F.

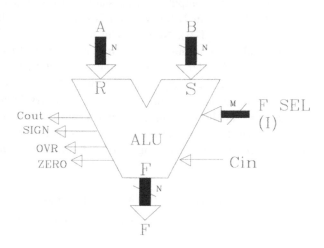

Figure 3-6: The Arithmetic/Logic Unit (ALU).

For a 4-bit wide ALU, we would need four complementers, four AND gates, four OR gates, four Exclusive OR gates, a 4-bit full adder, and a 4-bit full subtracter. Several multiplexers would also be required to handle the routing of the data to the network selected by F SEL and thence to the output F. Since the data paths are several bits wide, each multiplexer will be made using several copies, one for each bit in each data path. It is clear that this is a complex network to implement and some time needs to be spent in reducing it in order to minimize propagation delay and complexity.

The first minimization possible is associated with a property of the two's complement number in which we are implementing arithmetic. Recall that

$$F = A \text{ minus } B = A \text{ plus } (-B)$$

in any number system. In the two's complement number system

$$-B = \text{NOT}(B) \text{ plus } 1$$

where NOT(B) is the bitwise one's complement of the binary representation of positive B. This will cause the subtraction operation for F above to become

$$F = A \text{ plus } \text{NOT}(B) \text{ plus } 1$$

The subtraction operation has been replaced by addition done twice, once to add A and NOT(B) and then once to add the sum and 1.

Examination of the multibit FULL ADDER network in Figure 3-4 shows that multibit paths for A and B go into the network along with a single-bit path Cin. The multibit sum S and single-bit Carry Out are produced combinatorially; i.e., once the inputs are applied and a propagation delay is observed, the outputs will be stable until an input changes. Consider conducting the B data path through an N-bit wide one's complementer (here four inverters, one in each B bit-path) to form NOT(B). Now set Cin to 1 (High according to the symbol in Figure 3-4). After a suitable propagation delay time appropriate to the logic family in which the complementer and full adder are implemented, the *difference* is available on the Sum output of the Full Adder and the Carry Out holds the borrow, if any.

In order to interpret the meaning of the numbers on the Carry In and Carry Out lines, one should examine the value of the Full Adder output S and its meaning if Cin had been set to 0 instead of 1 as above. In other words, what does A plus NOT(B) plus 0 mean? If the Borrow In is assigned the values 0 meaning *True* and 1 meaning *False* then it is seen that

$$A \text{ plus } NOT(B) \text{ plus } 0 = A \text{ minus } B \text{ minus } 1$$

$$= A \text{ minus } B \text{ BORROW } 1$$

while

$$A \text{ plus } NOT(B) \text{ plus } 1 = A \text{ minus } B$$

$$= A \text{ minus } B \text{ BORROW nothing}$$

The automatic generation of the proper No Carry In and No Borrow In can be accomplished by generating a control signal called ADD/SUB* which will obey the following function table.

SUB/ADD*	Function
H	Subtract B from A with no Borrow In
L	Add A to B with no Carry In

Then, using the Full Adder network in Figure 3-4 and an Exclusive OR gate (with positive logic inputs), we can make the following replacements:

$$\text{Carry In} = SUB/ADD*$$

$$B0 = B0 \text{ ExOR } (SUB/ADD*)$$

$$B1 = B1 \text{ ExOR } (SUB/ADD*)$$

$$B2 = B2 \text{ ExOR } (SUB/ADD*)$$

$$B3 = B3 \text{ ExOR } (SUB/ADD*)$$

We have succeeded in replacing the separate addition and subtraction networks with a single addition network and a simpler Exclusive OR network.

The next simplification occurs when a truth table for the ALU is written and the standard minimization methods are applied to the outputs. This greatly reduces the number of gates required to implement the desired functions.

The flag outputs (i.e., the flags Cout, Sign, OVR, and ZERO) are generated combinatorially while the function F is being produced. They must be interpreted based upon the meaning of A, B, and F as arithmetic or logical variables. In the function table for the ALU given above, we can identify the first four functions as LOGIC functions. In these functions, there is no relationship between what happens in one bit position and any other bit position. Only Sign and ZERO have meaning for these types of operations. The ZERO flag is generated by the following function:

$$\text{ZERO} = \text{NOT}(F_n + F_{n-1} + ... + F_1 + F_0)$$

If all of the bits in the output function of the ALU are 0, then the ZERO flag will be *True*. This flag is useful both for arithmetic and logical operations, and its interpretation is not a function of which type of operation we are doing. The second "LOGICAL" flag is the SIGN, which is defined as

$$\text{SIGN} = F_n$$

The SIGN flag means NEGATIVE if *True* and the output function F is interpreted as a two's complement signed number. However, it always tells the value of the most significant bit of the function produced by the ALU whether that number is considered signed or unsigned, i.e., arithmetic or logical. It is useful for implementing tests of bits that have been isolated by moving them to F_n position via the shifter.

The flags that are normally considered to be arithmetic are the Carry Out (C) and two's complement overflow (V or OVR). The C flag is generated in the N-bit Full Adder that is the heart of the arithmetic network in the ALU, as we have shown above. Its generation is usually based on the Lookahead Carry Method. It can be interpreted as the Carry Out produced by adding two numbers A and B and the Carry In (Cin). In this capacity, the numbers being added are considered to be unsigned binary numbers, in which case the Cout flag also stands for *unsigned overflow*. It also is the Borrow Out, interpreted using negative logic, when the subtraction operation is implemented as above. The reason that two versions of the subtraction operation are implemented in the ALU is to give the designer a choice in the order of the operands so that this flag is more easily interpreted when it is thought of as the Borrow Out (e.g., in comparison operations). The values of the these two flags has no meaning after a logical operation and some ALU implementations will produce zeros on them in this case or will let them have any value that produces a simplification of the network. In either case it will be necessary for the designer to determine when the values of these flags are saved.

The two's complement overflow flag (V or OVR) is produced by the following function.

$$\text{OVR} = \text{Cout ExOR (Cin to Most Significant Bit)}$$

Its form was predicted in the discussion of two's complement overflow in Chapter 1. It, like all of the flags, is produced combinatorially, which means that it will be available at the same time the sum or difference is produced on the function output lines, F.

Interpretation of the result is performed by testing the OVR flag using logic during the controller states that follow the addition (subtraction) step. It is important to realize that the Full Adder, in which addition and subtraction of unsigned and two's complement binary numbers is performed, has no knowledge of the nature of the coding of the numbers passing through it. The designer imposes the meaning of the result after it is generated in the ALU when he provides states or hardware in which one or another flag is tested. If he tests the Cout flag when asking about overflow, then he has said that the F output is an unsigned number; and if he tests the OVR flag, then he has said that the F output was a two's complement number.

What about the MULTIPLY and DIVIDE operations — why are they not included in the ALU? Most combinatorial implementations of the multiply operation for integer numbers occupy more silicon than the entire ALU shown so far. It has proven more effective to implement other functions such as shifters and memory with the ALU shown above than to add other combinatorial operations. This may change in the future as technology improves, but we are currently in a period of implementing 32-bit wide ALU/memory/shifter networks for integer numbers that are very fast (about 50 nanoseconds) and flexible. If an architecture requires the addition of a MULTIPLIER, then one can be added as a separate device and operated by generating more control signals in the Controller. It may have access to the same internal buses available to the ALU if necessary.

The DIVIDE operation is used less often than the other arithmetic operations and is generally implemented by using an algorithm. A division operation between a constant and a data value can be replaced by the product of the inverted constant and the data value, thus eliminating the need for the division operation all together. Many times, dividing by a power of two can replace the actual division needed in which case the division can be replaced by a shift right by the power of the operation. If a barrel shifter is present, then this operation is very fast.

3.6. MEMORY ELEMENTS

Our descriptions of implementations of logic functions have included only combinatorial elements up to this point. We will now consider structures with which to build the memory blocks that are shown in Figure 1-2 in our implementable solution of the data transformations in Figure 1-1.

3.6.1. Flip-flops

The term "flip-flop" is used to denote a small class of networks that "remember" the last value of certain inputs even after these inputs change. The name comes from the two-state nature of their binary implementation in which the device is said to "flip" from one state to the other while "flop" implies the return to the original state. A mechanical implementation of this device clearly appears to "flip" and "flop".

There are several different designs that show this characteristic capability to "remember". Some of the more important flip-flops are shown in the following list along with their characteristic equations.

SR	$Q^+ = S + /R * Q$	Set/Reset flip-flop
JK	$Q^+ = /K*Q + J*/Q$	Clocked J-K flip-flop
D	$Q^+ = D$	Clocked D flip-flop

The "+" symbol on the output Q stands for the value of the output Q after the clock or input event. The SR flip-flop is set or cleared by synchronous inputs called S and R. An asynchronous version is frequently implemented on the output portion of a clocked flip-flop to provide a means to place a logical 1 or 0 in the flip-flop without reference to the clock. The output of the SR flip-flop is undefined after the active-edge of the clock when S and R are both True. This undesirable condition is eliminated in the next two flip-flops. The two inputs of the JK flip-flop, J and K control the state assumed by the output after the clock. All four combinations produce defined outputs including one that toggles at each edge of the clock. The D flip-flop on the other hand requires only one input, D.

The clocked flip-flops are so constructed that the duration of the clock pulse does not affect the operation of the element. The flip-flops are said to be *edge-triggered*, i.e., the output assumes its new state after a particular transition direction (e.g., Low-to-High) on the clock input. The time required from the triggering transition until the output assumes the new state is dependent only on the propagation delay through the circuit from the clock input to the Q output and not on any other input signal condition. This propagation time is dependent on the implementation technology.

Figure 3-7 shows the next state table, equation, and drafting symbol for the *edge-triggered D flip-flop*. This is an important flip-flop because of its simple next state control (i.e., the D input) and the edge-triggered property of its clock. Generally, if the D input is

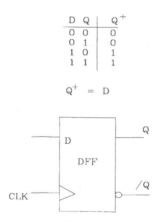

$$Q^+ = D$$

Figure 3-7: Edge-triggered D flip-flop.

valid at the instant the triggering event occurs on the clock input, the output will be set to the value of D. There is very little if any requirement for the data input, D, to be held after the clock transition. This makes the control of this structure much easier than other types of memory. All flip-flops that are used in the implementation of a state machine are built using the same technology as the combinatorial networks. This may raise their cost but it makes them as fast as the networks to which they will be connected.

Figure 3-8 shows a more complex structure consisting of a multiplexer and a D flip-flop. This structure is given the name of *enabled, edge-triggered D flip-flop*. Its characteristic equation in the form given in the figure begins to imply its usefulness as a device to be placed in the architecture of a state machine for control by a controller as shown in Figure 1-3. The output of the D flip-flop will assume the value placed on the DATA IN input to the two input MUX at the next rising edge of the clock if the control signal LOAD is *True*. Otherwise, the D input of the D flip-flop receives the old value of the Q output which is placed on the Q output at the next clock. The important feature of this structure that makes it so easily controlled is that the LOAD signal, and DATA IN if needed, must be stable only for some SETUP TIME before the clock transition, but need not remain stable after that time. Frequently, the same clock transition that serves to latch data into the architectural memories is the one that causes the controller to change to a new state. The control signals are no longer valid once the state change starts; therefore, this feature is very useful for memories in the architecture.

$$Q^+ = LOAD * DATAIN + /LOAD * Q$$

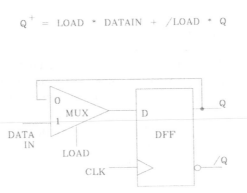

Figure 3-8: Enabled, edge-triggered D flip-flop.

Another method to accomplish the function of selectively loading a flip-flop is called a *gated clock*. In this case, a logic function is ANDed with the system clock of the D flip-flop. If the logic function "permits" the system clock to pass through to the flip-flop, then the data on the D input will be transferred to the Q output and "remembered". Otherwise, since there is no clock transition applied to the clock input, the Q output will not be changed. If the logic function granting permission to pass the system clock were implemented using gates with no propagation delay, then the system would work; however, all real gates have delay. One effect this has is that the input data may have changed during this delay. Another effect is that the various propagation paths in the "permission" network may not all be equal which will cause hazards, or glitches, that may cause the clock input to transition more than once or at a time far removed from an actual clock transition. The design of a system using a *gated clock* is possible; however, it requires that serious attention be paid to all possible (and impossible) signal combinations to determine if hazards will occur. This many times amounts to more time than the design phase can afford. The enabled D flip-flop completely removes this problem from consideration since there are *no* logic networks in the path of the system clock as it is distributed into the architecture. The enabled D flip-flop is generally available in each of the logic families as an integrated circuit and can be "made" when using various implementation technologies such as Programmable Logic Devices (PLDs), gate arrays, or custom logic.

3.6.2. Registers

Now we shift our attention to building more complex structures using the D flip-flop as a memory element in the same way that we built complex combinatorial circuits using the basic gates. Remember that the flip-flop structure is composed of the basic gates, as well.

The distinguishing feature between a flip-flop and a full adder is the layout of the interconnection paths and not the gates from which they are made.

Our first structure is a 4-bit parallel in, parallel out register shown in Figure 3-9. The 4-bit wide data path through the structure is made possible by placing four identical D flip-

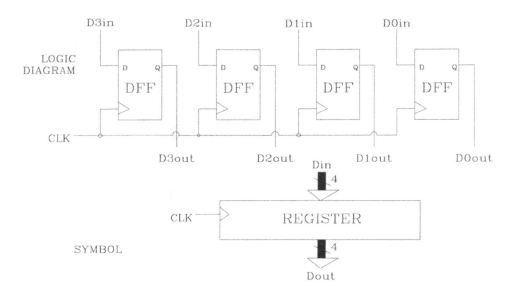

Figure 3-9: 4-bit parallel-input, parallel-output register.

flops side-by-side and connecting their clock inputs to a common control line called CLK. The circuit will require four separate 1-bit lines to implement the data input (D0in–D3in) and a similar output structure. An 8-bit register would be similarly constructed except that it would contain eight D flip-flops and require eight Data input lines as well as eight Data output lines. The meaning of the D3in line is established only when the register is connected to an element that places significance on a bit position, e.g., a 4-bit full adder. If the D3in line were connected to the most significant Sum bit (e.g., S3 in Figure 3-4), then the meaning of the D3out and the flip-flop from which it comes would be interpreted as the sign-bit of a 4-bit two's complement notation, etc.

While the register in Figure 3-9 is useful for implementing any memory that is updated every clock cycle, it does not lend itself to control any better than the plain D flip-flop from which it is made. The *enabled D flip-flop register* shown in Figure 3-10 places four enabled D flip-flops in a structure that preserves the multibit input and output data paths and provides a new control signal that operates on all four flip-flops simultaneously called LOAD. In this example, LOAD is a positive logic signal, but it is found in a negative logic implementation in some of the device families.

The 4-bit register in Figure 3-11 uses the original register structure in Figure 3-9 and adds a tristate output buffer to each flip-flop. The buffer structure obviously does a better job of isolating the flip-flop outputs from the devices to which they are connected and supplies added current for driving more loads than the register in Figure 3-9. However, an added level of control is also available through the tristate input on each buffer.

The output of the buffer will be the same as the output of the D flip-flop to which it is connected if the OUTPUT ENABLE signal is *True*. If this signal is *False*, then the current sourcing and sinking transistors are turned off and the flip-flop output is effectively dis-

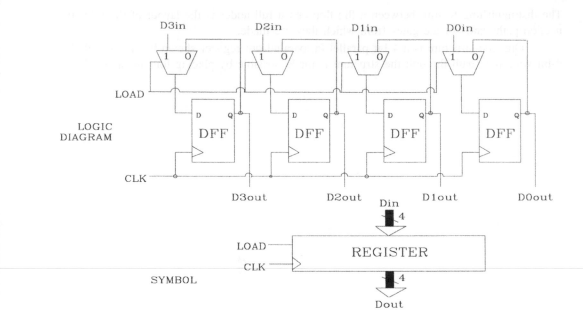

Figure 3-10: Enabled register.

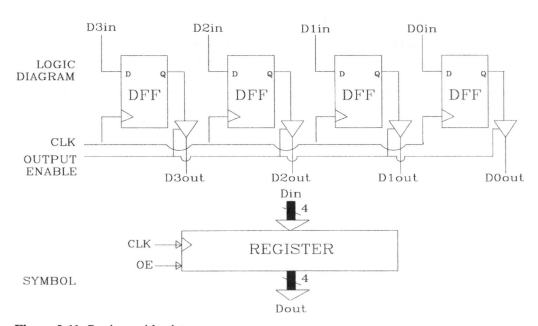

Figure 3-11: Register with tristate outputs.

connected from the output of the register. This allows the outputs of several registers to be connected to a common output structure called a BUS. In the example shown for a 4-bit register with tristate outputs, this implies that a 4-bit bus (consisting of four wires) could be connected to the inputs and outputs of several registers. Data could be moved onto the bus from a register by enabling its output buffers. When a clock signal was sent to another register on the bus, the data would be loaded into the register. This is called a register transfer and will be an important part of our design operations. Notice that the *gated clock* implied in the example would not be necessary if the tristate output buffers were incorporated in the register composed of enabled D flip-flops in Figure 3-10.

In this section, we have shown an important building block consisting of a one-dimensional array of D flip-flops called a *register*. The purpose of these structures is to "remember" groups of binary digits, i.e., numbers or values of variables. They can be extended to improve their control capabilities to make it easier to combine them to form more complex structures as we will now do.

3.6.3. Complex Register Structures

The first complex structure we will study will be used as a memory structure in our Register Arithmetic/Logic Unit (RALU) that forms a major part of the internal architecture of a computer. This structure is called a *multiport register* and is shown in Figure 3-12. A two-dimensional array of D flip-flops is constructed having an organization of 2 bits wide and 4 levels deep. The "two" direction can be thought of as a 2-bit version of the "register" described in Figure 3-9 above. Four of the 2-bit registers are arranged one-above-the-other in Figure 3-12. If one wanted to create an array of 8-bit registers, then the drawing need only be extended to the right by duplicating the array of D flip-flops. No new control structure would be needed.

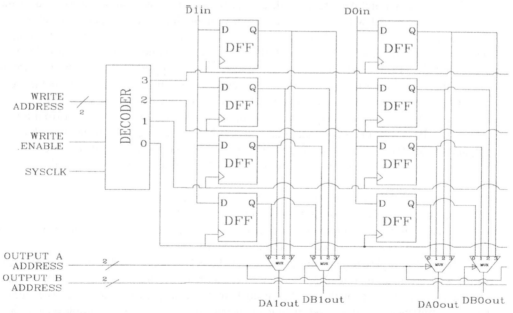

Figure 3-12: Multiport register (two 2-bit wide ports by four words deep).

Consider the four input multiplexers that are labeled with the symbols "DA0out" and "DA1out". The BUS called DA, of which these are two of the bits, will contain the outputs of the "register" (word) selected by the four-input multiplexer to which it is attached. The selection is made by the code placed on the control input called OUTPUT A ADDRESS. To "read" the top row of bits (i.e., the top register or word), the OUTPUT A ADDRESS lines both would be made 1. If these signals are interpreted as a single number, then the top register would be read onto the DA bus when OUTPUT A ADDRESS is 3. The bottom register contents would be placed on the DA bus when the OUTPUT A ADDRESS is 0.

To place the data on the Din BUS (i.e., denoted by D0in and D1in) into one of the "registers" or "words" in the structure, it is necessary to cause a positive (Low-to-High) transition on the clock input of the selected register. The decoder shown to the left in the figure is designed to use the WRITE ADDRESS to select one of the registers (00 points at the bottom one, while 11 points at the top one) to which to send the WRITE ENABLE, or clock, signal. In this example we have shown a *gated clock* in order to simplify the drawing. The structure can be built this way since the decoder is a small network that allows all hazards to be examined and removed.

If our network consisted only of the structures discussed thus far, we would call the network a *memory*. We have used an array of registers, some MUXes, and a WRITE operation decoder to build it. Notice that in this form data on the input bus, Din, can be written into a register at an address specified by the control input WRITE ADDRESS at the *same time* as data from a register specified by the control signal OUTPUT A ADDRESS can be placed on the output BUS, DA. This simultaneous READ and WRITE operation is lost in memories used in the external architecture of computers (Figure 1-6) since the WRITE and OUTPUT A ADDRESS are combined to minimize the cost of the structure. In most cases, the input bus and the output bus are the same in a bulk memory structure as well, in which case the WRITE ENABLE signal must be used to "turn off" the data output MUXes when a write operation is to be performed.

In the internal architecture we will generally not make the compromises used in the external architecture and will preserve the integrity of the input and output buses above. If we add another set of output multiplexers as is shown in Figure 3-12, we can create another output bus called DB, note DB0out and DB1out. A second output address control signal is added, called OUTPUT B ADDRESS. If our controller can supply all three addresses simultaneously, then we can write data on Din to one register while placing data from another register on the DAout bus and from a third register on the DBout bus in the same clock cycle. Notice that READing does *not* require a clock, but writing *does*. Data movement from a register to a bus (READing) is a combinatorial activity that will happen all during a clock cycle in preparation for the writing (i.e., data transfer to a register) that occurs at the clock transition. We have now arrived at the final form of our MULTIPORT REGISTER, a device that can perform three operations at once. Notice that it requires three buses and three control signals to support this operation. It may appear that we have created too many output buses since no apparent use for the second one comes to mind. Consider the full adder network. How many data inputs and outputs does it have, ignoring the carries? Could these be served by an array of registers configured as a multiport register?

The next complex register structure is called a STACK and is shown in Figure 3-13. It begins with an array of four 2-bit wide registers as before. The registers could be of any width (e.g., 4, 8, or more bits), but 2 is used here for clarity in the drawing. The output structure is implemented using a set of four input multiplexers with one multiplexer for each bit in one of the registers and one input for each register in the array. The MUX select signal is 2 bits wide and will be generated at the output of an UP/DOWN Counter which will be discussed below. An input bus called Din is connected as above to all of the registers. A decoder is used to gate the system clock signal such that only the register selected by the 2-bit wide address from the UP/DOWN Counter will receive the clock and thus load the data from the input bus. The array will be written into only when the signal STACK ENABLE* is *True*; otherwise, no new data is entered. The output bus DAout will always contain the data in the register pointed to by the output of the counter at all times regardless of the value

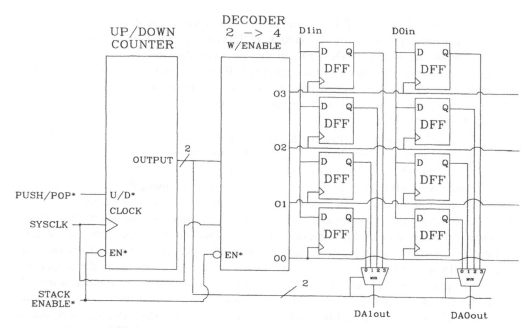

Figure 3-13: Stack (2 bits wide by four words deep).

of STACK ENABLE*. If the counter were ignored, then the structure described to this point, consisting of the register array, the output multiplexers, and the decoder, would be simply a single port register array or memory.

The addition of the UP/DOWN Counter to the structure implements a STACK. The counter obeys the following function table:

EN*	U/D*	Function of UP/DOWN Counter
L	L	Decrement counter on next positive clock
L	H	Increment counter on next negative clock

The active edge of the clock is a function of the counting direction so that the counter will not change while data is being written.

The STACK structure has as control inputs the STACK ENABLE* and PUSH/POP* signals. The data on Din can only be placed in a register in the STACK in synchronism with the clock when allowed by STACK ENABLE*. The following function table describes the operation of this structure.

Function of STACK

EN*	PUSH/POP*	Counter	Write	Read	Function
L	L	Decrement	Yes	Yes	POP
L	H	Increment	Yes	Yes	PUSH
H	X	Hold	No	Yes	HOLD

Notice that data is available from the STACK even when it is not enabled. This means that the value pointed at by the STACK POINTER (i.e., the UP/DOWN Counter) can be used without changing the POINTER. In this example, data will be placed on the stack whether the pointer is incremented or decremented. If we interpret incrementing the counter and writing as a PUSH, then decrementing the pointer is a POP. Notice that even if the data had not been written, it would have been lost since the next PUSH would have destroyed any data that was "POPped".

This stack structure contrasts with most stacks that you have encountered before. In most microcomputers, the stack builds (i.e., PUSHes) toward lower addresses. Also, in order to get a word from the stack, the data had to be POPped and was thus lost from the stack. While the data was still physically there it could not be reached using the stack structure and would have been destroyed by the next push. Most software stacks implemented in the computer environment have these same features so it will be necessary to justify these design differences when we employ this structure later. In microcomputer implementations it takes several machine cycles to change the stack pointer and move the data to or from the stack in memory. Here, the data in the stack can always be read, i.e., placed on the output bus. Data can be placed on the stack at the new address just produced by the counter at the falling edge of the clock because the data will be written when the clock rises at the end of the same cycle. This can be accomplished only if the elements in the stack (the registers, decoder, and counter) are faster than most other parts of the architecture. A POP operation is not used to get data off of the stack, but to repoint the stack pointer (throw away "exhausted" data).

In conclusion, we have discussed two complex structures of registers. Both structures did not in themselves modify the data placed in them, but served only to hold the data for later use. The multiport register and the stack contribute to the ease of use of the storage structures in an architecture by giving special control characteristics to them. This allows these characteristics to be optimized such that the devices are convenient to use with the anticipated controllers and are much faster than if the functions had been implemented using less sophisticated structures. In other words, a function that could be performed on general purpose hardware was implemented as a special structure to improve its speed and handling.

3.7. COMPLEX ARCHITECTURES

It is now time to combine several of the structures discussed above into even more complex architectures. The specific combinations have been dictated by commonly used operations such as accumulation. Our building blocks will consist of the ALU, register array, shifter, and data path. We will first consider units that implement arithmetic and logic functions on unsigned and signed binary numbers in integer or fixed point forms. Next, we will consider the operations that can be accomplished in one clock cycle to perform multiplication on this data type. Finally, we will outline the operations that can be performed on the floating point data type in this time.

3.7.1. The Accumulator

Figure 3-14 shows an ALU connected to an N-bit wide register. Both elements have been discussed in earlier sections and the *accumulator* represents the first of the structures formed by combining memory and data combination devices. The accumulator is literally a structure in which to perform the operation

ACC = ACC plus DATA.

Our version of this circuit is implemented using the ALU instead of the Full Adder. The ALU obviously reduces to the Full Adder when the Function Select input on the ALU is set to select addition. This equation describes an operation that is fundamental to many

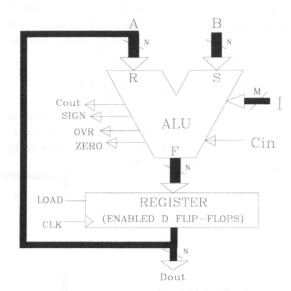

Figure 3-14: The accumulator.

algorithms, that of continued summation. We see it so often that it has been given the name above. We have combined two parts, one with three data ports and the other with two, which implies that there are several possible ways that interconnection can be made. Of the several combinations, only the two versions that connect the output of the ALU to the register input with the output of the register being connected to the ALU R or S inputs is called an accumulator. There are several important points to be made about this combination of memory, combinatorial elements, and DATA PATHs. First, the paths present are fundamental to the implementation of the desired function. There is a direct relationship between the placing of the paths and the function implemented. This was obvious in smaller structures shown earlier, but it now becomes so important that we should identify this step of design as the heart of architecture as a discipline.

The second point to be made is that the register *must* be edge-triggered since, when the clock input "ticks", it will be required to latch the new value generated by ALU even though its outputs will start changing at the same instant. Obviously, if the register outputs change, then the R input to the ALU will change, thus changing the ALU F output which leads to the input of the register changing. Any other clock behavior would cause the results in the register to be indeterminate, approximating the number of "sums" that could be performed while the register was "transparent" based on the propagation time of the ALU. A simple experiment will convince you of the problem. Consider implementing the ALU with a 74LS181 and the register with a 74LS374. Test the performance of this combination and then replace the register with a 74LS373 transparent D register. The second version will work only if the clock pulse is narrower than the propagation time through the network or if the structure is changed.

Data are applied to this version of the accumulator via the S port on the ALU and taken out on the Dout bus. It is up to the designer of the controller that supplies the ALU Instruction, I, and LOAD to determine which codes and what order to follow to implement

the accumulate equation. In words, one must supply, and hold for the duration, an ALU instruction that commands the ALU to ADD (e.g., FSEL = 4 from the ALU discussion above).
LOAD = *True* must also be applied and held. The data values to be accumulated are applied
through the S port to the ALU, one for each clock pulse. Notice that the only element to be
clocked is the REGISTER. The ALU does *not* receive a clock. Its outputs are *always valid*
(except for propagation delay). If 10 numbers are to be added, then each must be placed on
the B bus followed by a rising edge on the register's CLK input.

3.7.2. The Register Arithmetic/Logic Unit (RALU)

Figure 3-15 shows a more complex structure than the accumulator that nevertheless
contains many of its elements. The register in the accumulator could contain only a single
value. In the Register Arithmetic/Logic Unit (RALU) it is replaced by a register array implemented from the multiport register discussed earlier.

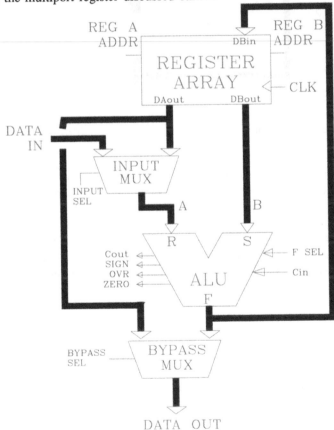

Figure 3-15: The Register/Arithmetic Logic Unit (RALU).

The ALU still has a path from its function output F to the input of the register array
just as in the accumulator. In this figure, the feedback path from the register array is via
DBout to the B input. This corresponds to the path from the register output to the R input
of the ALU in the accumulator. A path from the outside into the ALU is implemented via
the INPUT MUX connecting the data input bus DATA IN to the R side of the ALU. This
should be compared to the data input path B in the accumulator architecture (Figure 3-14).
With the elements described it can be seen that the accumulation equation can be implemented in an RALU.

There are some additional features that make the RALU more useful than the accumulator in implementing a variety of functions. The first exploits the fact that the register array is indeed an array of data. If the path from the DA port, one of the two ports on the register array via the input MUX to the R input of the ALU, is used, along with the DB bus into the S input of the ALU, then two data values in the register array can be combined by any one of the ALU functions selected with F SEL. The results can be saved in the register array and/or passed via the bypass MUX to the outside world. Notice that the saving of the result in the register array and its use outside are simultaneous and independent of each other. The last feature that is added in the architecture is the path from the DA port on the register array via the bypass MUX to the outside world.

We will now give some examples of the use of these paths. In these examples it should be noted that the REG A ADDR input points at the register in the array whose output is to be connected to the DA bus. On the other hand, the REG B ADDR input is used to specify the register whose output is to be placed on the DB bus going the ALU S input. The B address field in many RALUs is *also* used to specify the register to which the ALU F output is to be written when the clock input to the register array transitions. In some RALUs, a third address input is provided for optional use giving the appearance of a three address architecture. In this connection, remember that all operations are performed in *one* clock cycle. Data can be taken from a register pointed by REG A ADDR *and* one pointed at by REG B ADDR, combined in the ALU by one of several operations and the result placed in the register pointed to by REG B ADDR at the end of the clock cycle (the next rising edge).

The first example is the original one of accumulation except now we have a place for the input data to come from and for the result to be placed.

Reg(REG B ADDR) = Reg(REG B ADDR) plus Reg(REG A ADDR)

Here, the accumulator register is seen to be located where REG B ADDR points. If the controller supplies a new REG A ADDR each time the clock ticks and holds F SEL to select addition, then the data will "accumulate" for the number of registers specified by it.

The second example is given in a more general form that is intended to show that the fundamental pieces of operations can be performed in one clock cycle in the RALU.

Reg(REG B ADDR) = Reg(REG B ADDR) .OP. Reg(REG A ADDR)

The ".OP." stands for one of the operations that can be performed by the ALU, such as addition, subtraction, AND, etc. Many data transformations are sequences of steps using these operations. For example, a product can be performed, albeit slowly, by repeatedly adding the multiplicand to itself a number of times equal to the multiplier. So if the following sequence was performed,

HERE:

Reg(PROD) = Reg(PROD) plus Reg(MCAND)

Reg(MPR) = Reg(MPR) minus 1

IF (Reg(MPR) > 0) THEN GOTO HERE ELSE FINISH

FINISH:

This requires the controller to be able to test the results of an operation in order to implement the loop structure. The behavior of the controller will be discussed thoroughly in the next chapter; however, we can examine how the information is passed to it from the RALU where the data reside. Consider the arithmetic operation required in the IF statement above:

Reg(MPR) > 0

This is implemented by setting up the RALU to perform the following subtraction:

Reg(1) minus Reg(MPR)

The Reg(1) is intended to have the controller point, using REG A ADDR, at a register in the register array that has been previously initialized to one. The ALU is set to subtract that register containing the multiplier (MPR) from the register containing one. If the sign of the result is negative, then the operation is terminated by the controller. Notice that the register array write enable control would be set *False* during the comparison since there is no need to save the result produced by the ALU. The important information is in the sign flag that is tested by the controller, the subject of the next chapter.

It should be seen that the RALU can cooperate with the sequencing capabilities of the controller to produce complex sequences of operations that ultimately evaluate our data transformation equations with which we began Chapter 1.

3.7.3. Real RALUs

In order to implement the RALU structures that are used in current designs, we will introduce two more structures shown in Figures 3-16 and 3-17. We noticed in the preceding example that the multiplication algorithm used was one of the slowest possible. If we introduce a SHIFTER in the feedback path from the ALU to the register array, we will be able to exploit a much faster algorithm. This shifter is also important for bit isolation and for data reformatting. Our illustration here using the shift and add multiply algorithm is but one of a large number of applications for this network in our RALU architectures.

The shift and add multiplication algorithm is performed much like the one used when multiplying numbers by hand. In the case of decimal numbers, you form a list of partial products, each offset to account for the position of the digit in the multiplier. You start from the least significant digit of the multiplier and form the product of that digit with the entire multiplicand. The next partial product is formed the same way except that it is written down offset one digit position to the left since the multiplier digit had a weight of ten. This continues until the complete list of partial products has been produced. The column of partial products is then summed to create the final product.

The binary multiplication scheme runs the same way except there are several economies possible. The binary digits take on only two possible values, 1 and 0. The partial products are either equal to the multiplicand (i.e., if the multiplier bit is 1), or to 0 otherwise. This makes the actual multiplication to produce the partial products very simple — either copy down the multiplicand or do nothing. Offsetting of the partial products can be accommodated by shifting the multiplicand to the left 1 bit-position each time you try to add it to the list of partial products. The actual implementation of the operation is done by placing the multiplier in the least significant end of the double length field needed for the final product. Remember that multiplication of two three-digit numbers can produce a six-digit product.

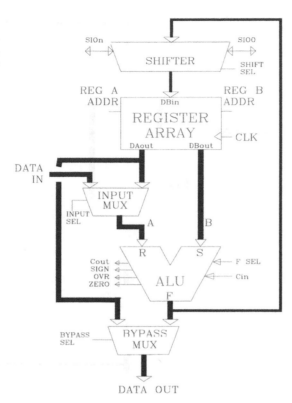

Figure 3-16: RALU with shifter — Version 1.

The digital world requires that we anticipate the largest possible data value in our algorithm. The most significant part of the product is cleared to 0. The algorithm then tests the least significant end of the multiplier and, if 1, adds the multiplicand to the most significant half of the product. To show this as a series of instructions, use the RALU in Figure 3-16. Consider that a D flip-flop input has been connected to SIO0 to capture the least significant bit of the multiplier as it is shifted out of its register in the register array. The controller can see this flip-flop and make control decisions based on its contents. Also, the Carry Out of the ALU will be captured in another D flip-flop which will can be right shifted into SIOn. Now consider the following sequence of steps expressed in PDL, a language described in Chapter 6.

```
LOOPAGAIN:
          Reg(MPR) = .SHIFT RIGHT. Reg(MPR);SIO0FF = msbit of Reg(MPR)
          IF (SIO0FF = 0) THEN GOTO NOADD
          Reg(PROD) = Reg(PROD) plus Reg(MCAND); CoutFF = Cout
NOADD:
          Reg(PROD) = .SHIFT RIGHT. Reg(PROD); SIOn = CoutFF
          Reg(COUNT) = Reg(COUNT) minus 1
          IF (COUNT > 0) THEN GOTO LOOPAGAIN ELSE END
END:
```

The two D flip-flops mentioned are identified here by CoutFF and SIO0FF. The semicolons ";" indicate that the events that follow on the same line happen during the same clock cycle or controller state. The variable COUNT is maintained in another one of the RALU

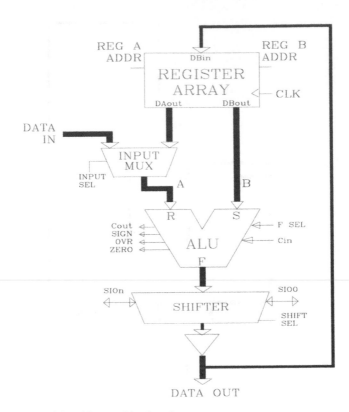

Figure 3-17: RALU with shifter — Version 2.

registers and contains a number equal to the number of bits in the multiplier when the algorithm starts.

We have implemented the multiply operation in fewer steps than was required in the repeated addition scheme shown earlier. Several improvements can be made, some now, some by adding sections to the RALU, and some in the controller. The first improvement would be to exploit the fact that both the multiplier and double-length product need to be shifted. The product is normally located in one register while the double-length product occupies two more. If the multiplier is placed in the least significant end of the product register pair, then it will not be overwritten by the accumulation of the multiplicand but *will* be shifted when the entire product is shifted. Thus, a three-register shift will be reduced to a two-register shift. This reduces the algorithm by one step for each bit in the multiplier.

The next improvement is made by recognizing that this double length shift operation appears in many other places as well, e.g., division, normalization, etc. If another register and shift network are introduced into the RALU, then the multiplier and least significant half of the product may be placed in it. The important thing about adding hardware is that if the data paths don't interfere, then the operation of the new network can occur at the same time as operations in the main structure. In this case, the upper half of the product is in one of the registers in the register array while the lower half (and multiplier) is in the new "shift register". A double-length single-bit shift can be accomplished in one clock cycle instead of the three with which we began. The only improvement left is to move the COUNT to a place where it can be modified and tested at the same time that something useful is going on in the RALU. We will reserve this modification until the next chapter.

In summary, a recent "generic" RALU is shown in Figure 3-18. Its architecture is derived from that in Figure 3-17 and the "Q-Register" and its shifter are added as separate and independent data paths to support multiplication and division as outlined above. A multiplex-

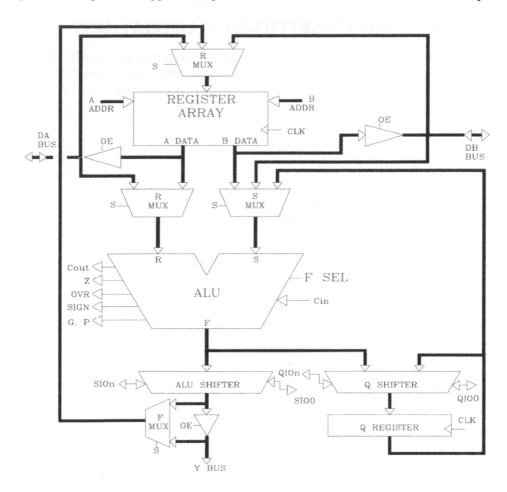

Figure 3-18: Generic modern RALU.

er is placed with its inputs straddling the output tristate bus driver. This allows the designer to protect the accumulator data path from data on the bidirectional Y-Bus.

The width of an RALU is determined by the number of bits in the data paths and the width of the registers, shifters, and ALU. At the earliest point that integration of a complete RALU could be done, the maximum width for all of these elements that could be placed on a single integrated circuit was 4 bits. If one needed a 16-bit wide RALU, then four of the 4-bit wide RALUs were placed side by side, creating a 16-bit RALU. Since each 4-bit RALU appeared to be a "vertical" slice down through a wider RALU, then the name "*bit-slice*" was applied to the idea of creating complex logical structures that could be paralleled to create wider structures. Notice that the "horizontal" connections involve mainly the Carry and shift bits. These slices implemented the carry generate and propagate terms, G and P, to improve carry propagation time across wide networks. Typical bit-slice RALUs of this period contained 16 registers so that 4 of these devices placed side by side formed the greater part of the internal architecture of a 16-register by 16-bit computer. Since the use of the registers was determined by the control codes supplied to the RALUs, the General Purpose Register (GPR) was the most common form implemented.

3.8. TABLE-DRIVEN ARITHMETIC OPERATIONS

With the advent of the single-chip combinatorial multiplier described in Section 3.9, the method of generating arithmetic operations using a table has become less important. However, it is still relevant in some circumstances. An example of the method is depicted in Figure 3-19.

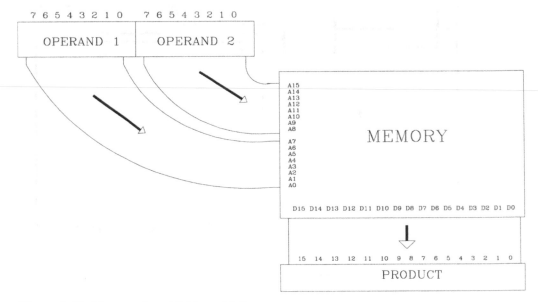

Figure 3-19: Memory-based 8-bit multiplier.

Two 8-bit numbers, Operand 1 and Operand 2, are connected to the address inputs of a memory. Each element of the memory contains, in this example, a product of two particular values of the two operands. The data output at any given moment is the product selected by an address formed by concatenating the two operands. The width of the memory is dictated by the precision of the result generated by the function contained in the memory, here 16 bits.

This method has long been available to the computer programmer where it provides a very fast, flexible method to generate functional results. In many cases, an algorithmic method can produce results with greater precision, but they are invariably slower. The problem with the table-driven method is that the memory size grows quite rapidly with the precision of the input operands. For the case involving integer multiplication shown in Figure 3-19, the depth of the memory is 2^{16} with a width of 16 bits. A memory to support the multiplication of two 16-bit operands would be 32 bits wide and 2^{32} "words" deep. While technology continues to increase the depth of memory at an astounding rate, one usually is faced with needs that outstrip the supply of parts.

Another shortcoming of this technique for the most common operations is that industry has many times produced minimized networks that are both faster and cheaper than a corresponding memory-based network. The speed factor is accounted for by noticing that a memory contains all possible Boolean product terms of its address inputs. The implication is that each product requires an AND gate that has as many inputs as there are address lines. The products that will cause a 1 to appear on a data output line must be ORed together. This implies that the OR gate for a particular output pin may have an appreciable number of inputs.

Since no technology supports AND and OR gates of this width, the functions are produced by applying the "associative" property for Boolean numbers to create groups of smaller terms that are small enough to connect to the actual gates available. The narrowing of the implementation of the Boolean function results in a greater number of gates in series in the signal path; hence, a longer propagation delay. At any given level of technological development, the specialized networks should be faster than the memory-based ones.

There are several places where the table-driven method is applicable. Since the function placed in the memory is independent of the structure of the memory, this method is appropriate for the functions where no specialized network has been developed. One very early use involved the generation of trigonometric functions. In this case, there is no second operand and the memory depth does not grow as rapidly as when there are two operands. The depth of the memory is governed by the precision of the desired results. This is a simple matter to consider if the function to be generated is linear; however, notice what happens for functions that have rapid changes in some part of their domain (e.g., the tangent) or are actually discontinuous. The table can contain a function whether it is continuous or not; however, the Sampling Theorem still applies to the interpretation of the result.

Another application of this method is for relationships that are not expressible as analytic functions. In this case, the table-driven method is the only one possible.

The last concern in this section is caused by the sheer size of a memory to hold a function to the precision dictated by the original problem one is trying to solve. If one can include the table in the architecture of a state machine, then many times the size of the memory can be traded against the speed of computation. This again is frequently done in computer programming where the table may contain very few samples or "seeds" and an algorithm refines the values to the desired precision. Two short examples to introduce this concept can be shown here.

In our multiplication example, we noticed that the depth of the memory doubled each time we increased the number of bits in one operand by 1. Presently, a table containing all of the values of a function of two 8-bit operands is relatively easy to make. In the case of commutative operations like multiplication, the depth of the memory is potentially only half as large. The actual precision can in some cases be extended by using short algorithms. In multiplication, for example, the 16-bit partial product formed by multiplying two 8-bit "digits" of 16-bit operands can be summed with the other three partial products to form a 32-bit product. Each partial product must be shifted to accommodate the value of the "digit" position as usual. This operation is faster than the single-bit shift and add method that will be discussed more thoroughly in Chapter 4.

As a second example, we observe that a table containing relatively few samples can be used with various interpolation algorithms to produce results having the desired precision. The algorithms, ranging from linear interpolation to curve fitting, are selected based on the nature of the function contained in the memory.

Once the controllers for state machines are examined in Chapter 4, we will proceed to a more thorough discussion of this problem.

3.9. MULTIPLICATION

The Digital Signal Processing (DSP) needs of the military have driven the development of the combinatorial multiplier. As will be shown in Chapter 7, the basic DSP equation is the arithmetic sum-of-products. Needless to say, the products need to be performed much faster than the sample rate of the system. Given a point in history, technology dictates the number of AND and OR gates having a particular number of inputs that can be integrated on a single piece of silicon. This ultimately determines the number of bits in the operands that can be multiplied in a given period of time. Figure 3-20 shows a symbol for a combinatorial multiplier. The symbol is intended to depict that a product of the multiplier and multiplicand is available after some propagation time determined by the physical structure of the transistors used to implement it. No clocks are needed.

The combinatorial multiplier is, in principle, a minimized network that implements the arithmetic product of two operands. The RALU accomplishes multiplication by providing two single clock operations that allow the multiplicand to be conditionally added to the developing product before it is shifted. This exploits the nature of the binary multiplication process shown in the following product truth table:

MPR	MCAND	PROD
0	0	0
0	1	0
1	0	0
1	1	1

The 1-bit product behaves the same as the "logical" product, or AND, of the inputs. This behavior allows one to express the multiply operation as a series of shift and conditional summations using the RALU.

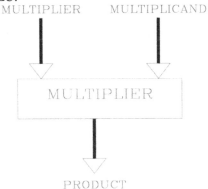

Figure 3-20: Combinatorial multiplier symbol.

A general multibit multiplication network would appear to require many wide AND and OR gates for its construction. This important network has been thoroughly examined during times when the number of gates and their width were severely limited. One particularly efficient implementation is shown in Figure 3-21. The partial products, A_nB_m, are produced by applying the product truth table above to each bit combination of the operands A and B as shown next:

Multiplicand A					A_3	A_2	A_1	A_0
Multiplier B					B_3	B_2	B_1	B_0
				A_3*B_0	A_2*B_0	A_1*B_0	A_0*B_0	
			A_3*B_1	A_2*B_1	A_1*B_1	A_0*B_1		
		A_3*B_2	A_2*B_2	A_1*B_2	A_0*B_2			
	A_3*B_3	A_2*B_3	A_1*B_3	A_0*B_3				
P_7	P_6	P_5	P_4	P_3	P_2	P_1	P_0	

Each column of partial products is then summed in single-bit Full Adders creating a bit of the final product, P_i. The carry is "saved" at each step and added in the next stage below, thus giving the adder its name, the Carry Save Adder.

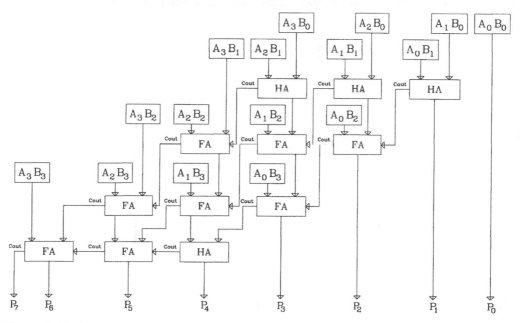

Figure 3-21: Combinatorial multiplier using the carry save adder.

Tracing the propagation path to any product term from any A_nB_m determines the propagation time of that path. The worst-case path for this example traverses the Carry In/Carry Out route for six full adders. Contrast this time with that incurred by using ripple carry adders to propagate the carry across each level. Notice that, as the number of input operand bits increases, the worst-case path length increases at a rate slower than would occur if ripple carry adders were used on each level.

Combining the multiplier with an accumulator creates an architecture that is well suited to performing the arithmetic sum of products operation. The Multiplier/Accumulator (MAC) architecture is shown in Figure 3-22. The multiplier and adder networks are combinatorial in nature and data propagation through them must be performed in one clock period, i.e., the path from the multiplier or multiplicand to the register data inputs forms the "critical path" in this device. Actual implementations of the MAC frequently include modifications to the adder network that allow additions or subtractions to be performed under control of a function select signal. The data types implied for Figure 3-22 are unsigned and two's complement binary numbers. The overall structure of a MAC is the same for any data type; however, the detailed data flow is not, as will be shown for the floating point data type in Section 3.10.

The accumulator register, implemented as an Enabled D register in Figure 3-22, is clocked every microcycle. Its contents will not be updated, however, until the Load Enable signal is exercised. This creates a convenient method to accommodate the very long critical path mentioned above, i.e., by allowing more than one clock cycle to elapse before loading the result.

As noted for the multiplier itself, the data path beginning with the product is double precision, i.e., twice as wide as the input operands. In real integer MACs, the entire adder/register structure is also maintained as a double precision path. For example, 16-bit MACs have 16-bit input operands and at least 32-bit wide accumulator structures.

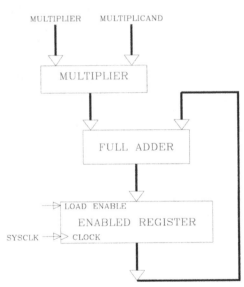

Figure 3-22: The Multiplier/Accumulator (MAC).

The width of the accumulator is usually made even greater to allow for the summing of a series of "full-scale" numbers of the same sign. Several more "guard bits" will be appended to the most significant end of the path. This results in the accumulator having a total of as many as 36-bit wide paths through it.

3.10. FLOATING POINT ARITHMETIC UNITS

Some general comments will be made at this time about Floating Point Units (FPUs) and how they support the major arithmetic operations. The basic operations performed by these architectures are addition, subtraction, and multiplication. Remember, these devices are combinatorial; hence, the data flow through them from input to output as opposed to the iterative multiplication methods supported in the RALU. Division does not lend itself to a noniterative method; however, the actual methods used (e.g., Newton-Raphson) do not require nearly the number of steps as the simple conditional subtraction "by hand" method.

Given the "flow through" overall design requirement, the floating point operands must be "separated" into fractional and exponent components. The major networks used will include shifters, adders, and multipliers. The role of the addition and multiplication networks is self explanatory, but the shift networks serve to align the binary point in the fractional part of the floating point number during addition, subtraction, and the normalization step following any operation.

Refer to Figure 3-23 during the discussion of combinatorial floating point multiplication that follows. Two "IEEE" single-precision floating point operands are introduced to the network as shown.

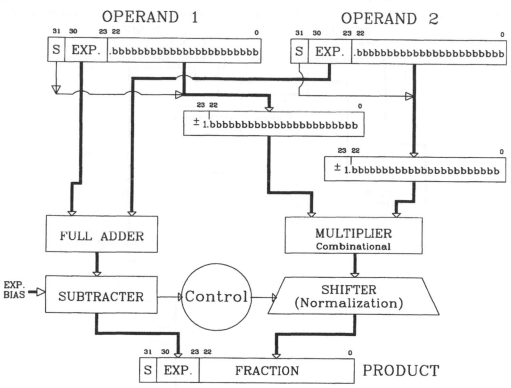

Figure 3-23: Combinatorial floating point multiplier.

The fractional parts are separated from their exponents and introduced to the right-hand data path. Notice that the "1" to the left of the binary point is restored to each fraction. Refer to Section 2.8.1 for a discussion of a normalized number, i.e., the reason for restoring the "1". The signed fractional parts have a range in value of $-2 <$ fraction < -1 or $+1 <$ fraction $< +2$. The range from -1 to $+1$ is excluded since the original operands were normalized. The fractions are combinatorially multiplied together using a network similar to that discussed in Section 3.9. The product is a double precision fractional number having a range of values from -4 to -1 or $+1$ to $+4$. All other values are excluded.

The exponents, meanwhile, are added. The bias value, $+127$, is subtracted from the sum to create a new biased exponent for the floating point product. Unfortunately, the fractional part of the product is not normalized; i.e., the most significant and only bit to the left of the binary point may not be "1". Recall the range of values established for the fractional part of the product in the previous paragraph.

Normalization of the product is accomplished by shifting the fractional part to the right while simultaneously decreasing the exponent. The amount of shifting (i.e., 0 or 1 bit-positions) is determined by the actual value of the portion of the fraction to the left of the binary point. All steps in the operation can be performed with combinatorial networks. As occurred for the product of integers, the data path for the product widens to double precision.

Floating point addition and subtraction are performed in a network like that shown in Figure 3-24. The operation is made considerably more complex by the fact that the binary points of the input operands must be "aligned" before the operation. Generally, the operand with the smaller exponent is denormalized by introducing it into the Operand 2 path.

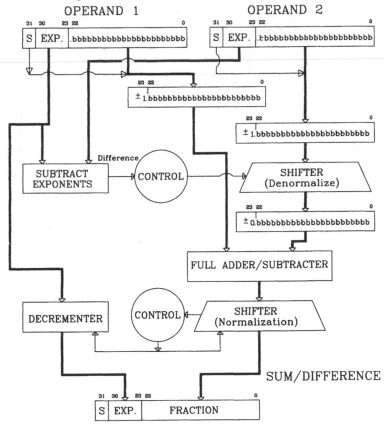

Figure 3-24: Combinatorial floating point adder/subtracter.

The fractional part of this operand is shifted to the right while adjusting its exponent until the exponents of both operands are equal. This is shown in Figure 3-24 by the path marked "Control" connecting the exponent comparator (i.e., subtracter) with the denormal-ization network. Since the amount of shifting may range from 0 to 31 bits, the denormalizer is built using a "barrel" shifter, i.e., a shifter that can combinatorially steer data to one of the 32 possible paths leaving the network. The example shows a 6-bit shift to the right caused by the exponent of Operand 2 being 6 less than that of Operand 1. It is important to emphasize that the denormalization shift is entirely combinatorial. No registers intervene in the path. The denormalized Operand 2 is created after a time delay that is independent of the shift amount.

Once the fractional part of Operand 2 is denormalized, thus "aligning" the binary points, the fractional parts are added or subtracted as determined by a function control signal. The result must now be normalized. A Boolean control network senses the distance required to shift the result to the left since its range is $-2.0 <$ result fraction < 2.0, i.e., the fraction may require shifting from 0 to 31 bit positions to the left. At the same time, the exponent must be decremented by the amount shifted to retain the value of the sum or difference.

Thus, the addition and subtraction of floating point numbers require even more combinatorial steps than multiplication. In order to create the flow through structure (i.e., no clocks) the designer must use wide-path barrel shifters, multipliers, and adders. The resulting designs require a large number of transistors to implement. The existence of modern FPUs that support the IEEE specification for single and double precision numbers is a tribute to current IC technology.

3.11. SOME REAL ARCHITECTURAL UNITS

The next several sections will include discussions of some real architectural units that allow the designer to build state machine solutions to serious compute-intensive problems. Most units shown are selected from a much wider field of parts to illustrate certain points and give specific examples that can be used in design examples. All of the devices shown are relatively complex, especially those with embedded "pipeline" registers. The serious designer should consult the manufacturer's data book and application notes when using the part in a specification. Several data books are listed in the bibliography.

The modern parts are frequently implemented in a high-speed CMOS family that is proprietary to the particular manufacturer. In general, the parts have propagation delays that are similar to those for the corresponding manufacturer's advanced TTL implementation, if one exists. The static power dissipation is much less, however. When the parts are operating at normal speed, power is consumed in charging the input capacitances on all Field Effect Transistors (FETs) in the device. Some devices, because of the very large number of transistors and high speed, may still require cooling, though not as much as that required by TTL implementations. The very high switching currents must be supported by careful signal path layout and adequate power supply wiring and bypassing. The size of the die on which the circuit is made has caused the manufacturer to bring several power and ground pins out from various positions on the die. This allows the circuit board designer to actually create the ground potential over the entire die since the traces on the die are too small for the currents involved.

3.11.1. Real Register Arithmetic Logic Units

Figures 3-25 through 3-29 show the current RALU offerings of two manufacturers, Texas Instruments (TI) and Advanced Micro Devices (AMD). There are two different integration philosophies used to produce devices that are both, when properly used, 32-bit wide RALUs. TI designers treated the RALU, its assorted shift extension multiplexers, and status registers as a complete physical structure.

The TI RALU, the 74ACT8832 shown in Figure 3-25, occupies a single, large "pin grid array" package capable of bringing three different 32-bit wide bidirectional data buses plus many control signals off of the chip. The device can handle operands and operations having widths of 8, 16, and 32 bits, updating the status register(s) correctly in each case. These configurations are the subjects of Figures 3-26 and 3-27. The overall configuration of the part is controlled by signals applied to the CFi inputs. To completely control the RALU from the microprogram, the designer must include not only the RALU instruction signals I7-0, but also the various signals associated with configuration and path control.

The register array contains 64 32-bit words that may be manipulated, within limits, as 8-bit, 16-bit, and 32-bit operand sources and destinations. The extra 4 bits beyond the basic 32-bit data path is used to contain a parity check bit for each byte of the 4-byte width of each register. There are three address buses, A, B and C, for the register array that allow, at the user's option, two or three address operations in a single clock cycle. The C address bus supplies the destination address for three address operations, while the B address bus serves for one source operand and the destination operand on two address operations.

Status information, such as Carry Out, Overflow, etc., is generated based on the configuration settings during the present microcycle. The correct flags may be gated to external status registers under control of the microprogram.

Several 1-bit registers are embedded in the design to make implementation of complete architectures simpler. These include the Divide/BCD flip-flops that streamline certain operations in order to reduce the number of clock cycles required for complex sequences related to division and format conversion.

The device also gives the designer many error checking aids, including parity generation and checking on each bus. A master/slave comparison structure that he can use to implement "fault tolerant" designs is also present. The designer can implement a major portion of the internal architecture of a computer using the single chip.

The AMD RALU chip set takes a different approach to the integration problem. The family of parts, designated Am29C300, includes the Am29C332 32-bit Arithmetic Logic Unit (ALU) shown in Figure 3-28 and the Am29C334 Four-Port Dual-Access Register File in Figure 3-29 among others. AMD chose to separate the functions so that major combinatorial elements occupied single chips with clocked elements concentrated on other chips. This allows the designer the freedom to place pipeline registers at any point in the data path or to use none at all. The configuration complexity is reduced somewhat while still allowing the flexibility presented by the TI approach. The trade-off is the increase in the number of chips in a "complete" system and the possibly slower data paths associated with taking a signal between chips.

The mask generator in Figure 3-28 is the mechanism that the AMD designers used to allow the microprogrammer to control the width of the operand path. The three major data paths are 32 bits wide with 4-bit extension buses to allow parity checking and generation on each byte. Only one bus, Y, is bidirectional while DA and DB are unidirectional as in the Am2903. A master/slave comparison structure associated with the ALU output supports "fault tolerant" design implementation.

Two structures increase the speed of certain complex operations over those shown for the "generic" RALU in Figure 3-18. The first of these is the 64-bit wide funnel shifter that can route a 64-bit data path formed by the concatenated DA and DB buses to any position relative to the ALU R input. This structure is particularly handy in scaling and normalization operations on physical data. The second structure is built into the ALU. It allows a 2×32 partial product to be formed instead of the 1×32-bit partial product generated by the "shift and conditional add" operation available on the older RALUs. A 32×32 multiplication can be performed in about 16 microcycles instead of the 32 required by an Am2903-based 32-bit RALU. An exotic bit field manipulation structure is also included to allow the performance of field logical operations on bit fields of arbitrary width and starting position.

The Am29C334 forms the register array with which the designer can configure a complete 32-bit wide RALU. Two devices are used side-by-side to implement a 64-word by 36-bit register array. The devices may also be "stacked" to increase the number of words in the register array. This, and the 18-bit width of the basic chip, shows the system design philosophy adopted by the AMD designers. The register array may also be used with 16-bit wide multipliers in designs that do not employ RALUs.

The address ports are more developed than those on the "generic" RALU. Each address port, A or B, has two addresses associated with it, one for reading and the other for writing. The device can perform two data reads and two data writes in a single cycle. This device can be integrated with the Am29C323 32-Bit Parallel Multiplier (with accumulator) and the Am29C327 Floating Point Unit, discussed later, to produce one- and two-level control systems having formidable power.

3.11.2. Real Integer Multipliers and MACs

Multipliers and multiplier/accumulators have been supplied by manufacturers for a long time. These devices were the central element in military digital signal processing applications so that their high cost was borne by the "nothing is too good for the troops" attitude of the times. TRW, Inc. was a pioneering company in the field with combinatorial units that ran hot and fast. In recent years, several companies have entered the field, first with TTL/ECL hybrids and then with high-speed CMOS implementations. It is the latter that we will feature in this section. Figure 3-30 shows the Texas Instruments 74ACT8836 Multiplier/Accumulator. This MAC follows the TI designer's philosophy of integrating as many features on a single chip and giving the designer control of them through configuration pins. Pipeline registers are present to increase the speed of those operations that benefit from such structures. In this manner, a new product can be accumulated every 36 nanoseconds. The designer is allowed to make some of the pipeline registers "transparent" for those problems that cannot be pipelined.

This device, like those offered by other current manufacturers, supports parity checking on the data buses and a master/slave comparator to support more robust designs. Since the accumulation path in any integer MAC is capable of overflow, most MACs have extension bits that allow the designer to implement a "soft landing" on overflow. Most output signals that "overflow" in analog situations merely "clip" at the voltage extremes provided by the power supply. In the digital world, the most significant bits will be lost if no further care is taken causing the output signal to "fold over". This distorted output is not only wrong, but *very wrong*. The TI MAC allows a certain degree of dynamic scaling and plenty of warning to enable the designer to more closely reproduce the correct output.

The MAC can also support double precision (i.e., 64-bit operands), multiplication as well as division. For the latter, the Newton-Raphson method is an iterative method that generates the reciprocal of the divisor by a converging series approximation.

Other manufacturers also make multipliers and MACs, including Cypress and Analog Devices. The second company's offerings include devices in which one or more of the three major data buses have been combined to reduce the number of pins on the chip. In one extreme case, a 16-bit MAC has been produced for attachment to a microprocessor bus as a memory mapped coprocessor as described in Chapter 7. The device is available with only a single bidirectional bus and a few control lines in a simple Dual In-line (DIP) packaging arrangement. The programmer can produce a very fast multiply/accumulation by moving two operands to the memory space occupied by the device.

3.11.3. Real Floating Point Units

We will conclude this chapter with the description of two modern FPUs, one by TI and the other by AMD. Both support the IEEE floating point format for single and double precision numbers. Internal data paths are wider than the 64-bit format in order to control round-off error during accumulations. The TI unit uses extensive pipelining and configuration control lines to provide a single integrated device that runs fastest in pipelined applications but still can be used for other situations. The AMD chip is configured more in the form of an RALU in which a parallel floating point multiplier is present along with the floating point ALU in a multiplier/accumulator architecture.

The TI 74ACT8847 FPU in Figure 3-31 is fabricated in TI's advanced 1-micron CMOS technology. Its data formats are compatible with the IEEE 754 Floating Point Standard of 1985. It supports combinatorial addition, subtraction, multiplication, and comparison on integer and floating point numbers. Division, square roots, and some format conversions are handled by short algorithms. It follows the TI pattern of parity checking on all three buses and the inclusion of master/slave comparison as aids in fault-tolerant designs. The device promises operations in a cycle time of less than 100 nanoseconds, especially in pipelined operation.

The Am29C325 Floating-Point Processor and the Am29C327 Double-Precision FPU comprise AMD's collection of CMOS floating point units. The Am29325 is also available in AMD's high-speed bipolar process. The '325 is organized as a three-port structure that can be configured as a traditional MAC or as a specialized MAC to support the power series expansion method shown in Chapter 7. It has registers on each port, the "transparency" of which is under control of the system designer. The data buses can be flexibly configured to allow integration into single, dual, or triple bus designs. Bus widths may be made 16 or 32 bits wide which means that the devices can be easily used as memory mapped floating point "coprocessors", i.e., as described in Chapter 7.

The Am29C327 is a more exotic architecture that expands the basic '325 data path by the addition of a register file and constants area. These are seen in Figure 3-32. Internal data paths are 64 bits wide and external ports allow some degree of flexibility in system integration.

SN74ACT8832A
32-BIT REGISTERED ALU

functional block diagram

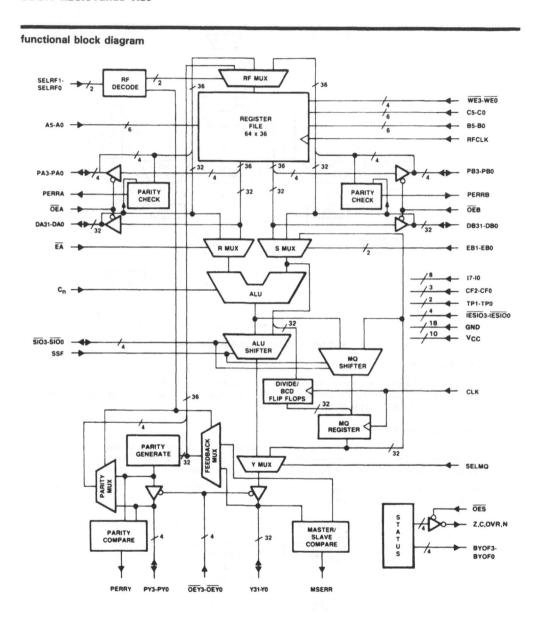

TEXAS
INSTRUMENTS
POST OFFICE BOX 655303 • DALLAS, TEXAS 75265

Figure 3-25: The Texas Instruments 74ACT8832 32-bit RALU.
(Reprinted by permission of Texas Instruments.)

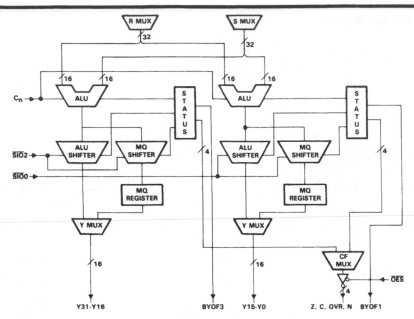

FIGURE 3. 16-BIT CONFIGURATION

ALU and MQ shifters

The ALU and MQ shifters are used in all of the shift, multiply, divide and normalize functions. They can be used independently for single-precision or concurrently for double-precision shifts. Shifts can be made conditional, using the Special Shift Function (SSF) pin.

bidirectional serial I/O pins

Four bidirectional $\overline{\text{SIO}}$ pins are provided to supply an end fill bit for certain shift instructions. These pins may also be used to read bits that are shifted out of the ALU or MQ shifters during certain instructions. Use of the $\overline{\text{SIO}}$ pins as inputs or outputs is summarized in Table 11.

The four pins allow separate control of end fill inputs in configurations other than 32-bit mode (see Table 6 and the functional block diagram).

TABLE 6. DATA DETERMINING $\overline{\text{SIO}}$ INPUT

SIGNAL	CORRESPONDING WORD, PARTIAL WORD OR BYTE		
	32-BIT MODE	16-BIT MODE	8-BIT MODE
$\overline{\text{SIO}}3$	—	—	Byte 3
$\overline{\text{SIO}}2$	—	most significant word	Byte 2
$\overline{\text{SIO}}1$	—	—	Byte 1
$\overline{\text{SIO}}0$	32-bit word	least significant word	Byte 0

TEXAS
INSTRUMENTS
POST OFFICE BOX 655303 • DALLAS, TEXAS 75265

3-17

Figure 3-26: The TI 74ACT8832 configured to use 16-bit operands.

(Reprinted by permission of Texas Instruments.)

SN74ACT8832A
32-BIT REGISTERED ALU

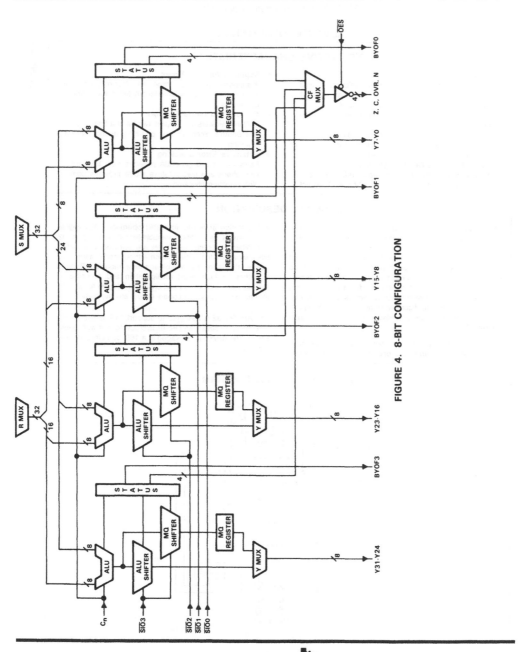

FIGURE 4. 8-BIT CONFIGURATION

TEXAS
INSTRUMENTS
POST OFFICE BOX 655303 • DALLAS, TEXAS 75265

Figure 3-27: The TI 74ACT332 configured to use 8-bit operands.
(Reprinted by permission of Texas Instruments.)

Am29C332

CMOS 32-Bit Arithmetic Logic Unit

ADVANCE INFORMATION

Am29C332

DISTINCTIVE CHARACTERISTICS

- **Single Chip, 32-Bit ALU**
  Standard product supports 110 ns microcycle time for the 32-bit data path. It is a combinatorial ALU with equal cycle time for all instructions.
- **Speed Select supports 80-ns system cycle time**
- **Flow-through Architecture**
  A combinatorial ALU with two input data ports and one output data port allows implementation of either parallel or pipelined architectures.
- **64-Bit In, 32-Bit Out Funnel Shifter**
  This unique functional block allows n-bit shift-up, shift-down, 32-bit barrel shift or 32-bit field extract.

- **Supports All Data Types**
  It supports one-, two-, three- and four-byte data for all operations and variable-length fields for logical operations.
- **Multiply and Divide Support**
  Built-in hardware to support two-bit-at-a-time modified Booth's algorithm and one-bit-at-a-time division algorithm.
- **Extensive Error Checking**
  Parity check and generate provides data transmission check and master/slave mode provides complete function checking.

GENERAL DESCRIPTION

The Am29C332 is a 32-bit wide non-cascadable Arithmetic Logic Unit (ALU) with integration of functions that normally don't cascade, such as barrel shifters, priority encoders and mask generators. Two input data ports and one output data port provide flow-through architecture and allow the designer to implement his/her architecture with any degree of pipelining and no built-in penalties for branching. Also, the simplicity of a three-bus ALU allows easy implementation of parallel or reconfigurable architectures. The register file is off-chip to allow unlimited expansion and regular addressability.

The Am29C332 supports one-, two-, three- and four-byte data for arithmetic and logic operations. It also supports

multiprecision arithmetic and shift operations. For logical operations, it can support variable-length fields up to 32 bits. When fewer than four bytes are selected, unselected bits are passed to the destination without modification. The device also supports two-bit-at-a-time modified Booth's algorithm for high-speed multiplication and one-bit-at-a-time division. Both signed and unsigned integers for all byte aligned data types mentioned above are supported.

The Am29C332 is designed to support 110-ns microcycle time standard speed, and 80-ns microcycle time with speed select. The device is packaged in a 169-lead pin-grid-array package.

SIMPLIFIED BLOCK DIAGRAM

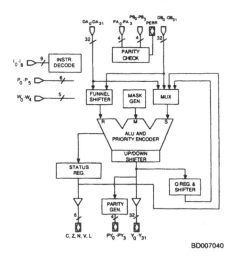

BD007040

2-38

Publication #	Rev.	Amendment
09288	C	/0

Issue Date: **January 1988**

Figure 3-28: The AMD Am29C332 32-bit arithmetic unit.

Am29C334

CMOS Four-Port Dual-Access Register File

Am29C334

PRELIMINARY

DISTINCTIVE CHARACTERISTICS

- **64 x 18 Bit Wide Register File**
  The Am29C334 is a 64 x 18-bit, dual-access RAM with two read ports and two write ports.
- **Pipelined Data Path**
  The Am29C334 can be configured to support either a non-pipelined data path (similar to the Am29334) or a pipelined data path.
- **Cascadable**
  The Am29C334 is cascadable to support either wider word widths, deeper register files, or both.

- **Built in Forwarding Logic**
  The Am29C334 provides simultaneous read/write access to the same address for double pipelined systems.
- **Byte Parity Storage**
  Width of 18 bits facilitates byte parity storage for each port and provides consistency with the Am29C332 32-bit ALU.
- **Byte Write Capability**
  Individual byte-write enables allow byte or full word write.

BLOCK DIAGRAMS

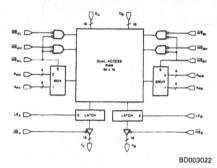

BD003022

Non-Pipelined Mode

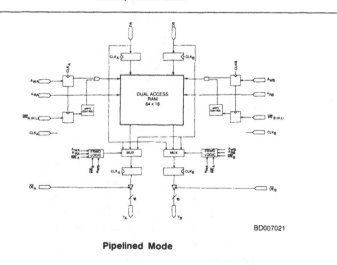

BD007021

Pipelined Mode

2-76

Publication #	Rev.	Amendment
08786	B	/0
Issue Date: **December 1987**		

Figure 3-29: The AMD Am29C334 four-port dual-access register file.

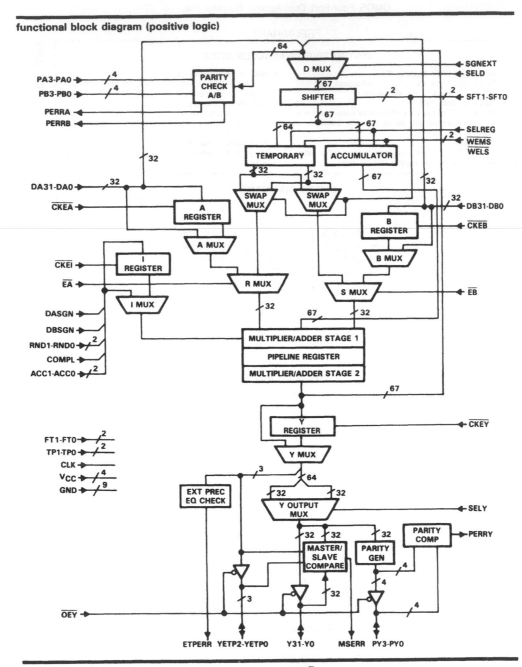

Figure 3-30: Texas Instruments 74ACT8836 multiplier/accumulator.

(Reprinted by permission of Texas Instruments.)

functional block diagram

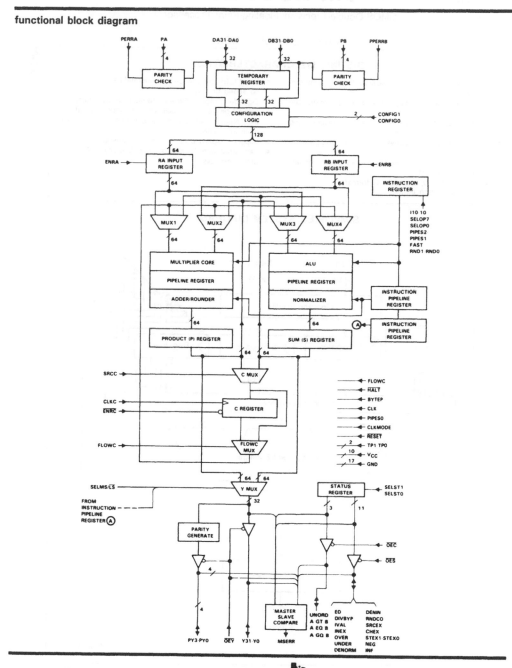

TEXAS INSTRUMENTS
POST OFFICE BOX 655303 • DALLAS, TEXAS 75265

Figure 3-31: Texas Instruments 74ACT8847 floating point unit.
(Reprinted by permission of Texas Instruments.)

Am29C327

CMOS Double-Precision Floating-Point Processor

ADVANCE INFORMATION

Am29C327

DISTINCTIVE CHARACTERISTICS

- High-performance double-precision floating-point processor
- Comprehensive floating-point and integer instruction sets
- Single VLSI device performs single-, double-, and mixed-precision operations
- Performs conversions between precisions and between data formats
- Compatible with industry-standard floating-point formats
  - IEEE 754 format
  - DEC F, DEC D, and DEC G formats
  - IBM system/370 format

- Exact IEEE compliance for denormalized numbers with no speed penalty
- Eight-deep register file for intermediate results and on-chip 64-bit data path facilitates compound operations; e.g., Newton-Raphson division, sum-of-products, and transcendentals
- Supports pipelined or flow-through operation
- Fabricated with Advanced Micro Devices' 1.2 micron CMOS process

SIMPLIFIED SYSTEM DIAGRAM

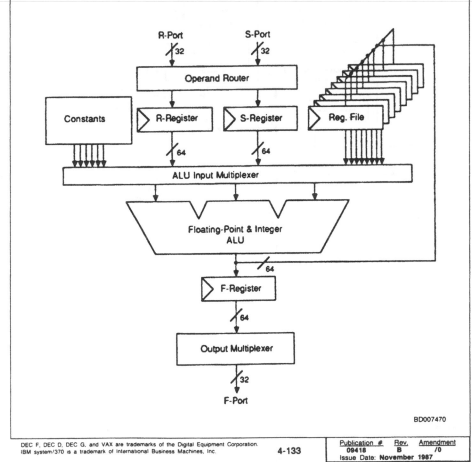

BD007470

4-133

| Publication # | Rev. | Amendment |
| 09418 | B | /0 |
Issue Date: **November 1987**

Figure 3-32: The AMD Am29C327 floating point unit.

EXERCISES

1. Draw a mixed logic diagram, including pin numbers, that implements Figure 3-10, the enabled D register, using the following parts: 74LS273 and 74LS157.

2. Consider that the enabled register "Load" ∫ signal drawn in Exercise 1 is being driven by a controller that obeys the following ASM diagram.

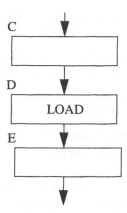

Use the logic diagram generated in Exercise 1 to demonstrate that the Data Setup Time (t_s) and Data Hold Time (t_h) for the 74LS273 will always be met in writing data to the register at the state D to state E transition regardless of the time delay between the clock edge and the "Load"∫ signal *True → False* transition. Refer to your Low Power Schottky TTL Device Databook for the necessary timing information. Note that the controller's clock also drives the register with the same phase.

3. Draw a logic diagram, including pin numbers, of a register with a gated clock implemented using a 74LS273. The register clock is driven by the following equation.

    LS273_CLK = SYSCLK * LOAD

   The equation is implemented using a 74LS08 positive logic AND gate. The signal, LOAD, is generated by the state machine controller used in Exercise 2.

4. Use the logic diagram produced in Exercise 3 to examine the constraints on the time delay in generating the LS273_CLK signal imposed by the Data Setup Time (t_s) and Data Hold Time (t_h) for the 74LS273. Examine the situation at the state D to state E transition when the SYSCLK edge goes *True* and the "Load" signal transitions from *True* to *False*.

5. Draw a logic diagram that implements the two-port read, one-port write register in Figure 3-12 using the enabled D register structure instead of the gated clock scheme shown. Implement the structure using suitable LSTTL parts (e.g., the 74LS377, 74LS139, and your choice of four input multiplexers). Warning: be sure that the address inputs of the multiplexers are independent for output address A and output address B. Include pin numbers on the logic diagram to show which multielement device was used where.

6. Draw a logic diagram that implements the stack in Figure 3-13 using the enabled D register structure instead of the gated clock scheme shown. Implement the structure using suitable

LSTTL parts (e.g., the 74LS377, 74LS139, and your choice of four input multiplexers). Include pin numbers on the logic diagram to show which multielement device was used for which purpose. Be sure to handle the clock phasing correctly, i.e., "POPing" the stack involves changing the stack pointer at the end of the system clock cycle (on the positive edge) while "PUSHing" the stack involves changing the pointer in the middle of the system clock cycle (on the negative clock edge) so that data may be latched on the next rising edge. This feature was introduced to allow the stack to operate in a single clock cycle.

7. Draw a logic diagram showing a funnel shifter for an 8-bit data path. Use a multiplexer with the proper number of inputs as the fundamental logic cell in the drawing.

    a. Show the values used on the data path extension inputs when 8-bit unsigned binary numbers are shifted left or right using the network.

    b. Show the values used on the data path extension inputs when 8-bit two's complement binary numbers are shifted left or right.

8. Draw a logic diagram of the networks needed to produce the three flags, Zero (Z), Sign (N), and Two's Complement Overflow (V), appropriate to the integer ALU.

    Hint: The most instructive way to proceed on the following problems is to analyze the procedures required to perform the operations requested and then to write a program in your favorite language that performs the task. The program's input and output should normally come from the "console" or "terminal" of the computer on which the exercise is performed. The data may be converted between the character format appropriate to the terminal and appropriate computation formats using the routines that you developed for the exercises in Chapter 2. The programs may be kept short and to the point by making them handle only a single operation that is the subject of the question.

9. Calculate the sums of the following numbers represented in the IEEE single precision floating point format:

    $$2.5 + 32.5 = ? \qquad -0.75 + 6.25 = ? \qquad 1.5 \times 10^4 + 7.25 \times 10^{-8} = ?$$

    Hint: You will find that the floating point conversion is more easily managed using a small program. Remember, you are trying to perform the operation as the computer does.

10. Calculate the products of the following numbers represented in the IEEE single precision floating point format:

    $$2.5 * 32.5 = ? \qquad -0.75 * 6.25 = ? \qquad 1.5 \times 10^4 * 7.25 \times 10^{-8} = ?$$

11. Design a memory-based multiplication network. Assume that each physical memory device is limited to 64 kilobytes. The input numbers use the 8-bit two's complement binary representation.

    a. Write a program to fill the array.

    b. How do you propose to move the array from its internal form in your computer to the physical memory (e.g., a pair of EPROMs)? See Chapter 5 Appendix C for a data representation that is compatible with most EPROM programmers.

Chapter 4

CONTROLLERS

4.1. INTRODUCTION

The controller in an Algorithmic State Machine (ASM), Figure 1-3, is a specialized architecture that determines the flow of states as a function of inputs, the present state, and outputs. It is formed using combinatorial elements, memory, and "data" paths similar to an architecture. A microprogrammed controller is the main focus of this chapter. In its fully developed form it is capable of performing any program control construct enjoyed by a person writing in a high-level language. In addition, it performs any of the constructs in a single clock cycle.

4.2. ALGORITHMIC STATE MACHINE DESIGN

In Chapter 1 we showed a method to design machines that were capable of producing a finite sequence of states, algorithmic state machines. Figure 1-5 summarized the three major graphic symbols used in this design method, the state box, the decision block, and the conditional output. While there are other notations in which to present a design (e.g., Mealy-Moore diagrams), this notation lends itself to the representation of complex decisions made using several inputs and in showing large numbers of outputs. The purpose of our controller designs is to generate control signals for a general purpose architecture. Most architectures require a large number of control signals; therefore, our design notation needs to accommodate this requirement as efficiently as possible. Unfortunately, the ASM diagram will also be lacking in this respect and we will be forced to change to another notation later. Until then, we will exploit the features of the ASM diagram.

The sequence of steps described in an ASM diagram bear a strong resemblance to the programming construct called a FLOWCHART. The important difference between an ASM diagram and a flowchart is that the state box in the ASM diagram has a duration of one *state time*, the period between two adjacent ticks of the system clock. Two outputs (e.g., Z0 and Z1 in the left part of Figure 1-4) occur simultaneously because they are shown in the *same* state box. In the same figure, Z2 does not occur at the same time as Z0 since it is in a different state box. Also, the output Z2 will turn on in the very next state *after* Z0 is on and will remain on for one state time only. The output Z0 turned on in the top state and remained on throughout the next state before it turned off. If this were a flowchart, then the only information carried is the ordering of the events Z0, Z1, and Z2. No information is imparted about how long they last. This implies that the actual implementation of an ASM *must* be made from an ASM diagram or similar method that imparts the same information. The flowchart, however, can be used as a preliminary statement of the general "flow" of the problem. It can then lead to the ASM diagram.

There are two general implementation methods for the ASM diagram. They may be called the *traditional* method and the *microprogrammed* method. In both methods, a means must be found to represent the state information in the ASM diagram. The *state of a system* is a unique description of the configuration of a system. The variables that completely describe a system have a particular set of values when the system is in a particular state. If even one variable has a different value, the system is *not* in the same state. The variables that completely describe a system are collectively given the name *state variables*.

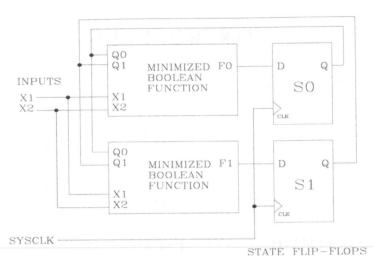

Figure 4-1: Traditional state machine implementation.

Notice that these ideas and terms can be used for continuous systems also, in which case the number of states is infinite even though the number of state variables may be finite; e.g., a charge-free point mass moving in a gravitational field requires four state variables, its mass and three dimensional coordinates. Luckily, our systems will have a finite number of states — remember the definition of an ASM. This allows us to assign a finite number of values to the state variables. If our ASM diagram had 10 states then we could use a 4-bit binary code to represent the value of each state. We could allot four 1-bit storage elements (e.g., D flip-flops) to "remember" the value of the state. These flip-flops are called the *state variable* flip-flops.

In this example, the four flip-flops could have handled an ASM diagram with as many as 16 states. As part of our design, we must examine the consequences of the flip-flops holding a value that is *not* part of our main design.

The *traditional* method of implementation uses dedicated Boolean networks to determine the next state, i.e., the new Q outputs, of the system. This is expressed in Figure 4-1. The actual logic performed in these networks embodies the algorithm expressed in the ASM diagram. If the ASM diagram is modified, even ever so slightly, to accommodate a design change, then all of the Boolean networks in the figure must be modified. While this is definitely a shortcoming in this "world of change", the Boolean networks themselves can be optimized for speed more effectively than in any other implementation. This means that the traditional implementation of an ASM is fundamentally capable of greater speeds than any other method.

Figure 4-2 shows two ways that ASMs can be implemented whether they are traditional or microprogrammed. The Moore machine at the bottom produces its outputs as functions of the state variables only. This means that a Moore machine ASM diagram would not contain the conditional output box shown in the right part of Figure 1-5 that enables Z2 and Z3 if X2 = 1 AND X1 = 0. The conditional output is possible only with the Mealy machine implementation shown in the upper half of Figure 4-2. Here, a single complex combinatorial network is used to combine the value of the current state with the inputs, X1...Xn, to produce the next state input for the state variable flip-flops and the outputs, Z1...Zn. In the Mealy machine, if an input is a term in the equation of an output and that input changes

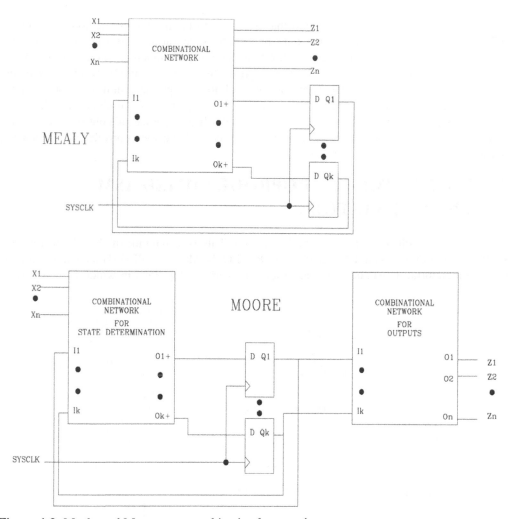

Figure 4-2: Mealy and Moore state machine implementations.

during a state in which the output is enabled, then the output will change. The major consequence of the conditional output is that the "on" duration of the enabled output is directly related to the input's temporal behavior. In systems in which the signals produced by the state machine are used to control an architecture, this timing behavior makes it very difficult for the designer to assure that the control signals meet the timing requirements of the architecture. Therefore, the hardware implemented controller that we use will be of the Moore type, whether the traditional or microprogrammed method is used.

In Figure 4-2, if the "combinatorial network for state determination" is implemented using dedicated, optimized Boolean networks, then the controller is said to be traditionally implemented. On the other hand, if the input network is implemented using a memory, then the controller is said to be microprogrammed. Notice that memories by way of the address decoder contain all possible MINTERMs and hence can implement all possible Boolean equations. The penalty for this flexibility is paid in the speed of the network. Obviously, if the dedicated network was implemented in the same technology as the memory, then because of its optimization the logical path length from input to output should be less; hence, the dedicated network should always be faster.

A third implementation method that is related to the traditional method is the ONE-HOT method. Its name implies that only one of the state variables is true at a time. This method will require more state variable flip-flops for a given implementation than one of the "encoded" methods above. Its use is predicated on the resulting simplicity of many of the output equations in small ASMs. One possible traditional implementation of a controller for our architecture could be based on a one-hot ASM that generates a multiphased clock that could be gated to the architecture in order to establish a sequence of control signals. This method is one of a group of traditional methods used to build the controller of a computer. It can result in a very fast machine.

4.3. ROM–LATCH MICROPROGRAMMED ASM IMPLEMENTATION

Now consult the upper half of Figure 4-3. This is a diagram of the Moore machine shown in Figure 4-2 with a Programmable Read Only Memory (PROM) substituted for the two combinatorial networks. There are four state variable flip-flops in which as many as 16

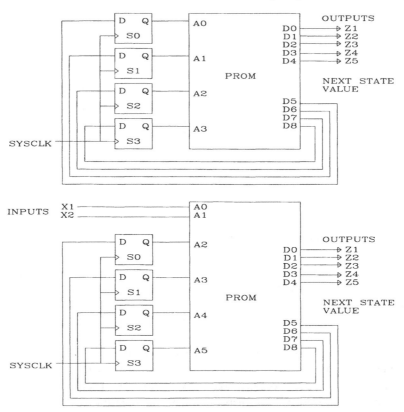

Figure 4-3: PROM-based state machine implementations

states can be binary encoded. There are five outputs that can each be a function of all of the state variables. There are no inputs which prevent this implementation from conditionally changing the flow of states as implied in the left part of Figure 1-5. This example begins our discussion of *microprogrammed* implementations.

Let's consider it in more detail. The PROM as shown in the drawing has four address lines and hence contains 16 words of 9 bits each. The words are connected to its outputs such that the bits at one end, D0...D4, are connected to the outputs Z1...Z5. The remaining 4 bits, D5...D8, are connected to the D inputs of the four state variable flip-flops. Each time a Low-to-High transition occurs on SYSCLK, the contents of the state variable flip-flops (i.e., their Q outputs) will be updated to the value contained in the data on the PROM outputs. Once the new state value is placed on the address inputs to the PROM, the data contained at that address will be made available on its data output pins, D1...D8. The next step must await the next occurrence of a Low-to-High transition on SYSCLK.

There is no maximum period for SYSCLK; however, the minimum period is set by the sum of the propagation delay time through the PROM plus the *setup time* of the state variable flip-flops. The setup time of an edge-triggered D flip-flop is the time required that the data on its D input must be stable before the clock transition occurs. For the LSTTL family, this time is about 20 nanoseconds. Both the propagation time for the PROM and the setup time for the state variable flip-flops are functions of the implementation technology (i.e., logic family) and range over a little more than two orders of magnitude for the currently active families.

If we place the following table of data at the addresses shown in the PROM in our example, we will implement an ASM that cycles through a sequence of five states just like a counter.

Address				Data								
A3	A2	A1	A0	D8	D7	D6	D5	D4	D3	D2	D1	D0
0	0	0	0	0	0	0	1					
0	0	0	1	0	0	1	0					
0	0	1	0	0	0	1	1					
0	0	1	1	0	1	0	0					
0	1	0	0	0	0	0	0					

The state variables S3, S2, S1, and S0 will assume the next value in this binary encoded list, 0, 1, 2, 3, 4, 0, 1, 2, 3, 4, 0, etc. each time SYSCLK ticks, assuming that they were initialized to 0 at the beginning. The outputs, Z1 through Z5, would be made to contain any arbitrary sequence merely by placing 1s or 0s in the data fields associated with D0 through D4, respectively. For example, if the system were in state 3, address 0011, and Z5 were required to be a 1, then the 1 would be placed in the D4 data column above at the address 0011. This production of a sequence of control signals using a memory is similar to playing a phonograph record. Any arbitrary sequence of sounds can be recorded on the record and all it takes to hear the desired performance is to "play" the record sequentially.

As it stands, this ASM implementation is useful only for producing preset sequences of outputs since as a controller it cannot change sequences based on an input. This is still a very simple, useful device. From it can be made a digital word generator of arbitrary width containing any desired sequence of output signals. The width of the output is increased by adding more PROMs in parallel using the same address lines. No more data outputs need to be devoted to determining the next state however. One could build a 16-state word generator having 12 independent 1-bit outputs by using two 8-bit wide PROMs and four D flip-flops. The word generator idea could be extended to the analog realm by converting the digital output to analog information. If the outputs, Z5 through Z1, were interpreted as a 5-bit two's complement binary number and the data region of the PROM contents contained the binary codes for the sine function, then this example would be an analog sine wave generator. Its frequency range would be set by SYSCLK and limited only on the upper end. One is not limited to placing a sine table in the PROM. Any arbitrary waveform can be encoded there, which yields a very flexible, small waveform generator.

The next step for our purpose is to adapt the microprogrammed ASM implementation to allow inputs that can be used to alter the sequence of states produced. The lower part of Figure 4-3 appears to be the same diagram as presented above. Detailed examination shows that the address inputs of the PROM have been "slipped upward" and reassigned so that the state variables are connected to more significant address lines than before, i.e., $S0 \rightarrow A2$, etc. Two inputs, X1 and X2, are connected to the lower address inputs A0 and A1, respectively. This structure can be used to implement an ASM diagram containing all of the elements in the right part of Figure 1-5, including the decision block and conditional output. Let's examine a "program" that could be used with this implementation.

Address						Data									
S3	S2	S1	S0	X2	X1	$S3^+$	$S2^+$	$S1^+$	$S0^+$	Z5	Z4	Z3	Z2	Z1	State/Out
A5	A4	A3	A2	A1	A0	D8	D7	D6	D5	D4	D3	D2	D1	D0	PROM
0	0	1	0	0	0	0	1	0	1						
0	0	1	0	0	1	0	1	0	1						
0	0	1	0	1	0	0	1	0	1						
0	0	1	0	1	1	0	1	0	1						

If the system is in state 0010 (i.e., $S3 = 0$, $S2 = 0$, $S1 = 1$, and $S0 = 0$) then the next one that is occupied by the system will be state 0101 regardless of the values of the inputs X2 and X1. This is caused by the fact that all four addresses have the same data in the next state determination field, D8...D5. This example shows the inputs being ignored for state determination purposes as shown in the left ASM diagram in Figure 1-5. The next example "program" will illustrate the generation of a conditional change of state.

Address						Data									
S3	S2	S1	S0	X2	X1	$S3^+$	$S2^+$	$S1^+$	$S0^+$	Z5	Z4	Z3	Z2	Z1	State/Out
A5	A4	A3	A2	A1	A0	D8	D7	D6	D5	D4	D3	D2	D1	D0	PROM
0	1	0	0	0	0	1	1	0	0						
0	1	0	0	0	1	1	0	0	1						
0	1	0	0	1	0	0	0	0	1						
0	1	0	0	1	1	0	0	0	1						

This table is generated using the decision block in the right half of Figure 1-5. The initial state is 0100 and the next state is dependent on the values of inputs X2 and X1. Notice that if $X2 = 0$ and $X1 = 1$, then the ASM diagram says that the system should go to state 1001 when SYSCLK next ticks. This is accomplished by placing the value of that state in the PROM data output field connected to the state variable flip-flop D inputs.

To illustrate the generation of unconditional outputs such as Z1 in state 0100 in the right part of Figure 1-5, we show the following fragment of the data in the PROM.

Address						Data									
S3	S2	S1	S0	X2	X1	$S3^+$	$S2^+$	$S1^+$	$S0^+$	Z5	Z4	Z3	Z2	Z1	State/Out
A5	A4	A3	A2	A1	A0	D8	D7	D6	D5	D4	D3	D2	D1	D0	PROM
0	1	0	0	0	0	1	1	0	0					1	
0	1	0	0	0	1	1	0	0	1					1	
0	1	0	0	1	0	0	0	0	1					1	
0	1	0	0	1	1	0	0	0	1					1	

Notice that the system is in state 0100 which sets addresses A5 through A2. Since the ASM specifies the Z1 output to be unconditional, then we place 1s in the D0 bit of all words having address inputs A5 through A2 equal to 0100 as shown. Since Z5 through Z3 are not shown in state 0100 in the ASM diagram which implies that they are "off", they will be set to 0s, omitted here for clarity.

Now consider the generation of the conditional outputs shown in the ASM diagram for state 0100 when X2 = 1 and X1 = 0.

Address						Data									
S3	S2	S1	S0	X2	X1	$S3^+$	$S2^+$	$S1^+$	$S0^+$	Z5	Z4	Z3	Z2	Z1	State/Out
A5	A4	A3	A2	A1	A0	D8	D7	D6	D5	D4	D3	D2	D1	D0	PROM
0	1	0	0	0	0	1	1	0	0			0	0		
0	1	0	0	0	1	1	0	0	1			0	0		
0	1	0	0	1	0	0	0	0	1			1	1		
0	1	0	0	1	1	0	0	0	1			0	0		

The columns for D2 and D1 driving outputs Z3 and Z2, respectively, are set to 0 except for the input combination X2 = 1 and X1 = 0. This corresponds to the address 010010. The D2 and D1 bit positions for that address are both set to 1 to cause Z3 and Z2 to turn "on" as required by the ASM diagram.

The complete contents of the portion of the PROM that will guide the state transition and outputs shown in the ASM diagram in Figure 1-5 is shown below.

Address						Data									
S3	S2	S1	S0	X2	X1	$S3^+$	$S2^+$	$S1^+$	$S0^+$	Z5	Z4	Z3	Z2	Z1	State/Out
A5	A4	A3	A2	A1	A0	D8	D7	D6	D5	D4	D3	D2	D1	D0	PROM
0	1	0	0	0	0	1	1	0	0			0	0	1	
0	1	0	0	0	1	1	0	0	1			0	0	1	
0	1	0	0	1	0	0	0	0	1			1	1	1	
0	1	0	0	1	1	0	0	0	1			0	0	1	

The columns for Z5 and Z4 contain zeros because these outputs are always "off", i.e., they do not appear in the 0100 state box or any conditional output connected to it. The outputs Z3 and Z2 have 0s except when the input condition connected to X2 and X1 is 10. The output Z1 is turned "on" unconditionally by the 1s placed in the D0 bit of all words with the address beginning 0100XX.

4.4. COUNTER–BASED MICROPROGRAMMED ASM IMPLEMENTATION

In the ROM-Latch ASM implementations, we had to specify the next state to be accessed in the ROM data of every state. In many cases, the next state will be the one whose state value is one greater than the current state. If this is recognized, we can begin a new series of implementations that are based on the Moore model in Figure 4-2 and borrow strongly from the traditional implementation method to determine the next state. Figure 4-4 shows the simplest version of the *counter-based microprogrammed ASM*.

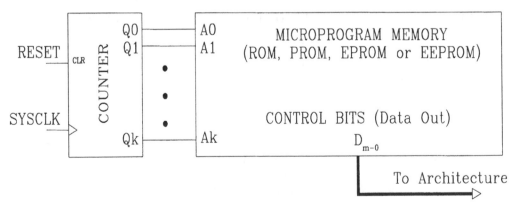

Figure 4-4: Counter-based state machine implementation — version 1.

A counter is a complex structure that includes the state variable flip-flops and input combinatorial network shown for the Moore machine in Figure 4-2. The output combinatorial network in that figure is implemented in Figure 4-4 with a memory. Borrowing the first table in the previous section and adapting it to eliminate the data fields associated with next state determination, we have:

| Address | Data |
A3 A2 A1 A0	D8 D7 D6 D5 D4 D3 D2 D1 D0
0 0 0 0	
0 0 0 1	
0 0 1 0	
0 0 1 1	
0 1 0 0	

The counter is made to count a binary sequence beginning with 0, i.e., reset. All data outputs from the memory are available for use in generating control signals. This machine, like the first one in the previous section, is of use only in generating a single sequence of control signals even though the actual values of the signals as a function of time are completely flexible. Once the memory is filled with the desired values, then its outputs will cycle endlessly until reset.

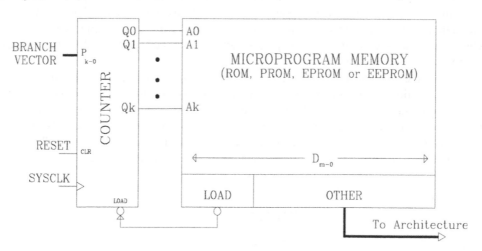

Figure 4-5: Counter-based state machine implementation — version 2.

Figure 4-5 begins the evolution of the counter-based ASM toward a generally useful design by introducing a mechanism whereby a count sequence can be changed to begin another sequence under control of the microprogram memory. Notice that most of the bits in the microprogram word are used to generate control signals for some architecture which is not shown. A 1-bit field, say PROM data output bit D_m, is connected to the load input of a counter that has the following function table:

UP COUNTER with PARALLEL LOAD

Load	Function
L	Place data on P0…Pk into outputs Q0…Qk
H	Increment Counter; $Q^+ = Q + 1$

We will begin our demonstration by using a fixed number applied to the parallel inputs (i.e., P0...Pk) of the counter. Let's make the branch vector = 0110 for this discussion. Now consider the following table of PROM data.

Address	Data
A3 A2 A1 A0	Dm D7 D6 D5 D4 D3 D2 D1 D0
0 0 0 0	1
0 0 0 1	1
0 0 1 0	1
0 0 1 1	0
0 1 0 0	1
0 1 1 0	1

The system will progress through the following sequence of states after reset: 0, 1, 2, 3, 6, 7, etc. The "program" itself determines the end of the current sequence and when to begin the next one. However, this point cannot be affected by the operations in the architecture unless they can supply a new branch vector to point to the sequence desired. This scheme is actually used in some computers; however, the task performed by the architecture is only one of vector storage.

To give our counter-based ASM a means for implementing decision-making, let us swap the location of the branch vector (external source) with the *Load* command (internal to the microprogram). This is shown in Figure 4-6. This seemingly small change allows us to implement all of the structures shown in the right ASM diagram in Figure 1-5 except the capability to test two inputs simultaneously and the conditional output. The following table contains the data for the microprogram memory for the example in Figure 1-5.

Address	Data									
	Branch Address				Other					
S3 S2 S1 S0	$S3^+$	$S2^+$	$S1^+$	$S0^+$	Z5	Z4	Z3	Z2	Z1	State/Out
A3 A2 A1 A0	D8	D7	D6	D5	D4	D3	D2	D1	D0	PROM
0 1 0 0	1	1	0	0	0	0	0	0	1	

We will specify an ASM diagram fragment that is similar to Figure 1-5 starting in state 0100. Following that state, we will place a decision block that has the following consequences:

IF (X1 .EQ. 1) THEN NEXT STATE = 1100 ELSE NEXT STATE = 0101

Notice in the data table above only one line of data, one address value, is used to contain the next state information, not two lines for each input required before. The only value placed in the data field associated with next state control is the branch address, the state to be entered if the condition is true. The "ELSE" situation is handled by the logic in the

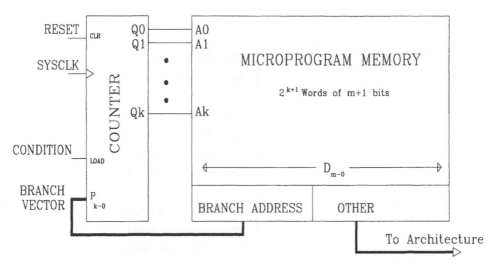

Figure 4-6: Counter-based ASM implementation with conditional branching — version 1.

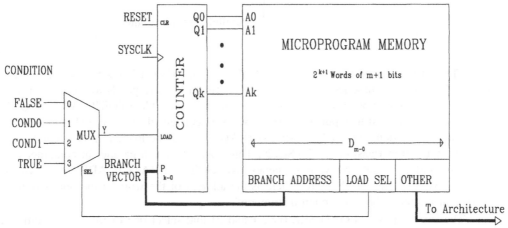

Figure 4-7: Counter-based ASM implementation with conditional branching — version 2.

counter, i.e., count up to 0101. If the condition is generated in the architecture under the control of the data outputs in the "OTHER" part of the memory, then flag testing and sequence modification can be accomplished using a controller implemented in this way.

The controller just described will run the IF...THEN...ELSE construct every cycle of SYSCLK. The only way it can be prevented from testing the condition input would be to add a circuit to disconnect it under control of the program in the memory. This is accomplished in Figure 4-7. The Condition Code Multiplexer (CCMUX) is placed between the condition input and the counter LOAD input. The MUX select line now receives its control signal from the microprogram memory. We have shown a 2-bit control line which allows us to select two conditional inputs and two other signals, one always *True* and the other always *False*. Here is a function table that describes the action produced by the two bit code LOAD SEL

LOAD SEL	COND0	COND1	Function
0	X	X	Increment the counter
1	T	T	Parallel load the counter
1	T	F	Parallel load the counter
1	F	T	Increment the counter
1	F	F	Increment the counter
2	T	T	Parallel load the counter
2	T	F	Increment the counter
2	F	T	Parallel load the counter
2	F	F	Increment the counter
3	X	X	Parallel load the counter

The application of a *False* to the CCMUX input when LOAD SEL=0 causes the counter to increment to produce the next state. This function corresponds to the normal flow of a program in a computer and has been given the name CONTINUE. When LOAD SEL = 3 the *True* applied to input 3 of the CCMUX is connected to the LOAD input of the counter causing the data field called "BRANCH ADdRess" to be loaded into the counter on the next Low-to-High transition of SYSCLK. This operation has a parallel in computer languages also and is known by various names like GOTO "BRANCH ADR" or JuMP to "BRANCH ADR". The branch address field is the address of the word in the microprogram memory that contains the next "microinstruction".

The two values of LOAD SEL (i.e., 1 and 2) that select one of the two conditional inputs implement the IF...THEN...ELSE construct as follows.

IF (COND0 .EQ. TRUE) THEN GOTO "BRANCH ADR" ELSE CONTINUE

and

IF (COND1 .EQ. TRUE) THEN GOTO "BRANCH ADR" ELSE CONTINUE

The flags COND0 and COND1 are 1-bit quantities created by a combinatorial operation elsewhere, usually in the architecture driven by the controller.

A new function table can be created that reflects the evolution of our interpretation of the LOAD SEL field. If we identify LOAD SEL as an instruction (INST) to the controller, then our function table looks like this:

INST code	Symbolic instruction
0	CONTINUE
1	BRANCH IF COND0 to BRANCH ADDR
2	BRANCH IF COND1 to BRANCH ADDR
3	JUMP to BRANCH ADDR

This simple controller consisting of a counter, CCMUX, and memory implements the primary program control instructions that we have used in all computer languages. We don't plan to stop here except to admire the simplicity of a hardware implementation of these important constructs.

We have created a little confusion in our move toward interpreting the memory data word as an instruction. It involves the time alignment of control signals to the architecture, in the field labeled OTHER, and the event we are specifying for the controller in the LOAD SEL and BRANCH ADR fields. As the controller is configured in Figure 4-7, the data in the "OTHER" field is active during the state that applies its address to the memory. Here is an example:

Address	BRANCH ADDR	LOAD SEL	OTHER
0100	0110	JMP	ADD

Interpreting the operation of the controller in terms of the function table immediately above, we find that the state of the system, $Q_k...Q_0$, on the output of the counter will change to 0110 on the next Low-to-High transition of SYSCLK. The architecture, using an RALU for example, will perform an ADD operation and place the results in its destination register on the same Low-to-High clock transition. Now assume that the following operation is performed instead:

Address	BRANCH ADR	LOAD SEL	OTHER
0100	0110	BRANCH IF C	ADD

Here, the Carry Out from the adder in the RALU is to be tested to determine the next state. The C flag will be *True* only if the ADD operation specified in the "OTHER" field causes an unsigned overflow. All signals beginning at the state variable outputs in the counter through the adder in the RALU via the C flag to the CCMUX must be stable before SYSCLK can be allowed to transition from Low to High. Data in the RALU which is selected by other fields within "OTHER" must be stable as well. This particular operation represents the worst case possible (i.e., WORST CASE DELAY PATH) through the entire network composed of controller AND architecture. Other operations, like AND, etc., in the RALU require considerably less time since they don't involve carry propagation.

In some operations, register transfers within the RALU do not pass through the ALU at all; thus, even less time is required. In a few cases, no architectural operations are performed, so only the time delay through the microprogram memory or in the counter sets a minimum period for SYSCLK. This wide range in the propagation delay requires that the period of SYSCLK be set to the slowest or worst-case delay. The consequence of this is that a lot of time is wasted when the quicker operations are being performed.

A cure for the problem can be made by breaking up the long propagation paths with registers. A register is placed on the output of the microprogram memory field "OTHER" as shown in the upper part of Figure 4-8, at A. This, when concatenated with the counter, is

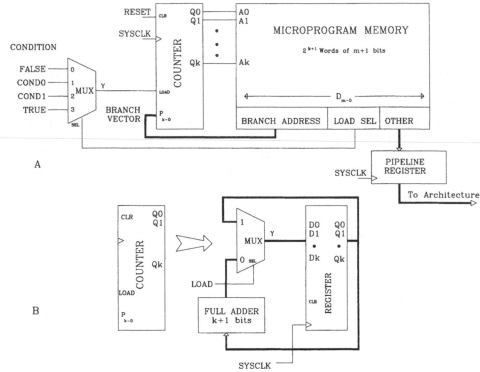

Figure 4-8: (**A**) Counter-based state machine implementation with pipeline register.
(**B**) Counter replacement.

called the *pipeline register*. In later versions we will replace the counter with a register without counting capabilities and the entire *pipeline register* will be just that, a register. Another small pipeline is placed on the output of the *adder* in the RALU. This will be known as the *status register* and it contains the various RALU flags (i.e., C, OVR, SIGN, and ZERO). Each register is updated on the same Low-to-High transition of SYSCLK. This structure does appear to slow the system since the single microinstruction above will require two SY-SCLK cycles to complete. The first cycle performs the ADD operation and the loading of the status register. The second cycle then implements the conditional branch since the C flag is only visible to the controller *after* the first clock cycle. This slowdown is compensated by the fact that the propagation time for signals associated with state changes (i.e., through the microprogram memory) is separated from that for architectural control. In fact, the two propagation times are now in parallel. The period of SYSCLK is still determined by the worst-case delay through the architecture, but only from the *pipeline register* output through the RALU to the Carry flag input on the *status register*. This time may be about half that in the previous example without the *pipeline register*.

The counter-based system is indeed faster since we have broken a long propagation delay loop that threaded both the controller and the architecture into two delay paths that can run in parallel. The price we paid for it is seen when one examines the microprogram execution of the ADD operation now *followed* by a conditional branch. This is left as an exercise that can be accomplished by writing a table of the contents of all elements in the controller and architecture. It will be seen that some very fancy control circuits must be included, or time wasted, to get the conditional branch to run. This feature is true for most pipelined systems, however. We will get around it by going back to the counter-based controller design and seeing how the problem can be avoided. Notice that a parallel-loading counter may be constructed as at B in Figure 4-8. The order of the MUX and incrementer in the feedback path is significant.

4.5. REGISTER–BASED ADVANCED MICROPROGRAMMED CONTROLLERS

Our first advanced controller is shown in Figure 4-9. The counter that caused our trouble has been divided into its constituent pieces, i.e., an incrementer and a register. The microprogram counter register, µPC, gives the name to this class of advanced controllers. Instead of taking the address into the microprogram memory directly from the state variable outputs as we did in the counter-based controller, we will interpose a multiplexer, MUX, between the state variables and the microprogram memory.

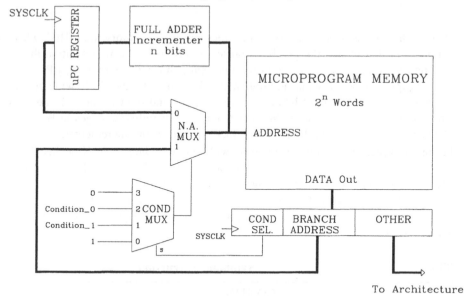

Figure 4-9: Register-based controller.

The path from the *Branch Address* field to the microprogram memory via the parallel load input of the counter is changed so that it passes through the MUX directly to the memory address inputs. This removes one level of clocked element from the path, thence the one clock cycle delay that we experienced in our counter-based implementation. The path from the µPC register outputs through the MUX to the µPC register inputs via the incrementer modifies the flow along the similar path in a parallel loadable counter. In the counter, the parallel data path OR the incremented register contents goes to directly to the register depending upon the value on the *Load* input. In our new design notice that the value applied to the address inputs of the microprogram memory will *always* be incremented and placed in the μPC register at the end of the current SYSCLK cycle no matter what the address source is.

The CCMUX, under control of the COND SEL field in the microinstruction, now determines which address source to use in the same sense that it did when it controlled the *Load* input on the counter. The important modification then is that the one clock period difference between a CONTINUE and a JUMP is removed in the PIPELINEd conditional branches. The order of the MUX and incrementer on the address path is what accomplishes this result. In other words, the components of a circuit are important, but the way they are wired can totally determine their behavior.

Let's examine the operation of this controller during the execution of an operation that adds two numbers in an RALU and conditionally branches based on the result in the C flag. We will still place the C flag in a status register making it impossible to do the conditional branch in the same cycle as the flag generation, i.e., the ADD. This time we will look at the microinstruction in the *pipeline register,* not in the microprogram memory.

OTHER	BRANCH ADDRESS	COND SEL	CYCLE #
ADD	X	CONTINUE	N
?	1001	BRANCH IF C	N+1

Assume that the address applied to the microprogram memory was 0100 when the SYSCLK transition that loaded the pipeline register occurred. The same transition that loaded the pipeline also placed 0100 plus 1 in the microprogram counter register, μPC. The incrementer is simply an M-bit wide addition network similar to the 4-bit wide full adder shown in Figure 3-2. If the A input bits are set to logical 0 and the Cin bit to 1, the network will continually produce S = B plus 1. In other words, S will follow changes in B by a logic family-related propagation delay. There is *no* clock applied to the incrementer.

In our example, the contents of the μPC will be 0101 when cycle N begins. The contents of the pipeline tells the architecture to add two numbers. The controller is instructed to continue to the next microinstruction. Using the function table below for this controller, we see that the COND MUX is made to point the address MUX at the μPC register when its COND SEL (i.e., INST) code is 3, meaning continue. Positive logic representation is used throughout Figure 4-9.

INST code	Symbolic instruction
3	CONTINUE
2	BRANCH IF COND0 to BRANCH ADDR
1	BRANCH IF COND1 to BRANCH ADDR
0	JUMP to BRANCH ADDR

At the end of clock cycle N, signified by the next rising edge of SYSCLK, the C flag from the adder in the RALU is clocked into the status register. The next cycle, N+1, begins when, due to the rising edge of SYSCLK in the last sentence, the next microinstruction appears on the output of the pipeline register. This microinstruction came from location 0101 since the μPC register was selected by the continue (INST = 3).

The controller part of the new instruction says to test the C flag and, if it is *True*, then fetch the next microinstruction from location 1001 in the microprogram memory. If it is *False*, however, use the microinstruction in the next memory location, i.e., 0110. For this example, assume that the C flag status register output is connected to the COND0 input of the CCMUX. To test this input, the actual INST code (i.e., value of COND SEL) is 2 from the function table above. Using positive logic, if COND0 is *False* (0), then the MUX will use the contents of the μPC register as the next address. We have shown above that when the current microinstruction from address 0101 was placed on the output of the pipeline register that this address plus one (i.e., 0110) was placed in the μPC register. Notice that due to the ordering of the incrementer and MUX in the address path to the microprogram memory, both CORRECT addresses are available at the end of the N+1 cycle. When SYSCLK goes High at the end of the N+1 cycle (the beginning of the N+2 cycle), the address selected by the C flag in the status register will be used to determine the microinstruction loaded into the pipeline at that time. The total propagation delay path starts at the status register and passes through the microprogram memory via the CCMUX and the address MUX to the data inputs of the pipeline register. Notice that the outputs of the μPC and the branch address field in the pipeline have been stable for most of the N+1 cycle period.

The next interesting point occurs if the branch was taken, i.e., if C flag was *True*. When the MUX placed the branch address, here 1001, on the address inputs of the memory, the incrementer placed 1010 on the inputs of the μPC. At the end of the N+1 clock cycle, the second microinstruction above caused the branch to location 1001 AND the 1010 was clocked into the μPC. If the controller instruction, INST, in the instruction at location 1001, which is now in the pipeline register, had been a "continue", the N+3 cycle would run with a microinstruction from location 1010. No clock cycles need to be rerun or skipped, all because of the flow of address signals in this controller. We have not added any new features to this controller, but we have eliminated some wasted cycles caused in the counter-based controller by conditional branching. This controller can perform all of its three instruction types in one cycle of SYSCLK. We will try to use this as a standard of performance for any other features that we introduce as this controller structure evolves.

The controller shown in Figure 4-10 borrows the BRANCH VECTOR from the counter-based controller in Figure 4-5. The branch vector is generated in the Second Address Register at the top of the figure. It and the branch address field in the pipeline register share the same n-bit wide bus that enters the address MUX. A virtual multiplexer is implemented using tristate buffers in the Second Address Register and the register containing the branch address field. The output enable control is part of the pipeline register in the *address select* field. Only the branch address field outputs in the pipeline register are tristateable, making their control from another part of the pipeline possible. If address select is High, then the Second Address Register output is selected by the next address MUX. If it is Low, then the branch address field is selected. Notice that either field can be employed in a conditional branch mode; however, they are each paired with the continue operation and not with each other. This allows an external agency to specify the beginning address of a sequence of microcode. No agency is shown in the figure that would cause the contents of this register to change. We will actually use this construct later when we implement an instruction set processor, ISP.

In summary, this controller can use an externally generated vector as well as branch addresses embedded in the microprogram to determine the state sequence. This vector can be employed conditionally.

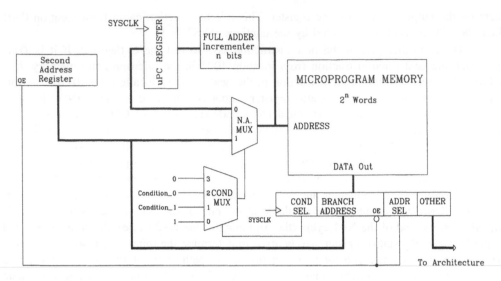

Figure 4-10: Register-based controller with virtual address multiplexer.

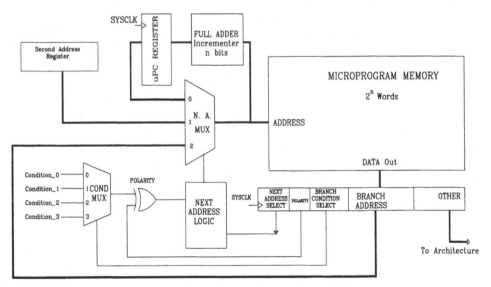

Figure 4-11: Register-based controller with condition code multiplexer.

The next controller structure, in Figure 4-11, expands the functionality of the address MUX and recognizes with new names some features that have developed relatively unnoticed so far. The address MUX receives another input which in this example implements the virtual MUX created by tristating the branch address field in Figure 4-10. To mark this point in evolving the controller, we will give the address MUX a name, i.e., NEXT ADDRESS MUX. While the name seems pedestrian, it clearly signifies the work that this element performs in our controller. Its physical structure consists of "n" copies of an "m" input MUX, one for each bit of each address path. Its function, as before, is to select one of the sources of address for the microprogram memory. Since there are more inputs on the next address MUX, we expect that the select function will require more bits. The number of inputs on the MUX will range between four and eight; therefore, the select code will be two or three bits.

The CCMUX will similarly evolve since the number of flags and other inputs is greater than those mentioned. Remember that this is the only path through which the controller can look at the architecture and the world beyond. We will usually not try to test more than one input at a time so that the path out of the CCMUX can be reduced to 1-bit wide. The CCMUX is pointed at one of the many sources of 1-bit inputs (flags, interrupts, etc.) by the *branch condition select* field in the pipeline register. This field must be specified by the microprogrammer to select the source of a condition code for which he is asking the controller to implement a conditional branch.

We will introduce a network, under control of the *polarity* field in the pipeline register, that will take the selected flag and logically invert it if desired by the microprogrammer. A function table for the polarity block in Figure 4-11 summarizes its operation:

Polarity control	Function
0	Pass 1-bit CCMUX output unchanged
1	Logically invert CCMUX output

The CCMUX and the polarity control allow the microprogrammer to point at a particular flag or other 1-bit input and supply its true or complemented form to the next block.

The NEXT ADDRESS LOGIC is a combinatorial network that translates codes in the *next address select* field of the microinstruction along with modified flag information into *next address MUX* selection codes. The actual transformation performed in this network establishes the relationship between the instruction codes and the instructions performed by the controller. A typical function table for the *next address logic* might look like this:

INST	CCMUX/POL	Controller function
0	X	CONTINUE
1	X	JUMP to Branch Address
2	T	JUMP to Branch Address
2	F	CONTINUE
3	T	JUMP to Address in Second Address Reg.
3	F	CONTINUE

Instruction code 0 causes the *next address MUX* to use the μPC contents as the source of the next microinstruction and ignores the polarity-modified CCMUX output. Notice that the contents of the *branch condition select* field in the microinstruction (pipeline register) does not matter. Instruction code 1 similarly ignores the CCMUX and points the *next address MUX* at the branch address field in the pipeline register for a microinstruction pointer source. A similar instruction could be provided to JUMP to the Second Address Register address.

Instruction codes 2 and 3 depend on the selected CCMUX input as modified by the polarity block for their outcomes. As shown, the following construct is implemented:

IF (CCMUX(COND SEL) .EQ. TRUE) THEN GOTO Branch Address
ELSE CONTINUE

If the polarity control is set to one as shown in its function table above, then the following construct is implemented:

> IF (CCMUX(COND SEL) .NE. TRUE) THEN GOTO Branch Address
> ELSE CONTINUE

or

> IF (CCMUX(COND SEL) .EQ. FALSE) THEN GOTO Branch Address
> ELSE CONTINUE

This allows the microprogrammer to phrase his conditional branch statements in a form appropriate to the interpretation of the input being tested.

The *next address logic* then allows simple codes represented only by the number of bits required by the number of controller instructions to be used to point the *next address MUX* at the necessary address source. This level of encoding can be performed in hardware or in software as we shall see later when we study the microcode assembler. Hardware encoding *may* reduce the number of bits in the controller field of the microinstruction.

In summary, this controller structure allows an almost "programming" type of interface to the implementation of a state machine. The pipeline register fields *next address select* and polarity can be thought of as an operation. The *branch condition select* allows the programmer to point at one of several test inputs and the *branch address* field specifies a possible destination for a control transfer. The symbolic constructs above are directly implementable on this structure.

The next controller, shown in Figure 4-12, retains all of the previous features including the *vector* address source. It adds a temporary storage location for a copy of the contents of the μPC register in the upper left-hand part of the drawing. The storage location is called the *subroutine return address register* and allows us to provide a one-level deep subroutine structure to our controller instruction repertoire.

In order to demonstrate the subroutine operation, we will examine the RETURN from a subroutine first. Assume that an address of a location in the microprogram has been placed in the *return address register* by some unknown (at this time) agency. The *next address logic* would translate the code for the RETURN instruction placed on its inputs by the *next address select* field in the pipeline register by pointing the *next address MUX* at the *return address register* as a source of address for the next microinstruction. This operation is identical to the JUMP instruction operation except that the address source is the return register instead of the branch address field in the pipeline.

A conditional return is easily implemented if the *next address logic* is made to use the 1-bit CCMUX/polarity output in its selection of the μPC, where *False* causes CONTINUE, or the return register, where *True* causes return from the subroutine.

The CALLing of a subroutine is a little more complex but will be made to take the same amount of time as the CONTINUE instruction, one clock cycle. Remember that the μPC register contains the address (plus one) of the instruction that is currently in the pipeline. If the controller part of that instruction CALLs a subroutine pointed to by the Branch Address field in the instruction, then the following two things happen *simultaneously*:

1. The *next address MUX* is pointed at the branch address field in the pipeline to handle to transfer to the subroutine
2. The address of the next microinstruction (return address) currently in the μPC is transferred to the return address register

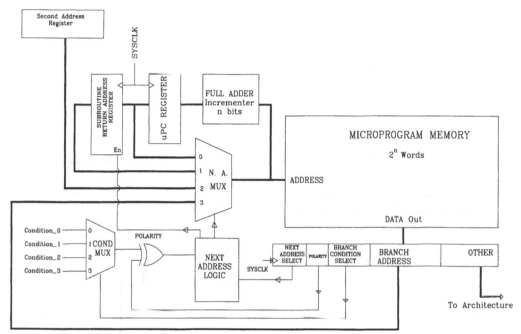

Figure 4-12: Register-based controller with subroutine capability.

All paths are established during the clock cycle in which they are commanded by the controller field in the pipeline register. When the end of the clock cycle comes, signaled by the Low-to-High transition of SYSCLK,

1. The return address register is loaded with the old contents of the μPC

2. The microinstruction specified by the branch address field (subroutine entry point) is placed in the pipeline register

3. The μPC is loaded with the address of the subroutine entry point plus one

Remember that the incrementer is combinatorial and is located at a point in the address flow of the controller such that it provides the incremented value of the current memory address to the controller by the end of the current clock cycle.

This controller is indeed powerful. It allows us to use the subroutine structure that we have available in all higher-level languages to save space and programming time even at the microprogram level. The architecture of the controller still remains very simple, consisting of MUXs, registers, specialized logic, and address paths. The actual flow of addresses is responsible for its time efficiency as well as its simplicity. The only shortcoming in this controller is the lack of a subroutine nesting facility. If one subroutine level is a good idea, then more nesting levels must be a better idea. We will add this facility in the next controller; however, it should be mentioned that because of the capability of the controller and architecture to perform their tasks *simultaneously*, the number of microinstructions in a program will generally be a great deal less than in a higher-level program.

The controller in Figure 4-13 retains all of the features found in the previous controller, but has its return address register replaced by a *stack* structure. This structure has been discussed previously in Chapter 3 and is displayed in Figure 3-8.

At this point we will show why we made the up/down counter used for a TOP OF STACK pointer change its value on the opposite phase of the clock when writing to the subroutine return address memory (FILE) occurred. Let's handle a subroutine RETURN first.

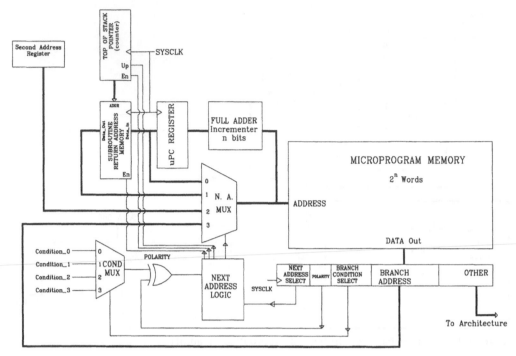

Figure 4-13: Register-based controller with nested subroutine capability.

Our design goal is to make all operations in the controller take the same period of time, one cycle of SYSCLK. In the case of a return, this is managed by leaving the top of stack pointer (up/down counter) pointing at the last entered return address. Since the memory (FILE) is *not* a clocked structure during a read operation, the last entered return address is always available on the input of the *next address MUX* for use at *any* time. A return operation, without an accompanying change of the stack pointer, is accomplished as in the previous controller by using a code in the *next address select* field of the pipeline that causes the *next address MUX* to point at its stack memory input (second from the right in the figure).

The RETURN operation can be made conditional as before by letting the *next address logic* decide between using the μPC output or the FILE output based on the polarity-modified, selected condition code.

A stack POP operation is accomplished by the *next address logic* through its control of the PUSH/POP* and STACK ENABLE inputs. The "top" RETURN address can be used *and* removed by combining the operation for a return described above with the stack pointer increment that implements a POP.

The subroutine CALL operation is built as before using the branch address field to contain a pointer to the subroutine entry. This time, however, the stack pointer must be decremented to the next empty location before the new return value can be placed in the FILE memory. The sequence of events is keyed to the phases of SYSCLK during the microcycle in which a CALL is requested. The sequence begins with:

1. Loading the microinstruction containing the CALL into the pipeline register on the RISING edge of SYSCLK

2. On the FALLING edge of SYSCLK at the midway point in the cycle, the stack pointer is decremented

3. At the end of the cycle (beginning of the next cycle), as SYSCLK RISES, the old contents of the μPC are loaded into the cell newly pointed at by the stack pointer

At the *same* time, the address of the second instruction in the subroutine is placed in the μPC by our friend, the incrementer.

How can this double operation be performed if we are running the state machine as fast as the technology allows? Remember that the period of SYSCLK was set by the propagation delay through the longest path in the system. After pipelining, this path appeared to be the one that started at the pipeline register, passed through the RALU, and ended at the C flag input on the status register. Even though this path is used only for arithmetic operations in the RALU, the worst-case delay is used in order to avoid varying the frequency of SYSCLK as a function of the microinstruction being executed. We need only to compare the worst-case propagation delay with the time required to increment the stack pointer and store data in the *file* memory. The pointer is an up/down counter optimized for speed in the technology of the controller. The *file* memory is relatively shallow, ranging from four words to as many as 32 in depth. It is also implemented in the fast technology of the controller and therefore would be expected to perform several times faster than the microprogram memory. Both the flow through the stack pointer and the *file* memory are independent of each other and are in parallel with the microprogram memory propagation path. So, for this structure, it would be expected that these elements would not increase the period of SYSCLK. The other paths are usually longer.

4.6. REGISTER–BASED CONTROLLER WITH ALL THE BELLS AND WHISTLES

The last controller version in this evolution will add two features to the previous ones. In the upper left portion of Figure 4-14 are two blocks, a down counter and a register. The Second Address Register is returned to a virtual MUX formed by its tristate outputs with the similarly equipped branch address field in the pipeline. This allows us to retain the four-input *next address MUX* without paying the time penalty caused by increasing its width. The *next address logic* is given the responsibility for converting instruction codes in the *next address select* field to controls for the new elements. The address register provides another place to store an address before it is used in the controller. The *next address logic* can implement JUMPs to this address or conditional BRANCHes using it along with the μPC. Other combinations are possible, including three-way branches.

The *counter* allows us to build counted loop control structures while simultaneously performing operations in the architecture. If you will review the last discussion of the shift and add multiply algorithm done in Chapter 2 using the RALU, you will notice that the body of the loop was managed by a counter initialized to the number of bits in the multiplier. In that example, the counter was located in one of the RALU registers and two SYSCLK cycles were needed to change it and get the zero flag through the status register. This program structure occurs so often that it is efficient to dedicate a structure in the controller to the task of loop counting. The counter like the register added above can be initialized from data in the branch address field in the pipeline register. This was one of the reasons for trying to employ a counter to handle the CONTINUE operations in the first place. It freed this field for more general use.

Two instructions are usually implemented for loop control, one to initialize the counter and the other to decrement the counter and do the conditional return to the head of the loop body. The counter can be initialized by having the *next address logic*, in response to the proper code in the *next address select* field of the pipeline, load the counter from the Branch Address field and use the μPC register as the source of the pointer to the next micro-

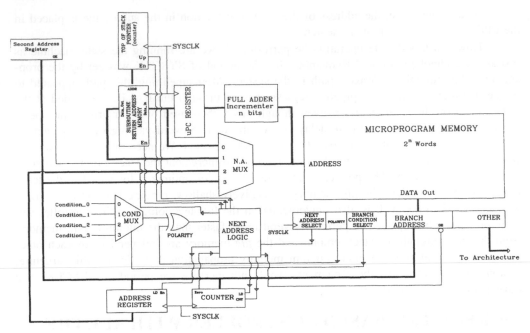

Figure 4-14: Complete microprogrammed controller.

instruction. This instruction can be thought of as LoaD CouNT and continue (LDCNT). Its operand is the value needed to initialize the counter; e.g., LDCNT 15 would place a 15 in the counter.

The next instruction actually manages the loop control and is placed after the body of the loop in the microprogram. It takes the symbolic form: DECrement Counter and LOOP to HEAD (DECLOOP HEAD). The counter = 0 flag (not in the status register) is pointed at by the CCMUX. The *next address logic* takes the next microinstruction from the location pointed at by the branch address field (the HEAD of the loop body). This is made condition-al on the counter = 0 flag, however, with the *True* case causing the μPC to be used as the address source (CONTINUE). This can be summarized as:

IF (COUNTER .NE. 0) THEN GOTO Branch Address (loop head)

ELSE CONTINUE

The counter is usually decremented at the end of the cycle at the same time (after) the address is selected based on it. This means that the counter will actually "run" the zero case. In order to compensate, the counter is usually initialized to one less than the desired count. This is a frequent situation for the DO LOOP implementation at the assembly lan-guage level.

Integrated circuit versions of the major parts of the controller have been made for several years now. The general trend has been to widen the address paths allowing deeper microprogram memories. Also, the return address stack is deepened to allow more subroutine nesting levels. The number of counters has been increased to improve the implementation of nested counted loops. In doing this, the separate register and counter shown in the figure are usually combined to save space on the silicon. This seems to cause only a small loss in func-tionality. In the next section we will discuss some of the real controller implementations in their integrated circuit form.

4.7. INTEGRATED CIRCUIT MICROPROGRAMMED CONTROLLERS

Now that we have studied the structure of a complete microprogrammed controller, we need to examine some of the integrated circuit (IC) offerings from various manufacturers. These ICs come in two different general forms: the first can be called *next address generators* and the second are complete integrated controllers. The next address generator integrates the following elements in Figure 4-14 onto a single chip.

- Next address multiplexer
- Microprogram counter register (μPC)
- Incrementer
- LIFO stack (FILE)
- Top of stack pointer
- Counter (for loops)
- Register (for addresses)
- Next address logic (in some cases)

Variations on the distribution of these sections will be discussed in the next section. The complete integrated controllers are a new line of parts by two manufacturers that result in an IC that contains all of the elements in Figure 4-14 except the Second Address Register. These devices are usually implemented in CMOS and are frequently mounted in Dual In-line Packages (DIP) with quartz windows. The microprogram memory is thus treated like an EPROM, i.e., it is programmable and ultraviolet (UV) light erasable. The device appears to the user as a logic block with several inputs to its CCMUX and several outputs from the "OTHER" field of its pipeline register.

4.7.1. Examples of the Next Address Generator

One of the older *next address generators* that is now available in gate array macro libraries is the Am2909 by Advanced Micro Devices, Inc. The device is a 4-bit wide "slice" through an N-bit wide *next address generator,* as shown in Figure 4-15.

It does not contain a loop counter or the *next address logic* of later units. However, this allows the designer to implement the *next address logic* in his own way to provide the control constructs appropriate to his needs. He can control the width of the instruction bit field in the microword by the method he uses to encode the *next address select* and polarity operations. The *next address logic* is implemented using a Boolean network, e.g., a ROM, that translates the *next address select* field or controller instruction code into the signals next address MUX select, file enable, push/pop, and register enable. If other elements (i.e., a loop counter) are added to the controller, then their control signals are also generated by the *next address logic* to create an integrated instruction set appropriate for the designer's tasks.

If his design requires counted loops, then the designer must add one or more counters suitable to the task such as one based on the 74LS169 in its "down count" configuration. The Terminal Count (TC) signal is used in the CCMUX/polarity structure to assist in loop termination.

Two features in the Am2909 that are not encountered in other next address generators are the array of AND gates connected to the "ZERO*" input and the array of OR gates connected to the OR control bus. The AND structure forces the Y bus, driving the microprogram memory address bus, to zero under control of the ZERO* input regardless of the next address MUX selection. This gives the designer a "hardware RESET" input to his state ma-

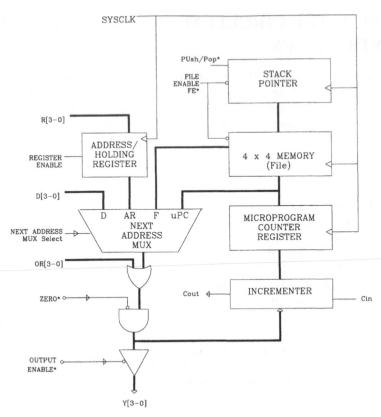

Figure 4-15: AMD Am2909 next address generator.

chine. The next microinstruction will be taken from microprogram memory physical address 0000.

The OR structure allows the designer to implement a function that he enjoys in the traditional state machine controller, i.e., the multiway branch in a single clock cycle. It accomplishes this by allowing the designer to provide another source of address that is ORed with the address selected by the next address MUX. To put this in context, imagine a controller having 12-bit wide address paths made using three Am2909 ICs placed "side-by-side". The Carry In of the least significant element is forced to 0, while the Carry Out is connected to the Carry In of the next more significant element. The incrementer is the only "horizontal" structure whose Carry signal must be propagated when expanding the width of the address paths. To implement a 16-way conditional branch instruction, assume that the 4-bit OR bus of the least significant Am2909 is driven by the output of 16-input/4-output priority encoder which in turn is driven by elements in the architecture or external to the machine. The upper 8 bits of the 12-bit Y bus will be derived from the branch address field of the microword by the next address MUX, i.e., a JUMP to branch address instruction is executed. The OR inputs of the two upper Am2909s will be held at 0. The resulting value on the Y bus, i.e., the address of the next microinstruction, will be formed by the concatenation of the upper 8 bits of the branch address field and the 4-bit code produced by the priority encoder as determined by events driving its inputs. Note: the lower 4 bits of the branch address field must be 0. The actual number of branch directions in a given clock cycle is under the control of the designer.

The last structure in Figure 4-15 that will appear in most other next address generators is the tristate buffer on the output of the Y bus. The reader may be surprised that one would want to disable the only signal driving the microprogram memory, i.e., the Y bus. During normal operations, the next address generator provides the address of the next microinstruction to the microprogram memory. There are two situations where normal operation is not used. To test the entire state machine, the designer might think that he can either execute his application program or some specialized test code to show that his machine works as specified. With a little reflection he should realize that this means exercising all possible instruction sequences in all possible orders with all possible data. This is not even a practical solution for a simple machine having a few states much less one appropriate to the capabilities we have discussed.

The most efficient test method involves placing the machine in a specific state, applying test input conditions, toggling the clock, and verifying that the correct next state was entered and the correct results were produced in the architecture. The tristate buffers on the Y bus allow the test operation to break the address path between the next address generator and the microprogram memory so that a state can be forced by driving the bus with the test generator. Once the machine is forced to a specific state, similar operations can initialize the rest of the architecture to continue the test. In the course of designing a system, the designer should keep in mind that he has a formidable debug and verification problem ahead. The tristate buffers on the Y bus contribute to the solution of this problem.

After the "bit-slice" next address generator discussed previously, AMD introduced a complete unit on a chip called the Am2910. We see in Figure 4-16 that the register and counter elements of Figure 4-14 have been merged to produce an address storage register and a down counter in a single element.

The two functions can obviously not be performed simultaneously. A single level of loop nesting can be done without adding any external parts. The next address logic, here called the *Instruction PLA*, is internal to the chip giving it a predefined instruction set having 16 members. The Condition Enable input (CCEN*) allows the microprogrammer to determine whether several instructions are sensitive to the Condition input (CC*). Thus, he can create a jump always from a conditional branch instruction, one of the 16 instructions in the basic set, by setting CCEN* to *True*. The next address logic (i.e., Instruction PLA) also supplies control signals to enable the pipeline register, MAP ROM (see Chapter 8) and a third address source onto the D bus by using PL*, MAP*, or VECT*, respectively. This allows the microprogrammer, through the proper controller instruction, to select the branch address field or external address registers as sources of possibly conditional branch addresses.

The stack was five levels deep in the first models of the Am2910 and was increased greatly as later models were produced. The stack is used to support subroutine nesting as indicated earlier. It also is used to contain branch addresses that point at the head of a loop body during looping. A stack FULL* signal indicates when stack overflow has occurred. This feature supports external testing as well as error tracing.

Resetting the machine is accomplished by a specific controller instruction Jump to Zero (JZ) which has a code of 0000. The reset operation is performed by forcing the Am2910 instruction bus to 0 by a signal (i.e., RESET) that is operated independently of the controller. One method to do this is to place the next address select field in a clearable register similar to the 74'273. The RESET signal is used to force the register contents to 0 which, by way of the JZ instruction, causes the *next address logic* to place an address of 0 on the Y bus. The first microinstruction after the RESET signal returns to *False* is taken from address 0 in the microprogram memory. As in the Am2909, the Y bus is tristate buff-

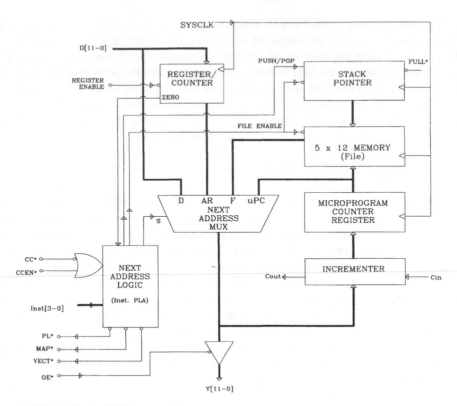

Figure 4-16: AMD Am2910 next address generator.

ered to provide for driving the microprogram memory by another address source (e.g., a test system).

The next address generator from Texas Instruments, Inc. (TI) is called the 74ACT8818 and is shown in Figure 4-17. Where the earlier AMD units were implemented using TTL technology, the 'ACT8818 is a high-speed CMOS realization. It is a very fast, modern unit having 16-bit address paths and several additional interesting features.

The 'ACT8818 has a structure containing all of the elements of a next address generator. There are two register/counter structures, A and B, with their associated buses, DRA and DRB, which serve the same function as the "Direct" bus in the previous examples, i.e., transport the branch address to the Next Address MUX. The TI "SN74ACT8800 Family Data Manual" (see Bibliography) lists the following features.

> "1. A 16-bit microprogram counter (MPC) consisting of a register and incrementer which generates the next sequential microprogram address
>
> 2. Two register/counters (RCA and RCB) for counting loops and iterations, storing branch addresses, or driving external devices
>
> 3. A 65-word by 16-bit LIFO stack which allows subroutine calls and interrupts at the microprogram level and is expandable and readable by external hardware.
>
> 4. An interrupt return register and Y output enable for interrupt processing at the microinstruction level
>
> 5. A Y output multiplexer by which the next address can be selected from MPC, RCA, RCB, external buses DRA and DRB, or the stack."

The presence of the two register/counters, RCA and RCB, makes implementing doubly nested loops more efficient. This operation will be encountered in the series expansion examples in Chapter 7.

The current version of the Advanced Micro Devices family of next address logic elements, the Am29C331, is summarized in Figure 4-18. While returning to a single loop counter, the device adds some decoding logic to create a complex conditional multiplexer that allows combinations of conditional inputs to be analyzed resulting in a "multiway" branching capability. The device is implemented in high-speed CMOS.

Both modern next address generators support methods to verify operations in such a way that supports "fault-tolerant" design principles. Such designs are required to run even if one or more parts fail. This capability is supported in the Am29C331 by using a master/slave error checking scheme in which two sequencers running in parallel continually check the addresses generated. The designer provides circuitry and code to cope with failures that are appropriate to his design.

4.7.2. Integrated Microprogrammed Controllers

The two examples to be discussed in this section are complete integrated controllers having all of the features and capabilities discussed in Section 4.6 and shown in Figure 4-14. The Altera EPS448 is presented in Figure 4-19 and the, now discontinued, AMD Am29-CPL151 is shown in Figure 4-20. Both devices are implemented in high-speed CMOS technology and present the designer with a logic block having approximately 8 conditional inputs and 16 architecture control outputs. The AMD chip contains 64 microwords in its microprogram memory, while the Altera device contains 448. Internal paths support the *next address select* and branch address fields of the microword so that all of the pins on the device, approximately 28, can be used to deal with the architecture of the machine. The principal difference between the devices is in how the width and depth of the microprogram is expanded from that contained in a single device.

The Altera device supports horizontal or vertical cascading to increase the size of the microprogram memory. Horizontal cascading to widen the microword is accomplished by placing the same controller program in each chip and using different architectural sequences for the section controlled by the chip. The AMD chip is expanded by using a mode control pin to couple the six address lines from the *next address MUX* to the lower architectural control pins. This address bus is used to drive external registered PROMs, thus arbitrarily widening the architectural control bus.

Both devices are available in erasable, reprogrammable form. Software may be prepared using the methods discussed in Chapter 5 or by using the manufacturers' own software support.

The devices support all of the control constructs described for the complete microprogrammed controller, including nested subroutine calls, counted loops, and other more exotic variations of the basic set. They represent a method with which the microprogrammer can prepare solutions that directly compete with those implemented using the traditional controller-based Programmable Logic Sequencers (PLS) and microcomputer solutions. Remember that single-level ASMs do not require instruction fetching as used in two level control situations based on microprocessors. These devices share a capability to run at high clock rates, currently above 25 MHz. The programming-like environment coupled with the capability to control multiple processes in the architecture make these devices very attractive as solutions to problems.

SN74ACT8818A
16-BIT MICROSEQUENCER

functional block diagram

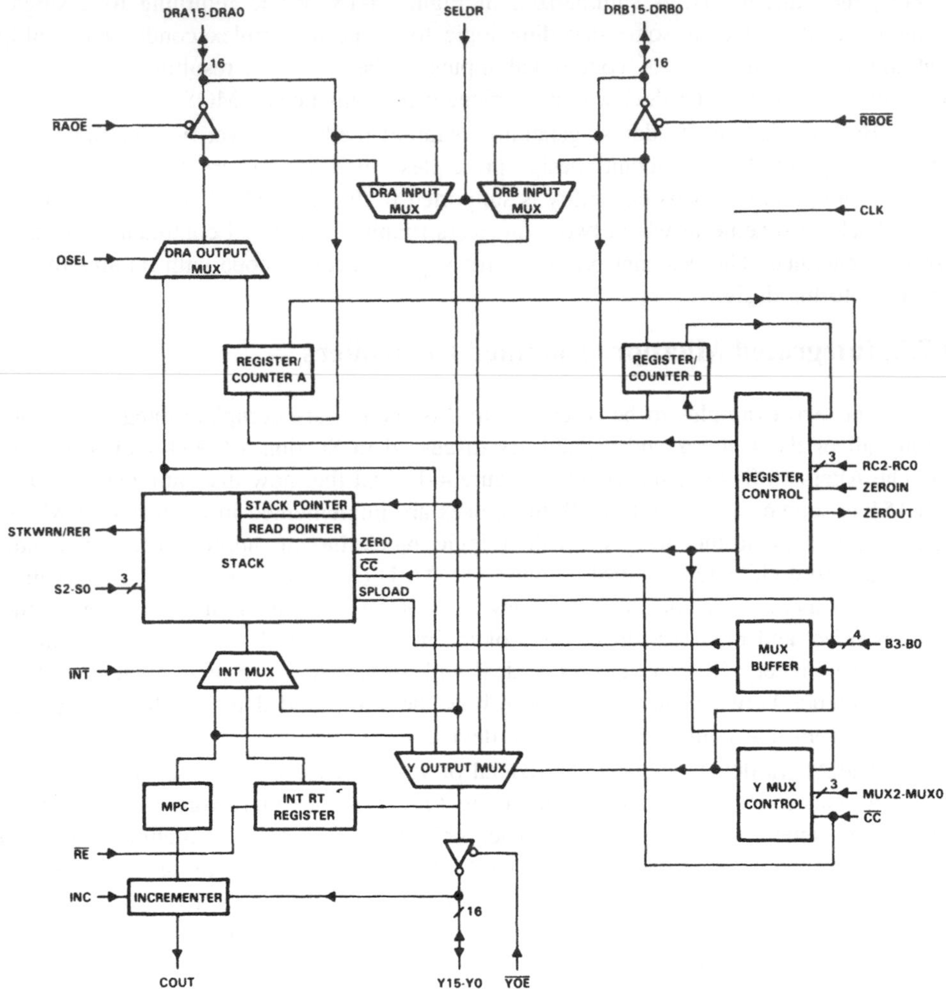

TEXAS
INSTRUMENTS
POST OFFICE BOX 655303 • DALLAS, TEXAS 75265

Figure 4-17: 74ACT8818 next address generator.

(Reprinted by Permission of Texas Instruments, Inc.)

Am29331

16-Bit Microprogram Sequencer

DISTINCTIVE CHARACTERISTICS

- **16-Bits Address up to 64K Words**
  Supports 80-90 ns microcycle time for a 32-bit high-performance system when used with the other members of the Am29300 Family.
- **Real-Time Interrupt Support**
  Micro-trap and interrupts are handled transparently at any microinstruction boundary.
- **Built-In Conditional Test Logic**
  Has twelve external test inputs, four of which are used to internally generate four additional test conditions.

- **Break-Point Logic**
  Built-in address comparator allows break-points in the microcode for debugging and statistics collection.
- **Master/Slave Error Checking**
  Two sequencers can operate in parallel as a master and a slave. The slave generates a fault flag for unequal results.
- **33-Level Stack**
  Provides support for interrupts, loops, and subroutine nesting. It can be accessed through the D-bus to support diagnostics.
- **Speed improvement with Am29331A (15% faster than Am29331)**

GENERAL DESCRIPTION

The Am29331 is a 16-bit wide, high-speed single-chip sequencer designed to control the execution sequence of microinstructions stored in the microprogram memory. The instruction set is designed to resemble high-level language constructs, thereby bringing high-level language programming to the micro level.

The Am29331 is interruptible at any microinstruction boundary to support real-time interrupts. Interrupts are handled transparently to the microprogrammer as an unexpected procedure call. Traps are also handled transparently at any microinstruction boundary. This feature allows re-execution of the prior microinstruction. Two separate buses are provided to bring a branch address directly into the chip from two sources to avoid slow turn-on and turn-off times

for different sources connected to the data-input bus. Four sets of multiway inputs are also provided to avoid slow turn-on and turn-off times for different branch-address sources. This feature allows implementation of table look-up or use of external conditions as part of a branch address. The 33-deep stack provides the ability to support interrupts, loops, and subroutine nesting. The stack can be read through the D-bus to support diagnostics or to implement multitasking at the micro-architecture level. The master/slave mode provides a complete function check capability for the device.

The Am29331 is designed with the IMOX™ process which allows internal ECL circuits with TTL-compatible I/O. It is housed in a 120-lead pin-grid-array package.

SIMPLIFIED BLOCK DIAGRAM

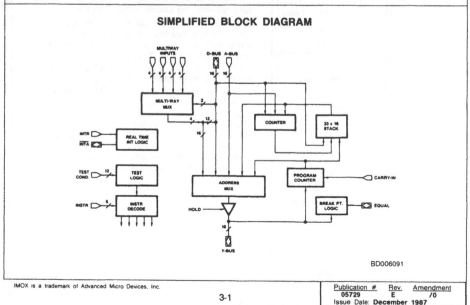

BD006091

IMOX is a trademark of Advanced Micro Devices, Inc.

3-1

Publication #	Rev.	Amendment
05729	E	/0

Issue Date: **December 1987**

Figure 4-18: AMD Am29C331 next address generator.

STAND-ALONE MICROSEQUENCER

EPS448

EPS448

FEATURES

- User-Configurable/Stand-Alone Microsequencer (SAM) for implementing high-performance controllers
- On-Chip reprogrammable EPROM Microcode Memory up to 448 words deep
- 15 x 8 Stack
- Loop Counter
- Prioritized, multi-way Control Branching
- 8 general-purpose Branch Control Inputs
- 16 general-purpose Control Outputs
- Cascadable to expand outputs or states
- Low-Power CMOS technology
- Footprint Efficient 28 pin 300 Mil DIP or 28 lead JLCC/PLCC package
- 25 MHz Clock Frequency

GENERAL DESCRIPTION

Altera's EPS448 (SAM) series of Function-Specific CMOS EPLDs are User-Configurable Microsequencers. On-chip EPROM (up to 448 words) is integrated with Branch Control Logic, Pipeline Register, Stack, and Loop Counter. This generic microcoded architecture provides an efficient vehicle for implementing a broad range of high performance controllers spanning the spectrum from basic state machines to traditional bit-slice controller applications.

The EPS448 has 16 output pins available in a 28-pin 300 mil DIP package as well as a 28-pin JLCC option. One-Time-Programmable plastic versions for SAM EPLDs are available to minimize volume production costs.

Programming the SAM device is accomplished on a standard Altera PLDS or PLCAD development system installed with the optional SAM+PLUS software package and device adapters. New users can purchase a separate PLDS-SAM development system with programming hardware included. SAM+PLUS allows designs to be entered in either state machine or microcoded formats. SAM+PLUS automatically performs logic minimization and design fitting for the device. The design may then be simulated or programmed directly to achieve customized working silicon within minutes.

Using a 1.0 micron CMOS EPROM technology allows SAM to operate at a 25 MHz clock frequency while still enjoying the benefits of low CMOS power consumption. This technology also facilitates 100% generic testability which eliminates the need for post-programming testing.

Ideal application areas for SAM include programmable sequence generators (state machines), bus and memory control functions, graphics and DSP algorithm controllers, and other complex, high performance machines. The devices may be cascaded easily to obtain greater output requirements (horizontal cascade) or greater microcode memory depth (vertical cascade) or both.

2

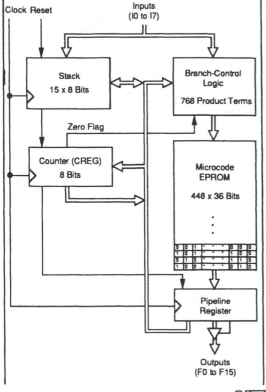

REV. 3.0

105

Figure 4-19: Altera EPS448 stand-alone microsequencer.
(Reprinted by permission of Altera Corporation.)

Final   **COM'L: -25/33   MIL: -25**

Am29CPL151H-25/33

CMOS 64-Word Field-Programmable Controller (FPC)

**Advanced
Micro
Devices**

DISTINCTIVE CHARACTERISTICS

- Implements complex state machines

- High-speed, low-power CMOS EPROM technology

- Functionally equivalent to the bipolar Am29PL141

- Seven conditional inputs (each can be registered as a programmable option), 16 outputs

- Up to 33-MHz maximum frequency

- 64-word by 32-bit CMOS EPROM

- Space-saving 28-pin OTP plastic SKINNYDIP® and PLCC packages and windowed ceramic SKINNYDIP package

- 29 Instructions
  - Conditional branching, conditional looping, conditional subroutine call, multiway branch

GENERAL DESCRIPTION

The Am29CPL151 is a CMOS, single-chip Field Programmable Controller (FPC). It allows implementation of complex state machines and controllers by programming the appropriate sequence of instructions. Jumps, loops, and subroutine calls, conditionally executed based on the test inputs, provide the designer with powerful control flow primitives.

Intelligent control may be distributed throughout the system by using FPCs to control various self-contained functional units, such as register file/ALU, I/O, interrupt, diagnostic, and bus control units. An address sequencer, the heart of the FPC, provides the address to an internal 64-word by 32-bit EPROM.

The Am29CPL151 is functionally equivalent to the Am29PL141 but is manufactured in CMOS technology and offers a space-saving 300-mil SKINNYDIP package. A pin-compatible larger FPC is offered as the Am29CPL154 with a deeper 512 x 36 memory and added flexibility.

This UV-erasable and reprogrammable device utilizes proven floating-gate CMOS EPROM technology to ensure high reliability, easy programming, and better than 99.9% programming yields. The Am29CPL151 is offered in both windowed and One-Time Programmable (OTP) packages. OTP plastic SKINNYDIP and PLCC devices are ideal for volume production.

SIMPLIFIED BLOCK DIAGRAM

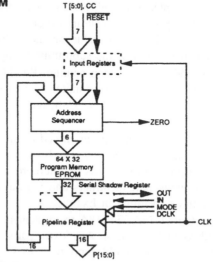

Publication # 10135   Rev. D   Amendment /0
Issue Date: November 1989

10135-001B

Figure 4-20: AMD AM29PL151 field-programmable controller.

EXERCISES

1. The most complex part about understanding a modern microprogrammed controller involves the way that its structure eliminates bad data in the pipeline during a branch instruction. The best way to study this exercise follows this procedure.

    a. Draw a fragment of an ASM diagram that contains a conditional branch to a state, e.g., 5 states beyond the current one, else continues to the next state. Use several outputs that are turned on in such a way that each state contains a different pattern. Your goal here is to identify when the pipeline contains bad information; i.e., information inconsistent with the ASM diagram.

    b. Implement tables similar to those shown in the text for the counter-based machine in Figure 4-7 and 4-8 and the register-based controller in Figure 4-9. Be sure to record the contents of the registers as well as the output of the incrementer and various buses.

    c. Examine the contents of the registers, including the pipeline where appropriate, during the execution of the branch and for the states immediately before and after. Do this once for when the branch is taken and once for when execution continues to the next instruction in sequence. You should see outputs misbehaving in the counter-based machine.

2. Perform an analysis similar to Exercise 1 during subroutine calling and returning using the machine in Figure 4-12. There are two instructions, Call and Return, and two cases for each when they are made conditional on the CCMUX input. Show that there are never any incorrect data in a register or on a bus during the execution of these instructions.

3. Examine the subroutine nesting capabilities of Figure 4-13 using the same analysis tool as in previous exercises. Here you will need to account for events on the falling edge of the clock as well as on the rising edge, i.e., two lines of information per clock cycle. The stack used is shown in Chapter 3 as well as having its operation described in Chapter 4. Show that there is never any incorrect data in a register or on a bus during the execution of the CALL instructions. What is the value on the address input to the microprogram memory when the first instruction of the subroutine occupies the pipeline register?

4. Redraw Figure 4-14 to add a second loop counter. Provide a truth table for the *next address logic* network that implements two counted loops, counter initializer, and the usual Continue, Conditional Branch, Conditional Call, and Return instructions using the new network.

5. Use the Am2909 to implement a microprogrammed controller similar to that described in Exercise 4. Any extra parts that are needed may be drawn from a suitable TTL databook. The controller should support 4096 words of microprogram memory. Draw a logic diagram of the network showing part identifiers and pin numbers. Prepare a truth table that implements the instructions specified in Exercise 4.

6. Use the Am29C331 or the TI 74ACT8818A to implement a microprogrammed controller similar to that described in Exercise 4. Any extra parts may be drawn from a suitable TTL data book. The controller should support 4096 words of microprogram memory. Draw a logic diagram of the network showing part identifiers and pin numbers. Prepare a truth table for the TI chip that implements the instructions specified in Exercise 4. You will need the appropriate data book for the selected part.

Chapter 5

ASSEMBLERS

5.1. INTRODUCTION

In the earlier chapters we discussed the controllers and architectural elements with which we will construct microprogrammed algorithmic state machines in the form shown in Figure 1-3. A microprogrammed controller contains a table of control sequences, i.e., the microprogram memory. One of the purposes of this chapter is to present a simple method for the preparation of the microprogram. In the context of preparing tables of instructions, we will also present a method to create assemblers for machines having a second level of control. The discussion of machines having a single level of control will commence in Chapter 7, while the second level of control will be added in Chapter 8.

Since we have adopted a "programmer's" model for the table of control sequences, we can extend this concept to include the creation of a language for writing the table entries. While the microprogram memory must contain data that can be interpreted by the hardware making up the controller and architecture (i.e., 1s and 0s) the microprogram can be expressed in a symbolic or human readable form. The translation from symbolic form to machine "readable" form is a well-understood algorithm in itself that can be automated. We will discuss the simplest of these algorithms, the assembler. The symbolic language will have the same human readable form as other assembler languages with modifications that support the capability of the hardware to perform more than one task in a single clock cycle. We will attempt to transfer any assembler language knowledge possessed by the reader into the understanding of an assembler for the first level of control.

Two assemblers will be considered in the microprogramming context. The first assembler is a widely used meta-assembler that was written specifically to be a "microprogram" assembler. Its concepts and methods for dealing with diverse architectures are worth studying even if the user does not have access to it. The second microprogram assembler is produced by adapting a "shareware" version of a second-level or microprocessor assembler for which the source code is available. The methods of adaptation may be extended to most major assemblers present on any computer system even if the source program for the assembler is not available.

The last subject covered in this chapter is to describe an assembler for machines having two levels of control (e.g., the computer). The design of such machines is covered in a later chapter. Since software support is needed for the second-level just as for machines with a single level of control, we will show how to adapt an existing second-level assembler for our new machine. Once the method is understood, the concept will be extended to produce an assembler for the first-level machine (microprogram) as well.

5.2. THE MICROPROGRAM WORD

The final controller discussed in Chapter 4, Figure 4-14, provided us with the program control constructs that we desire. They included the following operations: CONTINUE, JUMP, Conditional BRANCH, Subroutine CALL, Subroutine RETURN, and Counted LOOPs. The current value of the control signals is contained in the *pipeline register*. This register was added in order to break the propagation delay through the controller and architecture into two pieces that run in parallel. In so doing, we also created a structure in which part of the state of the system is stored. The pipeline contains all of the information about what the system will do at the end of the current clock cycle. If we place Figure 4-14 in the context of Figure 1-3 we would identify the path labeled "CONTROL" in Figure 1-3 as the field in the pipeline called "OTHER" in Figure 4-14. The path labeled "STATUS" enters the controller into the various inputs of the condition code multiplexer (CCMUX) in Figure 4-14.

At the beginning of a clock cycle, the pipeline register will be loaded with all of the bits in the microprogram memory having the address generated by the sequencer in the controller. This collection of bits is called the *microword*. While the "word" used in computer circles has a flexible meaning, it usually is a grouping of bits that have some logical connection to each other. If one is discussing integer arithmetic on the current generation of microcomputers, then a word is usually 16 bits long. The word is contrasted to the byte in that the byte now seems to be universally fixed on being a grouping of 8 bits. In many applications, a word may contain an integer number of bytes. In this application, however, the microword contains all of the bits necessary to control the architecture *and* the controller during a single clock cycle.

In order to simplify the preparation of tables of control information or "microprograms", it is useful to subdivide the microword into component structures. The first useful subdivision recognizes that certain of the bits are required to cause the controller to perform its program control function. The *controller field* is further subdivided into the *next address select*, *pol*, *branch condition select* and *branch address* fields, as shown in Figure 4-14. The architecture similarly has its own *architecture field* that can be subdivided into the various fields associated with each controlled element. If the architecture contained a Register Arithmetic/Logic Unit (RALU) similar to those in Chapter 3, then two or three fields might be needed to select the various registers to be used by the current instruction. Other fields contain codes to tell the ALU, shifter, or Q shifter which operations to perform. Any given microword contains many fields, each containing the number of bits appropriate for the control function they implement. In a given system configuration, the number of bits in the microword and the bit field groupings are fixed once the hardware is finalized. In a few cases, a field is shared by many elements in the architecture; however, it can serve only one purpose in each *microcycle* or tick of the system clock.

Remember that, from an electrical or digital design standpoint, each of the bits in the microword is a control signal traveling down a wire to its destination in the controller or architecture. Our grouping of these bits into fields is an artifice that will lead to a rational method for generating tables of control values for these signal lines based on programming methods with which we are familiar. We have intentionally changed the method by which these control signals are generated from using the *traditional* state machine implementation method and its associated Boolean networks to the *microprogrammed* or table-generated method of implementation. This has given us a structure that is much more flexible, albeit slightly slower to run than the traditional method.

One more step can be taken to cement our viewpoint that the bit fields that we have identified in the microword are indeed like the elements from which we write programs. We can assign them names or symbols just as we did in our high-level language programs. The *next address select* field in the controller really determines, with the help of the other fields, the next microinstruction that will be executed. This field appears to act like an "instruction" to the controller. The *branch address* field is obviously well named even though we may prefer to shorten it to *address*. This act of naming makes the next step obvious, i.e., set up an instruction format and some operations and start programming.

5.3. THE SYMBOLIC MICROINSTRUCTION FORMAT

Our goal is to prepare a file containing a set of sequential operations or statements in *symbolic* form. This is called the *source* program. It is human readable since it emphasizes a symbolic format as opposed to a list of numbers or control codes. We will then use this file as data input to a program called an *assembler*. The primary purpose of the algorithm performed by the assembler is to convert the symbolic statements in the source program into numeric values that are appropriate for placing in the computer's memory, the *object* program or file.

We will refer to the assembler language level of an instruction set processor (ISP) as a second, or macro-level. This term should not be confused with a *macroassembler*, a type of assembler that we will use later. First, we will discuss a *microcode*, or first-level, *assembler* that will assist in the preparation of numeric object files that can be placed in the microprogram memory.

The *microcode assembler* has an added component that is normally not explicitly used at the macro-level. When we use the microcode assembler, we are in the process of designing a microprogrammed state machine. Its architecture is in a state of flux and its "instruction set" has never been implemented before at the microcode or any other level. The number of bits in the microword as well as their detailed meaning will not be settled until near the end of the hardware/software design cycle. This requires that the microcode assembler be able to accommodate changes in the format of the microword as well as the addition of new "instructions" as the architecture evolves. The consequence is that the microcode assembler must have a method by which it can be told the microword format and the numeric "meaning" of a symbolic instruction. This feature is rarely found in macro-level assemblers since most of them are written specifically for an existing computer or ISP. Assemblers at either level that contain a capability to change the meaning of symbolic instructions and machine code formats are called *meta-assemblers* and our first microcode assembler, written by Microtec, Sunnyvale, CA, is of this type by necessity.

Our model for the microinstruction format will be based on the format of an assembler language instruction shown next.

Label:	Operation	Operands	; Comment

The label field starts in column one or is terminated with a colon, "∶". It has a "value" that is the *location* of this microinstruction in the microprogram memory. It will be used by the programmer as a destination in instructions such as JUMP or Conditional BRANCH. The comment field also has the same meaning as it did at the macro-level; i.e., it places the action performed by the microinstruction to which it is attached in the context of all of the microinstructions in the logical group in which it is embedded. The comment field begins with a semicolon "∶". Each field is separated by a *blank*.

The operation and operand fields must be reorganized since they must be able to accommodate the fact that many operations may be performed in the same microinstruction or microcycle. The second-level assembler was designed to assemble programs in which each line caused only a single operation to occur. An instruction could add two numbers, another instruction could test a flag and conditionally branch to another instruction's location, but the two operations would not be done simultaneously. Some computers implement decrement and branch instructions that require only one symbolic source statement; however, it should be noted that in most cases the actions carried out by the processor are sequential and not parallel.

To accommodate the *parallel* or simultaneous nature of operations in the controller and architecture, we will rearrange the operation and operand fields of the macro-level assembler source statement in adapting it to our needs. Since we can have many operations happening in the same microcycle, we will provide for many operation fields in the microinstruction. The operand field in the second-level assembler instruction was used to supply additional information to the processor concerning the location of data, etc. Therefore, we will attach an operand field to each of the operations that we have provided for in our microcode assembler source statement. To keep all of the operations separate, we will introduce an operation field separator symbol, the ampersand "&". The format of the microcode assembler source statement is then:

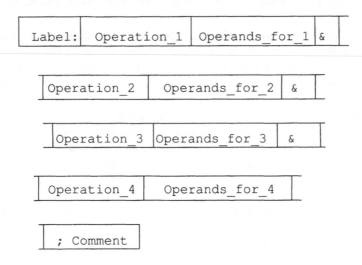

Each operation field might logically refer to one of the elements in the machine implementation such as the controller, the RALU, etc. The *machine code* format is much different since it is the actual arrangement of the bits in the pipeline register. Consider a simple machine that contains a controller similar to Figure 4-14 with an RALU such as the Am2903 or TI 'ACT8832 in the architecture. A microword (*machine code*) format might look like the following:

4	1	3	12
INST.	POL	CCMUX SEL	ADDRESS

4	4	1	1	1	9
REG A PTR	REG B PTR	EA	OEB	OEY	RALU INST.

The first line, containing 20 bits, will allow the controller to determine the microinstruction to execute in the next cycle. The second line, a continuation of the first line in the pipeline register, directs the RALU to perform a task that will be completed at the end of the current microcycle. The total number of bits in the microword for this simple machine is 40. The microword contains fields that vary in length from 1 to 12 bits. The order of bits in this example groups those control lines destined for the controller in a separate area from those for the architecture. This arrangement may not be possible in an actual system. The arrangement of bits in the microword may be totally disorganized due to hardware constraints. We expect the microcode assembler, *with our help,* to be able to take a symbolic statement of the form shown above and set the appropriate bit pattern in the machine format no matter what the organization of the microword.

5.4. THE DEFINITION PROCESS

The microcode assembler is organized in two sections. The first section, called the *definition* phase, tells the other section, the *assembler* phase, the organization of the machine instruction and the meaning of the symbols that we want to use. We will need to prepare a file of symbolic statements based on the assembler pseudo operations for EQUating (EQU) a symbol to a value and DEFining (DEF) a machine code format for an operation. This file is prepared once and after processing by the DEF program will remain in force for use by the assembler until the hardware is changed. If the file is designed correctly, all hardware dependencies will be concentrated in it. A hardware change should not directly require changes in any assembler source statement that doesn't explicitly use the modification. The arrangement of the bits in the pipeline can be changed without having to modify the assembler source program.

Another file, the *assembler source program,* will be prepared to contain the actions that are requested from the hardware. This file is very similar to any other assembler source program in that it explicitly contains the algorithms that the programmer is trying to implement. It also "knows" about the hardware except that the language that describes the operations desired is entirely symbolic.

Appendix A in this chapter gives a short description of the *definition* phase statements that accomplish the purpose of mapping the symbolic assembler statement onto the microword.

The main function of the definition phase is performed by the DEF directive in a process called *instruction definition.* The format of the *definition phase source* statement is the same as that for the macro-level assembler. A label *must* be present since the properties assigned in a DEF statement are associated with the name presented as a label. The operation symbol used in a DEF statement is DEF. The operand field may contain many subfields, all separated by commas.

There are three types of subfields in the operand field of a DEF statement; they are Constant, Variable, and Don't Care. The number of bit positions in a subfield is specified by a decimal number prefix to the field type specifier. The operand field must say something about every bit position in the machine- or microword. As an example, let us define an instruction for the controller of our hypothetical machine whose microword format was shown above and is repeated below.

```
CONTROLLER: DEF 4VH#E,1VB#0,3V,12VH#000,20X
```

This DEFinition statement, after processing by the definition phase, will tell the assembler phase that we have created an *operation* called CONTROLLER that has four operands corresponding to the bit fields INST., POL, CCMUX SEL, and ADDRESS in the microword format. Also, the last 20 bits, controlling the architecture, are skipped over and not associated with this operation. The word "operation" implies that the word "CONTROL-LER" is to be used in one of the operation boxes in the symbolic assembler source statement.

4	1	3 1 2	
INST.	POL	CCMUX SEL	ADDRESS
1110	0	XXX	000000000000

4	4	1	1	1	9
REG A PTR	REG B PTR	EA	OEB	OEY	RALU INST.
XXXX	XXXX	X	X	X	XXXXXXXXX

Notice that default values in the boxes above have been declared for three of the operand fields. The INSTruction field has been initialized to the 4-bit pattern 1110 which causes a CONTINUE operation if the controller had been implemented using an Am2910 sequencer. In this type of initialization, we have made the default value of the operand be the most commonly desired operation. If this had been an RALU, then we would have specified another type of initialization, one that left everything unchanged. Since the controller's action was initialized to a CONTINUE, then the CCMUX is not consulted, so we did not initialize the CCMUX SEL field.

The purpose of the constants supplied as initialization for variable-type fields is to place the hardware in the most commonly used or harmless state, as appropriate, automatically without our having to specifically consider this at the time of writing the assembler source program. We decide during the preparation of the definition source file the actual codes that will be needed for rendering the system harmless to itself and include them as default values in the appropriate fields. Our assembler will place either all 1s or all 0s in the fields that we leave completely to chance as in the case of the CCMUX SEL above. These values represent the "unprogrammed" state of the devices from which we implement the microprogram memory.

Since the 20 bits associated with the architecture were skipped over in this DEF statement, the assembler will allow us to write another DEF statement that deals with them. In general, while you are allowed to have many DEF statements dealing with the same bit field, you are allowed to use only one of the symbols so defined in a single assembler source statement. Since we want to have the architecture do something in the same statement that we tell the controller to perform, we must omit any reference to the architecture in the controller's DEF statements and vice versa.

If we now EQUate some other symbols to appropriate constants, we can use them along with this DEFined operation to write assembly language source statements for the controller.

```
JUMP: EQU       H#3
CALL: EQU       H#5
```

The assembler source statements could look like this:

```
        CONTROLLER                      ; st. 1
        CONTROLLER    JUMP   ,,,THERE    ; st. 2
THERE:  CONTROLLER CALL  ,,,SUB          ; st. 3
SUB:  ........                          ; st. 4
```

Statement number one, commented st. 1, uses the default value for its operands; hence, the controller will continue on to the next microinstruction in the sequence when the microcycle ends. No instructions to the architecture have been shown in this example. Statement number two sets the INST field to a binary 0011 from the EQUate above and uses the address THERE in the branch address field of the microword. The assembler keeps track of the number of microwords that it has assembled so that it knows the address of the line labeled THERE. The commas indicate that the default values for the operands in those positions are to be used. The only time commas are needed is to indicate when operands are skipped in getting over to one that is specified. The third statement will place a binary pattern 0101 in the INST field in the microword and use the address of SUB in the branch address field.

EQUates would normally be used to specify values for labels to point the CCMUX and the POL field. In Chapter 4, POL's function table was such that a 0 caused the CCMUX output to be taken as selected and a 1 caused the flag to be inverted. A suitable EQUate for this field is:

```
    NOT: EQU          B#1
```

Let's assume that the Conditional JumP (CJP) instruction for the sequencer chosen was prepared as follows:

```
    CJP: EQU          H#3
```

An assembler source statement like this could be written for the controller:

```
    CONTROLLER    CJP,NOT,Q#3,THERE
```

This would place a binary pattern 0011 in the INST field, a 1 in the POL field, a 011 in the CCMUX SEL field, and the address of the assembler statement labeled "THERE" in the branch address field. The statement effectively implements the following control structure:

```
    IF (CCMUX input 3 is NOT TRUE) THEN GOTO THERE
    ELSE CONTINUE
```

The POLarity control makes it much easier to write symbolic statements that reflect the logical intent of the programmer.

We can make the assembler source statement above contain only symbolic information, i.e., no explicit constants like the Q#3, by creating descriptive symbols with values set by EQUate directives. Here we would write:

```
CCMUX3: EQU      Q#3
```

This would transform the assembler source statement to:

```
CONTROLLER      CJP,NOT,CCMUX3,THERE
```

which very effectively expresses the intent of the programmer.

At the macro-assembly language level, we would have written a statement to perform this same control operation in the following form:

```
BRANCH  NOTCCMUX3,THERE
```

This syntax can be created using the DEF directive by regrouping some of the bit fields and assigning symbols to them. Consider the machine code format of the controller part of the microword modified as shown below.

4	1	3	12	
INST.	POL	CCMUX SEL	ADDRESS	Controller
0011	1	011	"THERE"	Part

4	4	1	1	1	9
REG A PTR	REG B PTR	EA	OEB	OEY	RALU INST.
XXXX	XXXX	X	X	X	XXXXXXXXX

The DEF directive is used to assign a value to the first controller field (i.e., the controller instruction) that will cause it to perform a conditional jump (CJP) operation. The POL and CCMUX SEL fields are grouped together into a single 4-bit variable field that is not initialized. The *address* field is made variable as before. The two variable fields are set up to truncate and right justify values that are placed in them using the ":%" modifiers.

```
BRANCH: DEF      H#3,4V:%,12V:%,20X
```

In the context of the assembly language source statement format that was shown near the beginning of this chapter, we have created an "operation" called BRANCH that has two operands as specified by the "V" fields in the DEF statement. In order to make the assembler source statement symbolic, we need to declare the values of some appropriate symbols as follows:

```
NOTCCMUX3:EQU   B#1011
```

These changes make the statement

```
        CONTROLLER      CJP,NOT,CCMUX3,THERE
```

completely equivalent to

```
        BRANCH          NOTCCMUX3,THERE
```

since the machine code generated for the controller field of the microword is identical in either case. The form chosen by the programmer is a matter of personal taste; however, his past experience with the macro-level assembler would probably influence him to chose the second form.

We could implement either the "branch if condition *True*" or the "branch if condition *False*" form of the control construct

```
        IF (CCMUX input 3 is TRUE) THEN GOTO THERE
        ELSE CONTINUE
```

```
        IF (CCMUX input 3 is "NOT" TRUE) THEN GOTO THERE
        ELSE CONTINUE
```

by defining a symbol in which the POL field was *False* and the CCMUX SEL field pointed at input three as follows

```
    CCMUX3:         EQU     B#0011
```

The binary form of the value is used here for clarity so that the POL bit, here equal to *False*, can be compared to the EQUate for the symbol NOTCCMUX3 shown above. Notice that we have not changed the actual bit patterns that are placed in the machine code format, but we *have* made the statement written by the programmer express his intent more clearly.

The "readability" of the assembler source program can be further enhanced by using the name of the source of the signal to which CCMUX input 3 is connected. For example, assume that input 3 is connected to the carry flag in the macro status register. An EQUate such as

```
    MCARRY:         EQU     B#0011
```

will establish the symbol MCARRY (i.e., macro carry flag) for use by the assembler. The program control construct above, in which CCMUX input 3 was tested, would then become:

```
        IF (MCARRY is TRUE) THEN GOTO THERE
        ELSE CONTINUE
```

The resulting assembler language source statement would have a form identical to that used at the macro-level; i.e.:

```
        BRANCH          MCARRY,THERE
```

Remember that this statement applies to the controller only and will share the assembler source statement line with operations controlling other parts of the state machine.

Another point to remember is that all symbols that were declared by using them as labels on DEF statements play the part of *operations* with respect to the syntax of the assembler source statement. The operand field consists of constants which may be declared using EQUate statements or variables that become known at assembly time. *No* symbol DEFined using the DEF statement may be used in the operand field.

5.5. THE ASSEMBLY PROCESS

Once the process of describing the physical structure of the microword to the assembler is complete, writing programs for the system can proceed. The definition process need be done only once and need not be repeated with each assembler program written. If the structure of the microword changes or if the hardware it controls is modified, then the definition source file should be modified to reflect the changes.

The assembler source program is the main document that describes to the assembler, and thence to the hardware, the algorithm that is the focus of the design. Its syntax is repeated here.

```
| Label: |   Operation_1 | Operands_for_1 | & |

    | Operation_2 | Operands_for_2 | & |

    | Operation_3 | Operands_for_3 | & |

    | Operation_4 |   Operands_for_4   |

    | Comment |
```

Each operation is a symbol that has been declared using a DEF statement in the definition phase. The operands are constants, symbols for constants, or labels within the assembler source program. If more than one operand is needed to specify an operation, a comma is used between each operand. An operation field is separated from its operand field by one or more blanks. No blanks may be placed within an operation or operand symbol since the assembler will interpret the blank as a field terminator. The various operation-operand fields are separated by the ampersand character "&". If the programmer decides to stretch the assembler source statement over more than one line, then a slash "/" is placed in column 1 to alert the assembler to the continuation. This increases readability and is recommended.

A comment may be placed at the end of the source statement by signaling its presence with a colon ";". However, the statements are usually too long for this. Comments should be placed between statements so that they can be used to more completely describe the context of the statement.

The detailed syntax of the assembler source statement is controlled by the definition process. The two contrasting pictures of the controller field given in the previous section illustrate this. Once the definition process is performed, various bit fields throughout the microword are collected together under the auspices of an "operation". Some of the bit fields are given values at definition time and others are made variable so that their values can be set at assembly time. The order of operands in a multioperand field is determined by the order of the corresponding bit fields in the microword. The BRANCH instruction above furnishes a good example. The DEF statement sets up two variable fields beginning in the fifth bit position from the left end of the microword. Looking at the microword format, we identify the 4-bit field with the POLarity-CCMUX SELect field and the 12-bit field with the branch address.

```
BRANCH: DEF    H#3,4V:%,12V:%,20X
```

The following controller portion of the assembler source statement shows the symbol (MCARRY, created for the POL and CCMUX fields first) followed by the symbol, THERE, indicating a location in the assembler source program.

```
        BRANCH  MCARRY,THERE
```

If they had been reversed, then the assembler would have placed the value for the address in the POL and CCMUX fields and the value for MCARRY in the branch address field. The *order* of the operands is determined by their *order* in the microword.

This is the only detail of the hardware microword structure that needs to remain visible to the microcode assembler programmer. The order of the operation-operand fields is entirely at his whim. A convenient way to write a microcode assembler source program is to produce a source line containing only the operation symbols separated by ampersands. Use the operation symbol for each of the physical microcode groupings that causes the associated hardware to "do nothing". The instruction for the controller in this case is CONTINUE and for the RALU is PASS with nothing stored. This statement is entered in such a way that there is plenty of space to the right of each operation in which to add operands. The statement is duplicated using the editor until enough are available with which to write the current module. Writing each line of the source program consists of modifying the operation, if appropriate, and adding any operands as required. If a portion of the architecture is not used in a particular instruction, then its operation field need not be modified. This method serves to simultaneously place each part of the structure in a "NO OPeration" state, makes remembering each field easier and speeds writing of the assembler source program. Since the architecture was just designed and is in a state of change, the programmer has no background of experience in writing for it. This source program structure helps build experience while handling the "NOOPs" needed in the unused parts of the parallel architecture.

5.6. OPERATION OF THE MICROCODE ASSEMBLER SYSTEM

Let's stand back and see how we will use the definition and assembly phases to produce machine code for our microprogrammed state machine. The flow of data through the various programs is shown in Figure 5-1. We will first prepare the definition source file using our text editor. The file "SWEET16.DEF" depicted just below the ".DEF" file symbol in the figure serves as an example. At this time, we will create the "instructions" or operations and operands that our assembler source program will use. This file will serve the assembler as a complete description of the machine code format and the values that we expect it to enter into the fields that we have called to its attention. The definition source file is processed by the Definition program, called "DEF2900", to create a machine-readable file for the assembler called the definition phase object file. This process need not be repeated until modifications are made to the format of the microword or to the hardware it controls.

We will also use our text editor to produce an assembler source file that contains the algorithm that we want to implement on our state machine. Two assembler source files, "SWEET16.ASM" and "SWMICTEST.ASM" are shown in Figure 5-1 near the ".ASM" file symbol. The source file is written in terms of the symbols that we have created in the definition phase.

FIRST LEVEL OF CONTROL

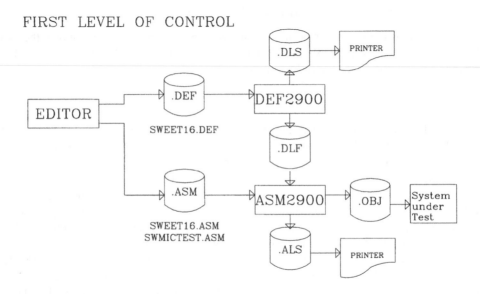

Figure 5-1: Use of definition and assembler programs.

The program should ideally contain only symbols, not values, since there are many design and implementation situations that may cause the value of a constant to change. If we have created a symbol using an EQUate statement, then we need to change only that statement and reassemble the source file instead of changing the value in all places in which it was used in the file. For this reason, symbols can be created using EQUates in the assembler source file as well as in the definition source file. One note of caution, however; the creation of a symbol requires that it appear as a label in one *and only one* statement, whether that statement is in the definition or assembler source files. Only symbols that have been created using the SET directive can be changed by placing them as labels on a new SET statement. This keeps you from inadvertently changing one of your operations, i.e., a label on a DEF statement, to stand for a location in the assembler program.

Once the assembler language source file is created, it is submitted for processing to the microcode assembler, here called "ASM2900". The assembler also uses the definition object file created above as input. It uses the two files to generate several output files that, in one form or another, contain the machine code for the system under development.

A listing file, indicated by the ".ALS" file symbol, is normally produced when commanded by the LIST directive. If no other changes are made, this file shows the assembler source statements with the machine code interspersed between each source statement. This is an especially useful form during the learning process and during the hardware testing phase since the programmer's intent in the source statement and comments is juxtaposed with the assembler's "understanding" of this intent. Needless to say, if a DEF statement is incorrect or an assembler source statement contains the wrong symbol, then the bit pattern produced will not contain the desired value.

By proper use of the operands on the LIST directive, the assembler output will also contain a symbol table, a machine language object file, and a cross-reference listing. The symbol table gives the name, type, and value of every symbol used in the definition and assembler source files. The cross-reference listing shows the line number in the definition or assembler source files where a symbol was defined, i.e., by a negative sign prefix, and every line number where that symbol was used in a source statement. This is valuable for removing conflicts caused by multiply defined labels.

The machine language object file, symbolized by ".OBJ" in Figure 5-1, contains the microprogram in a form that is readable to both machines and humans. Each bit in the microword is shown as a character 0, 1, or X depending upon whether its value is 0, 1, or "don't care". Each microword may occupy more than one line in the object file if it is too long to fit on one line. The microwords are arranged in blocks following a line that specifies the address in microprogram memory of the microword that immediately follows. The format of the address specifier is shown in the following example:

```
A 0010H
```

The "A" in column 1 indicates that an address follows. The address in this example is a four digit number, 0010, given in hexadecimal as specified by the suffix "H". The microword following this line in the machine code object file is intended to be placed at the address 0010 hexadecimal in the microprogram memory. The 1 after will be placed at 0011 and so forth. The address information is generated each time the ORG directive is used in the assembler source file.

The first line of the file has the form shown in the following example:

```
*$ 00064H 00156H 000D0H
```

The "*$" characters indicate that this line begins an object file. The next hexadecimal field specifies the number of bits in each microword which is 64 hex or 100 decimal in this example. The next field specifies the range of addresses or length of the data (e.g., 156 hex) in this file. The last field specifies the starting address of the first microword in the file (e.g., D0 hex). The last line in the file is always "$$", which signifies "end-of-file".

The object file can be directly loaded into the microprogram memory with the help of a development system. For our purposes, it is best that we consider the steps that must be performed before this file is in place in our microprogrammed state machine's control memory.

The steps taken will depend upon the implementation of the microprogram memory. If the memory is implemented using Programmable Read Only Memories (i.e., PROMs or EPROMs), this file must be "sliced" in the "vertical" direction parallel to the direction of increasing address. The slices must be the same width as the number of data outputs on the devices selected. The 100-bit wide microword specified above would occupy 13 devices having 8-bit data outputs. If the devices were placed side-by-side sharing a common address input, a single microword would be stretched across all 13 devices leaving 4 bits in the right most one empty. The next microword would be placed similarly at the next address above the first. When a new ORG-generated address specification line is encountered, the memories should be left "unprogrammed" up to the address specified by the ORG.

Slicing the object file and generating separate files for use by the PROM programmer is best done by a utility program. The programs available with the microcode assembler or in the development system can perform this task or the user may write them himself. This exploits the "machine" readable nature of this file.

If the microprogram memory is implemented using read/write memory (i.e., RAM) then a separate controller can be provided to transfer this file from the host system on which the assembly process was performed. This controller can be a microprocessor or a separately dedicated state machine. One of the first sequences implemented on the target system might be the loading program, in which case the target system could help in its own development. A microprocessor-implemented structure is attractive since it can greatly help in the development process in addition to loading the target program. If implemented correctly, it can load the pipeline, read the results of an instruction from the architecture, and report the results to the programmer via the host system interface. A complete development system environment can be placed in such a microprocessor.

5.7. AN ASSEMBLER FOR THE SECOND-LEVEL MACHINE — INTRODUCTION

Up to this point we have discussed the implementation of microprogrammed algorithmic state machines using a single level of control. In Section 1.4 it was noted that increasing the "power" of an instruction could make the production of control sequences more efficient. This led to the introduction of the second level of control or the Instruction Set Processor (ISP). We will digress from discussing the microcode assembler at this point to introduce a method for creating an assembler for the second-level machine. The method used will then be shown to be effective for creating an assembler for the first-level machine as well in the event that the Microtec assembler discussed above is not available to the designer.

The ISP consists of a set of state machine sequences with the desired sequence selected by an "instruction" fetched from another control store by the processor itself. The instructions are encoded numerically for compactness and ease of manipulation by the underlying first-level machine. Associated with the instruction set are a series of address modes or automatic sequences implemented in the first-level machine that assist in the location and transmission of data between parts of the ISP. A complete description of the ISP, therefore, includes the programming model (data locations), the instruction set, and the various addressing modes.

We are interested in examining methods for making the preparation of the "tables of instructions" (second-level programs) as easy and error free as possible. This can be achieved by the use of symbolic statements that are reduced by a fixed algorithm to the numeric codes (machine language) used by the ISP implementation. If each of the symbolic statements contains a command to the ISP requesting it to perform one of its sequences, then the symbolic language is called an *assembler language*. On the other hand, high-level languages generally allow statements that are similar to human language forms. In order for either type of symbolic language to be used, one must provide for the translation of the source program into a form executable by the ISP. The *assembler language* form is easier to support early in the development cycle of the ISP while the high-level languages appear when the ISP reaches maturity.

5.8. THE FORMAT OF AN ASSEMBLER LANGUAGE STATEMENT

As discussed above for the first-level machine, certain information must be present in *all* assembler language statements in order for the proper machine code to be produced for the ISP to execute. These can be summarized by reproducing the diagram of a single line of assembler language (the statement) similar to that shown for the microprogram. At the second level only, a single operation mnemonic is needed on each line since the second-level machine executes only one "operation" or sequence at a time.

LABEL:	OPERATION	OPERAND1,OPERAND2,…,OPERANDn	COMMENT

The four fields defined are the Label, the Operation Mnemonic, Operand, and Comment fields. The fields are generally separated by one or more blanks and no blanks are allowed to be embedded within a field. Some assemblers require that the Label field be placed starting in column 1 (with an alphabetic character) and/or terminated with a special symbol (e.g., a colon ":").

The Comment field is a string of text that is ignored by the assembler, but allows the programmer to explain the purpose of the present statement in the context of his program module. Since the operation field is usually self explanatory, it is pointless to reiterate it in the comment field. The programmer should use the comment field to establish the context of the statement by using constructs such as pseudocode or high-level language statements. As in the microcode case, the comment may be more complex than the space available on a single line. In this case, blocks of logically connected statements should be commented with lines that contain only comments. This greatly increases the readability of the program.

The Operation Mnemonic field is primarily used to contain a symbol that represents the name of one of the ISP sequences. The Operand field allows the programmer to supply the symbols that correspond to the rest of the information that the ISP needs to complete the requested sequence of operations.

The primary operation of the assembler algorithm is to translate the operation mnemonics into numeric codes "understandable" by the ISP. The ISP in turn will run the selected sequence of operations using the remaining information in numeric form in the operand field.

In addition to the operation mnemonics that are directly translated to numeric codes used by the ISP, there are codes that the assembler algorithm uses to aid the programmer. These are called *assembler directives* and allow the programmer to declare storage spaces, associate names with those spaces and place initial values in them. Other directives are used to control formatting of printed output and the inclusion of segments of the source program.

5.9. ASSEMBLER IMPLEMENTATION METHODS

The assembler algorithm may be implemented for our ISP in one of three ways. The first way involves writing a program that performs the actions implied above. This program is written in one of the languages available on an existing computer.

The second way is based on exploiting the common operations that all assemblers must perform. Most parts of the assembler algorithm are independent of the target ISP. These include address computation, source statement parsing, output generation, and others. The only real differences between all assemblers lie in the actual symbols and operand syntax that are to be translated and the resulting machine code to be produced. These differences can be concentrated into tables that can be changed each time the symbolic or machine instruction is changed. This type of assembler is called *table-driven* or a *meta-assembler*, which we first encountered for the microprogram above. They are extremely flexible and give the added advantage of maintaining a common "operating" environment regardless of the target ISP. This should be contrasted to the changes in operating environment experienced when the programmer moves from the VAX MACRO assembler to the IBM System / 370 assembler to the IBM PC MASM assembler. Once an assembler of this type is written in some language and run on a host computer, it is relatively simple to adapt it to a new or evolving ISP.

The third type of assembler exploits a feature found in many existing assemblers. This feature is called "macro expansion". The term "macro", as used in this section, should not be confused with the "macro-" (or *instruction set processor*) level of a machine. In this context, it refers to a text processing function that certain assemblers (called macroassemblers) perform before the actual step of converting symbolic assembler code to machine code.

It is possible to use this text processing feature in our application by creating tables of text substitutions, "macros", for the instructions of our ISP. The operation of this type of assembler is similar to the "table-driven" variety described above except that the assembler already exists in many computer environments. The adaptation process is much easier than either of the two forgoing methods.

5.10. THE CROSS ASSEMBLER

An assembler normally runs on the machine for which it translates symbolic instructions to machine code. Its algorithm is written in the language of its "host" machine. It is convenient in our work to configure a development system based on a computer that supports various software tools appropriate to our work. The host machine normally has editors, assemblers, simulators, and other tools that make our work possible.

However, since we are not particularly interested in developing an ISP that has the same instruction set as the host, we plan to create a *cross assembler*. A cross assembler is an assembler that is implemented in the language of one computer (ISP) and translates the symbolic assembler program to machine code for another ISP. It is necessary to be continually aware of the fact that an assembler embodies an algorithm which can be implemented (written) in many different languages and "run" on correspondingly many different machines. The target system's symbolic assembler source language and the corresponding machine language output serve only as input and output data to the assembler algorithm. Whether this language has anything to do with the host is based only on our needs.

In our application, we have developed the design of an ISP and we wish to write programs for it. The hardware we are designing may not exist yet, much less be in a state in which we can perform assemblies on it. Therefore, we want the assembler on the host to serve as a "cross" assembler for the system under design, or target.

We will choose method 3 mentioned in Section 5.9 in which the host macroassembler will be "converted" to perform as a cross assembler by using its macro (text processing) feature. We expect to be able to write many symbolic assembler source programs in the language of our "target" (newly designed) system. Each of these programs will accomplish some task appropriate to realizing the purpose for which our design was made. It is not necessary at this point to understand how each instruction in the new design is implemented, but only what it accomplishes with respect to the programming model (map of resources) of the target system. The programs may be written before any commitment to hardware is made and tried out using software simulation methods.

Certain information can be determined from simulation of the instruction set, i.e., the instruction set simulator. This includes an understanding of the efficiency of the instruction set, the performance of the instructions with respect to the expected algorithms for the target system, and even the ease of remembering the instructions and their address modes. If an instruction is hard to remember, then it probably would not be used even when its operation is needed.

Some things are not possible to determine in the simulator, however. These include a knowledge of the actual real-time performance of the system and other factors relating to performance of a physically implemented system. Most of the things that can be determined are worth the effort, however, and a cross assembler will greatly decrease the effort.

5.11. AN ASSEMBLER ALGORITHM DETAIL

In order for an assembler to convert an operation mnemonic in an assembler language source statement to machine code, it must search a set of three tables. One table contains all of the operation mnemonics and their associated machine operation codes for the sequences that the ISP can perform. Another table contains symbols and actions needed that are suggested by the assembler directives such as Declare Constant, Page Feed, etc. A third table contains names and text that should be substituted for the names when they are encountered in the assembler source program.

The third table is called a "Macro Definition Table" and is created by the application programmer. It may be placed at the beginning of the assembler source program before any other source statements or in a separate file that is "Included" at the beginning of the source program or in a separate Macro library that the assembler algorithm automatically searches in its quest for the resolution of an operation mnemonic. Most macroassemblers support all three methods of presenting the definitions of "macros" to the assembler.

An assembler must search these three tables in some order since a simultaneous search is impossible in a sequential algorithm. The best macroassemblers for our purpose search the tables in the following order: Macro Definition Table, Assembler Directives, and then Host Operation Mnemonic Table. If this order of search is performed, then we can use some of the operation mnemonics that are native to the host machine as our operation mnemonics.

For example, most machines use Load, Store, or Move as data transfer mnemonics. It would be convenient if we could use them also since our own automatic reflexes are to use the assembler language with which we are familiar. If our macroassembler searches the Macro Definition Table first, then we are free to use any symbols we choose including the ones with which we are familiar. We might not have to rewrite many of the programs that we need for our target system. We do not want to limit our target machine instruction set by this restriction, so this convenience should not be allowed to completely dominate our choice of instructions for our new ISP.

5.12. METHOD USED TO ADAPT A MACROASSEMBLER

In order to convert an existing macroassembler to serve as a cross assembler for our ISP, it will be necessary to study two types of assembler directives. These are the data declaration directives (e.g., ".byte" and ".word") and the macro definition directives (e.g., ".macro" and ".endm") used by the VAX MACRO assembler [DEC 88]. The actual spelling and syntax details are determined by the particular macroassembler we select but the general operation of all is similar to Miller [85].

Our examples for this section will be implemented using the VAX MACRO assembler. This assembler, representative of most major macroassemblers encountered, fulfills all of our basic needs and, in addition, it searches the Macro Definition Table first which allows us to use VAX operation mnemonics where it suits us. Remember, the program we are writing and assembling is never intended to run on the host computer itself, so the host assembler statements would have no meaning on our ISP anyway.

We will also need to know the assembler language syntax and machine language instruction format for the ISP we are developing. For the purposes of this chapter we will use an instruction from a computer called "Sweet16" that is described more completely in another part of this book. Our example is the instruction "ADIC R1,DATA". This is intended to be the assembler language syntax for an instruction that adds the immediate data to the data in the first operand along with the Carry Flag and places the result in the first operand. In this case, the first operand is one of the general purpose registers. The machine language format of the instruction is illustrated in the following diagram:

```
          15                8  7    4  3    0
          ┌──────────────────┬──────┬──────┐
Word 1:   │  Operation Code  │Reg 1 │Reg 2 │
          └──────────────────┴──────┴──────┘
          15                               0
          ┌─────────────────────────────────┐
Word 2:   │               Data              │
          └─────────────────────────────────┘
```

The instruction format is commonly used in many machines that support general purpose registers. The 4-bit width of the Reg 1 and Reg 2 fields that are used as register pointers in word 1 of the machine instruction allows the programmer to point at 1 of 16 registers with each register field. The immediate data is placed in the second word of the instruction. The "Reg 2" field is not used in this instruction.

In the simplest possible application of this method of creating a cross assembler, one would write a "Macro" for each machine language operation code that the ISP implements. A macro for the

ADIC R1,DATA

instruction has the format shown in the VAX MACRO example at the top of the next page. Note: In the VAX MACRO assembler, the ^x signifies that the number is in hexadecimal and the angle brackets <> serve as expression delimiters. The assembler directives ".byte" and ".word" tell the assembler to set aside a byte of storage or a word (16 bits) of storage, respectively. They further instruct the assembler to place the operand in the data space. The

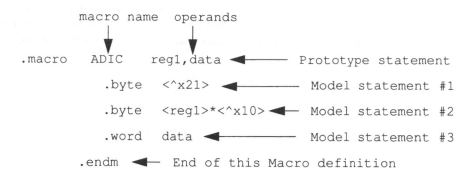

operand may be a number (here in hex) or an expression, i.e., <reg1> * <^x10>, but its result must evaluate to a byte or word of data.

The prototype statement is used to inform the assembler that the text that follows is a macro and that the name of this example macro is ADIC. The symbolic operands, reg1 and data, tell the assembler where to look on the source code line for the character strings that will be substituted during the "expansion" of the macro. As mentioned earlier, there will be one macro for each machine language operation code. All of the macros will be in a table (the Macro Definition Table) that may be placed at the beginning of the assembler source program, in a separate file that is "included" at this point or in a Macro Library. This library can be created on the VAX by using the LIBRARY program.

The assembler uses the above macro when it encounters any assembler language source statement that contains "ADIC" in the operation mnemonic field. For example consider the following:

```
ADIC   5,3
```

The assembler searches the list of macros for one named "ADIC". Upon finding it (above), it uses the position of the 5 and 3 in the operand field of the source statement to assign them to the symbolic operands, i.e., "reg1" and "data", respectively, in the macro prototype statement. Therefore, in this example, "reg1" is assigned the character string "5" and "data" is assigned the character string "3". Note that the operand field of the symbolic instruction could have contained symbols or strings of symbols (expressions) as well and the substitution would have been carried out just the same. The macro expansion is written here:

```
ADIC           5,3
    .byte    <^x21>      Expanded model statement #1
    .byte    <5>*<^x10>  Expanded model statement #2
    .word 3              Expanded model statement #3
```

The programmer wrote the first line and the assembler used the macro definition to expand the next three lines, i.e., model statements #1, 2, and 3.

The assembler will complete all macro expansions and then convert the directives to machine code. Notice that as far as it is concerned, the assembler is really preparing a large block of "data". This block of data does not contain any host machine language statements. It is not until the "data" is moved to the new ISP that it will be interpreted as instructions.

The example, upon assembly, creates a data space with the following format:

```
       1st byte      2nd byte
     ┌──────────┬──────────┐
     │   2 1    │   5 0    │  1st Word
     ├──────────┼──────────┤
     │   0 0    │   0 3    │  2nd Word
     └──────────┴──────────┘
```

Notice that the data area prepared by the assembler now contains numeric codes that are identical to that required by the machine code format shown above.

The general technique, then, involves the identification of the assembler directives in the selected macroassembler that define macros and serve to declare and initialize data storage areas. It is also necessary to study how to specify constants in various number systems. The hexadecimal notation used here is convenient since the first word of the machine code instruction generally maps directly onto hardware in the target ISP. The first word will be seen in binary or hex notation during software and hardware exercises and never in decimal.

The second word, e.g., "data", was specified in decimal. It will be necessary to understand how the assembler evaluates expressions since these serve as the method by which specific bits are placed in specific locations within the machine language instruction. In this example, the value of the symbol, "reg1", was placed in the left-most 4 bits of the second byte by multiplying ("*") it by hexadecimal 10 (decimal 16). Many assemblers contain expression operators such as shift left, shift right, AND, OR, as well as the multiply used here. These should be understood as well since they will be useful in creating the machine language instruction format desired.

5.13. EXTENDED USE OF MACROS

A macro may contain "instructions" that are defined in other macros. The first use of this feature in our application would be to reduce the repetitive copying of the model statements in the example macro definition. Consider another macro created as shown:

```
.macro    moderd    opnum,reg1,reg2,data  Prototype statement
          .byte     <opnum>                 Model statement #1
          .byte     <<reg1>*<^x10>>!<reg2> Model statement #2
          .word     data                    Model statement #3
          .endm
```

Note: The "!" indicates an OR operator to the VAX assembler.

The ADIC macro could be rewritten as follows:

```
.macro    ADIC      reg1,data             Prototype statement
          .moderd   <^x21>,<reg1>,0,<data>Model statement #1
          .endm                           End of this Macro definition
```

The example is the "Add the contents of reg1 to Immediate Data" instruction shown above. There would be SuBtract, AND, OR, and EXclusive OR instructions supported as well. The only difference between the ADIC and SBIC macro is the operation code (opnum) and the name of the macro. Since many instructions similar in format to ADIC will be defined, this economy in typing should result in fewer errors and more complete testing of the macro definition table.

5.14. PLACING THE ASSEMBLER PROGRAM IN THE TARGET ISP

When the macroassembler finishes its task, the machine code "data" will reside in an "object" file. The flow of data through the system of programs is shown in Figure 5-2. The macro definition file is shown by the ".MAR" file symbol to which is attached the file name "SW16MACRO.MAR". In most host environments, this file may be placed in a library by using a library manager program, symbolized here by the process "LIBRARY". The source program, indicated by two examples ("SW16MON.MAR" and "SW16TEST.MAR") next to the other ".MAR" file symbol, is assembled by the process labeled "MACRO". The Macro library, labeled "SW16MACRO.MLB", is used by the assembler to create the object file shown by the ".OBJ" file symbol. For most assemblers, including the VAX MACRO, the object file is intended to be linked with other programs before it will be executed on the host machine. The file is not in a form for easy movement to the target ISP. Fortunately, the program that performs the linkage on the host machine (e.g., VAX LINK) allows us to adapt the object form to any that we might require. Remember, the object file we have produced appears to the host as data not as a program.

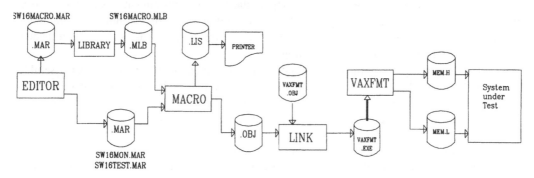

Figure 5-2: Assembling the second-level program.

In order to effect the reformatting of the object file, we must write a program that can perform the transformation of this "data" into a form that is more appropriate for importing into our target ISP. This requires three pieces of information. One is a knowledge of the form that the object file could take if it were linked with our "Transport" program. Second, we need to develop a transformation algorithm that we can write in an appropriate computer language. Lastly, we need to develop a message format appropriate for importing the data into our target ISP.

The simplest form that the data in our object file can appear in is as a table of integers. The example given above would adapt to appearing as a table of bytes or words. A program could be written that would pick up these bytes one at a time and send them to a file that could be loaded by the target ISP. The simplest program of this sort can be written in the assembler language of the host computer and placed at the end of the file containing the source program for the target ISP. After assembly, the object code file would contain the "data" block (the target ISP program) followed by the formatting program (host machine instructions).

Most host assemblers allow the programmer to supply the "entry point address" if the point at which execution is to begin is not at the first byte in the object file. Usually, this is done by adding an operand to the "END" assembler directive. Upon linking, the executable file (.EXE type files in the VAX) can be run and the "data" will be written out to a disk file in the host. This file can then be loaded by the various simulation programs for the target ISP and the performance of the target software could be studied.

This simple method has two possible disadvantages. One minor disadvantage involves the data format of the disk file. Each byte of data is in 8-bit binary form. This may cause difficulties in some systems if the data must be moved out of the host. Many systems do not routinely communicate using an 8-bit binary format. Some communications software programs (e.g., KERMIT, XMODEM and CrossTalk) support binary protocols so that the transmission of binary data between systems supporting these protocols is possible.

However, what happens if the destination of the data is the target ISP itself and the communications protocols have not been implemented yet? One intermediary for a transmission to a target that has no software running at the ISP level would be through an EPROM programmer. Many programmers do not support the binary communications protocols listed above but do support some older protocols that are within reach of beginning programmers. We will investigate these after we discuss another disadvantage of the use of an embedded host assembler program for data formatting.

The beginning programmer, especially, finds the thought of writing a host assembler program that includes writing to a disk file as a task that is beyond his capabilities. He should not take this view, but more understanding of the host system is required than some programmers have. Is there a method whereby the programmer's high-level language skills could be used to reformat the data and move it in a convenient form to the target ISP? Luckily, the answer is yes and it is found in a realization that the assembler object file is really DATA, as we have said all along. Furthermore, it is already organized in an array of "words". If we interpret the format of these words in terms of one of the data types in a high-level language and declare an array of that data type large enough to hold all of our program, then we are well on the way to solving our problem.

We will illustrate this method by writing a program, called "VAXFMT", in the FORTRAN language running on the VAX. The object code for "VAXFMT" will be linked with that produced by the assembler to produce an executable program labeled "VAXFMT" in the right-hand part of Figure 5-2. Executing "VAXFMT" causes the target system machine code to be written to one or more files in any format that we desire.

Consider the following declaration area at the beginning of a FORTRAN program:

```
INTEGER*2 ARRAY(1000)
COMMON /PROG/ARRAY
```

These statements inform the FORTRAN compiler that an array containing 1000 2-byte words is to be set aside and named ARRAY. Then this array is to be incorporated as a Named COMMON area called "PROG". Any other FORTRAN subroutine in this program can use the area called ARRAY by including similar statements at the beginning of the subroutine. The subroutine could use the common storage area even if it did not want to call the area ARRAY by just substituting the desired variable name for ARRAY in its version of the statements above.

The VAX MACRO assembler allows us to create a "program segment" for our "data" and assign a name to it by using the following statement:

```
.psect   PROG,pic,ovr,rel,gbl,shr,noexe,rd,wrt,long
```

The ".psect" operation mnemonic is an assembler directive that creates the program segment and the name assigned here is "PROG". The other parameters describe the segment to the LINKer program and are appropriate for our use in this application. Other assemblers on other host computers have similar directives (e.g., PSEG).

The VAX LINKer treats the name in our assembler object file the same as it treats the name in a Named COMMON area. So, in the case of our example, when the VAX links the object module for our FORTRAN formatting program and the object module of the ISP "program", it places the contents of the assembler program in the Named COMMON area called "PROG". The FORTRAN program can now refer to the words in the assembler program by the name ARRAY(I) where I is an integer index referring to the word desired. Note: the element ARRAY(1) refers to the first word in the assembler program.

If we now stand back and interpret our actions in terms of a FORTRAN program and its subroutines, we find that our assembler program is just a FORTRAN BLOCK DATA SUBROUTINE. This is FORTRAN's method of allowing the initialization of a COMMON area.

Our assembler program is probably not exactly 1000 words long, so we need a method to tell the FORTRAN program how long it is. The simplest method is to create another program segment in the assembler program that declares a single word containing the distance between the beginning and end statements of our assembler program. Pretend that the label on the first statement in our assembler program for the target ISP is

```
BEGIN:
```

and the label on the location after the last statement in our assembler program is

```
FINISH:.word 0
```

Let's create the following VAX assembler program segment:

```
.psect   LENGTH,pic,ovr,rel,gbl,shr,noexe,rd,wrt,long
.word    <FINISH-BEGIN>
.end
```

This program segment is placed immediately after the source code for our target ISP and the ".end" statement is the ONLY "end" statement in the entire program. Note: Most assemblers use the "end" statement to mark the physical end of a source program and require it to be the last line in a source file. The values of all symbols created between the first line of that file and the "end" statement are known to the assembler even if they are in different program segments. The program segment in the example is only one word long and it contains the length of our ISP assembler program in bytes.

The FORTRAN program is linked with this program segment by placing the following declarations in it:

```
INTEGER*2 LEN
COMMON /LENGTH/LEN
```

An integer variable, LEN, is created for the FORTRAN formatting program to use anytime it needs to refer to the length of our ISP assembler program.

The last detail of the FORTRAN program is now only a matter of the communications format chosen for the data. Several common formats (e.g., INTEL HEX and MOTOR-OLA HEX shown in Appendix C) consist of records that contain hexadecimal character equivalents of each 4-bit field in a word. Both formats came into being to allow the communication of 8-bit binary information by way of paper tape or other relatively unreliable paths. Each is record oriented with the record containing a record start character (":" for IN-TEL HEX), a byte count field, a starting address field, followed by several data byte fields. The concluding field is a form of error checking called the "checksum". All fields of the record are in hexadecimal character format.

The heart of the formatting algorithm consists of extracting the four hexadecimal digits from each word in "ARRAY" and converting them to the four characters that represent the hex digits ("0",...,"9","A",...,"F"). A file is written using the FORTRAN Formatted WRITE instruction to output each record.

The simulator and EPROM programmer will need to directly input this record type. If a format of this type is used as input to the simulator, then all required transformations of the program will have been tested in the simulator before it is entered into a hardware implementation. This removes one more possible source of error in the migration of the software into the hardware.

We have exploited one of the benefits of a "suite of programs" in a host computing environment. A "suite of programs" is a group of programs that share common data file formats. They usually do not perform separate parts of a task like a "system of programs" does, but are a group of tools. In this case, the FORTRAN compiler, the macroassembler, and the linker are used together to accomplish our task. Many potential host systems support similar language "suites", e.g., IBM System/370, IBM PC Microsoft FORTRAN, MASM, and LINK, and many others. This capability is one important factor in the choice of a suitable development system. It should be noted that other high-level languages on the VAX and IBM systems have similar relationships with the assembler as the one illustrated here. Other potential high-level languages to support the formatting function include C, PASCAL, and COBOL. The details of declaring a "common" area will vary but the intended function is supported in all of these languages.

5.15. A SHAREWARE ASSEMBLER

A macroassembler, called A68k, was written by Brian Anderson in 1985 for the Motorola MC68000 family of processors. It was converted to the C language and made to generate AmigaDOS linkable output by Charlie Gibbs in 1987. The source code, in C, for A68K is available through many "shareware" sources. The ready availability of its source program and low price, free to noncommercial users, make it an excellent choice as a vehicle to adapt for our uses. A compact user's manual is given in Appendix B. A flow diagram showing its use with files prepared for a project is shown in Figure 5-3.

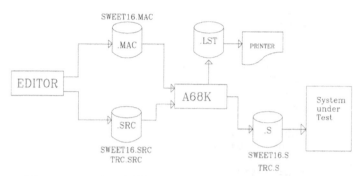

Figure 5-3: Assembly process using A68K as a microcode assembler.

The assembler will process records in a format defined for the MC68000 by Motorola. It produces output in S-record or AmigaDOS linker format (file extensions .S and .O, respectively). We will choose the S-record or "Absolute" form since it is most easily imported into our simulators, PROM programmers, and hardware. The S-record format is shown in Appendix C. The linking step is avoided since output in this form is absolute; i.e., all address references are satisfied at assembly time. If the designer wishes to use the AmigaDOS linker or write his own, he can have all of the normal programming environment frills including object libraries. A listing file, with the extension .LST, is also produced.

Our use of A68K will be to prepare both first- and second-level machine code for processors other than the MC68000. The task is done by employing A68K's MACRO feature. The MC68000 code-generating section can also be removed from A68K to produce a new uncommitted assembler. The change is simply made by modifying the "Instructions" function in the OPCODES.C module to always return *False* when queried about an MC68000 machine instruction. The new assembler is now unaware of the MC68000. The major instructions that remain are the Assembler Directives once the MC68000 code-generator has been disabled.

5.15.1. Examples of Microcoding with A68K

An example that uses the modified A68K to produce microcode is now shown. The demonstration file is TRACE.SRC that contains the microcode to implement the TRACE instruction to aid second-level debugging efforts. See Chapter 8 for a description of the microword for the Sweet16 computer.

```
; TRACE.src - source program for SWEET16 TRACE instruction
; This instruction produces an RALU register dump using the post processing
; report generator in the RALU module. A post processing status register
; dump is also produced.
; Opcode: $08
; Assembler syntax: TRACE - no operands.
```

```
; Generates an RALU register dump each time it is used.
; Created 8 Aug. 1989

TRC             macro
condeb          set     $0
aludeb          set     $1
srgdeb          set     $1
macdeb          set     $0
outdeb          set     $0
                endm


                NOLIST
                INCLUDE "sweet16.mac"
                DEFAULT
                LIST
                ORGA    $0


START           EQU     $000
FETCH           EQU     $004*16


                ORGA    $10
TRACE:

                JMP     ,FETCH
                CCMUX
                ALU
                AB
                MSRLD
                SHFT
                CNMUX
                REGMUX
                R2903
                DATAPATH
                NOREGWR
                MCMUX
                NOYBDEN
                NODBBDEN
                TRC
                endsc

                END
```

A macro, TRC, is defined that "SETs" some symbols to particular values. At the microcode level, we don't want to generate the machine code for a microinstruction until all bits of the machine code are known. So we defer this by "SETting" symbols to values that are later, at the end of the microinstruction, turned into machine code. The macro for TRC sets some symbols for the debugging fields of the microword to values that will cause a register dump at the end of the instruction. The rest of the instruction does nothing (the DEFAULT state). The TRACE instruction requires only one line of microcode to be generated to execute the action desired.

The macro definition file, SWEET16.MAC, is "INCLUDEd" but is not listed. The directive "ORGA" is a macro in the "include" file that multiplies the address operand by 16. We have observed that when left to its own devices, that A68K will generate 16 byte S-records. SWEET16 has a microword length of 100 bits (12.5 bytes). In writing the macro file, we have padded the remaining 3.5 bytes with zeros so that there is one microword per S-record. This made writing the memory loading module in the simulator much easier. The method used is shown in the macro "endsc" which *must* be invoked at the end of each microinstruction.

The microinstruction coding begins at the label "TRACE:". Each of the operators is a macro defined in SWEET16.MAC. The operators correspond to the operators used with the Microtek Microcode Assembler. Each operator may have operands, but none are specified here except the destination address for the JMP back to the FETCH sequence since the DEFAULT condition is taken. The macro "endsc" sets all symbols to the default state after they are used to generate code. This creates a situation akin to that specified in the DEF statements in the Microtec Microcode Assembler.

The listing file is shown next.

```
TRACE.SRC Page 1
Micro Program Cross Assembler
Modified 3/89 - M. Lynch Copyright (c) 1985 by Brian R. Anderson

AmigaDOS conversion copyright (c) 1987 by Charlie Gibbs.
Version 2.00 (March 11, 1989)

                1 ; TRACE.src - source program for SWEET16 TRACE instruction
                2 ; This instruction produces an RALU register dump using the post
                3 ; processing report generator in the RALU module. A post processing
                4 ; status register dump is also produced.
                5 ; Opcode: $08
                6 ; Assembler syntax: TRACE - no operands.
                7 ; Generates an RALU register dump each time it is used.
                8 ; Created 8 Aug. 1989
                9
               10 TRC macro
               11 condeb  set $0
               12 aludeb  set $1
               13 srgdeb  set $1
               14 macdeb  set $0
               15 outdeb  set $0
               16    endm
               17
             1024 ORGA $0
000000       1025+ ORG $0*NBYTE
             1026
000000       1027 START EQU $000
000040       1028 FETCH EQU $004*16
             1029
             1030 ORGA $10
000100       1031+ ORG $10*NBYTE
000100       1032 TRACE:
             1033 JMP  ,FETCH
000003       1034+con_inst set $3
000000       1035+i2910_oe set $0
             1036+ ifnc  ,
000001       1037+i2910_ccen set $1
000000       1038+i2910_ci set $0
             1039+ ifnc FETCH,
000004       1040+br_adr  set FETCH/NBYTE&$FFF
             1041+ endc
             1042 CCMUX
             1043+ ifnc ,
             1044 ALU
             1045+ ifnc ,
```

```
                  1046+  ifnc ,
                  1047 AB
000000            1048+i2903_ea set $0
000000            1049+i2903_oeb set $0
000000            1050+i2903_i0 set $0
                  1051 MSRLD
                  1052+  ifnc ,
                  1053+  ifnc ,
                  1054+  ifnc ,
                  1055+  ifnc ,
                  1056 SHFT
                  1057+  ifnc ,
                  1058+  ifnc ,
                  1059+  ifnc ,
                  1060+  ifnc ,
                  1061 CNMUX
                  1062+  ifnc ,
                  1063 REGMUX
                  1064+  ifnc ,
                  1065+  ifnc ,
                  1066 R2903
                  1067+  ifnc ,
                  1068+  ifnc ,
                  1069 DATAPATH
                  1070+  ifnc ,
                  1071+  ifnc ,
                  1072+  ifnc ,
000000            1073+maroe set $0
                  1074+  ifnc ,
                  1075+  ifnc ,
                  1076 NOREGWR
000001            1077+i2903_we set $1
                  1078 MCMUX
                  1079+  ifnc ,
                  1080 NOYBDEN
000001            1081+y_bbden  set $1
                  1082 NODBBDEN
000001            1083+db_bden  set $1
                  1084 TRC
000000            1085+condeb  set $0
000001            1086+aludeb  set $1
000001            1087+srgdeb set $1
000000            1088+macdeb set $0
000000            1089+outdeb set $0
                  1090 endsc
000100 36         1091+ dc.b
                   con_inst<<4!i2910_oe<<3!i2910_reg_ld<<2!i2910_ccen<<1!i2910_ci
000101 00         1092+ dc.b br_adr>>4
000102 4A         1093+ dc.b (br_adr&$F)<<4!ccmuxsel
000103 00         1094+ dc.b rama<<4!amux<<2!ramb>>2
000104 01         1095+ dc.b
                   (ramb&$3)<<6!bmux<<4!i2903_ea<<3!i2903_oeb<<2!i2903_oey<<1!i2903_we
000105 88         1096+ dc.b qrsel<<4!srsel
000106 88         1097+ dc.b qlsel<<4!slsel
000107 21         1098+ dc.b cnsel<<6!usrld<<5!mznld<<3!mcovld<<1!mdboe
000108 C8         1099+ dc.b i2903_i8765<<4!i2903_i4321
000109 57         1100+ dc.b i2903_i0<<7!dben<<5!marld<<4!maroe<<3!rd<<1!irld
00010A 30         1101+ dc.b mcmuxsel<<6!y_bbden<<5!db_bden<<4!halten<<2!condeb>>2
00010B 05         1102+ dc.b (condeb&$3)<<6!aludeb<<2!srgdeb
```

```
00010C 00  1103+ dc.b macdeb<<6!outdeb<<4!$0
00010D 00  1104+ dc.b $0
00010E 00  1105+ dc.b $0
00010F 00  1106+ dc.b $0
           1212
000110     1213 END
```

There are several areas to note in this listing. The addresses in the left-hand column are in terms of bytes. The adaptation of ORG to ORGA was required to keep this column meaningful. Notice that the microword occupies 16 bytes from 000100 through 00010F. This word fits exactly on one S-record, as we will see later.

We can see that macros like JMP, ALU, etc. do their work. In many cases, the conditional assembly instruction "ifnc" does not find an operand so the instructions in its domain are not produced. The TRC macro is present and causes several of the debugging symbols to be set for later use by "endsc". By the time "endsc" is expanded, *all* symbols have a value. The value was set by DEFAULT (a macro invoked by "endsc") or by one of the operator macros like JMP, AB, etc. The invocation of DEFAULT by "endsc" is not visible in the listing since it is bracketed by NOLIST/LIST directives.

The operation of "endsc" is seen in lines 1090 through 1106. Most of the operators available in this assembler are used to create the bytes of machine code. The symbols are shifted and pruned (by "&" operators) to fit in the position desired in the machine code word. While the format of the macro level machine instruction in ADDX, shown later, was mostly byte aligned, we have none of that at this level. Bitfields in the microword range in length from 1 to 12 bits, so a great deal of pushing and truncation is needed.

Machine Code S-Record File

The TRACE.S file is shown next.

```
S0030000FC
S21400010036004A0001888821C857300500000000E4
S804000000FB
```

The one word of microcode extends from the address field "000100" to the checksum "E4". The data bytes starting with "36" can be compared with the data in the expansion of "endsc" in TRACE.LST above. The zero padding on the end of the microword record is clearly visible.

S-Record Object File Dump Program: MCL

MCL is a utility program we wrote that reads the S-record file and produces a human readable version of the microwords in a file. It is an example of the many simple programs that the designer can provide to make his work easier and more error-free. It is invoked by typing:

```
mcl trace.s
```

The output will be sent to the screen. If output redirection is desired, typing

```
mcl trace.s >trace.mcl
```

under MSDOS or UNIX will send the output to the file "trace.mcl". A similar capability for VAX VMS can be provided. The output sample follows:

```
MCL Report for File: trace.s Page: 1

_____HARDWARE BITS_____SIMULATOR
_____
           C                                                  M
           A  C   A   B                                       C  D       SMO
    2      BD  M   M   M                   MM           MM     M  YB    2222TAU
    9   C  RDF U   U   U       Q S Q S C UMCD      DD AA    I  U  BB    9999ACT
    1   RC ARI X R X R X  OO   R R L L N SZOB  2 I BB RR    R  X  DD H  1100THS
    0 OLEC NEE S A S A S EEEW  S S S S S RNVO  9AN ED LO RW L  S  EE A  0033RII
      EDNI CSL E M E M E ABYE  E E E E E LLLE  OLS NI DE DR D  E  NN L  IIIIEND
    I **** HSD L A L B L ****  L L L L L DDD*  3UT *R ** ** * L  **  T  ONOOGEE
    +-+----+---+-+-+-+-+-+----+-+-+-+-+-+----+---+--+--+--+-+-+--+-+-------+
   010 |3|0110|004|A|0|0|0|0|0001|8|8|8|8|0|1001|C80|10|10|11|1|0|11|0|0001100
```

The data are much easier to read than the S-record. The column headings are read vertically. The 2903 instruction is parsed to match the data sheet. AM2903I bit 0 is the isolated bit in the 9-bit instruction field.

Microprogram Macro Examples

Several macros are extracted from SWEET16.MAC, the microcode "definition" file for the Sweet16 computer discussed in Chapter 8, to illustrate the details shown in the listing above.

Macro: JMP

```
; JMP:        DEF      H#3,B#0,1V:%B#1,B#1,B#0,12V:%H#000,80% UNCOND JUMP PL

JMP:          macro    \1,\2
con_inst      set      $3
i2910_oe      set      $0
              ifnc     \1,
i2910_reg_ld set      \1&$1
              endc
i2910_ccen    set      $1
i2910_ci      set      $0
              ifnc     \2,
br_adr        set      \2/NBYTE&$FFF
              endc
              endm
```

The first line shows the DEF statement used in the Microtec Microcode Assembler's DEFinition program as described in this chapter and in Appendix G of Chapter 8. The default for the first field on the left (H#3) is $3 in a field 4 bits wide (H). The default value ($3) is produced in the JMP macro by the "con_inst" field being SET to $3. The width and position of the field will be determined in "endsc" below. If there is a second operand (determined by the "ifnc \2," the branch address "br_adr" symbol will be SET to the value of that operand divided by the value of NBYTE which is the number of bytes in the microword (16 defined earlier in this file). The value produced is limited to 12 bits by the "&$FFF". Notice there are no blanks in the operand expression. If there had not been a second operand then the "br_adr" symbol would have been specified by the DEFAULT macro.

Macro: DEFAULT

```
DEFAULT    macro
           CONT     $1
           CCMUX    $A
           ALU      YBUS,LOW
           AB
           MSRLD    1,0,0,1
           SHFT     8,8,8,8
           CNMUX    0
           REGMUX   0,0
           R2903    0,0
           DATAPATH 0,2,1,3,1
           NOREGWR
           MCMUX    0
           NOYBDEN
           NODBBDEN
           DEBOFF
           endm
```

This macro is written in terms of the previously defined operation mnemonics (operators). In some cases, the operands (e.g., for the ALU) are given in terms of symbols that were previously equated. This is for human readability. In other cases, constants were used (e.g., the MSRLD operator). Some operators require no operands since they were defined in terms of constants themselves, like NOREGWR.

Macro: endsc

The next macro, "endsc", contains all of the actual code generation. No other macro actually sets aside space and initializes it.

```
; Make data formatter based on symbols defined above
; Equivalent to DEF statement - each ASM source line must conclude
; with this macro.
; The microword is 100 bits long and is padded to 128 bits (16 bytes)
; to make one microword fit one S-record line generated by A68K.

endsc macro
      dc.b con_inst<<4!i2910_oe<<3!i2910_reg_ld<<2!i2910_ccen<<1!i2910_ci
      dc.b br_adr>>4
      dc.b (br_adr&$F)<<4!ccmuxsel
      dc.b rama<<4!amux<<2!ramb>>2
   dc.b(ramb&$3)<<6!bmux<<4!i2903_ea<<3!i2903_oeb<<2!i2903_oey<<1!i2903_we
      dc.b qrsel<<4!srsel
      dc.b qlsel<<4!slsel
      dc.b cnsel<<6!usrld<<5!mznld<<3!mcovld<<1!mdboe
      dc.b i2903_i8765<<4!i2903_i4321
      dc.b i2903_i0<<7!dben<<5!marld<<4!maroe<<3!rd<<1!irld
      dc.b mcmuxsel<<6!y_bbden<<5!db_bden<<4!halten<<2!condeb>>2
```

```
dc.b  (condeb&$3)<<6!aludeb<<2!srgdeb
dc.b  macdeb<<6!outdeb<<4!$0
dc.b  $0
dc.b  $0
dc.b  $0
nolist
DEFAULT
list
endm
```

A total of 16 bytes are set aside and initialized by the "dc.b" directives. The lower case symbols like "br_adr", "rama", etc. have been specified by an operator invocation like

```
JMP     ,FETCH
```

or in the invocation of DEFAULT near its end. Since the "SET" directive has been used in each operator macro, the value of the symbol can be changed many times. The "EQU" directive would not have worked in this service. The assembler logical and arithmetic operators are used in the operand expressions to reduce the field widths and place them at the proper place in the microword. For example, "condeb" corresponds to a 2-bit field near the right end of the microword. The value of the symbol is truncated on the left by ANDing it with $3. Its position is at the left end of the byte shown so it is left shifted 6 bits to get it there. The other symbols are similarly treated and then ORed together to create the byte.

5.15.2. Examples Showing A68K Used as a Second-Level Cross Assembler

The macro facility of A68K may be used to help assemble second-level instructions as well. Two examples are given next for the ADDR and ADDX instructions.

```
; Add Register/Register (opcode = 00001000, 1 word instruction)
ADDR:     macro   \1,\2
          dc.w    ($08<<8)!((\1&$F)<<4)!(\2&$F)
          endm
; Add (RX) (opcode = 00101000, 2 word instruction)
; word 1 - like the RR instruction format
; word 2 - address in memory (symbolic)
ADDX:     macro \1,\2,\3
          dc.w    ($28<<8)!((\1&$F)<<4)!(\2&$F)
          dc.w    \3
          endm
```

The ADDR instruction, when implemented on the first-level machine, adds the contents of the register specified by the first operand, \1, with the contents of the register specified by the second operand. The results will be returned to the register pointed to by the first operand. The detail of its operation is unnecessary to know when creating the macro. The only important information needed is the format of the machine instruction and the number and types of operands.

The ADDX instruction performs the same operation as the ADDR, but its second operand is in main, or second-level, memory. The effective address is calculated by the underlying machine to be the sum of the contents of the second word of the instruction, operand \3, and the contents of the index register specified by second operand (i.e., \2).

The results of assembling a test program using this Macro file are given next.

```
         001000                65          ORG  $1000
                               66
                               67          ADDR 5,6
         001000  0856          68+            dc.w  ($08<<8)!((5&$F)<<4)!(6&$F)
                               69
                               70          ADDR 2,1
         001002  0821          71+            dc.w  ($08<<8)!((2&$F)<<4)!(1&$F)
                               72
                               73          ADDX 4,3,DATA
         001004  0843          74+            dc.w  ($08<<8)!((4&$F)<<4)!(3&$F)
         001006  1034          75+            dc.w  DATA
                               76
                               77          ADDX 1,2,DATA1
         001008  0812          78+            dc.w  ($08<<8)!((1&$F)<<4)!(2&$F)
         00100A  1040          79+            dc.w  DATA1
                               80
         00100C                81          DS.W 20
                               82
         001034  0019          83  DATA    DC.W 25
                               84
         001036                85          DS.W 5
                               86
         001040  414243444546  87  DATA1   DC.W 'ABCDEF'
                               88
         001046                89          END
```

The left-most column is the address of the first byte of data that appears in column 2. The data in column 2 is produced by the "declare constant" assembler directive "dc.w". The third column is the line number of the statement and the "+" sign to its immediate right indicates that the macro processor wrote the statement. Close examination of the operation codes produced in the second column will disclose the operation of the macroassembler.

The S-record format for the object file was chosen for this example. The records are given below:

```
S0030000FC
S21400100000856082108431034081210400000000005B
S2140010100000000000000000000000000000000000CB
S2140010200000000000000000000000000000000000BB
S21400103000000000000019000000000000000000000092
S20A00104041424344454610
S804000000FB
```

An absolute macroassembler like A68K makes the job of producing an assembler for the first and second levels of a machine relatively simple. The availability of the source code for A68K makes it possible to use any desired mnemonic for an operation in the target system without regard for the nature of the host machine.

5.16. CONCLUSION

This chapter has shown methods whereby the preparation of software for first- and second-level machines can be assisted during the design phase with a relatively small investment of effort. A commercial meta-assembler was illustrated with the intent of showing the various tasks that an assembler must perform to support assembly at the first level while the width and structure of the microinstruction word are in flux. A second-level assembler was created by using the macro facility of a common macroassembler to produce machine code for a newly designed Instruction Set Processor. An "absolute" macroassembler, A68K, was adapted to build an assembler for both the first- and second-level machines.

Skill that is developed in using the host macroassembler will manifest itself in the ability of the programmer to develop a natural assembler language syntax for the target ISP. Since Macros can contain Macros, new "instructions" can be implemented at the second, or ISP, level by writing them in the language of the target ISP itself. These new instructions can be used in programs as if they were part of the original "hardware" implemented instruction set. They will execute at a slower pace, but the programmer of ISP applications will, within that limitation, be able to think that he has the "extended" instruction set in which to program. If a "new" instruction is found to have great utility, or other endearing features, then it may be moved "down" to join the set of "hardware" implemented instructions in the first-level machine.

APPENDIX A. USER'S MANUAL FOR THE MICROTEC META-ASSEMBLER

Directives

Directives for use only in the Definition file:

WORD – This directive is used at the beginning of the definition file to specify the length of the microword. It remains in force for all assembler programs that use this hardware definition.

DEF – The DEF directive defines an operation mnemonic and associates the appropriate bit fields in the microword with it. It can be used to initialize some or all of the bit fields to appropriate default values and can designate some of the bit fields as operands that will be filled in at assembly time using information in the assembler source program.

Directives used in both the definition and assembler programs:

EQU – The EQUate directive is used to assign a value to a label. This assignment remains in force for the duration of both phases of microcode preparation.

SET – The SET directive works like the EQU directive except that the symbol that is assigned a value may have its value changed by a new SET statement later in the program.

LIST – This directive controls the printing of various types of output in both phases of microcode preparation. Its operands include:

E – list error containing source lines at the console

I – list instructions not assembled due to conditional assembly statements

Q – specifies that addresses be printed in octal in the listing

S – list the source text (default)

T – list the symbol table (default)

X – list the cross reference table

Use the following operand in the definition program only:

D – create a definition output file (default)

Use the following operands in the assembler program only:

B – list the object module separately in block mode

L – list the object code interleaved between the lines of source code (default)

O – produce an object module file (default)

V – complement each bit in the object module file

NOLIST – This directive uses the same operands as LIST and its action is the opposite of the meaning of that operand when used with LIST. For example, NOLIST S will cause the source text file to *not* be printed.

TITLE – This directive will place the text string that is its operand at the beginning of each page of the listing.

END – This directive specifies the end of the definition or assembler source program file.

Directives are also present to control whether a block of text in the source program is processed. These conditional assembly statements include IF, ELSE, and ENDIF.

Directive to be used only in the assembly phase:

ORG – This directive informs the assembler that the following instruction is to be assembled at the address specified by its operand. The operand must be a constant either symbolic or given by value.

Symbol Specification

In order to write symbolic programs it will be necessary to make symbols that represent the constants that we expect to used. *Constants* may be specified in one of four radices, i.e., binary, octal, decimal, or hexadecimal. This is specified by using the following prefixes on the value for the constant:

B#	Binary	1 bit per digit	B#0100
Q#	Octal	3 bits per digit	Q#53
D#	Decimal	–	D#21
H#	Hexadecimal	4 bits per digit	H#C2

The number of bits occupied by a constant is critical to the definition and assembly phases. The number of bits specified must "fit" the space allocated for it or instructions must be given that will indicate how to make it fit. No negative constants are allowed.

Modifiers are attached to the right end of a constant or variable field to specify the following operations:

* One's complement the constant

- Two's complement (negate) the constant

% Right justify the constant in the space provided

: Truncate a constant that is too long on the left

A number placed before the radix specifier on a constant or variable field specifies the actual number of bits associated with the field. For example, 5D#20: results in the bit pattern 10100, a 5-bit field formed by truncating the binary pattern for the decimal number 20 on the left. Both the definition and assembler phase will force you to pay attention to the width of a bit field since any overlapping of fields implies the generation of incorrect control signals.

Constants

3B#101 specifies a 3-bit constant field containing the binary number representation of 5.

6D#5% specifies a 6-bit constant field containing the bit pattern 000101, where the binary pattern for 5 is right justified in the 6-bit field with the upper bits padded with 0s.

8H#CC specifies an 8-bit field containing the binary pattern 11001100.

Variables

3V specifies a 3-bit variable field. The value placed in this field will be specified in the assembler source file.

5VH#3:% specifies a 5-bit variable field but assigns the pattern 00011 to it initially. The value may be changed by a statement in the assembler source program. If the value specified there has the wrong number of bit positions, then the value will be right justified and padded if too small or left truncated if too large.

Don't Care

3X specifies that the next 3-bit positions in the microword are to be skipped in this DEF statement. These bit positions are *not* to be associated with the operation symbol specified by the label for this statement.

APPENDIX B. USER'S MANUAL FOR THE MODIFIED A68K MACRO CROSS ASSEMBLER

Introduction

An overall description of the features of the assembler language supported by A68K is given in the sections entitled Symbols, Constants, Comments, and Assembler Directives. Since we have disabled the MC68000 instruction capability, we will *not* consider that aspect of A68K in our user's manual.

Symbols

Symbols (labels) may be 127 bytes long (1 line) and must begin with a letter. They may contain the following characters:

A through Z

a through z

_ (underline), . (period)

The label usually starts in column 1 and must be terminated with a colon, ":" if it does not. For text editing purposes, the label (except on a EQUate) should always be terminated with a colon. This allows one to use the Find operation in their editor to immediately go to a label instead of going to all of its occurrences in instruction operands.

Constants

Constants can be specified in one of three forms as follows:

1234 is a decimal constant

$1234 is a hexadecimal constant

'A', 'AB', or 'ABCD' are character constants

Note: Single Quotes (') can be used in a character constant by writing it twice, i.e., '' is a character constant containing one single quote.

Comments

Comment fields begin with a semicolon, ";". Any character may be used in the comment since the assembler totally ignores the field. Blank lines with no beginning ";" may be used as needed; they greatly increase the readability of the source program.

The MC68000 registers may not be used as a label or be redefined by an EQUate. They are D0, D1,,,, D7, A0,,,, A7, SP, USP, CCR, and SR.

Assembler Directives

The Assembler Directives and examples of their syntax for these assemblers are as follows:

Directive	Syntax	Examples
ORG	ORG expression	ORG $100

The ORG directive initializes the location counter in the assembler. The next instruction will assemble at the address specified. The ORG must be used at the beginning of the file for the S-record (-s) option.

EQU	label EQU value	A EQU 5

The EQU directive creates a symbol, "A" in this example, and assigns it a value. Any time the value is needed in an assembler program, the label "A" may be used instead. A symbol may be defined using an EQU only once in a program. See the SET directive.

DC	label DC<.length> value	B: DC.W 8

The DC directive sets aside space in memory for a constant and places the value in its operand field in that space at assembly time. While the value is normally considered a constant, it may be changed by the program. The number of bytes allocated is controlled by the symbol appended with the decimal to the basic DC directive. The choices available are ".B" to allocate one byte, ".W" to allocate 2 bytes (1 word on an even-byte boundary), and ".L" to allocate 4 bytes (1 long word also on an even-byte boundary).

DS	label DS<.length> <number of cells>	C: DS.L 5

The DS directive sets aside uninitialized memory space of the size specified by its operand. In the example, 5 long words are set aside with the label C attached to the byte at the lowest address. The length specifiers available for DC are applicable here.

EVEN	EVEN	EVEN

The EVEN directive adjusts the location counter in the assembler such that it is an even number (of bytes). Instructions in many 16-bit processors must be aligned (begin) on even addresses. Words and long words usually must be similarly aligned.

```
END        END                                    END
```

The END directive signals to the assembler that the end of the source program has been reached. It has no affect on the assembled program at "run time". If a linker is used, END may have an operand that specifies the address to be used as the entry point of the program.

```
XREF       XREF <label>                       XREF DATA2
```

The XREF directive signals the assembler and linker that a symbol used in the current assembler source file is defined in another file. The assembler will only flag symbols as undefined that are not used as a label (defined) in the current source file. The assembler places 0s in the address field of any instruction that refers to the label (here DATA2). This directive is only available of the AmigaDOS output file format (the -o option).

```
XDEF       XDEF <label>                       XDEF SUB1
```

The XDEF directive signals the linker that the label it defines is in the current assembler source module. When multiple object modules are linked together, the linker can quickly find the module containing the label requested in another object file. The linker cannot see any internal labels (those not flagged using the XDEF directive).

```
PAGE       PAGE                                  PAGE
```

The PAGE directive causes the current page to be ejected in the printing of the listing (.LST) file. The next line will be printed at the top of the next page.

```
LIST       LIST                                  LIST
```

The LIST directive causes the following lines to be printed in the listing file.

```
NOLIST     NOLIST                             NOLIST
```

The NOLIST directive causes the following lines to not be printed in the listing file. The pair of directives, LIST and NOLIST, are used to suppress printing of sections of the listing file.

```
SPC        SPC <number of lines>              SPC 5
```

The SPC directive causes the number of lines specified in the operand to be inserted in the listing file. In the example, 5 blank lines will inserted before listing is resumed.

```
TTL        TTL "title of this program"        TTL "Prog 5"
```

The TTL directive specifies a string given in its operand for the assembler to print on the first line of each page of the listing file. The quotes are necessary.

```
CNOP       CNOP (0,n)                         CNOP (0,2)
```

The CNOP directive instructs the assembler to skip over enough bytes necessary to assemble the next instruction on an address evenly divisible by "n". The symbol "n" may only have the values of 2 or 4.

```
INCLUDE   INCLUDE "filename.ext"   INCLUDE "SWEET16.MAC"
```

The INCLUDE directive instructs the assembler to read the file specified by the operand and assemble it. This statement is placed at the point in the program where the text in the specified file would logically go if the INCLUDE directive were not available. This directive may be nested, i.e., INCLUDE files may contain INCLUDE files. On the UNIX system, the filename field is case sensitive.

```
SET        label SET value A                      A SET 5
```

The SET directive performs the same task as the EQU directive except that a label so defined may be repeatedly redefined in the current assembly. This is one of the major directives that will be exploited in the use of these assemblers in support of "foreign" processors.

```
EQUR

REG

MACRO     label MACRO <operands>   ADD MACRO \1,\2
```

The MACRO directive is the vehicle by which we create an assembler for a target processor. It invokes a facility that is similar to a specialized word processor. The MACRO directive is placed on the first line of a body of text called a macro. The symbol in the label field is the "name" of the macro (here, ADD). Symbolic operands are defined in the operand field. This assembler requires that the symbolic operands take the form "\1, \2, etc.", where the number refers to the position of the operand in the list. These symbols can be used in the following text that forms the body of the macro. The last line of the macro is signaled by another directive, ENDM.

```
ENDM      ENDM                                    ENDM
```

The ENDM directive tells the assembler that the preceding body of text associated with the MACRO directive is terminated.

Conditional Assembly Directives:

```
IFEQ, IFNE, IFGT, IFGE, IFLT, IFLE, IFC, IFNC, IFD, IFND
```

These directives have the following syntax:

```
IF<COND> <expression>
```

If the expression, when evaluated at assembly time, meets the CONDition then the following text that precedes the ENDC directive will be assembled. If the condition is not met then the following text will be omitted from the assembly; i.e., no machine code will be generated from the text. All symbols in the expression must be known at assembly time, they cannot be run time dependent.

```
ENDC                                         ENDC ENDC
```

The ENDC directive terminates a body of text following a conditional assembly directive. It, with the IF <COND>, establishes the domain of the conditional assembly.

The following directives are available when the AmigaDOS object format option is used (-o):

```
CODE      CODE                                    CODE

BSS       BSS                                     BSS

DATA      DATA                                    DATA

SECTION   SECTION <section number>                SECTION 15

IDNT

DCB
```

Expressions

An expression is a combination of symbols, constants, algebraic operators, and parentheses that, when evaluated at ASSEMBLY time, provides a constant that is usable as an operand. Since blanks are used to demarcate the boundaries between the major fields (label, operator, operand, and comment) in an assembler source statement, no blanks can be embedded in an expression.

Operators available for use in expressions:

Unary:	.op.A
+	Positive A
-	Negative A
~	One's complement A

Arithmetic:	A.op.B
+	Two's complement summation of the operands A and B
-	Two's complement subtraction of the operands A and B
*	Signed multiplication of the operands A and B
/	Division of the operand A by the operand B

Logical:	A.op.B
&	Bitwise AND the operands A and B
!	Bitwise OR the operands A and B
>>	Logical Right Shift the operand A by the number of bits B
<<	Logical Left Shift the operand A by the number of bits B

The operator evaluation in a expression is in the following order:

1. Unary minus
2. Shift
3. AND, OR
4. Multiply, Divide
5. Add, Subtract

Special Symbols

Asterisk, *, when used as an operand will be treated as a symbol for the value of the location counter. The location counter is an internal variable that the assembler uses to keep track of the number of bytes allocated at any given instant during the assembly process. Its value can be seen in the left-most column of the assembly listing file, i.e., the one with the .LST extension. Notice its use with the SET directive in microprogramming applications. An example that stores two values of the location counter in two labels for later use in a program follows:

```
        org $200
L1      dc.l *
        ds.b $20
L2      dc.l *
```

The listing file that results has the following appearance:

```
000200                        16        org $200
000200  00000200              17 L1     dc.l     *
000204                        18        ds.b     $20
000224  00000224              19 L2     dc.l     *
                              20
```

The "back slash" followed by the "at" sign, "\@", when used in a macro, prevents multiple label definitions. This symbol stands for a counter that is incremented each time a macro is invoked. It is especially useful for implementing loops within macros that will be invoked more than once. An example macro showing the "\@" symbol attached to the label "test" follows:

```
delay       macro\1
test\@        dc.b\1
              dc.ltest\@
            endm
            org 0
            delay 5
            delay 5
            end
```

The absolute origin is set to "0" so that the address column in the following listing will reflect actual addresses. The macro is invoked twice with an operand of "5" each time. The label on the "dc.b" in each macro is different since the "\@" symbol was incremented each time the macro was invoked. Notice that this is shown in the data column for each appearance of the "dc.l" operation.

```
000000                        6            org 0
                              7
                              8            delay 5
000000  05                    9+test.001   dc.b   5
000002  00000000              10+          dc.l   test.001
                              11
                              12           delay 5
000006  05                    13+test.002  dc.b   5
000008  00000006              14+          dc.l   test.002
```

APPENDIX C. DATA TRANSMISSION FORMATS FOR OBJECT FILES

Motorola S-Record Format

The S-record data produced by the test program in Section 5.15.2 is presented below broken up into its logical sections.

```
S2  14  001000  08  56  08  21  08  43  10  34  08  12  10  40  00  00  00  00  5B
```

The first three fields define the record type (S2 → data), number of bytes in the record (14 hex), and the address of the first byte of data (001000). Data words have been separated to correspond to the 8-bit bytes counted by the first field. The final 8-bit field is the checksum, the sum of all bytes in the record beginning with the length field and concluding with the last data byte. The 6-digit address field, here 001000, is split into 3 bytes, 00 01 00, for purposes of checksum calculation. Any overflow beyond 1 byte in the checksum is discarded and the least significant byte is placed in the record.

The INTEL Hex Format

The INTEL Hex format contains the same information as the Motorola S-record except that there are some detail differences. A Hex record is shown next.

```
:10000000111F11001102790112FF3100490089001E
```

The record is now shown broken into its logical pieces.

```
:  10  0000  00  11  1F  11  00  11  02  79  01  12  FF  31  00  49  00  89  00  1E
```

The leftmost field, ":", indicates that this character position marks the beginning of an INTEL Hex record. The next field, here a hexadecimal 10 (16 decimal), contains the number of bytes of actual data within the data portion of the record. The next field is the 16-bit address, in hex, of the first data byte in the destination memory, here address 0000. The "00" in the next field specifies the record type, here a data record.

The data portion of the record begins with the first byte here containing "11" and terminates with the "00" in the 16th byte position. The last field is the checksum, the sum of all bytes in the record. The INTEL Hex checksum is two's complemented before being entered into the record. The system receiving the data merely adds all bytes, including the checksum; if the result is zero when all data in a record has been received, then the data has been transmitted correctly.

EXERCISES

Use the following machine code format for the next several exercises.

4	1	3	12
INST.	POL	CCMUX SEL	ADDRESS

4	4	1	1	1	9
REG A PTR	REG B PTR	EA	OEB	OEY	RALU INST.

Assume that the machine is implemented using a controller containing an Am2910 similar to Figure 4-14. The architecture fields in the second line command an Am2903. Obtain the appropriate data sheets for the parts so that the numeric values for the control codes are available. Appendices G and H at the end of Chapter 8 also specify these codes in a form compatible with the meta-assembler described at the beginning of this chapter. All operations requested of the hardware will involve these two parts only. These exercises are intended to deal only with the assembly language aspects of the problem and depend on your understanding the parts described in Chapters 3 and 4. Be sure to handle defaults in a manner consistent with the hardware and the problem.

1. Prepare a list of EQUates in the meta-assembler format that create symbols for the definition and assembly process for the RALU.
2. Create the DEF statement for the RALU fields that will allow you to write

   ADD R0,R1

   for the RALU part (i.e., the second line above) of the symbolic microinstruction when using the meta-assembler. It is desired to add the contents of R0 to R1 and return the results to R1. Prepare DEF statements in this form for the other standard ALU functions available in the AM2903.
3. The alternative way to command the RALU in the assembly process is to write a statement of the form:

   RALU R0,R1,,,ADD

   Using the meta-assembler format answer the following questions:
   a. For consistency with the use of the Am2903, what must the default values of EA, OEB, and OEY be if one desired to add the contents of R0 to R1 and place the results in R1?
   b. Write a DEF statement for the RALU "operation" with its associated EQUates so that you can use the form given in an assembler program.
   c. What else is needed to make the RALU "operation" handle the other standard ALU functions?
4. Write the complete set of DEF statements for all operations possible with an Am2910-based controller for each of the following formats:

a. The DEF statements should allow the assembler programmer to use the CON-TROLLER CJP,,,THERE form. Note that a single DEF statement and many EQUs are used in this form.

b. The DEF statements should allow the assembler programmer to use the

BRANCH   ,THERE

form. The first operand refers to the CCMUX for conditional branching.

5. Write a short microprogram that adds the contents of RALU registers 0 through 7 and divides the sum by 8 (using arithmetic shifting). The result is returned to R0. Note that a temporary register is needed. Use the meta-assembler format with both types of symbolic instruction formats given in Exercises 3 and 4. Take advantage of the loop counting capability of the controller in performing the multiple shift. Assume that the RALU is 16 bits wide.

6. Repeat Exercises 1 through 5 using the shareware assembler A68K. Take note of the difference in appearance of the complete assembly language source statement as compared to the meta-assembler. The statements should be identical except for the detail associated with the method each assembler uses to detect a new operation field.

7. Study an available macroassembler, e.g., one on your own personal computer or a workstation, minicomputer, or mainframe with which you have continuous contact. Obtain a book that describes the macroassembler that you have selected and answer the following questions about it.

a. What are the names and the syntaxes of the macro definition and macro end statements? Make a list of these in the form used in Appendix B of this chapter. Look up and list the same information for the byte and integer word data declaration, data space reservation, and conditional assembly assembler directives. You will also find useful the directives that work like the origin, end, set, equ, list, and include directives in A68K.

b. Determine how symbols are formed, how to specify constants, and how to write comments in the source program. Are blank lines tolerated? Must labels begin in column 1? Verify that blanks are used as field separators and that the assembler can use tabs for the same purpose.

c. Which symbols, if any, are reserved by the assembler for specific purposes? Make a list of the symbols, especially machine instructions for the host, that may not be used in your programs.

d. Are there any characters like the * and \@ in A68K that have special meaning? It is very useful to be able to specify the location counter value like * does. The symbol equivalent to the \@ allows labels within macros to be used repeatedly without incurring "multiply defined label" errors. It creates a new label each time a macro is invoked.

e. Find out how to prepare a source program, submit it to the selected host system, and get listing and data files produced. If the macroassembler produces relocatable code, as is probable, study the steps used with the VAX assembler in this chapter to determine how the machine code can be moved to the point of use.

8. Use the macroassembler studied in Exercise 7 to perform Exercises 1 through 5.

9. Use one of the assemblers discussed in this chapter or the one that you studied in Exercise 7 to write a cross assembler for the instruction set of a microprocessor that you have studied in another situation. The 8-bit microprocessors are the easiest to do; how-

ever, with the help of the conditional assembly directives, any instruction set may be handled. Try to write a single macro for each group of instructions, e.g., the arithmetic instructions having a certain common set of address modes. Then use your editor to duplicate the macro for all members of the set. The conditional assembly directives make it possible for the macroprocessor to do this for you at assembly time so that only one macro is actually provided for each group of instructions. Note that the meta-assembler is quite capable of this task, as well as the others discussed.

10. Identify a path, appropriate to your environment, that allows machine code to be moved to your "possibly hypothetical" target machine. This path may traverse a serial port in your personal computer or a telephone line from a remote host. Write a program that, like VAXFMT discussed in the text, takes the relocatable form of the object file, turns it into an absolute form, and formats it using one of the transmission forms appropriate to your situation or like those in Appendix C of this chapter. Describe a procedure for completing the transfer to your target hardware. This may involve an EPROM programmer or a microcomputer attached to your target.

11. The TI 74ACT8818 was shown in Chapter 4 to have many control input signals that go to specific elements within the device. Their are no translation effects from a Next Address Logic network. With the help of the data sheet for the component, write a set of DEF statements or macros that implement the set of instructions that were discussed for the controller in Figure 4-14. These include Continue, Jump, Conditional Branch, Call, Return, and Counted Loops. Assume that the controller is implemented using the TI device. Select the proper control lines for the device and connect them all to a suitably wide controller instruction field in the microword. The scope of this problem is limited to creating symbolic instructions in the list above for the TI device. This is a "software" version of Exercise 6 in Chapter 4.

Chapter 6

HARDWARE DESCRIPTION LANGUAGE
AND SIMULATION

6.1. INTRODUCTION

In the following chapters we plan to discuss the design of complicated systems using complex architectural elements. In many cases, the number of pins on each device is very great, ~200, and the systems composed of the devices involve an enormous number of interconnections.

To assist us, we will introduce a pair of design languages that will supplement the Algorithmic State Machine (ASM) diagram discussed in Chapter 1. The purpose of a design language is to allow us to describe our ideas in a complete, concise form. The form should take as little time to "write" as possible but allow us to embody our understanding of the problem specification. We hope to use the languages to help us "take notes" during the first, or discovery, phase of the solution. Most problems that we expect to encounter are far too complex to be understood in the course of one reading, but will require repeated readings and note-takings before our understanding is sufficient.

The first language, which we will call Program Design Language (PDL), is frequently used in the same context as the flowchart, i.e., to give a general description of an algorithm without seriously specifying the underlying hardware detail or timing. The second, the Hardware Description Language (HDL), parallels the ASM Diagram. It contains explicit timing and hardware resources information. The PDL and HDL are most easily used when one must specify many parallel operations, i.e., the flowchart and ASM Diagram boxes become too small for the number of entries. However, neither form shows complex program flow as clearly as the flowchart and ASM Diagram.

It should be obvious that the hardware fabrication stage cannot be started until the design has been thoroughly evaluated. We expect, through the use of the languages mentioned above, to be able to create a body of text that describes our hardware solution and the algorithm it supports. The remainder of the chapter will be used to describe methods to evaluate our design without having to build any hardware. Simulation is the attempt to evaluate a design by making and using tools that act like the design. The tools should be able to simultaneously act like the design and be relatively quick to produce and use. They can also be made to report information that is virtually impossible to obtain from the hardware itself.

This chapter will explore some of the simulation tools that can be made by the designer to support his own work. In some cases, it will be necessary to use more complex, general purpose tools provided by another source for those cases where the tool production would be too expensive. Several commercial software packages allow two very important factors (i.e., loading and propagation time) to be evaluated during simulation. The newest of these languages, VHSIC Hardware Description Language (VHDL) is required by the Department of Defense and defined in the "IEEE 1076-1987" specification (see Bibliography). It is intended to provide a complete and testable description of a system in a range of detail from the highest down to the gate level.

Simulators are used to understand all aspects of a design before the design is committed to hardware. Each level of simulation is used to reveal certain things about the design. The levels of simulation parallel the levels of control in a design.

A system with two levels of control [i.e., an Instruction Set Processor (ISP)] may be studied at the second level of control by the use of an Instruction Set Simulator (ISS). It is the purpose of this simulator to help the designer determine whether the instruction set is necessary and sufficient for his application. The simulator should allow sample applications to be written and executed. Needless to say, the applications may not run at full speed, but they can be used to study the instruction set. While application-dependent test data are normally designed to demonstrate particular points concerning the element under test, it is appropriate to operate on "real" data under some circumstances. The simulator should make this possible by being able to import "canned" real data acquired from outside the host system.

In some cases, the simulator's execution of the target algorithm itself meets all requirements of the problem specification, including speed. This allows a solution to be reached without the commitment to any specialized hardware. Applications written for general purpose computers fall into this category.

The simulator at the second level can be extended to provide a platform on which all aspects of the instruction set can be understood, including the algorithms on which the instructions are based.

Once the instruction set is characterized, or if a second level of control is not used, simulation should proceed to the first level of control. A simulator at this level allows the designer to map the selected algorithms onto the desired architecture. Any features described by the HDL in Chapters 7 and 8 should be demonstrated in simulation at this level.

For example, consider the mapping of the signed multiply onto a modern RALU. The fact that the multiplier will be placed with the lower half of the product in the Q register should appear in the simulation. A simulator at this level must allow the architectural elements, i.e., registers, combinatorial elements, and data paths, to be combined with the selected algorithms to interact with data in creating the desired solution.

In a simulation of a microprogrammed system, the designer should be able to write a microprogram embodying the algorithms running on the architecture under study. This microprogram should "run" on the simulation, allowing the designer to judge the accuracy and quality of his solution. Speed of execution of the final hardware-based solution can be determined from microcycle counts. Once the architecture has been refined to a workable level, the designer can perform a critical path timing analysis to actually determine the period of a clock cycle. This then leads to an understanding of the real time performance of his system.

The preceding paragraphs outline a general method of simulation that can allow the designer to evaluate the performance of his instruction set or the algorithms and the architecture on which they are to be implemented, as appropriate. The one area that is not covered cannot be addressed at present with simulators provided by the designer. The area of electrical simulation is entered when the designer wishes to evaluate the electrical behavior of his architecture. At the first level, he is particularly interested in real propagation delays and output loading. While the data books can give maximum values from which one can make critical path calculations, no information is given that would allow one to determine the presence of glitches (hazards) in combinatorial elements. The designer can build the system and run the software produced in the course of behavioral simulation. He can degrade (lengthen) the period of the system clock so that glitches that he assumes are present, but can't see, have settled. These may be caused by the variation in propagation delay due the variability of a parameter from part to part or due to variations in loading.

The alternative is to electrically simulate the architecture. This is an arduous task performed with the help of software provided by others. It does allow a thorough characterization of the system. It will not be covered in this book since our orientation has been to show the development and use of tools by the designer himself. Its place in the development of a system should come just before hardware fabrication. The time required to create a model of the desired system is longer than any of the steps traversed to this point; thus the time investment should be made only when all other means have been exhausted. In other words, why perform electrical tests on a system in which the algorithm does not work? The cure to algorithm and logical problems can be discovered far less expensively.

6.2. PROGRAM DESIGN LANGUAGE

Now that we have seen some of the capabilities of the hardware modules, we are ready to implement useful algorithms with them. Before this can proceed, we need a language in which to discuss the algorithms themselves. This problem has already been addressed in the computer programming field so that we may consider borrowing an algorithm description language from it. Two important types of language are *flowcharting* and *pseudocode*. These types are separated by the way they appear to the user: the first is graphical and the second is in the form of a written human language like English. The flowcharting method is the progenitor of the ASM design language presented in Chapter 1. It represents control structures (e.g., IF...THEN...ELSE constructs) very efficiently. However, when many operations are performed sequentially, it presents a cluttered and less readable appearance. The pseudocode type of presentation has the strength of presenting an uncluttered sequential flow while not having as clear a presentation of control structures. This type has received considerable attention as "structured" programming has increased in popularity.

Our algorithm description language is based on the pseudocode model. We will call this language PDL. Complex control structures will be accommodated by using flowcharting where appropriate. Generally, structures needing this treatment are found in the lowest "functions" in a structured program so both types of presentation will be useful.

PDL is very English-language oriented. At its highest level it is just a clear English statement of the algorithm. No storage locations, datapaths, or Arithmetic/Logic Units (ALUs) are defined. Borrowing the operator symbols from the next section, we will write a PDL description of the accumulation operation as follows:

$$A = A + B$$

If there are several B items to be accumulated into A then we could write:

$$A = \sum_{i=1}^{n} B_i$$

At some point, we would like to relate this algorithm-oriented description to a hardware-oriented description. This can be done using PDL by increasing the detail of what is used in the description. At some point, this language becomes too imprecise for our purposes and we must then move to the next language, HDL.

6.3. HARDWARE DESCRIPTION LANGUAGE

In order to describe the physical operations needed to implement the algorithms we wish to discuss below, we will introduce another language. The Hardware Description Language (HDL) is intended to describe data movement along signal paths (or arrays of signal paths called "buses"). The data may be combinatorially transformed during these transfers by elements along the path. While there are many HDLs differing in syntax and detail, their main reason for existence is to enable us to quickly and thoroughly describe the interaction of an algorithm with an architecture. If the HDL is thoroughly defined, it may serve as a programming language in its own right. This allows an architecture and algorithm to be converted by an HDL compiler into a simulation of the intended system. This powerful concept means that many of the labor-intensive tests of an algorithm/architecture combination can be completed quickly.

The HDL introduced here will serve as a language in which we can discuss the integration of algorithms and architectures. It will not be rigorously defined, but pointers will be given that will show the direction in which improvements need to be made to produce a rigorous (compilable) language.

As is true in any computer language, we need some nouns that define objects. Here, the objects are latches, registers, buses, and various combinatorial modules. It should be noted that registers can be thought of as arrays of latches. Also, buses are arrays of signal paths (or wires). This point can be exploited in the way we create objects.

We also need some operators. The first type we will consider are the assignment operators. The situation of placing data on a bus or on the inputs of a combinatorial element is different from the case where data is clocked into a latch. This difference is seen in the time required for the output of the object to assume the new state. If the object is a bus or combinatorial element, then the output will assume its new state after some propagation time related to the physical properties of the element. These properties include the output load resistance and capacitance as well as the logical complexity of the object. The latch, on the other hand, will not drive its output until its clock changes state. In our systems, the sensitive edge will most likely occur at the end of the system clock period. To discriminate between these two situations, we will use the following notation:

```
   =            (equal sign) combinatorial assignment
   :=           (colon, equal sign) clocked element assignment
```

Therefore, if A is the output of a latch and B is a wire (signal path), then wire B = A. Also, if C is the output of a buffer, then C = A. The other assignment operator is used when the signal path B drives the latch A, as in A := B. These operators emphasize that combinatorial elements (including buses) assume their new state after a physical delay, while latch (and register) outputs wait until their clock input has transitioned.

The functions provided by various logic elements are also available as logical operators. The following list is appropriate:

```
   &            Logical AND
   |            Logical Inclusive OR
   XOR          Logical Exclusive OR
   ~            One's Complement
```

These operators are used in a bit-wise sense connecting signal paths with latches. Some examples using a latch A and two signal paths B and C are as follows:

```
    A   :=   B  &  C
```

the output of latch A will be B .AND. C at the next sensitive clock edge.

```
    C   =  A  |  B
```

after a combinatorial delay the signal path C will contain A .OR. B.

```
    C   =  ~A
```

after a combinatorial delay the signal path C will contain the one's complement of the output of latch A.

If one defines register R as an array of latches, L0, L1, L2, and bus B similarly, wires P0 through P2, then the following examples are appropriate:

```
    R   :=   B
```

after the sensitive edge of the clock each latch in R will assume the state of the corresponding path in B. This is the same as saying

```
    L0 := P0; L1 := P1; etc.
```

In order for the symbolism to make sense, the number of latches in R must match the number of paths in B. It is assumed that the path number is in the same order and has the same values as latch numbering in the register, i.e., L0's input is connected to path P0. If this notation is observed, then the following examples are valid assuming B and C are two 16-bit buses and R is a 16-bit register:

```
    R   :=   B  &  C
```

The output of each latch in R is the bit-wise logical AND of the corresponding paths in buses B and C, i.e., $R(i) = B(i) \& C(i)$. where $R(i)$ is the ith latch in register R and $B(i)$ is the ith signal path in bus B.

A similar set of arithmetic operators is possible since one frequently finds integrated circuit logic elements of this complexity. A set of appropriate combinatorial arithmetic operators follows:

```
    +   The binary sum of two operands
    -   The two's complement difference of two operands
```

The two operations require Carry Input and Carry Output, as was seen in an earlier chapter. Two keywords are defined called Cout and Cin that are automatically used during the addition and subtraction operations. These normally would be connected to an appropriate source and destination for the carry.

In some architectures designed for speed, one finds a combinatorial multiplication circuit. If this is available, then a multiply operator is appropriate.

> * **The product of two operands**

Generally, the Register Arithmetic/Logic Unit (RALU) used in our designs does not support multiplication as a combinatorial operation. This causes us to not use this operator but to substitute an algorithm using the facilities available. The product itself can be unsigned or two's complement (signed) binary. Which one that is used must depend on the operands supplied. Our method of defining operands should allow us to discriminate between the two operations.

Division, whether binary or floating point, is generally performed as an algorithm, either based on the long-hand method or the Newton-Raphson method. For very small operands, a look-up table method based on a memory or minimized network may be applied. In this case, a binary division operator can be defined.

> / **The quotient of two operands**
> % **The remainder of the integer division of two operands**

In some applications, the necessity for speed has encouraged the production of floating point networks that can multiply and accumulate. The networks are combinatorial; however, registers are embedded in them to make some of the operations run in parallel (pipelines). The symbolization of these networks will remain separated; no single symbol will represent them due to the embedded registers. A macro facility allowing us to use a single symbol for their operation would be useful if this HDL were to represent a formal language.

Some additional operators to shift data left and right relative to its original position are appropriate. We will define only two operators as follows:

> << **Shift the operand left one bit position, e.g., A<<.**
> >> **Shift the operand right one bit position, e.g., A>>.**

These correspond to the action of the combinatorial shifters found in most RALUs. In some circumstances, one finds it necessary to employ barrel or funnel shifters, in which case the number of bit positions that can be shifted in one trip through the shifter may be specified. In this case, the notation is as follows:

> A>>N   **Shift A right N bit positions**
> A<<N   **Shift A left N bit positions**

The operand A in the examples must be a register or a bus since shifting would make no sense otherwise. In many cases, the "shift_in" and "shift_out" bits are of interest. These are handled by connecting latches or signal paths containing the desired bits to the shift input and shift output latches on the register being shifted.

From the previous discussion it can be seen that we need a means to define the variables (objects) that will be interconnected via the HDL. This definition phase is akin to the definition of variables in programming languages. Our words that define the properties of the variables are different from that for FORTRAN where we have the statement INTEGER*2 ARRAY(2,100), or C where the statement "int array[2][100]" means the same thing. Our properties are associated with whether the object is a clocked or combinatorial element and whether it is composed of several simpler elements like the register, etc. In declaring a register or bus, we must also specify the numbering pattern to be used for the elements. Furthermore, we might like to group some part of a complex structure under another name similar to the EQUIVALENCE statement in FORTRAN. As an example, assume that register RA is composed of 16 latches numbered 0 through 15. This might be a register that normally holds 16 bits (one word) of data. On some occasions, it is useful to refer to the "upper" or "lower" byte of data in the register; so let us define two more registers RU and RL that correspond to the upper and lower 8 latches in RA, respectively. Possible notation for this could read as follows:

```
define      REGISTER RA[15-0]
define      REGISTER RU[7-0] as RA[15-8]
define      REGISTER RL[7-0] as RA[7-0]
```

The command "define" will be used to formalize the variable definition phase of our HDL "program". The property "REGISTER" will serve to indicate that RA, RU, and RL should always be the recipient of a clocked assignment operation ":=". Further, their outputs will not assume the new value on their inputs until the next clock transition. The "as" operator establishes the correspondence between the latches making up RA and those making up RU and RL. The square brackets "[" and "]" will be used to denote an indexed quantity in the sense of the language C.

One could define RC as follows:

```
define    REGISTER RC[0-4] as RA[12-8]
```

This results in the correspondence in which RC0 is the same latch as RA12, while RC4 is RA8. A combinatorial element like a bus could be defined as follows:

```
define    BUS BA[15-0]
```

Similar partitioning could be done as in the case of registers.

In order to build systems, one must make connections. To describe a system of registers and buses, we will introduce another operator called "connect". It will be used after objects are "defined" and before any statements are made concerning register transfers. For example:

```
define      REGISTER      RD[0-4]
define      REGISTER      RE[15-0]

connect    BUS_OUTPUT    BA to REGISTER_INPUT RA
connect    BUS_OUTPUT    BA[15-11] to REGISTER_INPUT RD[0-4]
connect    REGISTER_OUTPUT RE to BUS_INPUT BA
```

In this example, the key words BUS_OUTPUT, REGISTER_INPUT, etc. are introduced. These key words along with the "to" operator are used in the "output" to "input" order to describe the system. If the element numbers are not given, it is assumed that the bus or register being connected have the same number of elements and that they are numbered in the same order. If this HDL were implemented as a formal language, this point would be checked by the compiler and a warning issued if this were not true. Notice that in the second "connection" example the ordering of the latches in register RD is opposite to that in the bus BA. The correspondence between elements is such that RD0 is connected to BA15. Any offset may be used to accomplish the actual wiring needed.

Using the definitions and connections for bus BA and registers RA and RE given above, we can now write an HDL statement that describes an operation involving these elements. Consider the simple assignment:

```
RA := RE
```

This means to transfer the 16 bits in RE to the 16-bit register RA such that RE15 goes to RA15, RE14 goes to RA14, etc. The next assignment would leave what bit pattern in RA no matter what the content of RE?

```
RA := RE & ~RE
```

Compare this with the following:

```
RA := RE XOR RE
```

If this were a formal language, we would expect the compiler to check that a path existed between the outputs of RE and the inputs of RA. Furthermore, a means should be made to verify that a combinatorial element capable of the logical operations requested lies on the path threaded by the bus BA.

Shifting can be handled now by introducing a means to indicate that several statements occur during the same clock cycle. We will use the line termination character used in the language C, the semicolon ";", to concatenate all statements that are to be performed during the same clock cycle. A "rotate", or circular shift operation involving the C_LATCH and the 16-bit register RA, defined above, is shown next.

```
define        LATCH C_LATCH
connect       C_LATCH_OUTPUT to RA15_INPUT
connect       RA0_OUTPUT to C_LATCH_INPUT

C_LATCH := RA0; RA15 := C_LATCH; RA[14-1] := RA>>
```

This statement creates a circular shift right of register RA through the C_LATCH. Notice that all inputs on the register and latch assume the new state after the propagation delay through the shifter and the interconnections. The outputs of the devices assume the new values at the end of the clock interval when the active edge of the clock occurs. This model works correctly for edge-triggered D flip-flop constructed latches and registers. The key here is the edge-trigger. The notation concerning the shifting of the inner bits of RA may also be written as follows:

```
RA[14-1] := RA[15-2]
```

It would be up to the compiler to examine that a combinatorial device in the form of a shifter lay along the path between the input and output of register RA.

Since many devices may be controlled during a single microinstruction, we will also introduce some punctuation. The curly brackets "{" and "}" will be used to enclose all statements that are to be processed during one clock cycle. This allows us to extend one microcycle over several lines using blank spaces as a means to enhance readability. The circular shift instruction above could be written as follows:

```
{
        C_LATCH := RA0;
        RA15 := C_LATCH;
        RA[14-1] := RA>>;
}
```

This usage is similar to the C language usage of both the semicolon and curly bracket.

In many circumstances, the operations within a microinstruction are dependent on a control function. For example, the output of a register can be placed on a bus only if an output control signal is true. To support this, we introduce a construct that is based on the conditional statements in high-level languages.

```
        IF (REG_OE == TRUE) {BA = REG;}
```

This statement can defer the control of an operation until "runtime", i.e., when data is being processed by the system under design. This form of the IF statement should not be confused with the program control statements introduced next.

The statements that control the flow of our HDL simulation are drawn from the capabilities of the controllers themselves. While we could describe a controller using the language introduced above, for architectures, we find that the clarity provided by submerging this detail under a common set of flow control instructions is desired. In this picture, the controller is a black box that furnishes architectural control statements in the order dictated by our HDL program. The statements will be modeled after the Am2910 controller for the present. These represent a useful basic subset of possible control statements. Other statements will be added as needed.

The syntax used will attach a single control instruction to a line of architectural control statements. Any controller, even the simple counter, can proceed to the next instruction. This is called

```
    CONTINUE
```

Associated with it is the unconditional GOTO instruction of the form:

```
    GOTO label
```

A conditional branch instruction is of the form:

```
    IF (condition) THEN GOTO label ELSE GOTO label
```

Remember these are instructions to the controller *not* the architecture; therefore, the action taken can only be another program control function. One must be careful in writing the condition to insure that only those tests supplied by the architecture are performed. The only conditions supported are the following:

```
        A == TRUE     The signal path or latch output is TRUE or 1
        A != TRUE     A is not equal to TRUE or 1, i.e., A is FALSE
        A == FALSE    A is False or 0
        A != FALSE    A is not equal to FALSE or 0,i.e. A is TRUE
```

The controller integrated circuits themselves *may* not implement the NOT EQUAL (!=) conditional test. This is frequently provided by a complementing circuit associated with the condition code multiplexer (CCMUX). The operand A must be a single-bit quantity. If a more complex function of various flags must be evaluated, then this should be done by Boolean networks in the architecture.

Other control constructs include:

```
CALL label        unconditional subroutine call
RETURN            unconditional return from subroutine

IF (condition) CALL label
IF (condition) RETURN

LOAD_COUNTER_AND_CONTINUE

LOOP_AND_DECREMENT_COUNTER
```

As a final example of the use of our HDL, let us create a text that will partially describe an Am2903 Register/Arithmetic/Logic Unit. First we must define the objects present.

```
define      REGISTER_ARRAY RAM[15-0][3-0]
define      COMBINATIONAL S_MUX[2-0]
define      COMBINATIONAL R_MUX[2-0]
define      COMBINATIONAL ALU[3-0]
define      COMBINATIONAL ALU_SHIFTER[3-0]
define      COMBINATIONAL Q_SHIFTER[3-0]
define      REGISTER Q[3-0]
define      BUS BA[3-0]
define      BUS BB[3-0]
define      BUS BDA[3-0]
define      BUS BDB[3-0]
define      BUS BRA[3-0]
define      BUS BRB[3-0]
define      BUS BF[3-0]
define      BUS BQS[3-0]
define      BUS BQ[3-0]
define      BUS BY[3-0]

connect     RAM_A_OUTPUT[3-0] to S_MUX_INPUT[2][3-0]
connect     RAM_B_OUTPUT[3-0] to R_MUX_INPUT[0][3-0]
connect     S_MUX_OUTPUT to ALU_S_INPUT
connect     R_MUX_OUTPUT to ALU_R_INPUT
connect     ALU_F_OUTPUT to ALU_SHIFTER_INPUT
connect     ALU_F_OUTPUT to Q_SHIFTER_INPUT
connect     ALU_SHIFTER_OUTPUT to BY_INPUT
connect     ALU_SHIFTER_OUTPUT to RAM_INPUT
connect     Q_SHIFTER_OUTPUT to Q_REG_INPUT
connect     Q_REG_OUTPUT[3-0] to S_MUX_INPUT[2][3-0]
connect     Q_REG_OUTPUT to Q_SHIFTER_INPUT
connect     BDA_OUTPUT[3-0] to S_MUX_INPUT[0][3-0]
connect     BDB_OUTPUT[3-0] to R_MUX_INPUT[0][3-0]
```

Several statements are given below to show the tristate buffer implementation method:

```
IF (EA == TRUE) {ALU_R = BDA; }


IF (OEB == TRUE && IO == TRUE)
{
                ALU_S = RAM_B;
                BDB = RAM_B;
}

IF (OEY == TRUE)
{
                RAM = ALU_SHIFTER;
                RAM[RAMB][3-0] := RAM;
                BY = ALU_SHIFTER;
}
```

6.4. SIMULATION OF THE SECOND LEVEL OF CONTROL

We will discuss machines having one and two levels of control in the following chapters. Our simulation discussion will begin at the second level of control. If there is no second level of control, obviously this level of simulation need not be performed.

Our second level of control will be associated with the ISP. We will create an instruction set, address modes, and programming model for a system that will solve our problem. In Chapter 8, we will study the second level of control and simultaneously learn how a particular ISP, a general purpose computer, works. In many applications we might find it useful to provide an instruction set that is not general purpose, but rather is directed explicitly toward our problem. In this case, the ISP that we produce would be called an Application Specific Instruction Set Processor (an ASISP). Even in the common situation in which we write a program for a computer that represents a specific solution to our problem, we find that we have usually created modules (functions or subroutines) that in themselves form a vocabulary in which to formulate our solution. What if these "words" were defined as "instructions"? We would find that our solution could be easily and efficiently given since the vocabulary is appropriate to our problem.

How does one discover the proper vocabulary in which to write solutions? In most scientific and engineering fields, the solution is couched in a mathematical language involving nouns, or symbols, that are the "lingo" of the field. The older high-level languages, FORTRAN and COBOL, derive their symbols and operators directly from the communities in which they were created, i.e., science and business. It was the job of the programmer in either language to create programs (i.e., solutions) in terms of the basic elements of the appropriate language. Over the years, the modularity of programs was recognized resulting in the explicit statement of the principles of structured programming. Once functions (words) were recognized and used, it became common to place the frequently used ones in dictionaries (i.e., function libraries) for use in other solutions. This has progressed at the present time to the form called "object oriented programming", where the task of dealing with data contained in an object is directly associated with the object itself. From the preceding, one can see that vocabularies are normally produced followed by a large number of solutions written in terms of the vocabulary.

There are two methods that can be used to create a simulator at this level. The first to be discussed will be based directly on the Fetch/Execute cycle performed by all ISPs. The second method will be a shorthand way to use the macro facility of an assembler running on a general purpose computer to produce a simulator. This method bears a strong relationship to the material discussed in Chapter 5 where an assembler for the second-level system-under-design was created from an existing macroassembler.

6.4.1. The Fetch/Execute Cycle-Based Simulator

The Fetch/Execute cycle-based simulator will be discussed first. A flowchart of this cycle is shown in Figure 6-1.

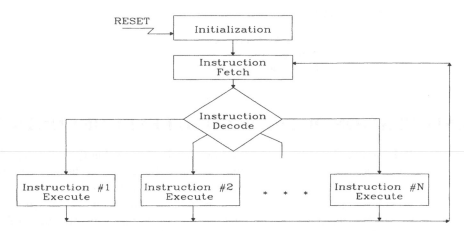

Figure 6-1: The Fetch/Execute cycle for the second level.

This simulator design is implemented as a program that contains modules to fetch the current instruction from an array simulating the second-level program memory. The instruction operation code (opcode) is decoded and control is transferred to one of several functions that simulate the action of the instruction. The goal of this simulator is to manipulate data in the programming model (i.e., the registers, program memory, and data memory) in the same way as the machine under design would. The algorithms in each of the execute boxes need only act so that the final results, as seen in changes in the data in the programming model, are the same as those expected in the final machine.

Generally, the algorithms are the same as those used in the hardware version. Since it is easy to create "registers" in the simulation, some of the details that are visible in the underlying architecture may be included. This really is not necessary at this level of simulation since it may obscure the things that need to be seen, i.e., the necessity and sufficiency of the instruction set design. The more detail added will cause a corresponding slow-down in both the evolution of the model and the speed of execution of the simulator. This degrades the speed of the development cycle itself.

Each execute module takes in application data and modifies it consistent with the instruction being studied. For example, consider the Signed Integer Multiply instruction, MPY R1,R2. Assume that incoming data is contained in registers R1 and R2 and the double precision signed product is to be returned to R1/R2. While one could write the shift/conditional add algorithm shown in Chapter 7 including the Q register, this does not contribute to the purpose of the task, i.e., the necessity of this instruction for the application. Only once its need is established should one consider studying the algorithm by which it is implemented. The actual code present in the Execute Function for MPY can be written directly in the high-level language of the designer's choosing, e.g.,

```
R1 = R1 * R2
```

Needless to say, R1 and R2 are memory variables and should be given the data type appropriate to cause the high-level language to perform the signed multiply. In some cases, the registers need to be able to hold both signed and unsigned quantities. Intermediate variables may need to be assigned in the execute functions to force the correct operations. Here, the C language is especially good since the designer can force particular operations directly by using CASTs. This same type of operation, albeit with different syntax, is available in most languages.

The previous paragraph implied that the simulator should be written in a high-level language, with C being suggested as a flexible choice. The first thought concerning the choice of language by one who is faced with writing a simulator is that the language of choice should be an assembler language on the computer available to the designer. The high-level language offers the designer a simple and usually more familiar environment in which to begin his simulation task. It is appropriate to use the high-level language early in the task when many modules need to be provided rapidly. Once the number of implementation choices are reduced, the speed of execution of the simulator can be addressed. The designer can apply the usual function timing methods to determine which modules should be "accelerated". He can use the high-level language's speed optimization algorithms first and then finally begin writing certain functions in assembler language.

There are trade-offs between a high-level language and an assembler that may affect the choice of language even at an early stage. The first concerns portability of the simulation. If it is important that the simulation source program be easy to transport to different development environments, then a high-level language like C that is noted for its portability is certainly necessary. On the other hand, the time investment by the designer in learning a language is enough that serious simulation would warrant the selection of one development environment for all of his work for many different projects. If this is a group effort, a network of similar systems would increase the productivity. Each member of the group would be responsible for one or more execute functions, including their creation and individual testing. Their results would be shared over the network when the total design was brought together.

A second consideration comes from the way that data is treated in high-level languages through the property of "typing". The "type" of a data value determines the operators that will be used to modify it. This was noted above in the signed vs. unsigned integer multiplication simulation. In assembler languages, there is little if any typing. This allows a single location to be used for any data type. For example, consider a 4-byte location called R1. This location could hold a 4-byte integer, signed or unsigned, as well as a single precision floating point number. The programmer is responsible for selecting the proper assembler language operator when modifying the data. This is the easiest method since it coordinates with the responsibility of the designer with respect to the architecture. The designer should not need to learn several assembler languages which would restrict the number of different development environments that he can use. If he does settle on assembler language as the proper choice for his simulation, he will find that he spends very little time trying to "force" a high-level language to do his bidding.

6.4.2. Macro-Based Simulator

In Chapter 5, we used a macroassembler to create an assembler for our yet unborn target ISP. This was accomplished by using the macro facility to substitute the desired target system machine code for target system symbolic statements in our assembler program. The workhorse instruction for this purpose was the assembler directive "declare constant"; e.g., ".byte" or ".word" on the VAX. The machinery in the assembler itself was used to parse our assembler source instruction in the target assembler language and then to search a list of macros for the one containing the machine code for our computer. It was also expected to keep track of addresses just as it would for its host machine. We embedded address references in our machine code by converting the relocatable addresses to absolute ones.

This time, we will create a simulator for the second level by replacing the "declare constant" directives with sequences of host machine instructions that perform the algorithm that is contained in our target machine's instruction. Using the signed integer multiply example above, the macro called MPY could be written directly in terms of the same instruction on the host machine if it had that instruction. If it did not have that instruction, some combination of host instructions would be used to create the desired result. The lowest level would be to recreate the shift/conditional add algorithm using the shift, add, and conditional branch instructions that are present in all computers. Remember, the goal of this level of simulation is to create only proper results on the data in the programming model.

The programming model itself is implemented through the declaration of storage areas in the assembler-based simulator. Space is set aside for program and data memory using the same directives used in the assembler in Chapter 7. For example, using the VAX assembler MACRO, the data area for a simulation of the computer example in Chapter 6 would appear as follows:

```
; Register locations
R0:         .BLKW 1         ;Specify a word to be used as register 0
... more registers similarly declared
SP:         .BLKW 1         ;Specify a word to be used as the stack pointer
PC:         .WORD 0         ;Specify a word to be used as the
                            ; program counter and set it to zero
                            ; to simulate reset
FLAGS:      .BLKB 1         ;A byte to hold the status flags
; Memory area
EPROM:      .BLKB 2048      ; 2 kbyte program data area
RAM:        .BLKB 2048      ; 2 kbyte RAM data area
```

The uninitialized storage allocation directive was used here. Registers could be implemented in the host processor's registers; however, many times these registers are being used to implement the simulations. If the target system's registers are allocated in memory just as the program memory is, then much register saving and restoration is avoided.

The simulation program contains small host machine language segments that simulate the target machine instructions. An example showing the implementation of the class computer's Branch on EQual instruction using the VAX MACRO assembler follows:

```
; MACRO FOR BEQ              (BRANCH ON EQUAL)
;          OPERATION: IF Z=1, THEN PC + d → PC
;          SYNTAX:       BEQ <LABEL>
;
            .MACRO        BEQ   DEST
                          BEQL  DEST
                          .ENDM BEQ
```

Another example for an arithmetic instruction (i.e., ADD using the Absolute address mode) follows:

```
;  MACRO FOR ADDA
;          OPERATION: SOURCE + DEST → DEST
;          SYNTAX:       ADDA SOURCE,DEST
;

           .MACRO      ADDA SOURCE,DEST
           PUSHR       #^M<R2,R3,R4,R5,R6>
           ADRCONV     SOURCE,DEST;CONVERT OPERAND ADDRESSES, IF NECESSARY
                       ;RETURN SOURCE IN R4, DEST IN R0
;
           ADDW        (R4),(R0)
;
           POPR        #^M<R2,R3,R4,R5,R6>
           .ENDM ADD.W
```

The macro ADRCONV is written in the host (i.e., VAX) assembler language to move the source and destination addresses from the target machine model to the VAX registers R4 and R0 where they will be used in the VAX add word instruction to actually access the data in the programming model. The VAX version of this macro is given next:

```
;          USED TO DETERMINE ACTUAL ADDRESS OF OPERANDS BY CHECKING
;          AGAINST THE ADDRESS OF UPPER AND LOWER BOUNDS OF THE
;          PROGRAM AS WELL AS THE LIMITS OF THE SIMULATED USER MEMORY
;
;                      REGISTER USAGE:
;
;                      R0: RESULT OF ADDRESS CONVERSION
;                      R1: NOT USED
;                      R2: INPUT ADDRESS
;                      R3: ADDRESS OF MEMORY
;                      R4: ACTUAL SOURCE ADDRESS
;                      R5: ADDRESS OF D0
;                      R6: END OF PROGRAM
;

           .MACRO  ADRCONV     SOURCE,DEST
           MOVAL   MEMORY,R3   ;ADDRESS OF MEMORY → R3
           MOVAL   D0,R5       ;ADDRESS OF D0 → R5
           MOVAL   FINI,R6     ;ADDRESS OF FINI → R6 (END OF PROGRAM)
;
           MOVAL   SOURCE,R2   ;ADDRESS OF SOURCE → R2
           JSB     ADDRCK      ;CALL SUBROUTINE ADDRESS_CHECK
                               ;ACTUAL ADDRESS → R0
;
           MOVL R0,R4     ;MOVE ADDRESS OF SOURCE TO R4
;
           MOVAL       DEST,R2;ADDRESS OF DEST → R2
           JSB         ADDRCK;CALL SUBROUTINE ADDRESS_CHECK
                       ;ADDRESS OF DEST → R0
           .ENDM       ADRCONV
```

The ADDRESS_CHECK procedure translates between relocatable addresses in the host system and absolute addresses in the target system.

```
;          PROCEDURE ADDRESS_CHECK (A: INTEGER)
;
;          CHECKS TO SEE IF THE ADDRESS IS IN PROGRAM SPACE OR
;          SIMULATED USER MEMORY. CHANGES USER MEMORY ADDRESS
;          TO ABSOLUTE ADDRESS BY ADDING MEMORY STARTING ADDRESS.
```

```
;
;              ASSUME ADDRESS TO BE CHECKED IS ALREADY IN R2
;--------------------------------------------------------------
;
ADDRCK:     CMPL      R2,R5      ;CHECK FOR ADDRESS < D0
            BGEQ      UPPER      ;R0 >= D0, CHECK UPPER BOUND
            JMP       ADJUST
UPPER: CMPL R2,R6                ;CHECK FOR ADDRESS < PROGRAM END
            BLSS      CHECKED    ;ADDRESS IS VALID PART OF PROGRAMMER'S
                                 ;MODEL OR IMMEDIATE DATA
                                 ;RETURN TO CALLING MODULE
;
ADJUST:     CMPL      R2,#^XFFFF;TEST UPPER RANGE OF USER MEMORY
            BGTR      ERR1       ;RANGE ERROR IF R2 > FFFFH
;
            TSTL      R2         ;CHECK FOR ADDRESS = 0000H
            BLSS      ERR1       ;RANGE ERROR IF R2 < 0
            ADDL3     R3,R2,R0;ADJUSTED ADDRESS → R0
            JMP       RETCHK
;
CHECKED:    MOVL      R2,R0      ;NO CHANGE- ADDRESS IN R0
;
RETCHK:     RSB                  ;RETURN WITH ADDRESS IN R0
;
ERR1:       PRINTCHRSMSG1
            DUMPLONG D0,D1,D2,D3,D4,D5,D6,D7
            DUMPLONG A0,A1,A2,A3,A4,A5,A6,A7
            DUMPLONG PRC,CCR
;
            JMP       RESET      ;SOFTWARE RESET
```

The target system application program being tested is located in the source program module after the macro definitions and the initialization code. A layout of the source code is shown on the next page.

This creates the effect that the target system memory area appears to contain a pro-

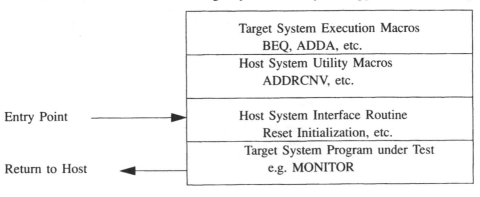

gram at the time the reset operation terminates. The bulk of the source code is written in the target computer's assembler language. The host computer's assembler processes the source code creating an object file that contains instructions for the host machine. After linking the object file with the run time library, the executable module is run on the host machine. Once the macro library is written, it can be used with many application programs intended for the target machine. These programs should run without modification when the target machine is actually built.

As usual, one must remember the goal set for this level of simulation. That is, how effective is the instruction set in writing solutions to the problems addressed by the target instruction set processor? One can also get a certain amount of experience with the algorithms that the target instructions are intended to use. The level of experience is limited since much of the underlying architecture that would make the instructions more efficient does not exist at this level of simulation.

6.5. SIMULATION OF THE FIRST LEVEL OF CONTROL

Simulation at the first level has three important goals. The first is to evaluate the actual algorithms that will be implemented on the target system's architecture. The HDL was introduced to provide a means to study the algorithm/architecture interaction. The second goal involves studying the components of the architecture and their control as they relate to the first goal. The third goal is to try to determine the actual circuit parameters such as timing and output loading that are encountered when the system must actually be constructed. In this section, we will address the first two goals in some detail. The third goal will be discussed only to the level of the principles of determining critical path timing by "hand" methods. The student is referred elsewhere for information concerning the machine simulation of the electrical behavior of circuits.

6.5.1. The Hardware Description Language as a Simulator

The HDL was introduced to serve as a method to describe architectures and the mapping of algorithms on them. The concept of an HDL has been present for some time as discussed in Shiva (refer to Bibliography). As was noted earlier, the rewards for producing a concise, thorough description of an algorithm and its hardware mapping are very high. Ultimately, the final design will be realizable in the most efficient version of the hardware. The subject of producing a simulator using the HDL description of an algorithm mapping was also implied. It is the purpose of this section to elaborate on the concept. We will restrict our view to possibilities and not write any specific example in this book.

As will be studied to much greater depth in the following sections, a hardware behavioral simulator consists of blocks of code (i.e., functions) that convert data and control inputs into data outputs that are identical to those of the hardware element being simulated. In later sections, the functions will be produced in an appropriate language "by hand". Once the functions are written, they will be "strung together" to create complete machines again "by hand". The subject of using an HDL and some sort of "compiler" will be treated now before the functions are produced in the later sections of this chapter.

Imagine that a particular version of HDL that has been so carefully designed that the various operators (e.g., +, - define, and connect) form a complete, unique description of an architecture. A program could then be written that would convert the symbolic description of this architecture/algorithm mapping into the appropriate functions that simulate the behavior of the actual algorithm running on the real hardware; i.e., the "HDL compiler" would produce a simulator. Without entering the detail that would require a book in compiler writing, it is our purpose here to show the elements and implications of such a compiler.

The HDL described is only one of many possible expressions of the language we need here. We will use it in our discussion with the realization that some terms may need to be added or modified to allow us to actually describe, without confusion, an architecture and its associated algorithm. In the HDL we described, there were two types of operators, some described the hardware and its interconnections (e.g., "define" and "connect") and others described operations to be performed on data passing through one of the elements (e.g., an ALU or multiplexer). Using the first type of operator we could write paragraphs that were symbolic descriptions of complex hardware structures, all ultimately based on the fundamental architectural elements introduced in Chapter 3, i.e., the memory, combinatorial (AND, OR, and XOR), and data path elements. We allowed our language to have enough latitude to describe more complex elements such as ALUs and MUXes at a higher level, thus making them fundamental elements, to save writing and better express our ideas. Our hardware elements appeared to us in this way so our language will contain "nouns" of that complexity. Define and connect are used to create "circuits" using the elements fundamental to our language. The HDL is given enough flexibility to describe arbitrarily wide data paths and their interconnection with parts of similar widths.

The second set of operators, including +, -, and MUX[], allow us to show how our algorithm would actually execute on the described hardware. This group of operators is also further divisible into constructs that operate on problem specific data (i.e., :=, +, -, etc.) and those that control the flow of the algorithm "program" (i.e., IF...THEN...ELSE, etc). While the controller could be separately described as an architecture that responds to control and data signals the same way as the architecture, it makes sense to speed up execution of the simulator to present these as simple control constructs and use them to directly control the flow of the simulation program. There is nothing in the language or "compiler" that prevents the expression of the controller in the same manner as the architecture however.

As shown early in the chapter, we are able to write "programs" using the HDL for our particular application. These programs have two major sections, the first contains paragraphs that describe the hardware to the compiler and the second contains paragraphs that describe the flow of control and data modifications as the algorithm performs its task on the hardware. A compiler for any language has the task of analyzing statements written in its vocabulary and producing machine executable code for the host expected to actually carry out the computations requested in the "high-level" source program. A "high-level" language, i.e., requiring a compiler not an assembler, uses human language phrases or sentences to express the thoughts of the programmer instead of the single-symbol/single-machine instruction correspondence in "low-level" languages. Our HDL is a "high-level" language.

To do its job, a compiler must be able to locate and identify the components of a sentence or statement. It must then, by using a set of rules called a grammar, make "sense" out of the statement; i.e., produce an equivalent set of machine instructions that will modify the contents of various variables in the way requested by the programmer. There are three steps then to the overall operation of a compiler. These steps are, in order of operation, analysis of a sentence (Parsing), make sense of the sentence from its pieces (Grammar), and generation of an equivalent set of host machine instructions.

The description of the language (e.g., our HDL described in Section 6.3) allows us to define the operators and the structure of our symbols (e.g., REGISTER). Also implied in the same section were the rules of use (i.e., grammar) that were acceptable. Thus, the statement

```
RA := RE & ~RE;
```

is a statement that ANDs the contents of register RE with the one's complement of itself and places the result in register RA at the end of the current clock cycle. We have referred to the fact that we defined registers RA and RE so that we would know that RA and RE are actually registers. The register assignment operator, ":=", if used correctly, has also informed us of this. The line termination character, ";", also tells us that there is no more information on the next line that affects this particular register transfer.

With a little work we could write down a set of rules that completely describes all of the allowed operations in our language. From these rules we can prepare a syntactical analyzer (i.e., a Parser) and a lexical analyzer based on our rules or grammar. Normally, the lexical analyzer will create a table that contains tokens for the operations and the symbols that undergo them. The order of the table is determined by the actual meaning (i.e., lexical analysis) of all of the sentences in the program. It is this table that is used by the code generator to produce an executable program to be run on the host machine.

Before we look at the code generator, let us see what else can be learned about our algorithm/hardware mapping expressed in HDL. Take the register transfer statement involving RA and RE above. We should have created a definition and connection section in the first part of our HDL program that defined RA and RE to be registers of a particular width. Further, we should have "connected" them together by a data path of the proper width and wiring relationship, i.e., a path from RE[0] through a multibit AND network to RA[0]. What if the compiler encountered the register assignment statement given above without having seen the appropriate declaration statement? We would expect our compiler to generate a warning appropriate to the problem. If RA was not declared as a register, or the declaration omitted, then should the compiler accept the register assignment operator, ":="? The FORTRAN compiler is expected to accept such situations and prepare data type translations so that the programmer can concentrate on the actual algorithm he is writing. Here, we not only want to understand how our algorithm works but whether we have provided the proper hardware and data paths for it to be performed in the way we specify. So a major purpose of the HDL compiler is to catch errors of this type and inform us of them.

The code generation section of the HDL compiler is responsible for assembling modules of machine code that represent the functional elements in our architecture and provide a mechanism to move data through them; i.e., run our algorithm on our specified architecture. In following sections you will be shown how to write modules in an appropriate language that describe the various fundamental elements ranging from the basic gates through ALUs and MUXes and including registers of various descriptions. It is the linking of these modules that we are concerned with here. The HDL compiler can simply produce a "main" program in FORTRAN, C, Assembler, etc. that calls the functions we will write below in the order dictated by the requirement that inputs to modules must be prepared from the outputs of other modules before the module is called. Since we are interested in microprogrammed state machines, the actual flow of control can be created by translating the second part of our HDL program as microcode that is placed in the microprogram memory of a suitable controller function. The input to the code generator is the list of tokens and variables produced by the syntactical analyzer step.

In concluding this brief introduction to HDL-generated simulators, we would like to note an alternative use for the HDL program. What if the description of our hardware/algorithm mapping expressed in HDL could be used to generate the interconnections in a gate array structure or by a silicon foundry to produce a custom IC that embodies our design? Notice that we have described our architecture in the first section of our HDL program. In our examples we have used a rather "high-level" description which was appropriate to our vocabulary and needs, i.e., the creation of a simulator. There is nothing that prevents this description from being thorough enough to serve as input to the various programs that are used to create custom ICs. Just as the HDL compiler above was used to create a microprogram embodying our algorithm, the code generation section can be used to emit statements in the appropriate language for one of the "silicon compilers" currently used to produce custom units. Students that are familiar with the various Programmable Logic Device languages such as PALASM from AMD can see conceptually how such statements can be produced.

What would this capability bring us? The fact that a single document exists (i.e., the HDL program containing our expression of the algorithm/architecture mapping that can be compiled using the HDL compiler for purposes of simulation and testing) gives us a much higher assurance that our design is correct and efficient. There are no "hand-assisted" stages that lie between our document and the finished product that can introduce random errors. The HDL and silicon compilers can both be "debugged" and verified using structured program development techniques, thus assuring their reliability independently of our design. The creation and testing of complex structures is then moved directly to the designer, where it belongs, without introducing random or untestable events in the path between design and "silicon".

6.5.2. Behavioral Simulation of Architectural Components

The term "behavioral simulation" used in the title of this section infers that we expect to write program segments, in a high- or low-level language on a host computer, that act like physical devices. We can satisfy the first two goals of simulation at the first level without including the electrical behavior of the parts. This simplifies the modules such that any designer can write useful simulations with little effort. The information gained from this type of simulation is great. We expect to determine the number and type of combinatorial and registered parts needed as well as the bus structure in our state machine design. The microprogram that embodies our algorithm can be tested in the simulation. This should reveal any deficiencies in the architecture or microprogram at a time when changes can be easily made.

A program that simulates a device consists of two major parts, the simulation module itself and a test driver. The test driver is the main program that applies values to the control and data inputs of the simulation module and reports the results produced. In reality, the test driver is actually a simulation of a simple controller. The simulation module contains the programming necessary to create the data transformation required by its control and data inputs. It should also contain a means for reporting the contents of any variables it has to the outside world; i.e., it cooperates with the test driver in reporting test results. A simple simulation program has the following organization:

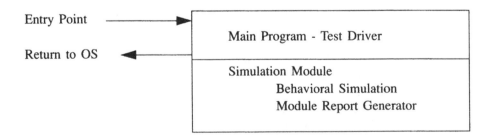

When we try to write programs containing several simulation modules we will find it convenient to add control inputs to each simulation module to control the report generator just as we control the behavior simulation.

The main program or test driver indexes through a large array of control information. It is just a simulation of the ROM-Latch controller discussed in Chapter 3. The control arrays may be implemented separately, one for each control line in the architecture. Each array should be of integer or unsigned data type and have enough elements to contain all of the test values used in the course of a "test run". The index to each array serves the part of the microprogram counter. For the purposes of testing a simulation module, it need only increment through the control arrays managed by a large "DO" loop. Once the simulation modules are drawn together into a system, the operation of the control section can be replaced by a simulation of the desired controller.

6.5.2.a. Simulation of Registers and Memory

Consider the register formed by an array of D flip-flops shown in Figure 3-9. We would like to write a program segment that "behaves" like the multibit register shown. We wish to describe the "storage" property of the register in our programming language. The entity in our programming language that also has the storage property is called a variable. Variables may be "typed" in most languages such that data having only a particular format can be stored in a variable having a particular type; i.e., integer data typed variables contain data having the two's complement integer format as described in Chapter 2. Our primary use of data typing is to inform the compiler about the type of arithmetic that it should implement on the data contained in the variable.

The register in Figure 3-9 has one clock input, four data inputs, and four data outputs. To simulate the register in Figure 3-9, all we need to do is to inform the compiler of the size and data type appropriate for our problem. There are two possible ways to simulate the register. In the first, we allocate a single "word" having integer data type for our register as follows:

FORTRAN	C
INTEGER*2 REG	int reg;

The integer data type is appropriate for the present since we will combine this register with an integer ALU later. If one wished to perform floating point operations using data from this register, then it would be simpler to use the corresponding floating point data type.

The storage space set aside in either language above was 16 bits. We can "align" our register to any point within the 16-bit word. The actual location can be determined after we place other elements (e.g., shifters) in our architecture.

The second method to describe a register is to allocate a single "word" to each bit in the register. Thus, an array in the programming language would correspond to a hardware register as follows:

FORTRAN	C
LOGICAL REG(4)	unsigned reg[4];

The unsigned integer data type available in C and the LOGICAL data type in FORTRAN are appropriate since the only values possible for each element of the array is a 0 or 1. This version has the advantage that it gives complete control over bit level operations which will be appreciated when we discuss combinatorial elements next. Its main disadvantage is its slow speed of execution which means that a large simulator may not be very useful. This is a good beginning method. The first method can be used as each block is optimized.

The clock input loads this register anytime it changes from Low voltage to High voltage. As shown in Chapter 4, all of the clock inputs on the registers in an architecture are tied together and driven by the system clock, SYSCLK. If the register shown in Figure 3-9 is used in an architecture, it will be updated on every clock cycle. The clock input therefore can be handled in one of two ways. The first way involves declaring an integer or logical variable that represents the clock signal. Cycling this signal corresponds to the clock behavior. This method leads to unneeded complication. The second way consists of implicitly representing the clock signal by letting the simulation program invoke the register module each time a clock cycle takes place. This method will be adopted here and will be explained more fully when the simulation program for a complete system is discussed below.

To simulate a register composed of an array of Enabled D flip-flops shown in Figure 3-10 requires another control signal, i.e., LOAD. Control signals can be declared to the compiler the same as the register itself. Either integer or logical variables are appropriate since the LOAD signal can take on only two values, either 0 or 1. If the clock signal is represented implicitly, as mentioned above, the logical operation performed by the Enabled Register simulation module can be shown as follows:

```
IF (LOAD = TRUE) THEN Contents of Register = Contents of Data Bus
```

Notice that the contents of the register will not change if LOAD is *False*, thus completing the behavioral simulation.

6.5.2.b. Simulation of Buses

Buses are simulated using programming language variables just as registers are. The problem here is that a bus or combinatorial network does not store information, so we will have to be careful in our interpretation of the time behavior of the variable. It is the accurate simulation of elements that do not have the storage property that causes the extreme complication of simulators that support the correct time behavior. We can satisfy the first two simulation goals by making two constraints on interpretation. The first is that the variable we use to represent an element without the capacity of storage accurately represents that element only for a very short time. The data placed on a bus may "decay" if the bus is not continually driven. The second is that we will place code in the simulation to detect whether an output actually drives a bus during the clock cycle.

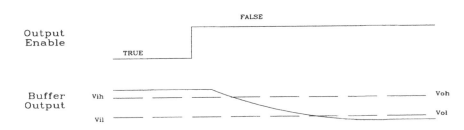

Figure 6-2: Tristate buffer output voltage as a function of time.

To make the time behavior of the data on a bus clearer, let us examine the behavior under a variety of conditions. Figure 6-2 shows an isolated wire, one signal in a bus, driven by a tristate buffer similar to that on the output of the register in Figure 3-11. The input to the tristate buffer is set to cause the wire to be at a "High" voltage when its output is enabled. The active low output enable signal for the register is shown as well. Observe the behavior of the voltage on the bus after the tristate buffer output enable signal goes false. Since the wire is isolated (i.e., not connected to any other inputs or outputs), the voltage will decrease to zero with a time constant that depends on the capacitance and the residual resistance of the wire to its surroundings.

Now consider a similar situation shown in Figure 6-3 where the wire is connected through a resistor to a "High" voltage; e.g., Vcc. Very little change would be seen if the buffer were driving the wire to a "High" voltage before the output was disabled. However, if the buffer were driving the wire "Low", the voltage would increase to Vcc after the output was disabled with a time constant depending on similar terms as above.

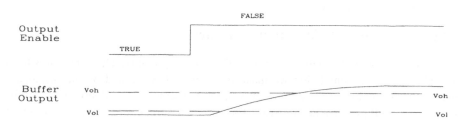

Figure 6-3: Tristate buffer output voltage as a function of time.

The input to a digital device discriminates between two ranges of voltages: those that it interprets as "High" and those that are "Low". In the first case, the wire will appear "High" to a device input for some time after the buffer is disabled, while the reverse occurs in the second case. Information seems to be "stored" on the wire for a short period of time. This storage method is actually exploited in certain devices, e.g., some microprocessors and all "dynamic" RAM. In our application we will interpret our simulation to represent events for a short period of time, i.e., one clock cycle. This means that if a bus is not driven during a clock cycle, the information that was on the bus in the previous cycle will still be present.

We can add a program segment in our simulator, not in the bus module, that detects whether the bus has been driven in the current clock cycle as mentioned above. The purpose of the segment is to generate a message that would alert us to the fact that the bus was not driven so that we can check our microprogram for the problem. Notice how this error checking technique is so much easier to apply in the simulator than in the actual hardware.

The behavior of the tristate buffer can be simulated with a line of code that looks very similar to that used for the Enabled D Register:

```
IF (OUTPUT ENABLE = TRUE) THEN Bus Contents = Register Contents
```

We do not update the bus contents but leave it as it was in the previous cycle if the output is not enabled. One is tempted to use an "ELSE" condition and place a non-data value like "HI-Z" in the variable representing the bus. However, we expect to place many registers, each with a tristate buffer on its output, such that they drive a single bus. The simulation program can not evaluate all of the register simulation modules simultaneously but must handle them one at a time. If the last module placed "HI-Z" on the bus it would destroy the data that an earlier module, by being enabled, had placed on the bus. This would not be an accurate simulation. Therefore, the simulation suggested above would be the correct method as long as we observe the limitations expressed in an earlier paragraph.

If we combine an Enabled D Register with a tristate buffer in such a way that the inputs and outputs of the Register/buffer were connected to the same bus, then the complete Register/buffer/bus module would appear as follows:

```
IF (LOAD = TRUE) THEN Register Contents = Bus Contents
IF (OUTPUT ENABLE = TRUE) THEN Bus Contents = Register Contents
```

While this simulation segment does not accomplish anything unless other registers are connected to the same bus, it does call attention to the problem of timing. What is the correct order of the statements in a simulation? This point will be covered in sufficient detail when we discuss systems of simulation modules below. At this point we can comment that the order shown here is backward.

6.5.2.c. Simulation of Combinational Elements

Combinational elements do not have the storage property. They share this feature with buses. Therefore, any simulation we make of them will be limited by the discussion on the subject given in the bus section above. Two important combinatorial structures will be shown here, i.e., the multiplexer and the ALU. The first is a simple structure, but the second actually is composed of several modules, one for each of the operations it can perform.

6.5.2.c.i. The Multiplexer

The MUX as described in Chapter 3 is a device having a control input that steers data from one of its data inputs to its output. We will first simulate a 1-bit-wide four-input multiplexer. To do this we need to declare four variables to hold the four inputs, one variable to hold the output, and one to hold the control signal. Since there are four inputs the control signal can take on four values; i.e., the least significant 2 bits contain the control value. One could declare two variables, one for each bit of the control signal, but this adds more complexity than is necessary at this point. The simulation appears as follows:

```
SWITCH (Control)
            CASE 0: Output = Input 0
            CASE 1: Output = Input 1
            CASE 2: Output = Input 2
            CASE 3: Output = Input 3
```

A "case" construct similar to that supported in most modern high-level languages is used here to avoid repeated IF_THEN_ELSE statements. The "Control" input can have one of four values. These are used by the programming construct as operands of the CASE commands to select the proper assignment statement for the output of the MUX.

The multiplexer could also be simulated by placing each of the inputs into a four-element array. The output is selected by using the control value as an index as follows:

```
Output = Input(Control)
```

This involves a two step procedure: loading the array "Input" and then assigning the correct element to the output. The case method appears to execute faster in many languages and will be the one preferred here.

Multiplexers are frequently used to steer words or "arrays" of bits. Data on several buses are steered by multibit MUXes onto a single output bus. Multibit MUXes can be built by using arrays of the 1-bit-"wide" multiplexers in hardware or as simulations. Another method of simulating multibit MUXes is based on the first method discussed in simulation of registers, i.e., allocate a variable to contain all of the bits in a bus. The switch or array method can then be implemented. This results in the faster simulation time for multibit structures.

6.5.2.c.ii. The Shifter

The shifter is potentially the most difficult element to simulate in a high-level language. The most direct method is based on declaring an unsigned integer variable, e.g., SHIFTER. In some high-level languages like C, the shift operation is defined allowing the operation to be expressed as follows:

```
shifter>>  - shift the location "shifter" right one bit
shifter<<  - shift the location "shifter" left one bit
```

Shifting involves more than just moving the contents relative to the fixed bit positions of a data path. There are both bits to be shifted into the data path and out of it. In principle, if one declares the variable for the "shifter" to be wider than the data path, then the "shift out" bit will be found 1 bit position to the "left" of the bits used in the data path simulation as a result of a shift-left operation. Using the 8-bit data path example simulated in a 16-bit unsigned variable from above, we find the shift out bit in the bit 8 position. Application of an AND with a binary mask of 0000 0001 0000 0000 allows us to isolate the shift out bit for use elsewhere in our simulation. The spaces were placed in the mask for readability. In languages that do not support bit manipulation, one can recover this bit after a lot of work using subtraction.

Recovering the "shift out" bit in the right-shift case is accomplished in a similar manner. However, the alignment of the data path within the variable used in the previous paragraph needs to be adjusted. To support recovery of the "shift out" bit in this case, it is more convenient to place the data path 1 bit position to the left of that used previously, as shown in an 8-bit example next.

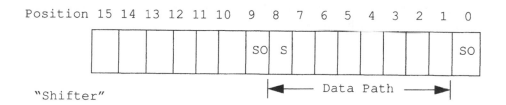

In this case, the "shift out" bit caused by shifting the variable to the right is found in bit position 0. It is found in bit position 9 after a left shift.

Shifting a bit into the data path is relatively simple if the language supports bit operations. In the case of a left shift, the number is shifted first as shown above. The bit to be "shifted in" resides in a variable declared similar to "shifter" having a value limited to 0 or 1. The "shift in" variable is then logical ORed with the shifter. The right-shift case is only slightly more complex. The bit to be "shifted in" is placed at the most significant bit position for the data path, e.g., in the bit 7 position above in the "shift in" variable. In our example, the numerical value of the "shift in" variable is either 0 or 128 (0080 hex) depending on whether the "shift in" bit is 0 or 1, respectively. After the "shifter" variable has been right shifted, the "shift in" variable is logical ORed as before.

To use languages that do not support bit operations, the writer has several possible solutions. For example, many modern versions of FORTRAN do not allow logical operators such as .AND. to be applied to the INTEGER data type in an assignment statement. Also, no shift operation is defined. Shifting can be accomplished by multiplication or division by a power of 2 if one carefully prevents the "shift out" bit from entering the variable's sign bit position (e.g., bit 15 above). The variable needs to be manipulated so that it always contains a positive number when seen from the host computer's viewpoint. Extraction of the "shift out" bits can be accomplished by subtracting suitable quantities. The runtime library for many FORTRAN compilers contains bit manipulation subroutines including AND, OR, and Shift among others. These are written in Assembler language for the host system, but do not require the writer to understand that language. They are heartily recommended over shifting using arithmetic methods under these circumstances.

Another method to circumvent the language restrictions in this application involves treating each bit in the data path separately, as was discussed for registers. Declaring the shifter as an integer array having as many elements as bits in the data path allows one to simply write a loop that passes the bits around among elements.

The "shift in" and "shift out" variables extend the width of the data path. The order of executing the assignment statements that move the "bits" is important. Each assignment statement is of the form:

```
SHIFTER(I) = SHIFTER(I-1) for a left shift
SHIFTER(I-1) = SHIFTER(I) for a right shift
```

A loop that shifts all bits in the SHIFTER array proceeds in a direction dictated by the direction. Shifting left begins at the most significant bit position where an assignment statement of the following form is executed:

```
SHIFT_OUT = SHIFTER(N)
```

where bit position N is the left-most bit in the data path. A loop manages the remaining bit positions proceeding toward the right-most end of the data path. The last bit position is updated from the "shift in" variable as follows:

```
SHIFTER(0) = SHIFT_IN.
```

To implement the right shift requires the order to be reversed.

In summary, shifting can be implemented using the host computer shift operations as provided by the compiler or runtime library. Data on the data path can be shifted by implementing the data path as an array of bits. Arbitrary shift amounts appropriate to the funnel and barrel shifters can be performed in either method whether the underlying host system directly supports multibit position shifts.

6.5.2.c.iii. The Arithmetic/Logic Unit (ALU)

The arithmetic/logic unit is described as a complex combinatorial element in Chapter 3. It normally supports the logical functions AND, OR, and Exclusive OR as well as the arithmetic functions of Addition and Subtraction of unsigned and two's complement signed binary numbers. The width of the operands, thus the data paths, through the ALU are under control of the ALU's designer. Integrated circuit fabrication technology limited the width of the early ALUs to 4-bit data paths while the current width is 32 bits. Even in the era of the 4-bit ALU "bit slice", most problems required wider data paths for their solution. This was accomplished by placing several identical ALUs "side by side". These points affect the choice of methods of simulation.

The ALU can be thought of as a device containing a group of sections, one for each of the operations it must perform. We will follow this idea in our simulation. The module is organized using the following control structure:

```
SWITCH (ALU Instruction)
        CASE ADD: Output = A plus B plus Cin
        CASE SUBTRACT: Output = A minus B minus Borrow
        CASE AND: Output = A AND B
        CASE OR: Output = A OR B
        CASE XOR: Output = A XOR B
        etc.
```

Within each section is all of the code needed to perform the operation on the operands. Also, that needed to manage the flags, Cout, OVR, SIGN and ZERO as shown in Figure 3-6 is included. If the ALU simulation is to create the bit slice structure, then it must implement the Carry Generate and Propagate terms as well. Once the ALU module is implemented, there are a number of optimizations possible especially in the area of evaluating the flags. For example, the ZERO flag calculation involves examining the bits of the ALU Function output to determine if they are all zero. This is shared in common with all sections (ALU operations); therefore, it can be moved outside of the individual sections and placed near the end of the ALU module.

A diagram of this organization along with the module reporting section mentioned above follows:

```
┌─────────────────────────────────────────────────────────────────┐
│  ALU - Behavioral Simulation Module                              │
├─────────────────────────────────────────────────────────────────┤
│  ADD Section                                                     │
├─────────────────────────────────────────────────────────────────┤
│  SUB Section                                                     │
├─────────────────────────────────────────────────────────────────┤
│  AND, etc.                                                       │
├─────────────────────────────────────────────────────────────────┤
│  ALU - Flag Setting Section                                      │
├─────────────────────────────────────────────────────────────────┤
│  ALU - Reporting Section                                         │
└─────────────────────────────────────────────────────────────────┘
```

The same two methods of simulation as described above are available here. The first method treats the entire data path, whether implemented using groups of narrower ALUs or one single-wide ALU, as an array of 1-bit elements. The second method involves the use of a single variable having a width greater than or equal to that of the data path being simulated. The second choice results in a faster executing simulation, while the first is easier to write in most high-level languages. We will let our ALU simulation cover the entire width of the data path in either case. In a later section we will revisit this question and explore the circumstances when the simulation module should have a width narrower than the data path similar to the "bit-slice" ALUs.

We will discuss the bit-by-bit simulation method first. While the execution time for this method is long, it allows us to manipulate the operand bits just as the hardware does. Using Figure 3-6 as a model to define the signal names and the ALU function table below, we will describe a few of the major points in this method of simulation.

```
ALU Function Table
        F SEL    |   Function
        0        |   Bitwise One's Complement the bits in A
        1        |   Bitwise AND the bits in A with those in B
        2        |   Bitwise OR the bits in A with those in B
        3        |   Bitwise Exclusive OR the bits in A with those in B
        4        |   Two's Complement ADD A to B with Cin
        5        |   Two's Complement SUBTRACT B from A with Cin
        6        |   Two's Complement SUBTRACT A from B with Cin
```

We need to declare the variables to be used for the operands, flags, and control signals. These will normally be the integer, or unsigned integer if available, data type of as narrow a width as supported by the compiler of the language being used. The unsigned data type can be used since each variable will contain only 1 bit taking on the values 0 or 1. The following examples in FORTRAN and C will suffice:

FORTRAN	C
INTEGER A(16),B(16),F(16)	unsigned a[16],b[16],f[16];
INTEGER F_SEL	unsigned f_sel;
INTEGER COUT,SIGN,OVR,ZERO	unsigned cout,sign,ovr,zero;
INTEGER CIN	unsigned cin;

The ALU instruction, F_SEL, is used to route program flow to the section of the ALU module that performs the selected operation. If the operation performed on a bit is table driven, as is appropriate here, then F_SEL can also serve to select the proper combination table.

Let us consider the logical operations first. The truth table for each single bit operation has two inputs and one output. It contains four lines, one for each input combination. Let's declare three two-dimensional arrays, one for each operation as follows:

FORTRAN	C
INTEGER ANDAR(2,2)	unsigned andar[2][2];
INTEGER ORAR(2,2)	unsigned orar[2][2];
INTEGER XORAR(2,2)	unsigned xorar[2][2];

The first index corresponds to the A input and the second to B. The contents of each element is the value of the desired function for that combination of inputs A and B as shown below for the "AND" array:

```
ANDAR        B
          |  0     1
      0   |  0   | 0
A     1   |  0   | 1
```

One such array is used for each function in the ALU module.

Instead of declaring three different two-dimensional arrays, let us combine them into a single three-dimensional array by using F_SEL as the index into the third dimension. From the ALU function table above, we see that F_SEL ranges from 1 through 3 for this set of functions. If we extend the three-dimensional array to include the case when F_SEL is zero (i.e., taking the complement of A) then we don't care about the value of B for that case. A version of this array is shown below:

LOGICAL FUNCTION

F_SEL	0		1		2		3	
	.NOT. A		A .AND. B		A .OR. B		A .XOR. B	
B	0	1	0	1	0	1	0	1
0	1	0	0	0	0	1	0	1
A 1	1	0	0	1	1	1	1	0

The arithmetic operations require an array with more dimensions. The arithmetic inputs include A, B, and Carry In, indicating that a three-dimensional array is needed for addition alone. If we also include F_SEL as an index to choose between addition and the two versions of subtraction, then we will need a four-dimensional array. At this point, two arrays of this type could be used, one for logical operations (i.e., F_SEL from 0 through 3) and one for arithmetic operations. Notice that F_SEL must be adjusted to the range 0, 1, and 2 for use with the arithmetic array. A single large four-dimensional array is possible in which the Carry Input term is neglected in the function output for logical operations similar to the treatment of B for F_SEL = 0 above. Remember that each element of the array represents the value of the function F.

The two subtraction operations are implemented by placing the results of the following functions in the corresponding elements of the array:

```
F_SEL = 5: F = /B plus A plus Carry In
F_SEL = 6: F = /A plus B plus Carry In
```

The arithmetic portion of the operation array for Cin = 0 has the following form:

ARITHMETIC FUNCTION

	F_SEL		4		5		6	
		B	A+B+Cin		/B+A+Cin		/A+B+Cin	
	A		0	1	0	1	0	1
Cin	0	0	0	1	1	0	1	0
0	0	1	1	0	0	1	0	1
	1	0	1	0	etc.		etc.	
	0	1	0	1	etc.		etc.	

A second array is needed to generate the Carry Out from the arithmetic operations. The ALU in most modern RALUs generates Carry Out equal zero for the logical operations as well as the correct carry out in arithmetic operations.

If the ALU being simulated represents a "bit slice", then the carry generate and propagate terms, G and P, must be created as well. Similar four-dimensional arrays are appropriate.

Once the operation arrays are created, the programming for 1 bit of the ALU is very simple. It is shown in FORTRAN as follows:

```
F = FunctionOpArray(F_SEL,Cin,A,B)
Cout = CoutOpArray(F_SEL,Cin,A,B)
```

Now, assuming the number of bits in the A, B, and F data paths is N, we can produce the complete output function array, F, by performing a loop that is executed N times beginning at the least significant bit location in A, B, and F. The Carry Input to the first iteration is the Cin bit to the ALU. The Carry Output from each iteration serves as the Carry Input to the next iteration. The last Carry Out is placed in the Cout variable for the ALU. These terms will be automatically ignored for logical operations because of the way the operation array was arranged.

The last step to be performed involves generating the flags for the entire ALU. The Cout flag was shown above. The sign flag is produced by copying the most significant bit in the F array [i.e., F(N)] to the flag. The zero flag requires examining all of the elements of the F array. The flag is set to 1 if all are 0. The two's complement overflow flag is easily created in this method. It is calculated just as it is in the hardware, i.e.,

```
OVR = Carry Out(N) XOR Carry In(N).
```

The second simulation method for the ALU involves allocating a single host computer "word" for all of the bits in an operand. As an example, consider the case when the data path through the ALU is 8 bits. We can allocate the memory as follows:

FORTRAN	C
INTEGER*2 A,B,F	int a,b,f;

The 8-bit data path is "right justified" within the variable. The control and flag allocations remain as before.

This method uses the arithmetic/logic instructions of the host computer to simulate the operations requested in the ALU module. Example lines are given below for C:

```
f = a & b; logical AND
f = a + b + cin; addition
```

This method has the advantage of greater execution speed over the first. The difficulty comes in two areas, forcing the type of the operation and recovering the flags. In a high-level language, the data type is used to tell the compiler the type of arithmetic operation to perform. While addition of two signed numbers results in the same functional output as adding two unsigned numbers, the overflow and carry flags have different meaning. Simulation using the C language allows one to specify the "unsigned" integer data type. The Carry and Overflow flags must be recovered by indirect methods. The modern versions of FORTRAN make it much more difficult to implement an assignment statement using the logical operators, as follows:

```
F = A .AND. B
```

The compiler is really expecting this operation to be part of an "IF" statement and many versions will flag it as an error.

The second difficulty, recovery of flag information, can strongly affect the choice of language used. In assembler language, the host computer generates the four flags that our ALU produces. A suitable conditional branching tree allows us to determine the values of the four flags directly. In a high-level language, we may need many lines of code to produce the flags. The format of the function output variable, F, for our 8-bit data path example follows.

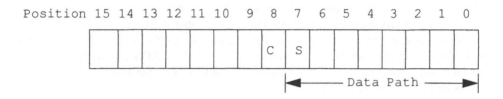

In this example, we can use the compiler's AND operation to examine bit 7 in the variable F to recover the sign flag. We can mask out the upper 8 bits and compare the remainder of the variable to zero to recover the zero flag. Since the data path occupies the right half of the variable, the carry output of an 8-bit operation will be found in bit 8 of the variable. Another AND operation will recover it. The overflow flag is more difficult. Given the organization of our data path within the function output variable F, we will have to examine the sign of our result compared with the sign of the input operands in bit 7 and produce an overflow flag compatible with the expected results. The compiler isolates us from computing this directly as in our first method or testing it as in the assembler language approach. We are reduced to using the tests that were shown in Chapter 2 to demonstrate two's complement overflow.

In summary, two methods for simulating the ALU have been discussed. The bit-by-bit method gives the writer the greatest control over the simulation but results in the slowest execution time. The word level simulation has fast execution time but is more difficult to implement in high-level languages. The behavioral module in either method should be accompanied by a reporting section under control of an additional input that allows varying amounts of information to be reported at execution time.

6.5.2.d. Simulation of An RALU

The representation of the behavior of a Register Arithmetic/Logic Unit (RALU) by a simulation module is our first chance to combine all of the elements that we have introduced. Many of the questions associated with creating systems will be dealt with here. We will use the RALU shown in Figure 3-17 as a model for our example. This RALU contains a register array, multiplexer, ALU, shifter, tristate buffer, and most importantly, buses. Each element except the register array has been discussed extensively above.

The register array is built in two logical sections, one responsible for creating the output buses DA and DB under control of addresses REG A ADDR and REG B ADDR. The other section is responsible for writing data into the register selected by REG B ADDRess. Either model for a register can be used, either bit-by-bit or one variable per register. Data declarations appear in C as follows:

Bit-by-bit   One variable per register

```
int reg_array[M][N];                    int reg_array[M];
```

Each output bus is created separately using whichever multiplexer model that is easily produced in the programming language of choice. Since the register array acts like a memory, it would appear to be most easily simulated using the array version of the multiplexer as follows:

```
da[j] = reg_array[i][j]; or da = reg_array[i];
db[j] = reg_array[l][j]; or db = reg_array[l];
```

The form depends on which model was used for a single register. The variables that represent the DA and DB buses are the same data type as that used for the registers within the array. Notice that the index "i" refers to the register within the array and assumes the value set by REG A ADDR while "l" is similarly determined by REG B ADDR. The index "j" in the bit-by-bit model must be cycled through from 0 to N-1, the number of bits in the data path, in order to complete the model.

Writing to the register array uses the REG B ADDR to point at the register to receive data as discussed in Chapter 3. In some RALUs, notably the Texas Instruments, Inc. (TI) series including the 74ACT8832, a third address is available for this purpose allowing completely independent three-operand operations. The register array shown in Figure 3-12 uses a write enable signal, WE, to allow register modification to occur on the next rising edge of the system clock. This is similar to the action of the Enabled D Register that we discussed earlier. The simulation code for 1 bit of the bit-by-bit version takes the following form:

```
if (we == TRUE)
{
    reg_array[l][j] = y_bus[j]; or reg_array[l] = y_bus;
}
```

The variable "y_bus" applies to the bidirectional data path that connects the input/output structure at the bottom of Figure 3-17 to the DBin input of the register array. As above, the variable "l" is controlled by REG B ADDR.

Building the RALU simulation module consists of arranging the simulation modules for the register array, input multiplexer, ALU, shifter, and tristate buffer in the order in which data produced by one is available to the next one below it in the direction of data flow through Figure 3-17. This means that the portion of the register array module that creates the DA and DB buses must come first. The section of this module that stores data in the register array must come after all other modules. The arrangement of modules is shown below:

Register Array: DAout, DBout
Input MUX: A
ALU: F
Shifter: Shifter Bus
Tristate Buffer: YBUS
Register Array: DBin

Notice that the tristate buffer module forms a two-input multiplexer between the shifter output bus and the bidirectional data path out of the RALU. Further, notice that while combinatorial elements can be placed in the order of data flow through the device, registers are more complex. Their output sections need to be executed first to place data on buses in order to "get things started". Their input sections are placed last to "finish up" the system clock cycle. Remember, this is the job that registers do for us. They break up long propagation paths to make our design jobs easier. We intentionally use them as starting and ending points for combinatorial events.

6.5.2.e. Simulation of Controllers

As was discussed in Chapter 4, controllers are a special interconnection of registers, combinatorial elements, and data paths that produce sequences of architectural control signals. The controller depicted in Figure 4-14 provides the microprogrammer with all of the common program control constructs plus some that are not so common. Writing a simulation program that behaves like a controller is accomplished just as was shown for the RALU above. One combines the simulations of the controller's components with interconnections (i.e., data paths) that allow microprogram addresses to pass through the network.

We will use the complete microprogram controller in Figure 4-14 as a model in our discussion. Variables to represent the registers including the microprogram counter register (μPC), the pipeline register, the LIFO stack, and the microprogram memory may be declared using the "word" format shown above since we need not manipulate individual bits in the controller. If the compiler supplies the unsigned data type, then it should be used since the controller's "data" are actually microprogram memory addresses (i.e., unsigned numbers).

The pipeline register and the various "data" paths contain fields of varying width that frequently do not fit any particular word size arrangement since they correspond to control signals for the architecture and the controller itself. If the language does not support arbitrary "word" widths, each field in the pipeline register should be assigned a single word. Several fields may be "packed" into a single computer word for compactness, but the repeated unpacking of the word in each microcycle will make the simulator run slower. The machine code presented by the microprogram assembler will most likely be in "packed" form. If space is available in the simulator, then, for execution speed's sake, the machine code may be unpacked once upon loading into the simulator's microprogram memory. The C language, among others, supports programming constructs called structures that allow the programmer to declare the number of bits in the various fields. The unpacking does not have to performed explicitly by the programmer, but by code embedded by the C compiler. Its execution speed lies somewhere between that of the programmer-created packed form and the totally unpacked form. An example of a structure definition is given below for some of the data paths and registers in the controller of the computer example found in Chapter 8:

```
/*
This structure defines the pin out
of the Am2910 Next Address Generator (Figure 4-16)
to the C compiler.
*/

struct f_io_2910 {
unsigned y: 12; /* output: microprogram address bits */
unsigned pl_x: 1; /* output: pipeline address enable */
unsigned map_x: 1; /* output: map address enable */
unsigned vect_x: 1; /* output: vector address enable */
```

```
 unsigned stack_full_x: 1; /* output: full */
 unsigned d: 12; /* input: direct input bits */
 unsigned i: 4; /* input: Am2910 instruction bits */
 unsigned oe_x: 1; /* input: output enable */
 unsigned rld_x: 1; /* input: register load */
 unsigned cc_x: 1; /* input: condition code */
 unsigned ccen_x: 1; /* input: condition code enable */
 unsigned ci: 1; /* input: incrementer carry-in */
};

/*
          This structure defines the internal registers
          of the Am2910 controller to the C compiler
*/

struct f_intern_2910 {
 unsigned regcnt : 12; /* register/counter */
 unsigned pc : 12; /* microprogram counter register */
 unsigned stack_pt: 3; /* stack pointer */
 unsigned stack[6]; /* five word stack */
};

/*
 This structure defines the pipeline bit fields
*/

struct pipe {
 unsigned is_2910_i : 4;
 unsigned is_2910_oe_x : 1;
 unsigned is_2910_rld_x : 1;
 unsigned is_2910_ccen_x : 1;
 unsigned is_2910_ci_x : 1;
 unsigned is_ba : 12;
 unsigned is_ccmux_sel : 4;
 unsigned is_2903_rama : 4;
 unsigned is_amux_sel : 2;
 unsigned is_2903_ramb : 4;
 unsigned is_bmux_sel : 2;
 unsigned is_2903_ea_x : 1;
 unsigned is_2903_oeb_x : 1;
 unsigned is_2903_oey_x : 1;
 unsigned is_2903_we_x : 1;
 unsigned is_qrsel : 4;
 unsigned is_srsel : 4;
 unsigned is_qlsel : 4;
 unsigned is_slsel : 4;
 unsigned is_cnsel : 2;
 unsigned is_usrld : 1;
 unsigned is_mznld : 2;
 unsigned is_mcovld : 2;
 unsigned is_mdboe_x : 1;
 unsigned is_2903_i : 9;
 unsigned is_dben_x : 1;
 unsigned is_dbdir : 1;
 unsigned is_marld_x : 1;
 unsigned is_maroe_x : 1;
 unsigned is_rd_x : 1;
 unsigned is_wr_x : 1;
 unsigned is_irld_x : 1;
```

```
unsigned is_mcmuxsel : 2;
unsigned is_y_bdbden_x : 1;
unsigned is_db_bdbden_x : 1;
unsigned is_mach_halt : 2;
unsigned is_db_2910_io : 2;
unsigned is_db_2910_in : 2;
unsigned is_db_2903_io : 2;
unsigned is_db_2903_in : 2;
unsigned is_db_stat_in : 2;
unsigned is_db_mach_in : 2;
unsigned is_db_outs_in : 2;
};
```

The structures do not actually allocate space in the examples shown. They merely assign names and sizes to elements in the structure; e.g., "regcnt" is a 12-bit-wide unsigned field in the structure called "f_intern_2910". Allocation of space is handled by statements of the following form:

```
struct f_io_2910 io_2910;
struct f_intern_2910 intern_2910;
```

The C programmer can then refer to the register counter in the controller as "intern_2910.regcnt" in an assignment statement like that shown next:

```
if ((io_2910.cc_x == 0 || (io_2910.ccen_x == 1))
        intern_2910.regcnt = io_2910.d;
```

The register/counter is updated from the controller's internal data bus "io_2910.d" if the conditional input "cc_x" is ZERO or the condition code enable, "ccen_x", is ONE. This statement has the same form as that shown for the simulation of the D type register in Section 6.5.2.a. above.

The controller in the computer example was simulated at the integrated circuit level. Since it contained an Am2910, as indicated above, one function in the simulation was devoted to the Am2910. As can be seen by comparing Figures 4-14 and 4-16, the Am2910 Next Address Generator, like the TI 'ACT8818, is a single integrated circuit that contains the following elements:

```
Next Address Logic
Next Address MUX
Incrementer
Microprogram Counter Register (μPC)
Stack Pointer (TOS)
LIFO Stack (FILE)
Register/Counter (the Register and Loop Counter combined)
```

Other functions dealt with the microprogram memory, conditional multiplexer, Instruction Register (IR), and the MAP ROM. A block diagram of "controller" the C function that simulates the Am2910 is given next:

Conditional Preprocessing Report	
A m 2 9 1 0 I n s t r u c t i o n	Jump to Address Zero Case
	Conditional Subroutine Call Case
	Jump to MAP ROM Address Case
	Other Functions
	etc.
Update μPC register and Register/Counter	
Conditional Post Processing Report	

The conditional pre- and post-processing reports are small functions that print varying amounts of information about the internal contents of the Next Address Generator. The amount reported, if any, is determined by a field in the microinstruction. This field would not normally be present in the hardware version of the device, so it is placed near the end of the microword where it can be removed without disturbing the hardware control fields. It allows the microprogrammer and hardware designer to see which registers change and whether those changes were the ones requested by the microprogrammer. The capability to turn off a report allows the microprogrammer to use debugged portions of the microcode to create a running machine in which he can debug newly written portions. Turning off the reporting mechanism allows the machine simulation to run as fast as possible. The use of this technique in the simulation of the computer example in Chapter 8 has allowed the designer to experiment with his second-level instruction set design and the implementation of new first-level sequences. A second-level MONITOR program gives him debugging capabilities at that level running on a microprogram that has all of its reporting turned off. He can write first-level sequences and, by turning on the reporting, he is able, by using a second-level test driver, to activate the sequences under test and watch them perform. The designer and programmer can "bootstrap" themselves into a running system without building the hardware platform.

The major section of the Next Address Generator simulation function consists of a "switch" statement that directs control to the "case" that is selected by the microprogrammer via the Am2910's instruction bus, i.e., io_2910.i in the first structure above. The operation of each case is handled by from one to four C statements that update the register or bus that is appropriate to the instruction selected. The Conditional Jump to Subroutine (CJS) is illustrated below:

```
case CJS :
        if((io_2910.cc_x == 0) || (io_2910.ccen_x == 1))
        {
```

```
                io_2910.y = io_2910.d;
                s_2910_push();
        }
        else
                io_2910.y = intern_2910.pc;
        break;
```

"CJS" is a symbolic constant "defined" to the C compiler that stands for the code of the conditional jump to subroutine instruction. Notice that if the conditional input is enabled (ccen) *and* the conditional input (cc) is *True* then the data output (Y) bus is updated with the contents of the (D) bus, i.e., the Next Address Selects the "Branch Address" field in the microinstruction as a pointer to the next microinstruction in the microprogram memory. The function "s_2910_push" is called to place the contents of the microprogram counter register in the stack to complete the subroutine call. If either condition is *False* then the next microprogram address is taken from the microprogram counter register. Remember "cc_x" and "ccen_x" are *True* when they are 0, i.e., active low, as symbolized here by the "_x" suffix in their names.

In all cases, the Am2910's Y bus (i.e., io_2910.y) is updated as described for buses in an earlier section. The microprogram memory module will use this information to select the next microinstruction with which to update the pipeline register.

In summary, the simulation of the Next Address Generator in a controller is very similar to the direct simulation of an ISP discussed in Section 6.4.1. It contains cases that update the proper registers or buses using techniques discussed in earlier sections. It is the use of the "data" contained in the registers or buses as addresses into the microprogram memory that makes this network appropriate to a controller and not an architecture. The elements of which it is constructed do not determine its nature, but their interconnections and the interpretation of the data they contain. The block diagram on the next page shows the various modules in the controller and their relationship to the data flow that concludes with the selection of the next microinstruction and its placement in the pipeline register for use in the next microcycle.

Extract Am2910.i and other enables from Pipeline
Collect all inputs to the Conditional MUX   **and prepare cc_x**
Invoke the Am2910 Controller Function   **to determine next microinstruction address**
Invoke Microprogram Memory Function   **to obtain next microinstruction**
Invoke Pipeline Register Function   **to update Pipeline Register**

The flow illustrates the action taken in the controller during a single microcycle, i.e., cycle of SYSCLK. The order of functions is important since the inputs to one module depend on the outputs of previous modules as *dictated by the system being simulated*. The actual implementation of the microprogram memory function and the pipeline register function, as shown below, is rather simple and gives us another look at integrating these modules into a system:

```
pipeline = mcm[io_2910.y];
```

The statement is written "in-line" in the system simulation module instead of within a function in order to improve speed by eliminating the function call. The variable "pipeline" is a specific instance of the C structure "pipe" given above. It is updated by using the "y" element of the io_2910 structure as an address into the array "mcm" which was declared as an array instance of the same "pipe" structure. The assignment statement in C represents the transfer of all of the bit fields addressed by the value of io_2910.y from the microprogram memory to the pipeline. While this is an economical form to write, it does involve execution time that is equivalent to moving one row of data values from a two-dimensional array equivalent to "mcm" to a one-dimensional array equivalent to "pipeline". The declarations of "mcm" and "pipeline" instances are shown next:

```
extern struct pipe mcm[2048]; /* the microprogram memory */
extern struct pipe pipeline; /* the pipeline register */
```

Both are declared external to the modules that use them since only a single copy of each must be shared among several functions. In the actual simulation, they were declared as global static variables similar to variables in a FORTRAN COMMON area.

6.5.3. Complete Algorithmic State Machine Simulations

Our technique to simulate complete microprogrammed algorithmic state machines will build on the methods introduced above to create the functional modules. We will combine modules for each integrated circuit with execution flowing in such a way that the data needed for input into a given module is made ready for it by modules ahead of it in the execution stream. The major program structure is organized around the events that can happen in a single cycle of the system clock, i.e., the physical SYSCLK. We will place pre- and post-processing "machine" reporting modules as the first and last modules, respectively, in the execution flow. The entire system module will be the body of a "DO" loop that can be executed a finite number of times so that the user can control the effects of "infinite" loops caused by errors in the microprogram or simulator itself.

The overall machine simulation program will also be made responsible for loading the microprogram and any other initialization needed to make the user environment, as well as the actual machine, work. We will try to load the actual microprogram format used in preparation of the machine code for introduction into the hardware version of our machine, e.g., our PROM programmer's data format. This step should allow us to test and remove data formatting as a cause of failure in the target hardware at an early stage. We would like to be able to say that once the microprogram runs correctly on our simulator, the only possible errors left are associated with physical problems, i.e., wiring *and* dead parts, instead of somewhere in our design and microprogram and wiring and dead parts. We at least will know where to look and will have provided test programs to use as tools as a result of our simulation.

6.5.3.a. A Simple RALU Driven by a Table-Based Controller

Our first simulated system is based on a state machine containing a simple RALU in its architecture similar to that shown in Figure 6-4. We will use a simple counter-based controller like that in Figure 4-4. This type of controller has been mentioned earlier as a good version of a "test driver" program for testing individual simulation modules. The statement, in C, that simulates the counter-based controller is shown here:

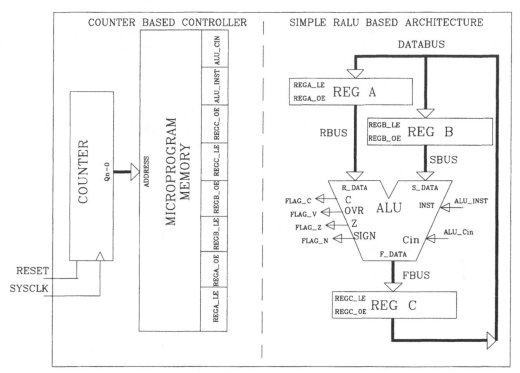

Figure 6-4: Simple RALU-based state machine example.

```
pipeline = microprog[i];
```

Both "pipeline" and "microprog" are instances of a structure "microword" that is declared similar to the one in the controller section. This version of the structure (shown below), however, contains only those fields appropriate to this machine:

```
struct microword {
        unsigned rega_le_x          : 1;
        unsigned rega_oe_x          : 1;
        unsigned regb_le_x          : 1;
        unsigned regb_oe_x          : 1;
        unsigned regc_le_x          : 1;
        unsigned regc_oe_x          : 1;
        unsigned alu_inst           : 4;
        unsigned alu_cin            : 1;
        unsigned alu_report         : 2;
        unsigned machine_report     : 4;
        };

struct microword pipeline;          /* pipeline instance of microword */
struct microword microprog[N];      /* N word microprogram memory instance */
```

Comparing the "microword" struct given above with Figure 6-4, we see that the first eight fields in the structure correspond directly to those that comprise the data outputs of the microprogram memory. The width of each field as specified in the structure is identical to that in the hardware. If languages other than C are used for the simulation or execution speed is a problem, then one of the standard data types (e.g., integer) can be used for each field. Two additional fields have been added to the "microword" to control reporting. The "alu_report" field is 2 bits wide to allow four choices of the contents of the ALU report, including no report at all. The "machine_report" field lumps the controller, registers, buses and their control signals together allowing the 4-bit report control field to select between the various amounts of information to be reported. The pre- and post-processing report technique is especially useful in pinning down errors within modules. One or the other can be disabled once the modules themselves are working, e.g., during microcode debugging.

An overall view of the simulation program for the Simple RALU Based Microprogrammed ASM is shown in Table 6-1. The organization shows two major sections, one to

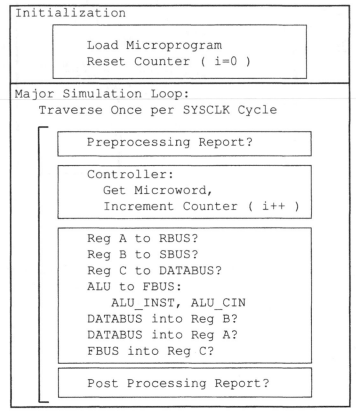

```
Initialization

        Load Microprogram
        Reset Counter ( i=0 )


Major Simulation Loop:
    Traverse Once per SYSCLK Cycle

        Preprocessing Report?

        Controller:
           Get Microword,
           Increment Counter ( i++ )

        Reg A to RBUS?
        Reg B to SBUS?
        Reg C to DATABUS?
        ALU to FBUS:
            ALU_INST, ALU_CIN
        DATABUS into Reg B?
        DATABUS into Reg A?
        FBUS into Reg C?

        Post Processing Report?
```

Table 6-1: Simulation program layout for simple microprogrammed ASM.

load and initialize the system and the second that embodies the actual simulation. One pass through the simulation section, the second major box, corresponds to the events that happen during one cycle of the system clock. The order of the smaller boxes is determined by the behavior of the actual hardware being simulated. The function that simulates an enabled D register with tristate buffers, as described in an earlier section, must be decomposed so that the "output" processing can be performed first to place the register contents on the respective buses. The contents of Register C must be placed on the DATABUS before Registers A and B are loaded. Remember the registers are edge-triggered at the end of the SYSCLK period requiring data to be present on their inputs before that time. In Table 6-1, the word "into" is used to correspond to the HDL register assignment symbol ":=", while "to" corresponds to "=", to emphasize the ordering required.

Once RBUS and SBUS have been updated, the ALU module can then be executed. The ALU creates the values for the FBUS and the flags under control of the ALU instruction, "ALU_INST", and the Carry Input, "ALU_Cin", which are derived from the microword.

There is some minor reordering of the modules that would not affect the outcome, such as swapping REG A to RBUS and REG B to SBUS. The two DATABUS to REG A or REG B lines could also be placed immediately after the REG C to DATABUS statement. The ordering of operations in the simulator is the major effort required of the simulator programmer. It is through this method that he represents the actual operation of the hardware.

6.5.3.b. Simulation of the Basic Computer Example

The basic computer example discussed in Chapter 8 represents a much more complex machine than that discussed above since it contains a second level of control with its attendant external architecture. The overall simulation program organization is still governed by the same rule as above, i.e., outputs of registers and memories must be placed on buses before combinatorial elements use them. Furthermore, all register and memory inputs must be prepared before the assignment statements that load them are executed. As the system increases in complexity, the ordering of modules gets more difficult. The program designer, however, need not take into account that the machine has a second level of control. In fact, the awareness of this organization will generally confuse him in creating the simulation program. He will erroneously try to include control functions handled in the first- or second-level programs in the simulation program itself. This should be avoided.

The controller for the Basic Computer Example, Figure 8-1, is based on the full-featured controller discussed in Chapter 4 and shown in Figure 4-14. The internal architecture contains a 16-bit RALU patterned after the "generic" RALU of Chapter 3 and two status registers, one for each control level. The RALU with its extended shift multiplexers is in Figure 8-2. The micro and macro status registers are in Figure 8-3. The external architecture that contains the second-level program and data memories along with the I/O controllers is shown in Figure 8-4. The complete computer logic diagram is presented in Figure 8-6 which is a composite of Figures 8-1 through 8-4. This overview provides a "road map" for the work ahead.

In this section, we will limit our discussion to the overall execution flow of the simulation program. The execution flow is shown in Table 6-2. The first major subsection loads the microcode and two second-level programs, a user interface/debugging program called SW16MON, and a second-level test program. The test program is invoked by SW16MON and is organized as a test driver with a subprogram under test. The subprogram may be the end-product of interest or may in itself invoke first-level sequences for testing through the instruction fetch sequence in the microcode.

The MAP ROM is loaded so that the connection between second-level opcodes and microprogram implementation sequence beginning address can be made more flexible for efficient use of the microprogram memory space. In the example computer, this initialization is fixed since each microsequence begins at an address that is twice the numerical value of the opcode. A more flexible method would involve loading a file derived from the microprogram's symbol table listing that contains the actual addresses of each sequence. The simulator's initialization section concludes with resetting the first-level machine and priming the pipeline register with the first microinstruction. It is up to the microprogram to initialize the second-level machine.

The simulator's initialization section concludes with resetting the first-level machine and priming the pipeline register with the first microinstruction. It is up to the microprogram to initialize the second-level machine.

The second major section of the basic computer's simulator consists of the simulation loop that is traversed once each cycle of the system clock. The number of times this loop is traversed is limited to restrict time spent on indeterminate loops. The simulation loop, as does the first-level machine, should run forever.

After the initial conditional machine reporting function that gives the preprocessing report, the pipeline is updated with the new microinstruction and a new feature is introduced, the machine conflict report. This specialized report generator examines all of the output enable signals in the pipeline and reports if two or more are active that would cause more than one device to drive a given bus. In our simple system, the conflict generates a report showing the location of the problem in the microprogram and terminates the simulation run. Other more complex simulators may invoke a microprogram editor to allow the user to examine and correct the problem on line.

The multiplexer preparation module contains a group of combinatorial elements, mostly multiplexers, that must be run to prepare outputs for various buses, the controller, and RALU. These include the shift extension, Carry Input, Register Address, and Conditional Input multiplexers. The CCMUX is run at this point (i.e., before the controller function) to select the proper status register output produced by the ALU in the last microcycle.

The controller and RALU modules are as described in earlier sections. The devices simulated are the Am2910 and Am2903, respectively. After the RALU is executed, the status register function is run. The first- and second-level status registers are each comprised of four 1-bit D flip-flops arranged in an extended version of the enabled D flip-flop shown in Figure 3-8. The simulator reflects the wiring shown in Figure 8-3 and not the overall status register manipulation philosophy which is under control of the microprogram. All flip-flops in the microstatus register may be loaded under control of the "μSRLD" signal in the microword. The more complex control structure of the macrostatus register is reflected in the grouping of the carry and overflow flag multiplexer select signals together. A similar grouping is shown for the "Z" and "N" flags. This allows the microprogrammer to update parts of the second-level status register separately or together. His actual microcode will reflect the flag setting strategy he has chosen.

Another new module is responsible for manipulating the devices and buses in the external architecture. The order of operations within this module is governed by the overall flow that we have seen in other modules, i.e., prepare inputs first and then calculate the output of a device. The flow in this module begins at the address bus and proceeds to the selected device where the operations that it can support are performed in connection with the external data bus. If the operation is an outbound data transfer from the CPU, the external data bus is updated from the internal data bus before any of the individual devices are simulated. On the other hand, if the operation is an inbound transfer, then the devices are simulated, thus updating the external data bus, before data is transferred from the external to the internal data bus.

The EPROM and RAM are simulated just like an array of registers except that if a "WRITE" is requested to the EPROM, a message is sent requesting confirmation from the user. This method allows the user to update the EPROM area while still protecting him from inadvertent mistakes.

The serial I/O controller simulation requires a little research into the less-used I/O functions associated with the simulation language. A serial input device looks like two input registers to the computer. One register contains bits that inform the second-level program

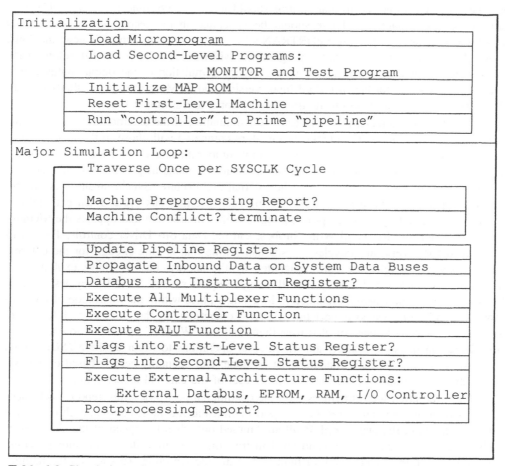

Table 6-2: Simulation program layout for the basic computer example.

that a character has been received and the second contains the received data. If the designer is not interested in the details of the serial port operation other than the location of its registers and status bits, he can simulate its operation rather simply. Serial input in the simplest case may be simulated using the formatted console data input routines (e.g., FORTRAN's READ or C's scanf statements). If these are used, then when the simulation program calls the input function, it will wait until the formatted data is available in the designated data input location that simulates the received data register in the I/O controller. The input status sensing contained in the second-level program will only be a formality. The simulation of the controller can be made more lifelike (i.e., the second-level polled I/O sense loop actually does the waiting for the input character) by using one of the host-system specific I/O routines that returns a character from the console without waiting or examining it. A flag is returned if no character is available. The simulation code then updates the status register variable with the results of the flag, and the data if appropriate. The desired function is system and language dependent and can be found after some research in the host-system runtime library. The simulator is less "portable" to new hosts; however, it does allow a closer simulation of the I/O functions.

Serial output to the console is easily simulated using the normal language output function, e.g., FORTRAN's "WRITE" and C's printf. Since serial I/O to the console is done

one character at a time, the format control should not be responsible for sending Carriage Return/Line Feed characters. These should be generated in the second-level program and sent as normal characters; i.e., use FORTRAN's "+" line control and omit C's new line, "\n". If normal line formatting is used, the programmer will not see a line of text develop on the console one character at a time as he would expect. Simulation of the output status flag is trivial in this case, as for the formatted input version above. More true-life methods also involve a search of the host-system runtime library for a useful output routine.

Another consideration in connection with simulation of the external architecture is the relative speed of events in the internal architecture compared with those in the external. The external architecture, because of a basic optimization rationale established for computers (i.e., that the memory section would contain as many locations as possible for the price regardless of speed costs) is slower than the internal architecture. Synchronization is accomplished by spending several microcycles during each external bus cycle. Each external bus cycle as orchestrated by the microprogram starts with a microcycle that updates the Memory Address Register, the Read and Write Strobes, and the Data Bus Driver controls. These values are held for several microcycles to establish a time that is long enough for the external memory to store or fetch the data designated by the address bus. We have exploited the knowledge that an external bus cycle will always require several SYSCLK cycles to place all of the code for the external architecture in a single place within the main loop of the simulator. Normally, one would expect the "READ" operations on the external architecture to come before the RALU function that uses them and the "WRITE" operations to come after. If the design in question actually allows external operations to occur at the SYSCLK rate, this division of the "external" module must be done. Our economy here is based on knowledge of the relative speed, hence less frequent use, of the external architecture.

The simulation loop concludes with the conditional post-processing report. Each module has its own pre- and post-processing report facilities as mentioned above, each controlled by its own fields in the microword. If all are "turned on", then the speed of the simulator is greatly reduced. Currently, the second-level instructions used to write the monitor program, SW16MON, have their report control fields set to disable reporting. This allows the monitor program and any second-level test program (e.g., the test driver itself) to run at "full speed". Microinstructions that implement second-level sequences that are being tested will contain report fields set to give the information desired by the user. The controller reports give a sense of the sequence's flow while the RALU reports show data modifications. By using this technique, the simulator creates an environment in which both first- and second-level programs can be developed.

6.5.3.c. Interconnection of IC Simulation Modules Using Pointer Arrays

In the simulation modules described above, the data produced by one device was sent to another device by an explicit assignment statement that contained the names of the two devices. For example, consider the following unconditional transfer of information between registers A and B by way of bus F:

```
fbus = rega_output;
regb_input = fbus;
```

This method of writing "freezes" a given design into a simulation program since each line contains some reference to the actual hardware and paths. Functions may be written that represent individual integrated circuits but their interconnection is not flexible since each "pin" must be connected explicitly to others as dictated by the schematic. One method by which a module could be connected flexibly to other modules would be to place the interconnections as variables in the argument list when the function is called. Each invocation of the function then describes the wiring for that device. For small devices such as registers, this may be appropriate but it exacts a heavy price on the typing and editing skills of the programmer for devices such as RALUs, Floating Point Units, and Next Address Generators. The simulator is to assist hardware design which requires that changes in the design caused by discoveries made during simulation be reflected in changes in the simulator. Finding and retyping all of the argument lists is not an efficient method.

The "structure" construct in the C language gives us a method to produce a very flexible "wiring" technique for our simulators. The structure, introduced above, is described by Kernighan and Ritchie (see Bibliography) to be a "collection of one or more variables, possibly of different types, grouped together under a single name for convenient handling." We have exploited the structure in our earlier discussion to create "variables" that have appropriate groupings of signals of varying widths united by a common theme, e.g., the pipeline. We described the interconnections between all elements in our basic computer example in the form of various structures. The interconnections between each "pin" on a device and any other "pin" was accomplished using an assignment statement. The structures, declared "globally", were visible within all functions. To control the ALU within the RALU function, we merely used the "RALU instruction" element of the pipeline structure. Unfortunately, this meant that if we wanted to modify the source of this information, we would have to edit the actual RALU function itself to modify the element used.

The next step toward flexibility is to consider passing the names of the structures that describe the interconnections for a given IC as arguments to the function describing the IC. The flexibility gained by this method is reflected in the fact that one copy of a function that describes an IC can be used to simulate all of the ICs of that type in the design. A design having eight enabled D registers with tristate output buffers need have only one copy of the simulation module. The simulation module itself can be built using one multiplexer and one D flip-flop module, for the enabled D flip-flop, and a tristate buffer module. To build an entire 8-bit register would only require a function that called the three 1-bit modules eight times in the proper order using the appropriate structure in its argument list. There is a trade-off being made. It is one of execution speed. A large design may require a rather fast computer for running the simulation in a timely manner. The actual balance is rather dynamic and thus no hard rules can be given here. As the simulation modules are written, the programmer can determine if this method can be used to save writing. Notice that the method discussed is simply that of structured programming.

Let's extend our interconnection scheme one step further. In the previous paragraph, we passed the names of structures that described the sources and destinations of data for a particular connection of an IC to the function that simulates the IC. We will now pass only one pointer to a list of pointers that describe the interconnections for one "instance" of the IC. Kernighan and Ritchie (see Bibliography) describe a pointer for the C language as "a variable that contains the address of another variable". The concept is derived from assembler language and is a very powerful one. An instance of an IC corresponds to a particular occurrence of a given device in a circuit: e.g., the symbol U5 on a schematic signifies the fifth integrated circuit in a design which we can determine to be a 74LS00 from the parts list. In the basic computer example structures for the pipeline, Am2903 "pins", Am2903 register, status registers, and buses must be made known to the RALU simulation function. Calling the RALU simulation in C using this concept would appear as follows:

```
ralu(&pipeline,&io_2903,&reg_2903,&statreg,&buses)
```

The C operator, "&", stands for "the address of"; i.e., we are sending the address of each of the structures to the "ralu" function. This has reduced the amount of typing in the argument list but has not relaxed the rigidity with which the simulator is "wired". You can see why we chose to make the various structures "global", and thus visible to all functions.

Instead of sending pointers to a set of data containing structures organized like the data storage organization of the circuits they represent, let us create a single pointer that points to a "wiring" structure. The wiring structure is itself a structure of pointers, each pointing at the appropriate element of a data storage structure. We have, essentially, introduced indirect addressing by using the "wiring" structure as an intermediary. This technique may slow execution but is a very useful organizing construct. Let's describe a simple 8-bit register having a tristate output buffer similar to Figure 3-11 (e.g., 74LS374). A structure that gives form to the variables used to describe the data storage of our device follows:

```
/* Describe an 8 bit tristateable register similar to the 74LS374 */
    struct f_reg8ts {
            unsigned      inreg       :8; /* data input pins of register */
            unsigned      creg        :8; /* contents of register */
            unsigned      outreg      :8; /* output pins of tristate buffers */
            unsigned      oe_x        :1; /* output enable line - active low */
    };
```

Each occurrence of the 74LS374 in our schematic will be simulated by the following:

```
/* Declare two instances of registers of form reg8ts */
    struct f_reg8ts rega;          /* one instance of an LS374 */
    struct f_reg8ts regb;          /* second instance of an LS374 */
```

We now have two registers in our design, "rega" and "regb". If we want to refer to the contents of register "rega", then we would say "rega.creg". This format may be used just like any other variable in an assignment statement.

The actual simulation function for an 8-bit register with tristate outputs would look like the following:

```
reg8ts(pd)
struct f_reg8ts *pd;
{
        (*pd).creg = (*pd).inreg;
        if ((*pd.oe_x) == 0) { (*pd).outreg = (*pd).creg; }
}
```

The name of this function is "reg8ts" and it needs the pointer argument, pd, when it is called. The argument "pd" is a pointer to a structure that points to the variables that are needed by the function. The variable "(*pd).creg" is an alternate form of pointer addressing given by Kernighan and Ritchie (see Bibliography) that we will employ here for its explicitness. The symbol can be read as designating the "creg" element of the structure pointed at by the pointer "pd" (i.e., *pd). This fact is made known to the compiler by the data declaration "struct f_reg8ts *pd;" at the beginning of the function. The declaration says that "pd" is a pointer to a structure described by the structure "f_reg8ts" given above. Notice that the function does not have to know, at the time it is compiled, which data storage structure is to be used, but only that it will have the form given by "f_reg8ts". Each time it is called during execution, a pointer to the desired data storage structure such as "rega" will be sent in its argument list. The following program segment initializes register "rega" to hexadecimal "FF" and transfers the data under explicit control to register "regb":

```
      rega.creg = 0xff;/* initialize "rega" contents */
      rega.oe_x = 0; /* enable the output buffers of "rega" */
  /* run a register transfer */
      reg8ts(&rega); /* simulate "rega" to place contents on output pins */
      databus = rega.outreg; /* move data from "rega" output pins to databus */
/* build machine, i.e., interconnect ICs */
      regb.ireg = databus; /* data bus to "regb" input pins */
      reg8ts(&regb); /* place data into "regb" contents */
```

The variable "databus" is declared as an unsigned integer just like the contents "creg" of the "reg8ts" structure. Normally, control of the register output enable signals, i.e., rega.oe and regb.oe, would be accomplished from the pipeline which is not explicitly shown here.

Now what do the wiring structures look like?

```
/*
      Describe an 8-bit register wiring structure for an enabled D register
 with tristate buffer output.
*/
      struct f_reg8ents {
      unsigned char *datain; /* pointer data input source */
      unsigned char contents;/* contents of register */
      unsigned char *dataout; /* pointer to output data destination*/
      unsigned char *le_x; /* pointer to load enable line */
      unsigned char *oe_x; /* pointer to output enable line */
      };
```

There are 4 pointers and 1 data storage cell declared in the structure. The wiring for each occurrence of the device in our schematic will be simulated by the following:

```
/* Declare two instances of registers of form reg8ents */

 /* wire the first instance of register */
      struct f_reg8ents rega_wire =
            {&databus_0, 0x00, &databus_1, &pipeline.rega_le_x,
            &pipeline.rega_oe_x};

 /* wire the second instance of register */
      struct f_reg8ts_wire regb_wire =
            {&databus_1, 0x00, &databus_0, &pipeline.regb_oe_x,
             &pipeline.regb_oe_x};
```

Notice that a structure called "pipeline", declared elsewhere, has elements "rega_oe_x" and "regb_oe_x" to serve as data storage for the register output control signals.

The actual simulation function "reg8ents" is given next:

```
reg8ents(pd,dir)
struct reg8ents *pd;
char dir;
{
      if (dir == IN)
      {
          if (*(*pd).le_x == 0)
          { (*pd).contents = *(*pd).datain; }
      }
      if (dir == OUT)
      {
```

```
            if (*(*pd).oe_x == 0)
              { *(*pd).dataout = (*pd).contents;}
        }
}
```

The argument "dir" is a way to allow the function reg8ents to be placed where needed by the designer and still allow all of its parts (i.e., input and output assignment statements) to be located in one "register" function. The two values for "dir" are "defined" to be
the constants "IN" and "OUT" symbolizing the transfer of data into and out of the register,
respectively. This notation contributes to a structured program but definitely slows execution.
Its employment is therefore conditional on the speed of the host system.

The significant pointer addressing feature is shown by the notation "*(*pd)" which
signifies using the pointer "*pd" to point at another pointer which in turn points at the desired data storage location. The following program segment shows the transfer of data from
"rega" to "regb" under control of the current values in the "pipeline" structure:

```
/*
      If the rega_oe_x element of the pipeline structure permits then
 place the contents of register "rega" on "databus_1"
*/
      reg8ents(&rega_wire,OUT);
/*
      If the rega_le_x element of the pipeline structure permits then
 place the contents of "databus_1" into register "regb"
*/
      reg8ents(&regb_wire,IN);
```

A simulation involving three enable registers with tristate output buffers connected to
a single databus is given next. Under control of a microprogram loaded during the course of
initialization, data can be transferred between any pair of registers. The definition of the wiring structures is given below; however, the program requires that the reg8ents function be
included before it will run.

```
/*
Program to simulate three 8-bit enabled D registers having tristate output buffers. All
register inputs and outputs are connected to a common databus.
Arbitrary transfers between the registers may be accomplished under control of a micro-
program loaded in the "micromemory" by a user-written "load_microprogram" function. The
microprogram machine code should be prepared by a microprogram assembler like those
described in Chapter 5.
Initial contents of registers are determined by the "initialize_register" function
.Reporting functions, "pre_processing_report" and "post_processing_report" should be
prepared by the user to present the contents of the registers, buses, and control signals
in the desired format.
Be sure to #define the various constants, e.g., N and NUMBER_OF_STEPS.
*/
/* Declare three instances of registers of form reg8ents */
 /* wire the first instance of register */
      struct f_reg8ents rega_wire =
            {&databus, 0x00, &databus, &pipeline.rega_le_x,
            &pipeline.rega_oe_x};

 /* wire the second instance of register */
      struct f_reg8ts_wire regb_wire =
```

```
                {&databus, 0xFF, &databus, &pipeline.regb_oe_x,
                &pipeline.regb_oe_x};

  /* wire the third instance of register */
        struct f_reg8ts_wire regc_wire =
                {&databus, 0x80, &databus, &pipeline.regc_oe_x,
                &pipeline.regc_oe_x};

/* define the pipeline structure */
struct pipe {
        unsigned rega_le_x: 1;
        unsigned rega_oe_x: 1;
        unsigned regb_le_x: 1;
        unsigned regb_oe_x: 1;
        unsigned regc_le_x: 1;
        unsigned regc_oe_x: 1;
  }

/* define storage structures */

        /* storage based on pipe data format */
struct pipe pipeline; /* storage for pipeline register */
struct pipe micromemory[N]; /* storage for microprogram memory */

        /* miscellaneous storage */
unsigned char databus; /* define storage for the databus */
unsigned int upc; /* microprogram memory address */

main()
{
/* Initialization Section */
        load_microprogram();
        initialize_registers();

/* Main Simulation Loop */
        for (upc = 0; upc < NUMBER_OF_STEPS; upc++)
        {
                pre_processing_report();

         /* simple counter based controller */

                pipeline = micromemory[upc];

        /* conditional transfer of register contents to databus */
                reg8ents(&rega_wire,OUT);
                reg8ents(&regb_wire,OUT);
                reg8ents(&regc_wire,OUT);

        /* conditional loading of databus contents into registers */
                reg8ents(&rega_wire,IN);
                reg8ents(&regb_wire,IN);
                reg8ents(&regc_wire,IN);

                post_processing_report();
        }
}
```

Notice that a single copy of the register simulation module is used for all instances of the register. The wiring is clearly exposed within the wiring structures which are the only elements that need to be changed in the event of a design change in the hardware. The simulation functions may be compiled and placed in a library in the same sense that ICs are described in a databook. This is a concept espoused for general programming purposes by those interested in Object Oriented Programming and the name has been used by Cox in his discussion of Objective C (see Bibliography).

A particularly interesting case can be made for building complex simulation functions using this method. Consider the Arithmetic/Logic Unit (ALU). One can simulate an 8-bit-wide unit using the word method discussed above. This is the method where the entire data path is aligned within one word of the host language. The execution time is faster; however, flag recovery is more difficult. Consider that a mix of methods, possibly including assembler language, is used to produce an 8-bit-wide ALU simulation function called "alu8". Furthermore, let us assume that an "alu8" prototype structure having the following form has been defined:

```
struct alu8 {
        unsigned char *inst;/* point at pipeline for instruction */
        unsigned char *r_data; /* point at alu_r input bus */
        unsigned char *s_data; /* point at alu_s input bus */
        unsigned char *f_data; /* point at alu_f output bus */
        unsigned char *c_in;/* point to source of carry input */
        unsigned char *flag_n;/* point to storage for sign flag */
        unsigned char *flag_z;/* point to storage for zero flag */
        unsigned char *flag_c;/* point to storage for carry flag */
        unsigned char *flag_v;/* point to storage for overflow flag */
        };
```

A particular example of a simple 8-bit ALU called "alu8" is given next:

```
alu8(pd)
struct alu8 *pd;
{
        unsigned r_si; /* sign bit for r path */
        unsigned r_lo; /* least significant seven bits for r path */
        unsigned s_si; /* sign bit for s path */
        unsigned s_lo; /* least significant seven bits for s path */
        unsigned f_si; /* sign bit for f path */
        unsigned f_lo; /* least significant seven bits for f path */
        unsigned cin; /* carry in to alu */
        unsigned c6; /* carry out of lower seven bits */
        unsigned c7; /* carry out of eighth bit position */

        r_si = 0;
        s_si = 0;

        if ((*(*pd).r_data & 0X80) != 0) { r_si = 1;}

        if ((*(*pd).s_data & 0X80) != 0) { s_si = 1;}

        r_lo = *(*pd).r_data & 0X7F;
        s_lo = *(*pd).s_data & 0X7F;
        cin = *(*pd).c_in & 0X01;

        switch(*(*pd).inst) /* process one of the ALU's instructions */
```

```
        {
      case R_PLUS_S_PLUS_CIN:
           f_lo = r_lo + s_lo + cin;
           c6 = 1; if ((f_lo & 0X80) == 0) { c6 = 0;}
           f_si = r_si + s_si + c6;
           c7 = 1; if ((f_si & 0X02) == 0) { c7 = 0;}
           break;

      case S_MINUS_R_MINUS_CIN:
           f_lo = s_lo + (~(r_lo) & 0X7F) + cin;
           c6 = 1; if ((f_lo & 0X80) == 0) { c6 = 0;}
           f_si = s_si + (~(r_si) & 0X01) + c6;
           c7 = 1; if ((f_si & 0X02) == 0) { c7 = 0;}
           break;

      case R_MINUS_S_MINUS_CIN:
           f_lo = r_lo + (~(s_lo) & 0X7F) + cin;
           c6 = 1; if ((f_lo & 0X80) == 0) { c6 = 0;}
           f_si = r_si + (~(s_si) & 0X01) + c6;
           c7 = 1; if ((f_si & 0X02) == 0) { c7 = 0;}
           break;

      case R_PLUS_CIN:
           f_lo = r_lo + cin;
           c6 = 1; if ((f_lo & 0X80) == 0) { c6 = 0;}
           f_si = r_si + c6;
           c7 = 1; if ((f_si & 0X02) == 0) { c7 = 0;}
           break;

      case S_MINUS_CIN:
           f_lo = (~(s_lo) & 0X7F) + cin;
           c6 = 1; if ((f_lo & 0X80) == 0) { c6 = 0;}
           f_si = (~(s_si) & 0X01) + c6;
           c7 = 1; if ((f_si & 0X02) == 0) { c7 = 0;}
           break;

      case R_AND_S:
           f_lo = r_lo & s_lo;
           f_si = r_si & s_si;
           break;

      case R_OR_S:
           f_lo = r_lo | s_lo;
           f_si = r_si | s_si;
           break;

      case R_XOR_S:
           f_lo = r_lo ^ s_lo;
           f_si = r_si ^ s_si;
           break;
        }

/* place result on output bus */

      *(*pd).f_data = (f_si << 7) | (f_lo & 0X7F);

/* flag setting */

      f_si = f_si & 0X01; /* isolate sign flag */
```

```
*(*pd).flag_n = f_si; /* place the sign flag where schematic says */

if ((f_si | f_lo ) == 0) /* set the zero flag */
     { *(*pd).flag_z = 1;}
else
     { *(*pd).flag_z = 0;}

if (*(*pd).inst >= R_AND_S) /* set the carry and overflow flags */
     {
                         *(*pd).flag_c = 0;
                         *(*pd).flag_v = 0;

     }

else
     {
                         *(*pd).flag_c = c7;
                         *(*pd).flag_v = c7 ^ c6;

     }
}
```

Once the "alu8" function is created, ALUs of any larger size may be created by invoking it with the correct "wiring" structure. Consider the following two structures that create two 8-bit ALUs to serve as the upper, "alu_h", and lower, "alu_l", halves of a 16-bit-wide ALU:

```
/*
      Declare a single 16-bit alu.
      Make it using two instances of an 8-bit "alu8".
      Use the Carry In - Carry Out connection.
*/
struct alu8 alu_h =
      {
             &pipeline.alu_inst,&r_bus_h,&s_bus_h,&f_bus_h,
             &hc,&mn,&mz,&mc,&mv
      };
struct alu8 alu_l =
      {      &pipeline.alu_inst,&r_bus_l,&s_bus_l,&f_bus_l,
             &pipeline.alu_cin,&mn,&mz,&hc,&mv
      };
```

The wiring of each instance of "alu8" shows the instruction coming from the "alu_inst" element of the "pipeline" structure similar to those shown above. Data enters and leaves the instances of "alu8" via the upper and lower halves of "r_bus", "s_bus", and "f_bus" which are designated by pointers in the instance structures above. The Carry Input to "alu_l" is derived from the pipeline register, while that for "alu_h" comes from the Carry Output of "alu_l" via a location called "hc" (i.e., half carry). The four output flags, "mn", "mz", "mc", and "mv" are produced in each instance of "alu8. If a status register function was present, it would be "wired" to the flag outputs of "alu_h" to capture the flags resulting from 16-bit operations. The 16-bit zero flag must be produced by "ORing" both copies of the "mz" flags, i.e., z = mz_h | mz_l. Notice that the status register could be dynamically controlled by the microinstruction to capture either the lower ALU's flags or the total flag group to support both 16- and 8-bit operations as is done in the TI 'ACT8832 RALU.

In summary, system wiring can be accomplished in a direct manner using global data areas. The unfortunate side effect is that the wiring is embedded in each simulation function. Part of this shortcoming may be removed by passing data areas to the simulation functions as members of an argument list. This is rather bulky and does not reach the full potential given by the third method, i.e., passing a pointer to a list of pointers that describes the wiring for a particular instance of an integrated circuit. The pointer method yields separate descriptions of the device, i.e., a device-dependent template such as the "alu8" structure, and the wiring, e.g., the "alu_h" and "alu_l" instances. A single function (e.g., "reg8ents") can be used to contain the actual simulation code for all instances of that type of device in the design. This gives the designer a process whereby he can create a library of devices that can be incorporated in several designs. The actual design only needs to contain the wiring structures and a simulation loop body that invokes all of the functions in an appropriate order. The detailed operation of the module is not as important as the effects produced at its "pins", just like the actual IC.

6.5.4. Simulation of the Electrical Behavior of Architectures

The moment-by-moment electrical behavior of each pin on an integrated circuit is a very complex job to simulate. In the digital domain, the information one hopes to obtain at this level is divided into two areas. The first relates to the time required to propagate signals around the circuit. The second involves the load driving capabilities of devices used in the design. Several commercial software packages allow this level of simulation to be performed. Various professional Hardware Description Languages allow the two most important factors to be evaluated, i.e., loading and propagation time. The newest of these languages is VHSIC Hardware Description Language (VHDL) defined in "IEEE 1076-1987". The student may also use analog device simulation methods at the transistor level. The practicing design engineer has had the electrical simulation capability for several years in the analog field and some serious efforts have been made to bring this capability to the digital field. The major hurdle to overcome in simulating the electrical behavior of digital circuits is related to the sheer magnitude of the number of transistors in the major devices. The program execution time becomes so long that this level of knowledge can be obtained more quickly for complex circuits by actually constructing them. Besides using ever faster computers to run the simulations, some exotic methods have been employed, including embedding physical copies of very complex devices within the simulation.

It is beyond the scope of this book, and most designer's programming skills, to provide simulation at this level. Commercial software should be employed for these purposes. However, there are methods that can be done "by hand" that can assist in determining some overall aspects of the two pieces of information needed about the physical circuit, i.e., timing and output loading. These will be discussed briefly in the following sections.

6.5.4.a. Critical Path Determination

The basic timing information desired about a state machine design is used to determine the period of the system clock, SYSCLK. The determination of the "critical path" provides this information. The "critical path" is the signal propagation path through a network that takes the longest time and hence is critical upper limit to the frequency of SYSCLK. In our discussions of controllers in Chapter 4, we encountered this term when we introduced the pipeline register to divide the signal propagation path that had originally threaded through the controller and architecture. By introducing the pipeline register, we cut the propagation path in two allowing events to proceed in parallel in the controller and architecture.

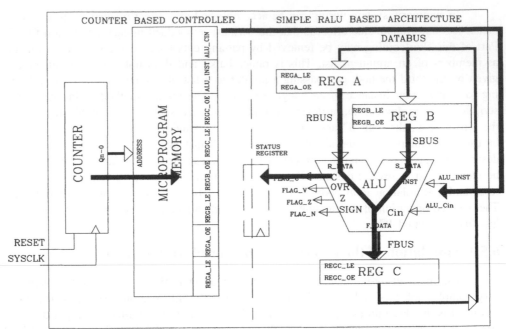

Figure 6-5: Propagation paths through the simple RALU-based state machine.

Figure 6-5 shows the simple RALU-based state machine discussed above. On it are traced several propagation signal propagation paths. Several more exist, including one from the output of REG C to the inputs of REG A and REG B. To find the critical path, the designer traces all propagation paths on a diagram of his system. The only requirement for the degree of detail in the diagram is that he can identify the actual "pins" on the integrated circuits that lie along the path. Once all paths are traced, he then sums the propagation delays along the path of interest as it threads the ICs along it. A control signal propagation path generally starts at the pipeline register if one is present, as it normally will be in a complex machine. Data propagation paths start at a register that serves as the source of data for the path being examined. All propagation paths terminate at the input of the next register they encounter.

Thus, the data path in Figure 6-5 that originates at the output of REG A terminates at the input of REG C. The time required for data to propagate along this path is determined by the ALU instruction; i.e., an add operation takes longer that a logical operation. An equation for each path can be written that has the following form for the example path:

$$t_p = t_{p\ REGA\ output} + t_{p\ alu\ (R-F)} + t_{p\ REGC\ setup}$$

If one considers the problem in more detail, then one should also examine several data combinations since many devices have different propagation times depending on whether an output transitions from LOW to HIGH or HIGH to LOW, i.e., t_{pLH} or t_{pHL} respectively. Another cautionary note is that complex combinatorial elements like the ALU have propagation delays that depend on the operations that they perform and data that is being used.

A control path is shown in Figure 6-5 that originates in the pipeline register in the controller and enters the ALU at the ALU_INST and ALU_Cin inputs. Assuming the registers are already enabled (i.e., RBUS and SBUS are stable), then the control path becomes two data paths, one that terminates in the Status Register and the other in REG C. The propagation time along the path depends on the ALU instruction. If the Status Register had not been used, this path would have terminated inside the controller, i.e., ultimately at the pipeline register's input for a more complex controller than that shown.

In the simple controller shown in Figure 6-5, a propagation path exists from the active edge of SYSCLK, through the counter and microprogram memory, to the inputs of the pipeline register. Since a pipeline and status register are used, signal propagation along this path can occur in parallel with data propagation in the architecture.

After all paths are identified and calculated using the integrated circuit manufacturer's data books, the "critical path" is the path having the longest propagation time. The propagation data used from the data book should be the values that make the path being examined the longest, i.e., "worst-case" analysis. Another version of "worst-case" analysis is to use data that minimizes register set-up time with respect to the active edge of the system clock. Our designs have used devices that do not contain any combinatorial circuits in the clock path, thus minimizing the effects of this problem. Since one particular path has the longest time, a time by the way that is set by the performance of addition along the REG A→ALU→REG C path, then that time sets the maximum frequency for SYSCLK. Some designers consider that, since the logic operations along that path take so much less time, one should just change the period of SYSCLK to accommodate the slow instructions. This is a possible solution, but most designers reject it as unnecessarily complex. The introduction of the pipeline register compensated for a similar problem. Solutions that retain a constant SYSCLK period generally result in more reliable systems and ones that are guaranteed to meet the Sampling Theorem that is the basis of all digital designs.

The laborious task of identifying and calculating the propagation times for all paths within a system can be improved with experience. The ALU is a notorious culprit that turns up in most critical paths. It should be suspected from the beginning of the analysis. The path from REG C via DATABUS to the inputs of REG A and REG B in Figure 6-5 contains no separate combinatorial elements; hence, it would be expected to have one of the shortest propagation times in the system. The designer can overlook it in favor of analyzing the paths that pass through the ALU. Any memory element has the potential for being a major contributor to propagation delay along paths that thread it. The microprogram memory is probably the single slowest item in the controller. However, while the register array in the architecture is generally fast compared to the RALU, a memory like those used in the external architecture of the computer is, by trade-off, slower than any element in the internal architecture. It was so slow that we actually used a form of clock rate control to compensate. The designer can speed up this analysis through experience by carefully examining only those paths that thread slow elements.

6.5.4.b. Output Loading

Analysis of the design to determine the loading on any output must wait until the hardware elements of the design are known; i.e., a relatively firm circuit diagram must exist. The "critical path" analysis discussed above depends on knowing that all outputs are driving loads of less than those indicated as test limits by the manufacturer. Each data sheet specifies the important parameters that form the basis of the tests the manufacturer used in determining the parameters. The two most important terms are the loads, or resistances connected between the output and the power supply, and the capacitances between the output and the rest of the world. These terms define a transmission line that will be excited by voltage transitions on the output.

The output pin of the device drives the input pins of other devices. The load resistance and capacitance in the test data must be distributed among the device inputs being driven. In many cases, the input capacitance of a device from one of the standard logic families is dominated by the capacitance of the packaging as opposed to that of the input structure itself. An input capacitance of 10 pFd is frequently found for digital parts. The load resistance presented by an input depends on the logic family to which the part belongs. It is characterized through the V_{il} @ I_{il} and V_{ih} @ I_{ih} input parameters in the DC Characteristics chart in the data sheet. As in all digital circuit designs the sum of all input currents that an output must drive when the inputs are driven LOW (i.e., the total I_{il}) should be less than the current that the driving device can "sink" (i.e., I_{ol}). If this limit is not observed, then the output voltage of the driving device, V_{ol}, will increase, possibly to a point above the highest voltage that an input will think is LOW. A similar restriction is in effect for inputs that are driven high. The failure to observe this requirement may result in the driven device thinking that its input has not been driven HIGH.

These DC characteristics are especially important for Transistor–Transistor Logic (TTL) and Emitter-Coupled Logic (ECL) devices that draw current through their inputs at all times. The time-dependent behavior of the currents is also important in these devices as well as those in the Complementary Metal Oxide Semiconductor (CMOS) family. The input capacitance along with the wiring between the driver's output and the input in question forms a tuned circuit, actually a low-pass filter. Rapid changes in the driver output voltage during a state change will appear, at best, to an input as a "slowly" changing voltage "low-pass filtered" by the wiring inductance and capacitance. In the worst case, the interconnection circuitry will ring at a frequency determined by the inductance, capacitance, and resistance along the path. The "ringing" will cause the input to be at a voltage different than the output for some time after the output changed state. In some cases, the input voltage may go below ground or above the positive power suppy voltage (V_{cc}). These conditions generally inhibit high-speed operation since the system clock must be slowed to allow the voltage transients to subside before the system state is latched into registers. The effect of wiring is to increase the propagation time through elements along a propagation path. Normally, short, direct leads and adequate local power supply filtering minimize this effect. In some cases, especially ECL circuits, serious attention is required to prepare the signal paths as transmission lines while constructing printed circuit boards. This aspect of circuit analysis dominates very high speed systems. Obviously, one does not wish to debug a design simultaneously with analyzing the physical questions. This point should place the design analysis steps in earlier sections into perspective. One should not create hardware until the logical design has been completed and evaluated carefully.

6.6. CONCLUSION

A side effect of the function encapsulation discussed in Section 6.5.3.c. is seen as one studies the more complex devices that are appropriate for application in specialized machines. Floating Point Units (FPUs), for example, are not easily studied using the manufacturer's data books. Generally, the designer needs access to a unit placed in an environment in which he can manipulate it to understand how it works. This is especially true of devices that contain embedded pipelines such as the TI 74ACT8847 FPU.

Manufacturers frequently produce prototyping boards that can be placed in a computer, such as the IBM PC, to allow the designer to exercise the part until he is ready to incorporate it in his designs. This method is derived from one used successfully in the microcomputer field. Most student engineers have encountered one of the many microcomputer "evaluation" boards. In some respects, the availability of these boards has directly affected the growth of a given microprocessor. The current versions of boards that serve the microprogrammed state machine field are rather expensive. They are capable of running fast and they do provide, through their host PC, a good development environment.

There is an "in between" step possible that falls between reading the data book and purchasing an evaluation board. The component manufacturers or third parties may release simulation models along the lines of those shown above. The designer may then have the best of both worlds. He can combine the model within his evolving design or run it with a simple "counter-based" controller to study how the unit works. Once the design evolves far enough that the data paths and other components approach a running system, he may begin supplying microprograms that address the problem for which the design is being made. A separate design path can be developed in which algorithms and their interaction with the new design are studied. Even though the solutions run slowly, the behavioral simulation gives enough insight to establish the correctness of the hardware design. Once the real hardware is produced, the designer will have produced proven software with which to exercise the system while he finishes his timing evaluations. We have had an experience in this connection with Texas Instruments, Inc. while designing a multi-FPU data flow structure and found the results well worth the effort. The final hardware has progressed with few surprises and the evaluation of actual algorithm mapping on the simulation model led to some changes in the design that resulted in a much more efficient machine.

EXERCISES

1. Write a program that simulates the action of each of the 74LS273, 74LS373, 74LS374, and 74LS377 registers. Connect the inputs of each register to an 8-bit wide databus, named DBUS, in your program. Each time the simulation code is executed (once through the loop), the edge-triggered clock input will be activated; hence, it need not be separately simulated. This causes the 74LS273 and 74LS374 to always be loaded from the bus every time the code for the simulation of the particular device is encountered. Be aware that the 74LS373 and 74LS374 are each composed of two elements, the register itself and a set of tristate bus drivers. The program should simulate the performance of the devices for all possible control input conditions and three possible values of data on the bus. The overall flow of the program should be in the form of a loop with one combination of inputs used during each pass through the loop. The input data and conditions may be contained in several one-dimensional arrays or one multidimensional array. Each pass through the loop should cause the internal contents of the registers and the data on the output pins to be displayed in a meaningful way along with the input data and conditions.

   Hints:
   a. The names of the output enable lines for the two registers that have them should be OE373L and OE374L.
   b. The name of the clear input on the 74LS273 should be CLR273L and the load input on the 74LS377 should be LD377L.
   c. A TABLE with headings and line numbers should be printed showing all of the inputs and appropriate variables and their contents for each register. This TABLE should look very much like a Truth Table.
   d. The register contents for each device should be labeled C273, C373, C374, and C377, while the output pins of each device should be OUT273, etc.

e. If your programming skills are minimal, you might find it easier to make each variable (e.g., C273) a one-dimensional array with the same number of elements as the width of the device. This means that each operation on a device will require a loop to manipulate each bit. If your skill permits, you may use the integer (or Logical or Boolean, where appropriate) data type to represent each of these variables. Use the smallest word width possible in your language that will contain these quantities. The complication with the second method comes in implementing shifters and ALU's since you must use wider words and extract bits from appropriate places.

f. The clock inputs on the 273, 374 and 377 are edge-triggered and should not be shown explicitly in the table since they are assumed to have occurred each time the line of simulation code is executed. The clock of the 373 is quite different since this is a "transparent" register. Make it appear explicitly in the table as a control input.

2. Write a module that simulates the action of an ALU that implements the following table of functions:

```
I    |                              Function
-----+-----------------------------------------------------------------------
  0  | Add inputs A and B with Carry In; produce outputs F and Carry Out
  1  | Add input B to Carry In; Produce outputs F and Carry Out
  2  | Subtract input B from A with Borrow; produce F and Borrow
  3  | Subtract input A from B with Borrow; produce F and Borrow
  4  | AND inputs A and B; produce F
  5  | OR inputs A and B; produce F
  6  | EXOR inputs A and B; produce F
  7  | One's Complement A; produce F
```

The basic ALU functions are ADD, AND, OR, and COMPLEMENT. Implement subtraction using the two's complement and add operation (e.g., F = A + /B + Carry In). The various flavors of adding and subtraction are implemented by selecting the proper inputs within the ALU module (dedicated MUX's).

Note: This ALU should produce the flags SIGN, ZERO, Carry Out, and two's complement Overflow.

ALU TEST PROCEDURE: Test the module by using 3 different data values and both values for Carry In (where appropriate) for each Function. One method to do this is to provide a test table (vector) with 6 rows (one for each function) and 3 columns (data A, data B, Carry In). This forms one test of all functions. Repeat it with new data for as many combinations as required.

3. Write and test a module that performs the operation of a shifter. The shift network should be able to move all bits in an operand one bit position to the left (or right) relative to their original position or pass the operand through unchanged. The bit shifted out should be caught in a variable called "SO" and the contents of a variable called "SI" should be shifted in. Note: SO and SI contain only one significant bit, but must be represented in your program like any other integer. Shifting to the left is equivalent to multiplying by two; however, some compilers have runtime libraries that have a set of functions to perform shifts on integers.

The shifter may also be simulated using an array. Each element of the array is used to contain 1 bit of the shift path. Instead of using the host computer's shift or multiply function, you may transfer data between elements of the array with the direction specified by the shift control input.

The shifter has three functions controlled by a 2-bit control signal S as follows:

```
S   |      Function
--------+----------------------------------------------------------------
  0   | Do not shift; SO=0
  1   | Shift left one bit position; SO = ms bit; SI into LS bit
  2   | Shift right one bit position; SO = ls bit; SI into MS bit

  3   | Do not shift; SO=0
```

Test this module with the data used in Exercise 2 for all values of the function code, S, and shift in bit, SI.

4. Write and test a module that performs the function of a 4-input multiplexer (MUX). The MUX inputs are connected to the outputs of four multibit "registers" as implemented by variables A, B, C, and D (all integers). Implement the MUX using an array or CASE statement. The select lines to the MUX are in an integer called SEL. Test this module using different data values for A, B, C, and D and all of the possible values of SEL. The output of the MUX is an integer called MUX4.

5. Take the modules for the 74LS377, the ALU, and the shifter to implement a single module that simulates the architecture in the figure below:

The data paths, register, ALU, and shifter are each 16 bits wide.

Notes:

a. Make a list of all control lines and their names. The names must be unique since they will be labels in your program.

b. Design an array that implements the following table:

```
Index   |                      Elements
--------+-----+----+----+----+-------+-------+--------+--------+-----
 ICOMND | R0LD R1LD R2LD R3LD MUXASEL MUXBSEL ALUFSEL SHFTCNTL etc.
--------+-----+----+----+----+-------+-------+--------+--------+-----
   0    |     |
 
   1    |  etc.
```

Each row represents one command (microinstruction) to the architecture. Along the row are elements for each of the control lines listed above. The array will have as many dimensions as implied by the number of control signals for each command. The variable, ICOMND, is an index into the table and should be allowed to have a range from 0 through 9.

i. Each row contained in the array will be indexed by ICOMND such that all of the control signals to cause a complete operation corresponding to one cycle of SYSCLK will have the same value of ICOMND. Another index can be used to point at the control signal within a command so that each module (ALU, REGister, etc.) can access its command.

ii. The order of use of a given control line is important since the expected data must be on the input of an element before that element is simulated. The data is made available (e.g., a MUX pointed at a register containing the source data) during the processing of the same command line (value of ICOMND) but in an earlier module. This order is

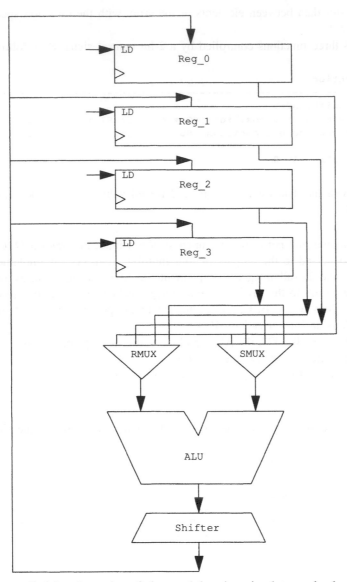

controlled by the order of the modules that simulate each element in the simulation program.

c. Write a program that simulates the architecture described above using the modules that you designed and tested in previous exercises. Be sure the order of the modules is such that data will flow as indicated by the architectural drawing above. The order of the control lines in the control table above is not important within a given value of ICOMND since each module knows the control lines it needs. The program should be organized in such a way that the events that will happen in one cycle of SYSCLK will be performed under the control of the Table in part (2) for a given value of ICOMND. The index is incremented and then the program runs again. The body of the simulation program is in the form of a loop that is executed as many times as there are different values of the index (ICOMND).

6. The control array that you designed in Exercise 5 above should be made to contain the codes necessary to implement the following algorithm:

    REG0 = (REG0 - REG1 + REG3) * 4

    Initially the registers should be set to the following data:

    REG0 = $FFEE_x$, REG1 = 0123_x, REG2 = $FF00_x$, REG3 = $5F00_x$

    Print out the contents of all simulation variables at the end of each pass through the simulation loop. Be sure the program labels each variable. The output would look best if data is printed in hexadecimal.

7. Refer to Sections 6.4.1, 8.2.1, and 8.7.3. Write a program that simulates the Fetch/Execute Cycle described in Figure 6-1. Implement the execution phases of the instructions in the Basic Instruction Set using the Immediate and Absolute address modes for all appropriate instructions. Assume that the programming model for the second-level processor you are simulating has 14 general purpose registers, one Subroutine Return address register, and one second-level Program Counter. The registers, program and data memory and databus are 16 bits wide. Each of these elements should appear as a 16-bit integer in your program, i.e., in C you might declare the PC as

    unsigned int PC;

    Use a machine instruction format that has an 8-bit operation code, two 4-bit register specification fields, and a 16-bit integer field for the immediate data or absolute address. Remember that the PC and SP are treated modulo-16; i.e., they wrap around when incremented or decremented past either end of memory.

8. Use an available macroassembler to write simulation macros in terms of the host machines assembly language for the programming model given in Exercise 7. Provide test programs and data to assemble with the macro file that demonstrate the correct execution of the simulator.

9. Select a real RALU and prepare a simulation module for it. It is best to begin with a simpler RALU like the Am2903. Refer to the appropriate data book to determine how the various pins, especially the flags, react to the control signals. Prepare test data and verify that your module performs as specified in the data sheet.

10. Use the schematic capture and simulation capabilities of a design program on a development system to set up a test program that verifies the performance of an RALU module in the provided library of models.

11. Build your own version of a real RALU module using models in a design system on a computer available to you. Draw the logic diagram using basic parts, e.g., registers, MUXes and adders. Provide test data to show that your model works like the data sheet provided by the manufacturer of the real RALU that you copied. Try to pick a relatively simple RALU for this purpose.

12. Design and simulate a 8 × 8 combinatorial multiplier. Write the simulation module such that your block "acts like" the multiplier; i.e., you are not interested in the internal structure. Provide a test program that verifies the operation of your multiplier module.

13. Design and simulate an integer MAC that performs a 16 × 16 multiplication and sums the products with previous ones into a 32-bit register. Use a single copy of the 8 × 8 multiplier block from Exercise 12 in the MAC architecture. Place it in such a way that a controller can use it to perform the 16 × 16 multiplication and produce a 32 bit product. You may use more adders but no more multipliers in this architecture. Produce suitable overflow flags. Provide a test program that verifies the operation of your MAC module.

14. Use a suitable schematic capture and simulation program to implement the 4 × 4 Multiplier architecture based on the Carry Save Addition technique described in Chapter 3. Use basic structures like AND gates and adders to implement the network. Set up test vectors that allow you to demonstrate that the circuit works. If the simulation program provides timing facilities, evaluate the worst-case propagation delay for the network.

15. If you have access to one of the workstation-based design systems that has models for the more advanced modules discussed in Chapter 3, then explore the TI 74ACT8847 FPU by implementing a simple counter-EPROM-based control word generator to stimulate the FPU. Use the TI data book and its example programs to stimulate the device and examine the results produced. This is an efficient way to study the more complex devices discussed in Chapter 3.

Chapter 7

ALGORITHMS AND ARCHITECTURES

7.1. INTRODUCTION

Having laid the groundwork in the preceding chapters, we will now discuss the design of Microprogrammed Algorithmic State Machines using a single level of control. Each of these designs is driven by an important algorithm that the reader may have encountered in the course of writing programs for computers.

Most of the algorithms shown deal with numerical calculation. They have affected the design of the architectural components and controllers discussed earlier. The algorithms are important in their own right and the reader has probably encountered them in other contexts. The important area of graphical display applications is not covered in this chapter. While the area of generating high-end computer graphics displays has been traditionally dominated by machines of this type, their complexity does not lend itself to the scope of this book. The flow of data into and out of the state machine examples is not dealt with explicitly. This portion of the design is application specific. Data flow control would normally be supervised by additional microprogram sequences in the controller running the algorithm, i.e., data flow and algorithm controls would be coordinated for maximum efficiency.

All of the applications shown in Section 7.2 are produced using an Algorithmic State Machine composed of a controller and Register Arithmetic/Logic Unit (RALU). The last two sections show the use of the more complex architectural components in unusual environments. Section 7.3 discusses the replacement of sequences performed in one architectural component with single clock cycle operations performed by another more complex component. Coprocessors extend the instruction set of a host computer. Section 7.4 describes two implementation methods for coprocessors. Memory-mapped coprocessors consist of a very complex arithmetic component used as an architecture in a state machine in which the controller is simulated by a computer. This surprising juxtaposition results in a fast, cost-effective improvement in the performance of the host computer. The conventional coprocessor is shown as a microprogrammed state machine attached to a host Central Processing Unit (CPU).

A specific design example is shown in the last section of the chapter. It is a loosely attached coprocessor that involves multiple processing elements with a single microprogrammed controller to explore various dataflow architectures.

7.2. DESIGNS USING ONE LEVEL OF CONTROL

We will now study the implementation of several common algorithms in a Microprogrammed Algorithmic State Machine having a single level of control using the languages that were introduced above. The architectural elements discussed in Chapter 3 will be controlled directly by microinstruction sequences installed in the microprogram memory of a controller from Chapter 4. The complete algorithm resides in the controller's microprogram memory. This is to be contrasted with the case where a set of general purpose sequences is placed in the controller's memory and a second level of control is established such that the algorithm is expressed in terms of the second level of control. This case will be considered in Chapter 8.

Many applications are possible using a single level of control. In all cases, these applications can operate at a higher speed than those implemented with a second control level. A large class of problems that can be executed only on the very fastest computer, or not at all, can be solved using dedicated solutions similar to the examples given in this chapter.

Most people consider that a dedicated solution, while having faster execution time, is slower to develop and less flexible in case the needs of the solution change. It is the purpose of this chapter to show that for many cases this is not true. Chapter 7 discusses several standard mathematical algorithms and their implementation. These implementations can be thought of as building blocks that can be incorporated into more complex designs. The generality of the hardware building blocks in Chapters 3 and 4 causes most variations in the solution of problems to be reduced to changes in the microprogram. The flexibility of a programmable solution is present in our case as well as in the general purpose computer. In other cases, small changes to data paths are needed. If the description and simulation of a solution can be performed at a high level, the designer can anticipate some of the hardware flexibility needed for an array of solutions.

The first set of solutions is implemented on a state machine having a ROM/Latch controller. This simple controller is given "instructions" by using the power of the microcode assembler. Once the machine is defined to the assembler, the designer has as simple a programming environment as on any computer. The assembler chosen for this task will be the Macroassembler as opposed to the Meta-Assembler since it provides more flexible control of memory allocation. The next several solutions will be based on common arithmetic algorithms implemented in a microprogrammed algorithmic state machine of the form shown in Figure 7-7. This simple RALU-based state machine represents a pedagogical platform for the examination of the accumulation, multiplication, division, sum of products, and power series algorithms in the various parts of Section 7.2. In most real applications, these algorithms would be combined with others to complete a system that could acquire data from a host, or the real world, transform it, and present the results back to a host or directly to the user. The author isolates the algorithms for study here for the sake of simplicity. Interface methods are discussed in the context of coprocessors in Section 7.4.

7.2.1. Some Introductory Designs

As an introduction to using the tools described earlier, we will show two small solutions to a classic problem — the traffic light controller. It is assumed that the reader has experienced such a controller and has possibly studied it as an example in other state machine and microcomputer design areas. Our example will be implemented using the ROM/Latch controller shown in Section 4.3 and Figure 4-3. The purpose of this exercise is to demonstrate the creation of a programming environment for a machine that does not appear to have any. By using the features of the microprogram assembler introduced in Chapter 5, we intend to create a "language" with which we can represent our solution as expressed in Algorithmic State Machine (ASM) diagrams or Hardware Description Language (HDL).

Let us first state the specification of the problem. A street intersection is shown in Figure 7-1. Traffic is controlled in the intersection by light signals for cars and walk signals for pedestrians. Light signals are paired across the intersection such that lights and walk signs having the "1" subscript control traffic in the North-South direction, while those having the "2" subscript control traffic in the East-West direction. Sensors (i.e., TRIP_1 and TRIP_2) are placed in the street to determine the presence of cars. The sensors are paired in the same way as the lights.

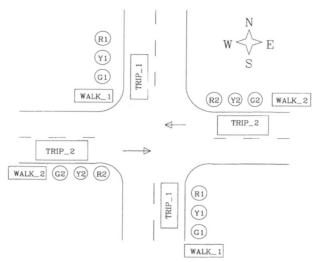

Figure 7-1: Road intersection.

The rules of operation, beginning with traffic stopped in the North–South direction, are as follows:

1. Lights RED_1, GREEN_2, and WALK_2 are ON while all other lights are OFF.

2. After 4 units of time, the WALK_2 light should begin blinking OFF/ON, each state requiring one unit of time. The WALK_2 light blinks twice and is then extinguished. The TRIP_1 sensor is enabled.

3. The controller should wait until TRIP_1 senses the presence of a car. The YELLOW_2 light turns on for one unit of time.

4. Finally, the RED_2, GREEN_1, and WALK_1 lights are turned on.

5. The E-W sequence, similar to that for the N-S direction, continues.

6. The entire sequence repeats endlessly.

A portion of the ASM diagram is shown in Figure 7-2. The RESET condition sends the machine to State_0 (A) where the traffic in the North-South direction is stopped. After four state times, the WALK_2 light begins to blink, turning OFF for one state and then ON for the next. This action takes a total of four state times. TRIP_1 sensing begins in State_8 (I) where the ASM remains until a car is sensed attempting to move in the N-S direction. The yellow light for the E-W traffic is turned on while the N-S red light is retained. The cycle continues with a similar set of states to allow N-S traffic to move.

We will implement our first solution using the ROM/Latch machine shown in Figure 7-3. Each state will represent one unit of time, thus setting the period of SYSCLK. The number of states needed will be greater than 9 (see Rule 2) and less than 32 requiring five state variables if we use binary state coding. The two inputs TRIP_1 and TRIP_2 will be applied to the least significant address lines of the memory (here shown as two 8-bit wide EPROMs), while the state variables are applied to address inputs A2 through A6. A register

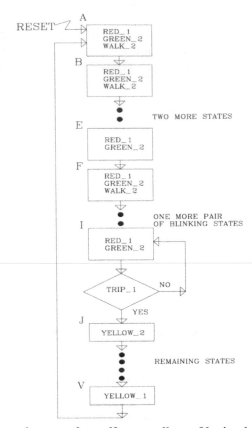

Figure 7-2: ASM diagram fragment for traffic controller — Version 1.

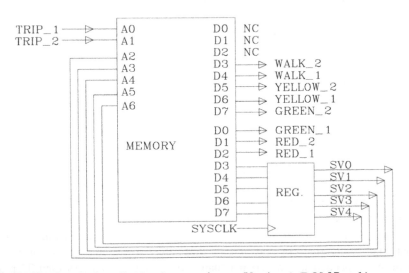

Figure 7-3: Traffic signal controller implementation — Version 1 (ROM/Latch).

(REG) is placed on the output of the memory to hold the state variable values between clock ticks. The light controls, RED_1, RED_2, etc. are not latched to make programming easier. The inputs, TRIP_1 and TRIP_2, should be latched or synchronized also (not shown here) to prevent transitions during the setup time just before the transition of the clock for the state variable register.

Two fragments of memory data are given next. The first shows the memory contents used to support the transitions from State_0 (A) to State_1 (B) and from B to C. The controller function in a ROM/Latch state machine implementation is accomplished by using the ROM data fields, here labeled State V, to contain the value of the next state to be reached based on the conditions of TRIP_1 and TRIP_2.

```
                          S                 Y Y
                          t         G G e e
                          a         r r l l W W U
                          t     R R e e l l a a n
                          e     e e e e o o l l u
                                d d n n w w k k s
                          V                     e
                                1 2 1 2 1 2 1 2 d
  State     Addr(words)   +--+-+-+-+-+-+-+-+-+-+
   000          000       |01|1|0|0|1|0|0|0|1|7|
   000          001       |01|1|0|0|1|0|0|0|1|7|
   000          002       |01|1|0|0|1|0|0|0|1|7|
   000          003       |01|1|0|0|1|0|0|0|1|7|

   001          004       |02|1|0|0|1|0|0|0|1|7|
   001          005       |02|1|0|0|1|0|0|0|1|7|
   001          006       |02|1|0|0|1|0|0|0|1|7|
   001          007       |02|1|0|0|1|0|0|0|1|7|
```

In the preceding fragment, program flow "CONTINUED" from State_0 (A) through State_1 (B) to State_2 (C) regardless of the values of TRIP_1 and TRIP_2. Thus, the STATE V column shows that the next state desired for all combinations of TRIP_1 and TRIP_2 (address inputs A0 and A1, respectively) to be one plus the value of the present state. The memory is organized such that each state requires 2^n "words" to determine the next state, where n is the number of conditional inputs; here n = 2, thus four words of memory are required for each state.

The next fragment shows the memory contents at State_8 (I).

```
008   020   |08|1|0|0|1|0|0|0|0|7|
008   021   |09|1|0|0|1|0|0|0|0|7|
008   022   |08|1|0|0|1|0|0|0|0|7|
008   023   |09|1|0|0|1|0|0|0|0|7|
```

Here, the next state conditions reflected in the STATE V column are determined by the value of TRIP_1. If TRIP_1 is *False*, the ASM diagram shows that the next state time should also be spent in State_8 (I) while State_9 (J) is occupied otherwise. Notice that TRIP_1 is connected to memory address input A0. The entries for addresses 20 and 22, corresponding to TRIP_1 being *False*, are State_8; while for addresses 21 and 23, the entries are State_9.

Creating an assembler language for this network requires that each microinstruction specify the contents of four "words", one for each exit possibility. For our example we will use the MACRO assembler, A68K, instead of the dedicated microprogram assembler in Chapter 5. The microcode assembler is normally used when a single microinstruction requires only one word of machine code. The other assembler is easier to use in this case. The process of creating a language is carried out in two steps. The first step assigns values to symbols as a result of an instruction in the program. The second step places the values of the symbols in the correct location in the machine instruction word.

We will begin our example by creating the second step. The format of the machine instruction word is shown next.

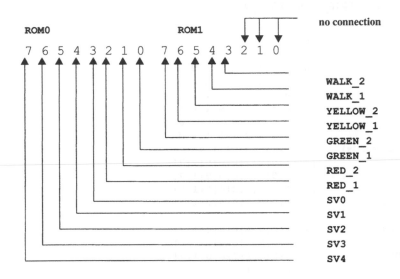

For each machine word, we need to create 2 bytes of data, one for the left EPROM (ROM0) and one for the right EPROM (ROM1) since many commonly available EPROMs are 8 bits wide. To do this, we will create a macro that will be invoked in each microinstruction after the symbols have been set. We call the macro "endsc", implying that it *must* be used at the end of every microinstruction or else data will not be created. The text of this macro is given as follows:

```
; endsc macro combines all the previously defined macros
; and creates an assembled word
endsc macro
; Word Number 0 - TRIP_1 and TRIP_2 are FALSE
 dc.b by0_nxt_st<<3!red1<<2!red2<<1!green1
 dc.b green2<<7!yellow1<<6!yellow2<<5!walk1<<4!walk2<<3!$7
; Word Number 1 - TRIP_1 is TRUE and TRIP_2 is FALSE
 dc.b by1_nxt_st<<3!red1<<2!red2<<1!green1
 dc.b green2<<7!yellow1<<6!yellow2<<5!walk1<<4!walk2<<3!$7
; Word Number 2 - TRIP_1 is FALSE and TRIP_2 is TRUE
 dc.b by2_nxt_st<<3!red1<<2!red2<<1!green1
 dc.b green2<<7!yellow1<<6!yellow2<<5!walk1<<4!walk2<<3!$7
; Word Number 3 - TRIP_1 and TRIP_2 are TRUE
 dc.b by3_nxt_st<<3!red1<<2!red2<<1!green1
 dc.b green2<<7!yellow1<<6!yellow2<<5!walk1<<4!walk2<<3!$7
 nolist
 DEFAULT
 list
 endm
```

The "endsc" macro declares constant byte (dc.b) storage for a total of four words, one for each exit combination, or 8 bytes. The declare constant word (dc.w) assembler directive would have been effective here except that the line of text was too long to easily print. The fortuitous alignment into a "computer word" normally would not occur. Symbols for each of the lights are present (e.g., red1, red2, and walk2). The "<<" operator tells the assembler to shift the value in that symbol to the left the number of bits specified; e.g., red1<<2 yields 4 if red1 is *True* (1). The shifting moves the values of the symbols to their correct place in the physical word.

Four symbols are needed to specify the next state for each combination of the inputs, TRIP_1 and TRIP_2. The symbols, by0_nxt_st through by3_nxt_st, are set as a result of commands given in the program. A short segment of the program showing the microinstruction for State_0 (A) is given next.

```
; Run East-West Traffic

A:      CONT       ;State_0
        RED_1
        GREEN_2
        WALK_2
        endsc
```

Some of the commands that make up a language appropriate to our problem are now apparent. The ASM diagram shows the outputs RED_1, GREEN_2, and WALK_1 to be *True*. They are reflected here as commands within the State_0 (A) microinstruction. Each microinstruction requires several commands, one for each element of the architecture (here the traffic lights) and one for the controller to determine the next state. The "CONT" command reflects the information in the ASM diagram, indicating that a transition to State_1 (B) is to be made regardless of the values of TRIP_1 and TRIP_2; i.e., "CONTinue" to the next state. The "endsc" command causes the assembler to perform the operations requested in the commands.

Now let us consider how the symbols like red1 and red2 within the endsc macro are set by a command like RED_1 in our program. We create another macro called RED_1 as below:

```
; LIGHT MACROS

RED_1:   macro
red1     set        $1
         endm
```

The assembler directive "set" is similar to "EQU" in that it assigns a value, here hexadecimal one, to the symbol red1. The difference is that the assembler allows this to be done again and again without causing an error. In most programs, a symbol such as CR, which stands for the "Carriage Return" character, is assigned a value at the beginning which should not be changed. The EQU directive is normally used since the programmer would like to be told if he inadvertently changed the value, hence the code, for a carriage return command by re-EQUating it. Here we want to use the "set" directive because we plan to change the value of a symbol in each microinstruction.

The assembler reads our program file and, as is discussed more fully in the second part of Chapter 5, translates our RED_1 command using the statements within the RED_1 macro. Once all of the symbols like "red1" are specified, the endsc macro is invoked to create the actual machine instruction by "ORing", i.e., the "!" operator used in the endsc macro, the shifted symbol values together.

The ASM diagram protocol tells us that an output is *False* if its name is not present in an output box in the diagram; see Chapter 1. We can create the same effect in our programming language by having endsc invoke another macro called DEFAULT after the symbols are used to create the machine code. The DEFAULT macro is shown next for this example.

```
; set up default values
DEFAULT    macro
red1       set        0
red2       set        0
green1     set        0
green2     set        0
yellow1    set        0
yellow2    set        0
walk1      set        0
walk2      set        0
cont_st    set        (*/8)&$1F
           endm
```

The effect of this macro is to set all of the symbols *False*. If a symbol is not turned to *True* by a command in a microinstruction, its value will be *False* when endsc is executed as the last command.

The last set of macros is the most important in that they create the control constructs of our programming language. To support the program for the ASM diagram shown in Figure 7-2, we need four control instructions, CONTinue, JMP, TEST_1, and TEST_2. The "TEST" instructions should sense the appropriate input and return to the current state until the input becomes *True*. The JMP instruction needs an operand that specifies the state that is the destination of the "jump". Macros to set the values of by0_nxt_st, etc. are shown next.

```
; NEXT STATE MACROS
CONT:   macro
by0_nxt_st set cont_st+1
by1_nxt_st set cont_st+1
by2_nxt_st set cont_st+1
by3_nxt_st set cont_st+1
        endm

JMP:   macro \1
by0_nxt_st set (\1/8)&$1F
by1_nxt_st set (\1/8)&$1F
by2_nxt_st set (\1/8)&$1F
by3_nxt_st set (\1/8)&$1F
        endm

TEST_1: macro
by0_nxt_st set cont_st
by1_nxt_st set (cont_st+1)&$1F
by2_nxt_st set cont_st
by3_nxt_st set (cont_st+1)&$1F
        endm

TEST_2:  macro
by0_nxt_st set cont_st
by1_nxt_st set cont_st
by2_nxt_st set (cont_st+1)&$1F
by3_nxt_st set (cont_st+1)&$1F
        endm
```

Several points must be noted here. The symbol "cont_st" contains the value of the state code for the present state until the end of the expansion of "endsc" in each microinstruction. The symbols are masked to the length of the fields they are intended to fill by the AND (&) operation. The state variable field in this example is 5 bits long. This prevents the number from overwriting adjacent fields. Conditional assembly permits one to extend this concept to actually flag symbols that are out of range.

The JMP macro carries one operand signaled by "\1". Whatever symbol (i.e., the address of a state) is found at macro expansion time in the operand field of the JMP command will be used everywhere the "\1" appears in the body of the macro. Notice that the value of the symbol is shifted right by 3 bit positions, i.e., divided by 8. Each state requires 8 bytes, which the assembler counts, to create a single state. We are adapting the byte addressing of the assembler to our need to address states.

The two TEST commands combine commands to continue in the lines where the test condition is *True* with the value of the present state location otherwise. They need no operands since the destination of the exit is either the present state or the next state. A program fragment is shown next that depicts the use of the TEST_1 instruction.

```
; Sense the TRIP_1 sensor
; Loop in state I until a car comes
I:         TEST_1 ;State_8
           RED_1
           GREEN_2
           endsc
```

The flow of control in State_8 (I) in the ASM diagram shown earlier is generated by this set of commands. An excerpt from the assembler listing below shows the action taken by the assembler.

```
            464 ;  Loop in state I until a car comes
            465
            466 I:    TEST_1      ;State_8
000008      467+by0_nxt_st set cont_st
000009      468+by1_nxt_st set (cont_st+1)&$1F
000008      469+by2_nxt_st set cont_st
000009      470+by3_nxt_st set (cont_st+1)&$1F
            471       RED_1
000001      472+red1 set $1
            473       GREEN_2
000001      474+green2 set $1
            475       endsc
000040  44  476+     dc.b by0_nxt_st<<3!red1<<2!red2<<1!green1
000041  87  477+     dc.b green2<<7!yellow1<<6!yellow2<<5!walk1<<4!walk2<<3!$7
000042  4C  478+     dc.b by1_nxt_st<<3!red1<<2!red2<<1!green1
000043  87  479+     dc.b green2<<7!yellow1<<6!yellow2<<5!walk1<<4!walk2<<3!$7
000044  44  480+     dc.b by2_nxt_st<<3!red1<<2!red2<<1!green1
000045  87  481+     dc.b green2<<7!yellow1<<6!yellow2<<5!walk1<<4!walk2<<3!$7
000046  4C  482+     dc.b by3_nxt_st<<3!red1<<2!red2<<1!green1
000047  87  483+     dc.b green2<<7!yellow1<<6!yellow2<<5!walk1<<4!walk2<<3!$7
```

Notice that each command, here shown in capital letters, results in the evaluation of a macro. The value chosen by the assembler is shown in the left-hand column. The value of cont_st was 8 in line 467 which resulted in by0_nxt_st having a value of 8. RED_1 causes the symbol red1 to take on the value of 1. After all command macros are evaluated, the endsc macro is invoked. The symbols are shoved over to their appropriate positions by the shifting action and a byte of memory data is created by the dc.b instruction. These values are seen on the lines from 476 through 483 in the second column next to their address in a byte-wide memory space.

To move this list of data bytes to the appropriate left- or right-hand EPROM, we must take the assembler output, shown below in S-record format, and "split" it into two streams for the EPROM programmer. Most EPROM programmers have this function; however, it may be performed by a small utility program written by the design engineer if necessary.

S-Record File:

```
S0030000FC
S2140000000C8F0C8F0C8F0C8F148F148F148F148FF3
S2140000101C8F1C8F1C8F1C8F248F248F248F248F63
S2140000202C872C872C872C87348F348F348F348FF3
S2140000303C873C873C873C87448F448F448F448F63
S21400004044874C8744874C875027502750275027 93
S214000050582758275827582763176317631763177B7
S2140000606B176B176B176B1773177317731773175B
S2140000707B177B177B177B1783078307830783070B
S2140000808B178B178B178B1793079307930793077B
S2140000909B179B179B179B179B079B07A307A307FB
S2140000A0A847A847A847A847004700470047004773
S804000000FB
```

For the purposes of comparing the S-record data with that shown in the memory contents table above, we will take the sixth S-data record and break it up into its logical sections.

S2 14 000040 4487 4C87 4487 4C87 5027 5027 5027 5027 93

The first three fields define the record type (S2 → data), number of bytes in the record (14 hex), and the address of the first byte of data (000040). Data words have been separated to correspond to the 16-bit words needed in our state machine. The final 8-bit field is the checksum, the sum of all bytes in the record. Refer to Appendix C of Chapter 5 for more detail on the S-record format.

Now let us take the first data word and separate it into a form comparable with that shown in the memory contents table and the assembler listing fragment for state I above.

4487 → 0100 0100 1000 0111 **binary**

01000 1 0 0 1 0 0 0 0 111 **binary**

It is grouped according to the logical fields for our machine; i.e., the left-hand field, 01000, corresponds to the 5-bit state variable field.

08 1 0 0 1 0 0 0 0 7   memory table format.

For ease of debugging, a utility program may be written that performs the reformatting of the S-record into the memory contents form.

We will need one more macro to accomplish some housekeeping for the assembler. The macro is called RESET and it is invoked immediately after the origin (ORG) statement at the beginning of the assembler source program. Its purpose is to initialize the symbols associated with state control; e.g., cont_st. It will also be used to specify the initial values of the lights as well. The RESET macro performs the work of the DEFAULT macro at the beginning of the program. It is shown next:

```
; RESET macro -- All outputs are set False.
;                       Primary purpose of RESET is to initialize
;                       cont_st.
; The RESET macro must be the first statement in the program
;  after the origin statement (ORG).
RESET:                  macro \1
red1                    set $0
red2                    set $0
green1                  set $0
green2                  set $0
yellow1                 set $0
yellow2                 set $0
walk1                   set $0
walk2                   set $0
cont_st                 set (*/8)&$1F
                        endm
```

We have shown in the previous example that we can create a programming environment where none existed by the judicious use of a macro assembler. A program written in our "language" can be made to correspond directly with the ASM diagram, thus allowing our programs to be concise, complete, and very easily written. We notice in our solution above that time periods are set by stepping through several identical states. While the large number of bytes available in an EPROM actually encourages this solution to event timing, we will use this opportunity to make a small addition to the architecture of our machine to assist in this effort. It will allow us to illustrate the control of a more complex architecture than simple signal lights.

A counter, modeled on the SN74LS163 binary up-counter, is introduced into the architecture of our state machine implementation to help us create a counted-loop control construct. Our ASM diagram takes the form shown by the fragment in Figure 7-4. The program segment, written in the problem-specific language introduced above, is shown to the left in the figure.

The loop run by our counter replaces states 0 through 9 in our earlier ASM diagram. The counter is initialized in State_0 (A) so that it reaches terminal count (15) after being incremented 5 times (a total of 10 state times). The two states within the loop body are responsible for blinking the WALK_2 light.

The problem specification has been changed slightly to cause the light to blink for the entire period before TRIP_1 sensing begins. Notice that the counter is incremented within the loop body; i.e., permission for the counter to increment is given by the programmer. For economy of illustration, we will further reduce the number of outputs from our state machine by eliminating the WALK_1 signal. Running the E_W direction a fixed amount of time allows us to replace the TRIP_2 input with the terminal count output of the counter, CNTR_TC.

The implementation of our state machine is shown in Figure 7-5. The 74LS163 up-counter signals when its Q outputs are all equal to 1 (i.e., count = 15) by setting the TC output *True* (1). Our ASM will sense the CNTR_TC output by using the memory address A1 input replacing the TRIP_2 input in the previous example.

The counter's functions are controlled by several inputs including PE (parallel enable) and CEP (Count enable parallel). We have chosen to use only these signals since we do not need to clear the counter. Data may be loaded into the counter from the microprogram via the D3–D0 bit positions in the right-hand EPROM. We therefore have created a programmable counter that can be initialized to a particular value, held, or incremented under program

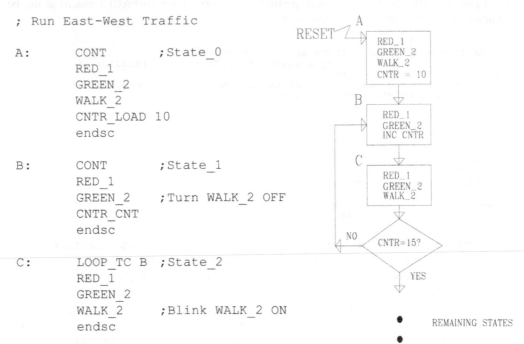

```
; Run East-West Traffic

A:      CONT        ;State_0
        RED_1
        GREEN_2
        WALK_2
        CNTR_LOAD 10
        endsc

B:      CONT        ;State_1
        RED_1
        GREEN_2     ;Turn WALK_2 OFF
        CNTR_CNT
        endsc

C:      LOOP_TC B   ;State_2
        RED_1
        GREEN_2
        WALK_2      ;Blink WALK_2 ON
        endsc
```

Figure 7-4: Program segment and corresponding ASM diagram for counter-assisted traffic signal controller.

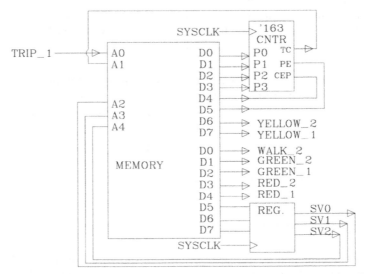

Figure 7-5: Traffic signal controller implementation — Version 2 (counter assisted).

control. The remainder of the machine implementation is the same except that we have taken advantage of the reduced number of states by using only three state variables. The complete bit map is shown below.

We will introduce two groups of instructions, one to control the counter itself and another to take advantage of the control construct it makes possible. Counter control instruc-

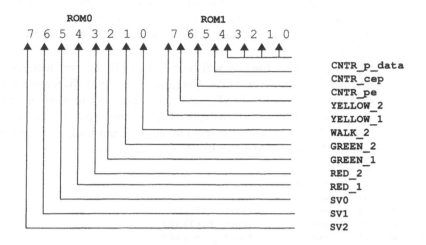

tions should allow the programmer to load an arbitrary data value presented on the P inputs, increment, or hold the counter. These operations are possible using a combination of the PE and CEP inputs that can be determined from the SN74LS163 datasheet. In creating our language, we could make PE and CEP explicit commands in the same way as we did the light controls (e.g., RED_1). This method did not improve the readability of our program, so we chose to create a set of commands as follows:

    1. CNTR_CNT
    2. CNTR_LOAD AMOUNT

These commands set values for both PE and CEP and have the following form:

```
; Counter Control Macros

CNTR_CNT:       macro
cntr_cep        set         1
cntr_pe         set         1
                endm

CNTR_LOAD:      macro       \1
cntr_cep        set         1
cntr_pe         set         0
cntr_data       set         \1&$F
                endm
```

Notice that the operand for CNTR_LOAD is placed in a symbol, cntr_data, that will be used to initialize the counter in endsc. The first word of the endsc macro takes the following form:

```
dc.b    by0_nxt_st<<5!red1<<4!red2<<3!green1<<2!green2<<1!walk2
dc.b    yellow1<<7!yellow2<<6!cntr_pe<<5!cntr_cep<<4!cntr_data
```

Masking of the 4-bit counter data field is used to prevent the programmer from using a value that would overflow into the counter control fields. Notice that the next state control portion of the word (i.e., by0_nxt_st) is shifted to the left end of the word, a 5-bit shift, since it is now only a 3-bit wide field.

The new program control instruction will be called LOOP_TC and requires one operand, the destination if the condition is *False*. Comparing the ASM diagrams for the two versions of this example shows that the TEST_1 instruction does not need an operand since the two possible destinations are the present state, held in the symbol "cont_st", or the present state plus one. LOOP_TC allows the programmer to jump back over an arbitrarily long loop body. A macro that creates the LOOP_TC instruction is shown next:

```
LOOP_TC:        macro       \1
by0_nxt_st      set         (\1/8)&$7
by1_nxt_st      set         (\1/8)&$7
by2_nxt_st      set         (cont_st+1)&$7
by3_nxt_st      set         (cont_st+1)&$7
                endm
```

We use the operand as the next state (suitably adjusted to convert byte addresses to states by the >>3 construct) if the Terminal Count output of the counter is *False*. Counter_-TC is applied to the memory address A1 input. The "cont_st+1" is used if TC is *True*; i.e., the counter has reached 15 decimal or F hex.

The counter is initialized in State_0 (A) to 10 decimal in the program in Figure 7-4 since it is incremented once in a loop body containing two states. The total number of state times before TRIP_1 sensing begins is to be 10. Blinking the WALK_1 light is accomplished by one of the state pairs used in the earlier version. The counter is incremented in State_1 (B) and tested in State_2 (C). Incrementing could be accomplished equally well in C since both counter and state register are rising-edge-triggered, i.e., no race condition is possible.

In the remaining part of the program, TRIP_1 sensing is accomplished with the same instruction used in the earlier example. Other combinations of the TRIP_1 and Counter_TC testing are possible, allowing the designer to create an instruction that is especially well tailored to his problem.

We will conclude our simple traffic controller example by moving to a controller based on the Am2910 in a circuit shown in Figure 4-14. The new version of the controller, in Figure 7-6, is substantially different than the ROM/Latch versions above. The four word structure needed for each microinstruction in order to build the control functions is now replaced by hardware consisting primarily of the next address multiplexer in the Am2910. We also have moved the counter from the architecture of our state machine into the controller. Since only a single controller command can be used in a microinstruction, we must make at least a little change in our microprogram.

Our goal for this version is to illustrate the power of the definition, or macro, file in allowing us to adapt an existing program to the new hardware. We will attempt to use the same microprogram used in the second version, i.e., the one with the loop counter. If we are able to exploit the capabilities of the macro file to its fullest, then we should be able to use the original source program file with little or no editing.

The Am2910 Next State Logic Device, in Figure 7-6, furnishes several major parts of a fully developed microprogram controller. The microprogram memory and pipeline register must now contain all of the architectural control signals (i.e., the lights) but also many more lines associated with the controller function. These include the "instruction" fields, 2910_i

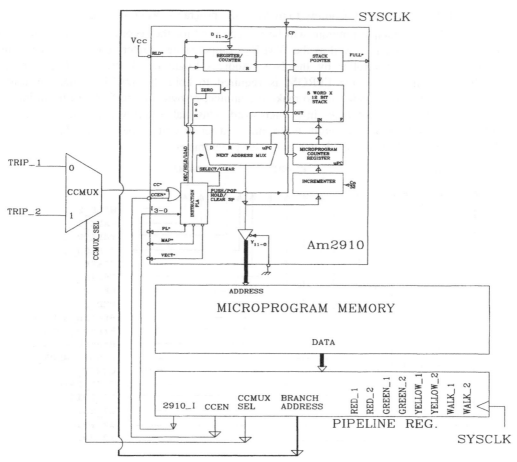

Figure 7-6: Traffic signal controller implementation — Version 3.

and cccn, as well as the *conditional multiplexer* control field, CCMUX_SEL, and the branch address field. We will make the "pin" assignments shown below and at the top of page 238.

EPROM Address/Data Line Definitions:

1. *Memory Address:*
   A0:2910_Y0, A1:2910_Y1, A2:2910_Y2, A3:2910_Y3, A4 - An:0
2. *EPROM #1:*
   D7:2910_I3, D6:2910_I2, D5:2910_I1, D4:2910_I0,
   D3:2910_CCEN, D2:CCMUX_SEL, D1:Branch_Addr_3, D0:Branch_Addr_2
3. *EPROM #2:*
   D7:Branch_Addr_1, D6:Branch_Addr_0, D5:RED_1, D4:RED_2, D3:GREEN_1,
   D2:GREEN_2, D1:YELLOW_1, D0:YELLOW_2
4. *EPROM #3:*
   D7:WALK_1, D6:WALK_2, D5-D0:no connect.
5. *EPROM #4:*

D7-D0:no connect. Dummy-to make utility programs easier to write.

While our example program will be the one used for the second version, we have restored the TRIP_2 input as well as the WALK_2 output since counter control is now accomplished within the controller by the Next Address Logic or "Instruction PLA" in the Am2910. A third 8-bit wide EPROM is required to hold the WALK signals. A fourth EPROM is added to the definition to make the number of bytes in the S-record be an integer multiple of the number of bytes in a microword. This makes utility programs such as the EPROM data splitter and S-record unassembler easier to write. The fourth EPROM is not physically included in the circuit nor is it ever programmed.

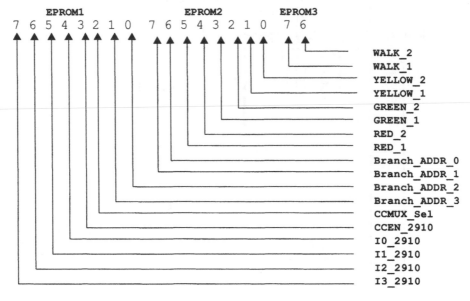

We must now adapt the macro file used for the previous version to the new hardware. The next table shows a subset of the instructions supported by the Am2910.

2910_i	Mnem.	Name	Reg/Cntr Contents	Fail		Pass		Reg/Cntr
				Y	Stack	Y	Stack	
3	CJP	Conditional Jump	X	PC	Hold	D	Hold	Hold
9	RPCT	Repeat BR Addr,	<> 0	D	Hold	D	Hold	Dec
		Cntr <> 0	= 0	PC	Hold	PC	Hold	Hold
12	LDCT	LD Cntr & Continue	X	PC	Hold	PC	Hold	Load
14	CONT	Continue	X	PC	Hold	PC	Hold	Hold

The subset chosen will be used to directly implement the control instructions that were used in the second version of our example. These were CONT, TEST_1, JMP, and

LOOP_TC. More instructions implemented by the Next State Logic in the Am2910 can be found in *The Am2900 Family Data Book* by AMD (see Bibliography). The instructions include other looping and branching combinations. The subset reproduced here allows us to build the instructions used in our version 2 program.

The macros for the program control instructions are now presented.

```
; NEXT STATE MACROS
; 2910 built-in instructions

CONT          EQU    14 ;Continue
CJP           EQU    3 ;Conditional Jump to address
                       ;in Pipeline (Branch Address)
LDCT          EQU    12 ;Load counter and continue
RPCT          EQU     9 ;Repeat loop, counter <> 0
                       ; address of head of loop in Pipeline

JMP:          macro \1
I_2910     set    CJP
CCEN_2910  set    1 ;False
branch_addr set  \1/4&$F
              endm

; Test TRIP_1 and loop in place until TRUE (0)

TEST_1:    macro
I_2910     set    CJP
CCEN_2910  set    0 ;True
CCMUX_SEL  set    0 ;Point at TRIP_1
branch_addr set  cont_st
              endm

; Test TRIP_2 and loop in place until TRUE (0)

TEST_2:    macro
I_2910     set    CJP
CCEN_2910  set    0 ;True
CCMUX_SEL  set    1 ;Point at TRIP_2
branch_addr set  cont_st
              endm

; Loop until terminal count
; - also decrements counter inside 2910

LOOP_TC:    macro \1
I_2910      set    RPCT
branch_addr set  \1/4&$F
              endm

; Counter Control Macros - now part of controller

CNTR_LOAD: macro \1
I_2910      set    LDCT
branch_addr set  \1&$F
              endm
```

Symbols for the Am2910 instructions to be used are EQUated at the beginning allowing succeeding code to be more readable by the programmer. The first thing to notice is that since the Am2910 itself implements all of the control constructs, each microinstruction re-quires only a single, multibyte word. The controller portion of the word consists of fields for the 2910 instruction, the Condition Code Enable input (CCEN), the Conditional Multiplexer Control signal (CCMUX_SEL), and the Branch Address. The rest of the microword contains fields for all of the architectural control signals (e.g., RED_1).

The JMP instruction is constructed using the Am2910 Conditional Jump to Branch Address instruction, CJP. The CCEN input to the Am2910 is set to False to make the con-dition appear always *True* (Pass) to the controller. The branch_addr symbol is set to one fourth of the byte address specified with the help of the assembler through operand one, "\1". TEST_1 and TEST_2 use the same Am2910 instruction. The operand is already known since it is the present state specified by the symbol cont_st which was set to the assembler's location counter in DEFAULT. The CCEN input is enabled and the CCMUX_SEL field points the CCMUX at TRIP_1 or TRIP_2, respectively.

LOOP_TC is built using the RPCT instruction which decrements the counter and tests it against zero in the same clock cycle. The loop body is executed when the counter is zero. When the counter "underflows", control passes to the next state after the one containing the LOOP_TC instruction. The branch_addr symbol is set to the address of the head of the loop body.

Counter initialization is now made a program control function since the Am2910 LDCT instruction simultaneously loads the counter and CONTinues to the next state in the sequence. The combination of functions that were handled in both the controller and archi-tecture in the previous version requires a minor modification to the actual program.

The segment of the program from version 2 is shown next with the modifications.

```
; Run East-West Traffic

A:                     ;State_0
     RED_1
     GREEN_2
     WALK_2
     CNTR_LOAD 4
     endsc

B:                     ;State_1
     RED_1
     GREEN_2           ;Blink WALK_2 OFF
     endsc

C:   LOOP_TC B         ;State_2
     RED_1
     GREEN_2
     WALK_2            ;Blink WALK_2 ON
     endsc
```

The modifications consist of changing the counter initialization value to accommodate a down-counter that must underflow, and the removal of all the CONT instructions. Removal could have been accomplished by making the CONT macro empty; however, a smoother method that is appropriate to any of the previous versions is used. The DEFAULT macro sets the I_2910 symbol to 14, the CONT instruction code as follows:

```
; Set up default values
; Lights are OFF

DEFAULT    macro
I_2910     set   CONT
CCEN_2910 set   1 ;False
```

```
        red1      set    $0
        red2      set    $0
        green1    set    $0
        green2    set    $0
        yellow1   set    $0
        yellow2   set    $0
        walk1     set    $0
        walk2     set    $0
        cont_st   set    */4 ;convert byte address to state
                  endm
```

If the I_2910 default value is not changed in the program by using another control instruction, then the CONT value will be assembled in the microcode. This method could have been used in any of the other examples, but explicit use of CONT was intended to re-inforce the "multiprocessing" properties of this type of system. Does this controller need a RESET macro as used earlier to initialize any variable symbols at the beginning of the assembler program?

The endsc macro is now shown to illustrate the code generation adaptations.

```
; endsc macro combines all the previously defined macros
; and creates an assembled word

endsc macro
      dc.b  I_2910<<4!CCEN_2910<<3!CCMUX_SEL<<2!(branch_addr>>2&$3)
      dc.b  ((branch_addr&$3)<<6)!red1<<5!red2<<4!green1<<3!green2<<2
              !yellow1<<1!yellow2
      dc.b  walk1<<7!walk2<<6!$3F
      dc.b  $FF
      nolist
      DEFAULT
      list
      endm
```

The only new point shown here is the method used to split the value of the branch_addr symbol across byte boundaries. This particular split could have been avoided by using the dc.w or dc.l directives, but the method will find application in certain cases. Remember that the fourth byte is in an EPROM that will not be programmed and is present to make other utility programs easier to write. Its presence accounts for the "divide by 4" operation on addresses in many of the macros.

The assembler listing for the program fragment is now given to show how the various macros are used to create the machine code.

```
000000      179 A:                ;State_0
            180        RED_1
000001      181+red1    set $1
            182        GREEN_2
000001      183+green2 set $1
            184        WALK_2
000001      185+walk2   set $1
            186        CNTR_LOAD 5
00000C      187+I_2910 set  LDCT
000005      188+branch_addr set  5&$F
            189        endsc
000000 C9   190+  dc.bI_2910<<4!CCEN_2910<<3!CCMUX_SEL<<2!(branch_addr>>2&$3)
```

```
000001  64  191+  dc.b  ((branch_addr&$3)<<6)!red1<<5!red2<<4!green1<<3
                        !green2<<2!yellow1<<1!yellow2
000002  7F  192+  dc.b  walk1<<7!walk2<<6!$3F
000003  FF  193+  dc.b  $FF
            208
000004      209 B:              ;State_1
            210        RED_1
000001      211+red1 set $1
            212        GREEN_2      ;Turn WALK_2 OFF
000001      213+green2 set $1
            214        endsc
000004  E9  215+  dc.bI_2910<<4!CCEN_2910<<3!CCMUX_SEL<<2!(branch_addr>>2&$3)
000005  64  216+  dc.b  ((branch_addr&$3)<<6)!red1<<5!red2<<4!green1<<3
                        !green2<<2!yellow1<<1!yellow2
000006  3F  217+  dc.b  walk1<<7!walk2<<6!$3F
000007  FF  218+  dc.b  $FF
            233
            234 C:     LOOP_TC  B       ;State_2
000009      235+I_2910          set      RPCT
000001      236+branch_addr     set   B/4&$F
            237        RED_1
000001      238+red1 set $1
            239        GREEN_2
000001      240+green2 set $1
            241        WALK_2       ;Turn WALK_2 ON
000001      242+walk2 set $1
            243        endsc
000008  98  244+  dc.bI_2910<<4!CCEN_2910<<3!CCMUX_SEL<<2!(branch_addr>>2&$3)
000009  64  245+  dc.b  ((branch_addr&$3)<<6)!red1<<5!red2<<4!green1<<3
                        !green2<<2!yellow1<<1!yellow2
00000A  7F  246+  dc.b  walk1<<7!walk2<<6!$3F
00000B  FF  247+  dc.b  $FF
```

The S-record output has the following form:

```
S0030000FC
S214000000C9647FFFE9643FFF98647FFF30E43FFFE9
S214000010E8E13FFFCA583FFF99983FFF38123FFF7D
S804000000FB
```

The machine code is much more readable in a form produced by a simple utility program written to separate the bit fields along the lines discussed above. Unassembled machine code is shown next.

```
            2 2 C B         Y Y
            9 9 C r      G G e e
            1 1 M a      r r l l W W
            0 0 u n R R e e l l a a
              c x c e e e e o o l l
            i c s h d d n n w w k k
              e e
              n l A 1 2 1 2 1 2 1 2
State Addr  +-+-+-+-+-+-+-+-+-+-+-+-+
000   000   |C|1|0|5|1|1|0|0|1|0|0|0|1|
001   001   |E|1|0|5|1|1|0|0|1|0|0|0|0|
002   002   |9|1|0|1|1|1|0|0|1|0|0|0|1|
```

```
003   003   |3|0|0|3|1|0|0|1|0|0|0|0|
004   004   |E|1|0|3|1|0|0|0|0|1|0|0|
005   005   |C|1|0|9|0|1|1|0|0|0|0|0|
006   006   |9|1|0|6|0|1|1|0|0|0|0|0|
007   007   |3|1|0|0|0|1|0|0|1|0|0|0|
```

Now compare the AMD Am2910 Next Address Generator in Figure 4-16 with the TI 74ACT8818 Next Address Generator shown in Figure 4-17. The TI and AMD parts are quite similar with respect to the method by which they generate addresses since the data paths through the Next Address MUX (2910) and the Y OUTPUT MUX (8818) are virtually identical. While the TI controller also includes an extra loop counter which allows nested counted loops, the main difference lies in the absence of a Next Address Logic element, i.e., the Am2910's Instruction PLA. As would be expected, this allows the designer more freedom in the construction of control instructions over the 16 provided by the Am2910. This freedom is accompanied by a cost; i.e., the extra width required in the microword to accommodate the extra control lines.

The designer has the choice of keeping the complete freedom at an extra cost in parts or reducing the microword width by the addition of an external Next Address Logic network with the concomitant loss in flexibility. This choice is possible in all parts of the state machine and represents a range of control field encoding described by the words "horizontal" (no encoding) to "vertical" (maximum encoding) control. Obviously, the horizontal control results in the most microprogram flexibility but at the cost of the widest microword.

We will show a method using the macro method introduced above to create the logical simplicity of the vertical control structure on a machine with horizontal control. The Am2910, as a vertical control device, will be used to contribute an "instruction set" to a controller based on the TI 74ACT8818. The following macro file defines a group of macros used in the solution to the earlier problem using the Am2910:

```
; Macro file to create 2910 micro instructions
; 2910 built-in instructions
cjs2910   EQU 1       ;cond jump subroutine to place
cjp2910   EQU 3       ;cond jump place
push2910  EQU 4       ;push mpc and continue
rfct2910  EQU 8       ;decrement count, repeat loop if counter <> 0
rpct2910  EQU 9       ;decrement count, jump to place if counter <> 0
crtn2910  EQU 10      ;cond return from subroutine
ldct2910  EQU 12      ;load counter and continue

;continue macro
CONT      macro
          endm

;jump to place
JMP       macro       \1
          I_2910      cjp2910
          PLACE       \1
          endm

;conditional jump to place
CJP       macro       \1
          I_2910      cjp2910
          CCEN_2910
          PLACE       \1
          endm

;load counter and continue
LDCT      macro \1
          I_2910 ldct2910
```

```
              COUNT \1
              endm

;decrement counter
;jump to place if counter <> 0
RPCT        macro        \1
              I_2910 rpct2910
              PLACE \1
              endm
```

Comparing the I_2910 symbol in the earlier solution (e.g., LOOP_TC macro) with the I_2910 macro above shows that we have created a set of "instructions" like the JMP and RPCT that are similar to the JMP and LOOP_TC. "Underneath" the I_2910, COUNT and PLACE macros must be the translation to the TI 8818. These translation macros are shown next:

```
I_2910 macro \1

; conditional jump
              ifeq          \1-cjp2910
              DBUS          ,,PIP2D
              MUX           OP6
              CCMUX         ,POLLO
              endc

; conditional subroutine call
              ifeq          \1-cjs2910
              DBUS          ,,PIP2D
              MUX           OP6
              CCMUX         ,POLLO
              STACK         OP5
              endc

; push µPC on stack and continue
              ifeq          \1-push2910
              STACK         OP6
              endc

; loop if count is not zero
              ifeq          \1-rfct2910
              MUX           OP0
              CCMUX         ,POLLO
              STACK         OP2
              RC            ADBH
              endc

; jump to place if count is not zero
              ifeq          \1-rpct2910
              DBUS          ,,PIP2D
              MUX           OP4
              CCMUX         ,POLLO
              RC            ADBH
              endc

; conditional return
              ifeq          \1-crtn2910
              MUX           OP2
              CCMUX         ,POLLO
              STACK         OP3
              endc

; load counter and continue
              ifeq          \1-ldct2910
              DBUS          ,,PIP2D
              RC            ALBH
              endc

              endm
```

```
CCEN_2910 macro
          CCMUX      EXCPT0,POLLO
          endm

PLACE     macro      \1
          ifnc       \1,
          DATAM      (\1/NBYTE)&$ffff
          endc
          endm

COUNT     macro      \1
          DATAM      \1&$ffff
          endm
```

The programmer may use the LOOP_TC macro from the earlier solution or the RPCT macro to control his counted loops. Either way, the final solution can be expressed in terms of instructions that are appropriate to the original problem. The translation to the TI part is handled once the macro file containing I_2910, PLACE, and COUNT macros is created. We have the programming simplicity of a group of control constructs that are appropriate to our problem without the clutter of constructs that might be useful in other solutions. The flexibility of having the other constructs can still be obtained by creating more macros.

To summarize the three solutions to this simple problem, we wish to make these notes. We have shown how to create a programmed system where none appeared to exist. The simplicity of the program segments that we have employed should be convincing evidence of the power of the method. After the short time required to prepare the macro or definition file, the designer need not be concerned with the location or control behavior of any part of his implementation beyond that presented in simple, readable commands that he created. Programs written in his language should be inherently easy to read by anyone if he chooses the words to fit his problem, not the parts within his implementation. Notice that as long as the hardware structure does not change that a large number of different programs can be written without changing the definition file. Changing the hardware a little (i.e., changing the number of bits in a symbol) causes only the endsc macro to need editing. All programs should still produce correct machine code on the new machine. Larger changes such as adding the counter may still be done transparently if none of the older symbols were removed as we chose to do in our example. Their locations within the microword are handled entirely in the definition (macro) file, specifically in endsc, and not in the program. The creation of new control instructions is entirely a matter of editing the definition file. In our example, we could have created many instructions that take advantage of all possible combinations of state exit paths. In addition, another input can be easily handled by devoting 8 words, 16 bytes, per state. The seemingly large change caused by introducing the advanced controller in version 3 was handled almost entirely by changes in the macro file.

7.2.2. Accumulation

The microprogrammed state machine shown in Figure 7-7 will be used to implement the next several solutions. It is composed of a controller based on Figure 4-14 and the "generic" RALU in Figure 3-18. The three bidirectional data buses, Y, DA, and DB, are brought out of the state machine for connection to the "real world" through digital or possibly analog inputs and outputs. A pipeline register holds the current microword while the controller fields are used to determine the next microinstruction. The combinatorial propagation delay from the pipeline through the controller is in parallel with that through the architecture. A status register is placed between the flags produced by the architecture (e.g., ALU Carry Out) and the controller's Conditional Multiplexer to complete this division of propagation times. If the status register had not been introduced, then the total propagation time from the ALU Func-

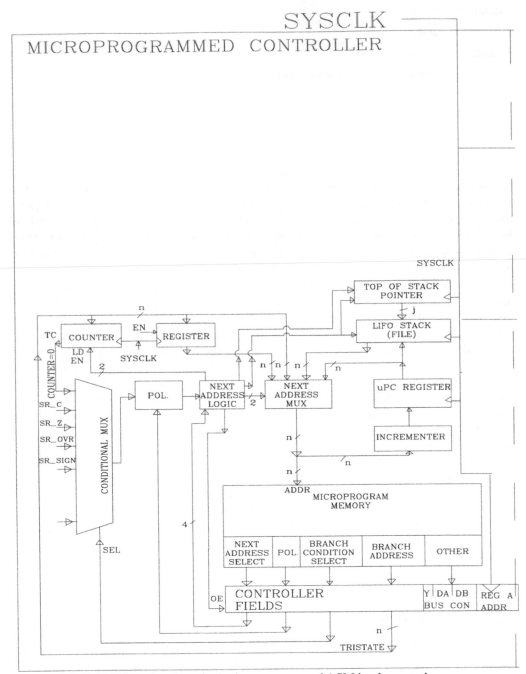

Figure 7-7: RALU-based pedagogical microprogrammed ASM implementation.

tion Select or Carry Input Multiplexer via the ALU – the controller – the Conditional MUX – to the pipeline, during conditional branching instructions, would have determined the period of the system clock, SYSCLK. The controller and architecture are designed and integrated in such a way that combinations commanded by the pipeline during a given period of SYSCLK are latched into registers at the end of the clock period, i.e., the rising edge. With

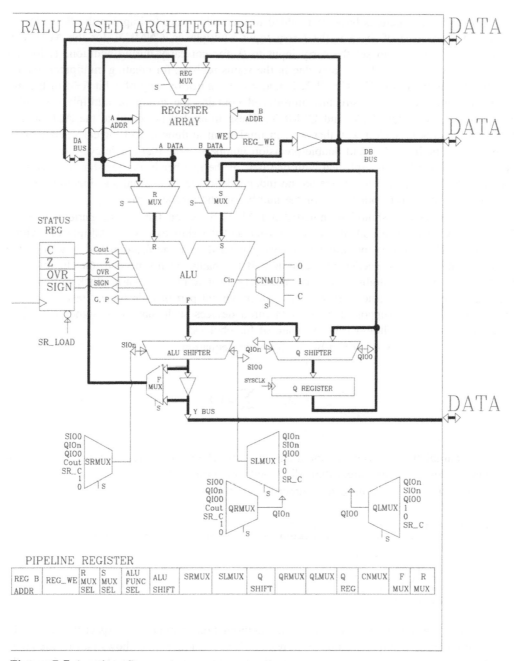

Figure 7-7 (continued).

the pipeline and status register in the design, the maximum SYSCLK rate is determined by the longest propagation time in the controller or architecture, not the sum of the two. As was discussed in Chapters 3 and 4, the longest propagation time in the controller is through the microprogram memory and in the architecture is through the ALU along the carry path during arithmetic operations, with the latter dominating for comparable technologies.

Two other structures have been added to the architecture to improve the flexibility of the RALU. The first of these is the Carry Input Multiplexer, CNMUX, for the ALU. Our choices here allow us to set the carry input to 0, for not carry during addition, 1, for not borrow in subtraction, and the Carry flag in the status register, for creating multiprecision additions and subtractions. The CNMUX is under our control by way of a bit field in the microinstruction. The second structure introduced contains the group of multiplexers called SRMUX, SLMUX, QRMUX, and QLMUX. These multiplexers complete the ALU and Q Shifter structures in such a way that we can implement arithmetic and logical shifts as well as rotates on both single- and double-precision numbers. Note that single-cycle double-precision shifts must be accomplished with one half of the number in the register array and the second half in the Q register. There are no independent data paths to allow single-cycle double-precision shifts for both halves of the number in the register array.

In summary, it should be noted that RALUs based on the generic pattern given in Figure 7-7 may include both the status register and the shift extension multiplexers within the RALU. This reduces the amount of wiring needed in an application and is a desirable feature if the paths are needed in the design. If the paths are not needed, the flexibility of the RALU is reduced slightly such that the designer may have to waste clock cycles overcoming the paths. This selection process is an important part of a designer's task. The various philosophies are supported by manufactured devices such that the designer is able to produce an efficient solution if he is aware of his choices.

Accumulation is described by the following equation:

$$F(x) = \sum_{i=1}^{n} x(i)$$

Accumulation, in itself, is important only for calculating averages; however, it is fundamental to the remaining algorithms discussed in Section 7.2.

In PDL, the algorithm can be described as follows:

```
Place zero in the Accumulator, F.
Initialize the Count to the number of elements to be accumulated, n.
LOOP:
          F = F plus x(i)
          Decrement Count
          IF (Count .GT. 0)
          THEN GOTO LOOP
          ELSE END
```

Since the RALU was designed with accumulation in mind, we expect that it will be able to implement a single accumulation per clock cycle. Figure 7-8 shows part of the architecture section of Figure 7-7. Overlaid on the drawing are traces that show the data paths followed by data originating in the register array. If the number of bits in the operands match the width of the register array, then a single clock cycle will be needed for each accumulation and the Carry Input will be fixed to zero.

The HDL description of the process is as follows:

```
{
    Reg[Accumulator] := Reg[Accumulator] XOR Reg[Accumulator];
    SR_C := Cout;
}
```

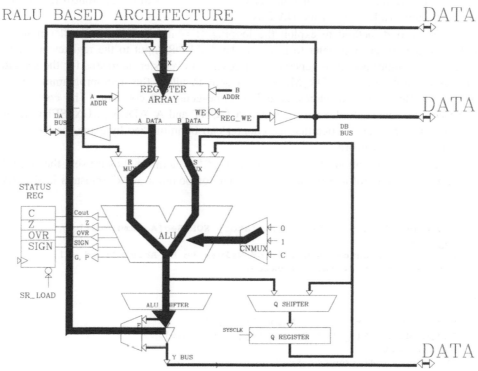

Figure 7-8: Accumulation data paths in the RALU.

```
{
   Reg[Accumulator] := Reg[Accumulator] plus Reg[first.operand] plus CNMUX[0];
   SR_C := Cout;
}
{....................}
{..other operands....}
{
   Reg[Accumulator] := Reg[Accumulator] plus Reg[last.operand] plus CNMUX[0];
   SR_C := Cout;
}
```

The address of the accumulator is placed in the REG B ADDR field since this field is also used to point at the destination of the results in many RALUs. The operand "x" is pointed at by the REG A ADDR field which changes in each "statement" of the HDL. The REG_WE field in the microinstruction is true in any statement in which data is written into the register array; i.e the operator is ":=". In each case, the Carry Output is captured in the C flag of the status register.

Notice that the looping capability of the controller is not employed since the microinstruction REG A ADDR field is being changed to point at other "x"s. Two methods may be used to make the microprogram shorter. First, if the engineering problem itself presents data to the state machine one sample at a time, the DA Bus data path can be used instead to pass the "x" value into the system. Each accumulation line would appear as follows:

```
{
   Reg[Accumulator] := Reg[Accumulator] plus DA_BUS plus CNMUX[0];
}
```

Control lines originating in the pipeline as separate bit fields in the microinstruction are used to control data availability on the DA Bus.

The second method to exploit the looping capabilities of the controller requires the addition of a register or counter to generate the A ADDR input to the register array. Figure 7-9 shows a counter added to generate a succession of register addresses for the second operand. A multiplexer, A_ADDR_MUX, is present to allow the microprogrammer to select between counter-generated addresses and pipeline-generated constants. The control signals for the counter and multiplexer (i.e., LOAD/COUNT and A_ADDR_MUX_SEL) are driven by the pipeline register. Notice that the counter can be initialized by using the REG_A_ADDR field in the microword.

An HDL description of the accumulation of "n" items is shown using the controller's counter for loop control and the address counter for pointing at "x" registers in the RALU.

```
{
    Reg[Accumulator] := Reg[Accumulator] XOR Reg[Accumulator];
    SR_C := Cout;
    Controller.Counter := Number of Values to be Accumulated - 1;
    A_ADDR_Counter := initial value;
}
LOOP:
{
    Reg[Accumulator] := Reg[Accumulator] +
    Reg[A_ADDR_MUX[A_ADDR_Counter++]] + CNMUX[0];
    SR_C := Cout;
    IF (Controller.Counter != 0)
    THEN
    {
        { Controller.Counter := Controller.Counter - 1;}
    GOTO LOOP
    }
    ELSE
    {
        { Controller.Counter := Controller.Counter - 1;}
    CONTINUE
    }
}
```

The first "statement" enclosed by the "{}" clears the register within the register array selected as the accumulator by Exclusive-ORing it with itself. The Carry flag is cleared in the status register and the counter in the controller is initialized with the number of iterations for the loop. Notice that each function within the statement is possible in the same clock cycle since there are independent data paths available.

The second statement delineated by the outer pair of "{}" contains instructions for both the controller and architecture. The accumulator register is summed with the register array operand pointed at by the address counter in the architecture. The symbol A_ADDR_Counter++ is borrowed from the C language to indicate the post-incrementation of the address counter. This model is consistent with a counter based on the 74'169. The controller counter is simultaneously decremented and tested against zero to determine if looping should stop. When the rising edge of SYSCLK occurs at the end of the cycle, the loop control will execute as will the storing of the accumulated result. The body of the loop will require only a single clock cycle for each number to be accumulated.

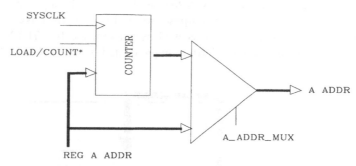

Figure 7-9: Register array address generation detail.

The concept presented by the method in Figure 7-9 is adaptable to larger memories than the register array of the RALU. If such a memory is used, its address lines will be driven by a larger field in the pipeline similar to REG_A_ADDR, as will its write enable signal. The data bus for the memory is then attached to the DA Bus for the RALU.

7.2.3. Multiplication of Unsigned Binary Numbers

The RALU contains an ALU that can perform the sum or difference of unsigned and two's complement binary numbers. It, however, cannot combinatorially multiply. An algorithm was discussed in Chapter 3 that would use the RALU to perform multiplication. With the introduction of controllers in Chapter 4, we are ready to discuss an Algorithmic State Machine (ASM) that can multiply two unsigned binary numbers.

In PDL, the algorithm can be described as follows:

```
Place zero in the double width Product.
Initialize the Count to the number of bits in the multiplier.
LOOP:
   IF (the Least Significant Bit of the Multiplier is 1)
       THEN add the Multiplicand to the upper half of the Product.
       ELSE do nothing
   Shift the double width Product to the right 1 bit position.
   Shift the Multiplier to the right 1 bit position.
   Decrement Count
   IF (Count > 0)
       THEN GOTO LOOP
       ELSE END
```

The multiplicand is a single-precision number (e.g., 16 bits) while the product must be double width (e.g., 32 bits wide). A total of three single-width numbers, i.e., the upper and lower halves of the Product and the Multiplier, must be shifted. The accumulation of the Product above must be done to its upper half, otherwise the accumulating partial products will be lost in the right shift. Since there is no overlap between the current bit positions in the Multiplier that are active (i.e., don't contain shifted in zeros) and the accumulating partial products, we can place the Multiplier in the lower half of the Product and zero only its upper half. The two shift statements above become the following:

```
Shift the double width Product/Multiplier right 1 bit.
```

This is as far as we can go with the description of the algorithm until we begin its mapping onto the hardware. For this purpose, we will use an ASM like Figure 7-7, in which the controller can run counted loops as well as perform the IF...THEN...ELSE construct. The architecture, shown in Figure 7-10, is implemented using an RALU that includes the Q

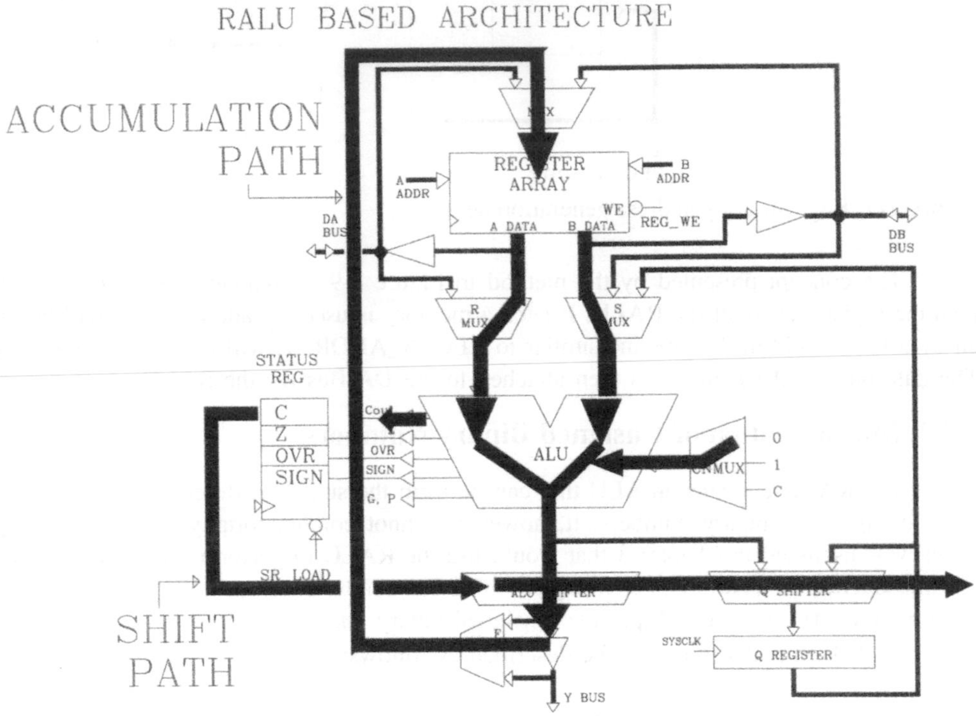

Figure 7-10: Data flow detail for multiplication in the pedagogical ASM.

register, its shifter, and data paths. Choices include the Am2901, Am2903, TI AS888, and TI ACT8832. Each has a different width and number of registers in the register array. If we choose to perform a multiply algorithm involving 16-bit unsigned numbers, we would use 4 Am2901 or Am2903 bit slices. We would require 2 TI 'AS888 bit slices and only one TI 'ACT8832 RALU. These choices assume that we want each add/shift operation to handle the entire width of the Multiplicand and Multiplier. If we are content with handling a "slice" of these operands at a time then the number of the narrower parts would be reduced accordingly. The partial products are created and then summed at the end of the operation. The choice of the width of the data path through the RALU is based on the speed required in the solution. At any time one can trade an increased number of control states for a reduced number of architectural parts.

The PDL description of our algorithm now takes the following form allowed by the increase in register and data path resources:

```
Upper Half of Product ← 0
Q Register ← Multiplier
Controller Counter ← number of bits in Multiplier
LOOP:
 IF (Multiplier Least Significant Bit = 1)
```

```
      THEN Add Multiplicand to Upper Half of Product.
         Shift Product/Q Register right 1 bit.
      ELSE
         Shift Product/Q Register right 1 bit position.
   Decrement Counter
   IF (Counter > 0)
    THEN GOTO LOOP
    ELSE GOTO END
```

The PDL description has not mapped the algorithm onto the hardware to the extent that the strengths of the hardware are exploited. This is best shown using the HDL description that reflects the data paths in the architecture and the actual controller instructions.

```
{
   Reg[Product.upper] := Reg[Product.upper] XOR Reg[Product.upper];
   Carry.Latch := Cout;
}
{
   Q.Register := Reg[Multiplier] >>; MPR.LSb.LATCH := QIO0;
   Am2910.Counter := Number of Bits in the Multiplier - 1 ;
}
LOOP:
{
 IF (MPR.LSb.LATCH == 0)
   THEN { GOTO SHIFT }
   ELSE
   {   Reg[Product.upper] :=
       Reg[Product.upper] + Reg[Multiplicand];
       Carry.Latch := Cout;
   }
   CONTINUE
}
SHIFT:
{
   Reg[Product.upper] >>; Q.Register >>;
   Reg[Product.upper][15] = Carry.Latch;
   Q.Register[15] := Reg[Product.upper][0];
   MPR.LSb.LATCH := Q.Register[0];
   IF (Counter != 0)
   {
       THEN {Counter := Counter - 1;}
       GOTO LOOP
   }
   ELSE
   {
       Counter := Counter - 1;}
       CONTINUE
   }
}
```

Several points need to be made to show the mapping of the algorithm onto the hardware. The first is associated with control of the major shift-and-conditional-add loop. The construct beginning IF (Counter...) and ending with the last bracket after the ELSE is implemented in the controller. A counter and its zero flag are provided in most modern controller devices, including the Am2910 and TI 74ACT8818. The counter can be initialized, decremented, and tested independently of the architecture. The test is performed during the present state and the next state is determined by the value tested before the sensitive edge of the clock. The counter itself is decremented regardless of its present state. The testing of the counter contents is done before the counter is changed at the next clock edge. This means that the state when the counter is zero will be performed (i.e., the counter will "roll over" — contain all 1s) when the loop terminates.

The second point relates to the parallel nature of the shift paths for the Register Array and the Q Register, as shown by the heavy horizontal arrow in Figure 7-10. These paths, taken together, implement a double-precision shift in one clock cycle. The arrow pointing to the left from the ALU Carry Out shows the transfer of the carry out from the previous addition, contained in the Carry.Latch in the Status Register, into the most significant end of the Product. The transfer of the least significant bit in the middle of the Product to the Q Register is also shown. Notice that the least significant bit of the Multiplier, also in the Q Register, is available to be tested by the controller to determine whether an accumulation is to be done. We have introduced a 1-bit latch, i.e., the MPR.LSb.Latch, to hold this bit. When a summation and a shift are needed, two clock cycles (states) are required because of this pipelining.

Since the controller runs in parallel with the RALU, the loop control can be performed during the same state in which shifting is performed. Notice how the program control statements are placed among the architecture statements in the single construct labeled SHIFT. This block of instructions down to the last bracket occurs during a single clock cycle.

The initialization section, above the LOOP label, takes two clock cycles since the zero is created in the ALU for clearing the upper half of the Product. The ALU lies along the transfer path over which the Multiplier travels to the Q-Register. Notice that the Q Shifter can be used during the cycle to "preset" the MPR.LSb.LATCH via QIO0. The Carry.LATCH is cleared by the Cout bit from the ALU during the Exclusive-OR function. The behavior of Cout is dependent upon the RALU, but is given in the appropriate data sheet.

The major loop requires two clock cycles regardless of whether the summation was performed. Since the loop is performed as many times as the number of bits in the multiplier, there is room for further optimization. The conditional summation was implemented in this example by performing a conditional branch to the SHIFT label if the bit previously shifted out of the Multiplier is zero. Most of the current RALUs provide an internal control path that allows the testing of the least significant bit of the Q Register by the ALU function select input. The ALU, when informed by the RALU instruction provided by the controller, decides to add and shift or just shift based on the value of the least significant bit of the Q Register. It is obvious that the importance of the multiply algorithm under discussion influenced this design point. With this improvement, the major loop labeled LOOP requires only one microinstruction or one clock cycle for each iteration of the loop. It can now be written as follows:

```
LOOP:
{
   IF  (Q.Register[0]  ==  1)
       THEN
       {
           {Reg[Product.upper]  :=
            Reg[Product.upper] + Reg[Multiplicand]};
       }
   }
   Reg[Product.upper] >>;  Q.Register >>;
   Q.Register[15]  :=  Reg[Product.upper][0];
   Reg[Product.upper][15]  :=  Cout;
   IF  (Counter != 0)
```

```
      THEN
      {
          { Counter := Counter - 1; }
          GOTO LOOP
      }
      ELSE
      {
          { Counter := Counter - 1;}
      CONTINUE
  }
}
```

Notice that we have been able to eliminate the MPR.LSb.LATCH and the Carry.-LATCH since internal data paths have provided connections between the Cout and the most significant bit of the ALU_F bus as well.

In summary, we have shown the mapping of the multiplication algorithm for unsigned numbers onto a modern RALU. In so doing, we have made use of improvements provided by the manufacturers to reduce the number of clock cycles within the major loop to one per multiplier bit.

7.2.4. Multiplication of Two's Complement Binary Numbers

When the multiplication of signed binary numbers is contemplated, it is seen that the sign of the product depends on the signs of the multiplier and multiplicand. A similar problem was confronted in the case of addition and subtraction. This resulted in the selection of the two's complement number system since it handled the sign of the result "automatically". Once this number system is selected for signed numbers, we must determine how best to continue using it in the rest of our arithmetic operations.

The first possible algorithm is one that determines the sign of the product directly from the signs of the inputs by the following algorithm written in PDL:

```
IF (Sign of Multiplier == Sign of Multiplicand)
   THEN Sign of Product = positive
   ELSE Sign of Product = negative
IF (Sign of Multiplier = negative)
   Make Multiplier positive (two's complement MPR)
IF (Sign of Multiplicand = negative)
   Make Multiplicand positive (two's complement MCAND)
Perform Unsigned Multiply Algorithm
IF (Sign of Product == Negative)
   Make Product = negative (two's complement PROD)
```

This is much more complex than the corresponding algorithm used for addition and subtraction. To continue the actual multiplication, the negative number, if present, is converted to positive by taking its two's complement; i.e., $M = /M + 1$. Then, the unsigned binary multiplication algorithm discussed above is used. The product must be returned to its correct form if it should be negative. The two's complementation requires only one clock cycle for each number that must be converted. This adds at most only three clock cycles to the unsigned multiplication operation described above. Since this was reduced to 18 clock cycles (i.e., 2 for initialization and 16 for the multiply) for a 16-bit multiplier, the entire operation is probably an acceptable compromise under some circumstances.

As often happens, a solution to the sign problem can be achieved by reexamining the number system, the algorithm or the functions performed in the RALU. The unsigned multiplication algorithm not only produces the incorrect sign for the product, but also the wrong product if carried out on a two's complement number. A cure for this problem is based on "correcting" the product when the sign bit of the multiplier is processed. The first n - 1 steps of a two's complement multiply using an "n" bit multiplier proceed as for the unsigned multiplication. The multiplicand is accumulated into the upper half of the product area and the product/multiplier is shifted to the right to accommodate the weight of each binary digit in the multiplier. However, instead of shifting the Carry Out of the previous conditional summation into the most significant end of the Product, the algorithm requires that the sign of the summation be shifted instead. This is further conditioned by the two's complement overflow as will be shown below.

During the last step of the algorithm, the multiplicand is subtracted from the accumulating product if the multiplier is negative. The designers of the RALUs that we use have addressed the problem. If we examine in detail the function performed by the ALU, Q shifter, and ALU shifter in the Am2903, for example, we find the following functions provided during the unsigned multiply instruction to the RALU:

Unsigned Multiply Special Instruction for the Am2903

```
ALU Function: IF (MPR.LSb == 0)
   THEN ALU.F.Bus = S + Cin
   ELSE ALU.F.Bus = R + S + Cin
ALU Shifter: ALU.F.Bus >>; ALU.F.Bus[MSb] = Cout
Q Shifter: Q.Bus >>
```

Note: External connections are made to send ALU.F.Bus[0] to Q.Bus[MSb].

The Am2903 and TI 74ACT8832 also provide a pair of two's complement multiply instructions. The first, called Two's Complement Multiply, is executed n - 1 times where n is the number of bits in the multiplier. The second is executed once as a final-stage correction and is called *Two's Complement Multiply Last Cycle*. A table of the functions provided is given below:

Two's Complement Multiply Special Instruction for the Am2903

```
ALU Function: IF (MPR.LSb) == 0)
        THEN ALU.F.BUS = S + Cin
      ELSE ALU.F.BUS = R + S + Cin
ALU Shifter: ALU.F.Bus >>; ALU.F.Bus[MSb] = Sign XOR OVR
Q Shifter: Q.Bus >>
```

Two's Complement Multiply Last Cycle Instruction for the Am2903

```
ALU Function: IF (MPR.LSb) == 0)
        THEN ALU.F.BUS = S + Cin
        ELSE ALU.F.BUS = S - R - 1 + Cin
ALU Shifter: ALU.F.Bus >>; ALU.F.Bus[MSb] = Sign XOR OVR
Q Shifter: Q.Bus >>
```

The contrast between these two instructions is seen in the operation performed by the ALU as it produces a combination of its R and S inputs. If the least significant bit of the multiplier, in the Q register, is 0, then the "pass" function is performed since Cin is set to 0 all during the multiply algorithm. However, during the last cycle of a two's complement multiply, the multiplicand, pointed to by S, is subtracted from the product. This corrects the product that was developed in the previous n - 1 cycles.

Notice the quantity that is shifted into the most significant bit of the ALU F bus. In the unsigned case, it is the carry out of the summation. In the signed case, it is the Sign Exclusive ORed with the Overflow flag. This effectively causes the creation of the correct sign on the product "automatically".

The effect of including these RALU instructions is that the two's complement binary multiply requires only one more microinstruction than the unsigned binary algorithm. The Two's Complement Multiply Instruction replaces the Unsigned Multiply RALU instruction used in the previous algorithm. This instruction is run n - 1 times. The Two's Complement Multiply Last Cycle RALU instruction is placed immediately following the major loop and is executed once. The total number of clock cycles for the two types of multiplication algorithms remains the same.

Any modern RALU supports signed multiplication in this way, including the TI 74AS888 and 74ACT8832 and the AMD Am29332.

An example of the operation performed is useful at this point. We will first show the multiplication of a pair of 4-bit two's complement number using this method in long hand.

```
   +5   0101      Multiplicand (MCAND)
X  -3   1101      Multiplier (MPR)
  ----  --------
  -15   11110001  Product
```

The expected product of -15 in decimal is shown in two's complement representation. It occupies 8 bits since the number of bits in the product is the sum of the number of bits in the multiplier and multiplicand. Our signed multiplication algorithm will consist of three (n - 1) steps of conditional addition. The last step, corresponding to the partial product for the sign bit, will consist of subtracting the multiplicand from the left half of the product accumulated to that point.

```
1st Partial Product        0000 0101      from MPR bit 0
2nd Partial Product        0000 0000      from MPR bit 1
3rd Partial Product        0001 0100      from MPR bit 2
                           ---------
Sum of Partial Products    0001 1001      PP1 + PP2 + PP3
```

Now we correct this partial product if the sign of the multiplier is negative, i.e., sign bit is 1. The correction is performed by subtracting the multiplicand from the accumulated partial products by two's complementing it and adding it to the position where it would have aligned if this were just an unsigned multiplication. The two's complement of the multiplicand is /(0101) + 1 or 1011.

```
  0001 1001      PP1 + PP2 + PP2
+ 1101 1         Aligned -MCAND
  ---------
  1111 0001      Corrected Product
```

Since we must implement the alignment steps using shifts when we map this algorithm on the hardware, we will have to shift the accumulating partial products to the right during the operation. The same multiplication will now be performed using the instructions given above. The algorithm begins with the most significant half of the product in one of the registers in the register array and the multiplier in the Q register. It will end with the double-width product in the register array and the Q register. The multiplicand is in another register in the register array.

```
        Prod.H          Prod.L              MCAND
      |      0     |    1101     |      |    0101     |
                MPR
```

```
PP1        0101  1101      LS bit of MPR was 1; No shift yet
PP1>>      0010 1 110      Sign XOR OVR = 0 was shifted in on left
PP2        0010 1 110      No add since LS bit of MPR was 0
PP2>>      0001 01 11      Shift in 0 on left
PP3        0110 01 11      LS bit of MPR was 1
PP3>>      0011 001 1      Sign bit of MPR is only bit left in Q
PP4        1110 001 1      Subtract MCAND
PROD>>     1111 0001       Shift in Sign XOR OVR = 1 since Sign is 1
```

The final result appears in the register array and Q register as follows:

```
   Register Array          Q Register
  |     1111     |        |     0001     |
  |     0101     |
```

There are several details to observe. The first is associated with the use of the ALU to perform addition and subtraction during the algorithm. The actual ALU operation always includes the Carry Input, Cin. In all steps except one, the Cin = 0. The last step of the Two's Complement Multiply requires that the Cin = 1 if the multiplier sign is negative. This is a consequence of the way the subtraction of the multiplicand from the product is implemented. Remember that the ALU actually one's complements the number being subtracted and the Cin is expected to contain the 1 needed to create the two's complement. In this case, the subtraction operation depends upon whether the most significant bit of the multiplier is 1 (i.e., is negative). If it is not, then the double-width product is only shifted right.

The second detail involves creating the correct bit to shift into the most significant bit of the product. The unsigned multiply required that the Carry Out, Cout, of the conditional summation be shifted into the product. This accommodated the unsigned overflow (i.e., carry out) from the most significant half of the product at any point in the accumulation of partial products. Notice that the modern RALUs provide an internal path for this carry during the execution of the Unsigned Multiply RALU instruction. Our first example was provided with an external path through the Carry Flag Latch. In the case of two's complement multiplication, the bit that is shifted into the product depends upon the sign of the multiplier and the multiplicand. The bit is created by the following combinatorial term:

```
Shift_In_bit = Sign_of_the_ALU_output XOR ALU_Overflow_Flag
```

In our example above, the multiplier sign bit was 1 which caused the ALU to subtract the multiplicand from the accumulated product. The sign of the result of this subtraction was negative while there was no two's complement overflow, thus creating the 1 that was shifted into the product.

Mapping the signed multiplication algorithm onto the machine shown in Figure 7-7 is identical to the mapping shown for unsigned multiplication. The design of the modern RALU has been adapted to contain the ALU instructions and internal control paths described above. No external latches are used on the ALU Carry Out or the Multiplier's Least Significant Bit since these form internal control signals used by the added instructions that determine whether the ALU adds the Multiplicand to the Product or not before shifting. Figure 7-10 shows the data flow paths as before.

7.2.5. Division of Unsigned Binary Numbers

The binary integer division operation may be accomplished using one of several methods. For smaller numbers, in the range of 8 bits, a table look-up method may be used. The table is normally implemented using a Read Only Memory (ROM). For larger numbers, there are several algorithmic methods related to long division. Each of the algorithmic methods requires that the divisor and dividend be aligned such that the divisor is greater than the dividend. This is a special case of "normalization" mentioned in connection with floating point numbers in Chapters 2 and 3.

The first of these methods relies on subtracting the divisor from the dividend. If the result is greater than 0 a 1 is entered in the corresponding place in the quotient. On the other hand, if the result is negative, the divisor is added back into the dividend, i.e., restored. Shifting the divisor toward the less significant end of the dividend, to the right, prepares for the next iteration. When a number of iterations – equal to the difference in the number of bits in the dividend and the divisor – is complete, a quotient and remainder will have been developed. The *"restoring" division algorithm* is the name of this method which reflects the fact that an addition takes place.

The second, and most commonly used, method is called the *"non-restoring" division algorithm*. Instead of adding (i.e., restoring) the divisor to the dividend if the result of subtraction produced a negative number, the divisor is shifted right one position and then added. This replaces an add–shift right–subtract sequence with a shift right–add sequence resulting in each iteration requiring one fewer step.

There is another general method if the divisor is a constant. The constant can be stored as its inverse (i.e., 1/K) and the divide operation is then replaced with a multiplication. Needless to say, this is the desired method in this case since the multiplication operation may already be provided for other reasons and, where speed is of the essence, a combinatorial multiplier may be used resulting in single state operation.

7.2.5.a. Table Look–Up

The division operation may be accomplished by "looking up" the quotient and remainder in a table. The index into the one-dimensional array that embodies the table is formed by concatenating the divisor and dividend. A similar operation was described for multiplication in Chapter 3; see Figure 3-19. The number of bits in the quotient depends on the number of effective bits in the divisor and dividend. In the table look-up method, we will need to provide the same number of bits in the quotient as we do for the divisor and dividend data path. All possible combinations of effective divisor width are present.

As an example, we can use the ROM based multiplier in Figure 3-19 as the hardware basis for our divider. Two 8-bit operands, the divisor and dividend, are concatenated and used to address the memory. One 8-bit half of the memory is now used to contain the quotient while the other 8-bit half contains the remainder. The number of words in the memory is 2^{16} since division is not commutative. A simple program may be written for the host computer to generate the table which is then transmitted to a PROM programmer. A partial listing of the table for unsigned binary division is given below:

```
A15...Address......A0D7..........D0 Data D7........D0
```

Dividend	Divisor	Quotient	Remainder	Error
0000 0000	0000 0000	xxxx xxxx	xxxx xxxx	1
0000 1111	0000 0011	0000 0101	0000 0000	0
0000 1111	0000 0100	0000 0011	0000 0011	0
1000 1111	0000 1000	0001 0001	0000 0011	0
1111 1111	1111 1111	0000 0001	0000 0000	0

An important fatal error in division operations is caused by dividing by zero since the quotient is undefined. The first entry in the table is one of 256 entries that cause this problem. An "Error" bit is included in the data produced by the memory to signify to the controller that such an event occurred. It is up to control algorithm to cope with the problem since the significance and correction technique is algorithm sensitive.

One convenience of using the table look-up method is that positioning of the quotient and remainder relative to the system data paths is virtually automatic. No normalization or alignment steps are needed since the results can be placed at their correct positions within the memory.

The example has shown the division algorithm implemented for 8-bit unsigned binary integers using the table look-up method. If the designer required that signed binary integers be supported, then the table entries for the quotients would be changed to reflect the sign of the result. Thus, in line 4 of the table, the dividend represents -143 as a two's complement pattern. The quotient should be the integer part of -17.875, i.e., -17 which is 11101111 in two's complement. The table look-up method is appropriate for "small" numbers and represents a way to calculate results rapidly. The minimized networks that have been developed for multiplication are generally not available for division since so many algorithms employ division by constants.

7.2.5.b. Restoring Division Algorithm — Conditional Subtract and Shift

Let's look at an exercise in which we divide two unsigned binary integers (143d by 7d) "by hand". The expected result is 20d with a remainder of 3d. The steps are shown on the next page.

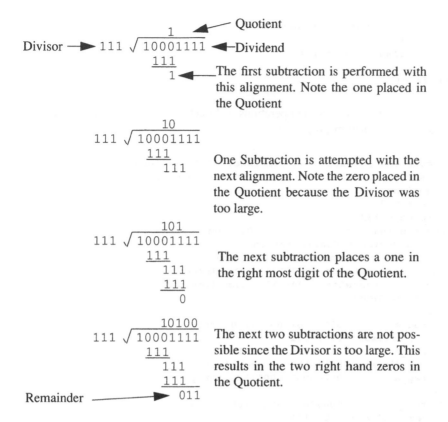

Let us now map the algorithm demonstrated above onto the RALU in the pedagogical ASM in Figure 7-7. We will restrict our "instruction set" to subtraction, addition, and shifting in order to emphasize the fact that we cannot see the results of a comparison test until the next "clock cycle", i.e., the status flags are pipelined. This implies that, at times, the result of a subtraction will "overflow" and must be restored before the next step. The data path width is fixed by the hardware exposing a restricted region of the dividend to the subtraction and restoration operations. The active area of subtraction is limited by the width of the divisor. In most systems, the divisor and dividend contain the same number of bits and the dividend, padded with 0s on the left, will undergo double-precision shifting in the course of creating the "developing remainder" in each step of the algorithm. The divisor, 00000111, is placed in a register in the register array, here denoted by "reg_divisor". The dividend, 10001111, is placed in the Q register, here "qreg", while another register in the register array, "reg_a", is cleared to 0 to receive the developing remainder. The Q register concatenated with reg_a in the register array will be double-precision shifted to the left using the Q shifter, the ALU shifter, and the SIO0 and QIOn data path.

We can summarize the algorithm as adapted for the RALU using the following pseudo-code:

```
for (n = 1; n <= NBITS; n--)
{
        shift_aq();
        sub_a_divisor();
        if (reg_a[msb] == 1)
        {
                QIO0 = 0;
```

```
                add_a_divisor();
        }
        else {QIOO = 1;}
    }
    shift_q();
```

The following table shows the operations, subtract "-∫, add "+", shift "<", and the register contents as the algorithm progresses:

```
Compute the Quotient and Remainder of an Unsigned
Divisor and Dividend using the Restoring Division
 Algorithm
divisor = 7
dividend = 143
Expected results: quotient = 20_d; remainder = 3_d
Initial Register Configuration:
reg_a         contains:00000000    zero
reg_divisorcontains:00000111    the divisor
qreg          contains:10001111    the dividend
Operation Sequence:
Step: 1
 <  reg_a:          00000001 00011110 :qreg
 -  reg_divisor: 00000111
    reg_a:          11111010 00011110 :qreg
 +  reg_divisor: 00000111    reg_a[msb] = 1
    reg_a:          00000001 00011110 :qreg
Step: 2
 <  reg_a:          00000010 00111100 :qreg
 -  reg_divisor: 00000111
    reg_a:          11111011 00111100 :qreg
 +  reg_divisor: 00000111    reg_a[msb] = 1
    reg_a:          00000010 00111100 :qreg
Step: 3
 <  reg_a:          00000100 01111000 :qreg
 -  reg_divisor: 00000111
    reg_a:          11111101 01111000 :qreg
 +  reg_divisor: 00000111    reg_a[msb] = 1
    reg_a:          00000100 01111000 :qreg
Step: 4
 <  reg_a:          00001000 11110000 :qreg
 -  reg_divisor: 00000111
    reg_a:          00000001 11110000 :qreg
  reg_divisor: 00000111    reg_a[msb] = 0
    reg_a:          00000001 11110000 :qreg
Step: 5
 <  reg_a:          00000011 11100001 :qreg
 -  reg_divisor: 00000111
    reg_a:          11111100 11100001 :qreg
 +  reg_divisor: 00000111    reg_a[msb] = 1
    reg_a:          00000011 11100001 :qreg
Step: 6
 <  reg_a:          00000111 11000010 :qreg
 -  reg_divisor:00000111
    reg_a:          00000000 11000010 :qreg
  reg_divisor: 00000111    reg_a[msb] = 0
    reg_a:          00000000 11000010 :qreg
Step: 7
 <  reg_a:          00000001 10000101 :qreg
```

```
  -  reg_divisor: 00000111
     reg_a:       11111010 10000101 :qreg
  +  reg_divisor: 00000111     reg_a[msb] = 1
     reg_a:       00000001 10000101 :qreg
Step: 8
  <  reg_a:       00000011 00001010 :qreg
  -  reg_divisor: 00000111
     reg_a:       11111100 00001010 :qreg
  +  reg_divisor: 00000111     reg_a[msb] = 1
     reg_a:       00000011 00001010 :qreg
  <  reg_a:        00000011 00010100 :qreg
The Computed Results are:
qreg         contains: 00010100    the quotient
reg_a        contains: 00000011    the remainder
reg_divisor contains: 00000111    the divisor
```

The first subtraction is tried using the developing remainder, in reg_a, and the divisor, reg_divisor, both in the register array. The subtraction is accomplished in the ALU using the function F = R plus NOT S plus 1. A borrow can be detected by observing the most significant bit of the developing remainder since the numbers are considered to be unsigned. The first subtraction produces a borrow, as signaled in the most significant bit (msb) of reg_a, and a restoration is needed. At the beginning of Step 2, the dividend is shifted to the left one bit position and the quotient bit just developed is shifted into the least significant end of the Q register via QIO0.

The shift-subtract-restore operation proceeds until Step 4 where the developing remainder is finally larger than the divisor. No correction is needed and a one is shifted into the least significant bit (lsb) of the Q register, q_reg, at the beginning of Step 5. The subtraction in Step 5 results in a borrow, thus a restoration is needed to conclude the step. The last shift in Step 8 shifts only the quotient register to complete the operation.

Even if the programmer takes advantage of the data paths present in an RALU, the subtract, add, and shift operations may not be grouped because of the order of operations and the buffering of the status that determines the need for a restore operation. The next algorithm will be adapted by RALU designers to remove this problem. The number of steps to execute the restoring algorithm varies between one and two times the number of bits in the divisor.

7.2.5.c. Non-Restoring Division

In the previous example, we notice that if a borrow is generated in the subtraction, the developing remainder must be restored by adding the divisor back into it. The total number of steps in the restoring algorithm is related to the number of digits in the divisor. A technique, called the *non-restoring division algorithm*, was developed early in the history of the computer. It reduced the number of steps required in the algorithm to the number of digits in the divisor. The heart of this method is summarized as follows:

1. Subtract the divisor from the dividend
2. IF the result does not generate a borrow, THEN
   a. Record a 1 in the quotient
   b. And shift the dividend left 1 bit position
   c. And subtract the divisor from the dividend
   ELSE

      d. Record a 0 in the quotient
      e. And shift the dividend left 1 position
      f. And add the divisor to the dividend
   3. Repeat step 2 until the quotient has been developed

Step 2 is identical to that performed in the restoring method described above when no borrow is generated. Step 2f specifies that one half of the divisor is added to the dividend. The restoring and non-restoring methods produce the same result after Step 2 since adding the divisor to the dividend, shifting, and then subtracting the divisor is arithmetically equivalent to shifting the dividend and adding the divisor as described by Stone in his *Introduction to Computer Architecture* and Rafiquzzaman in *Modern Computer Architecture* (see Bibliography). The only shortcoming of the method occurs when the operation in the last step causes a borrow. Under these conditions, the remainder has a sign opposite that of the quotient. An extra step is executed that "corrects" the remainder by performing the needed addition.

The main body of the algorithm is summarized using the following pseudo-code:

```
for (n = 1; n <= NBITS; n++)
{
        if (reg_a[msb] == 0)
        {
                shift_aq();
                sub_a_divisor();
        }
        else
        {
                shift_aq();
                add_a_divisor();
        }
        if (reg_a[msb] == 1){QIO0 = 0;} else {QIO0 = 1;}
}
```

The divisor and dividend are unsigned binary numbers each containing the same number of bits as specified by "NBITS". The dividend is initially placed in the Q register (qreg) and is shifted into the accumulator register (reg_a) in the course of the division operation. The "shift_aq" operation performs a double-precision left shift of reg_a and the qreg. The msb of the qreg is placed in the lsb of reg_a during the shift operation. Depending upon the value of the msb of reg_a, the "sub_a_divisor" and "add_a_divisor" steps subtract or add, respectively, the divisor to reg_a, thus generating a new "remainder" at each step. Since the ALU employed performs arithmetic upon unsigned or signed binary numbers, the borrow can be detected by sensing the msb of the register in which accumulation occurs.

We will repeat the example used in the previous section using the non-restoring algorithm. The dividend is placed in the Q register as an 8-bit wide RALU, "qreg" in the algorithm above. The divisor is in a register, "reg_divisor", in the register array positioned such that it can be moved to one of the ALU inputs. The register in which the remainder is developed, "reg_a", is also placed in the register array. The contents of the Q register are shifted serially into the "reg_a" register using the Q shifter, ALU shifter, and the SIO/QIO data paths during the course of the algorithm. The following table depicts the register contents and steps performed. The "+", "-" and "<" characters indicate the execution of the "add_a_divisor", "sub_a_divisor", and shift operations.

```
Compute the Quotient and Remainder of an Unsigned Divisor
 and Dividend using the Non-restoring Division Algorithm
divisor = 7
dividend = 143
Expected results:quotient = 20_d or 14_x;remainder = 3_d or 3_x
Initial Register Configuration:
reg_a           contains:00000000    zero
reg_divisor contains:00000111    the divisor
qreg            contains:10001111    the dividend
Step: 1     reg_a[msb] = 0
 <  reg_a:       00000001 00011110 :qreg
 -  reg_divisor: 00000111
    reg_a:       11111010 00011110 :qreg
Step: 2     reg_a[msb] = 1
 <  reg_a:       11110100 00111100 :qreg
 +  reg_divisor: 00000111
    reg_a:       11111011 00111100 :qreg
Step: 3     reg_a[msb] = 1
 <  reg_a:       11110110 01111000 :qreg
 +  reg_divisor: 00000111
    reg_a:       11111101 01111000 :qreg
Step: 4     reg_a[msb] = 1
 <  reg_a:       11111010 11110000 :qreg
 +  reg_divisor: 00000111
    reg_a:       00000001 11110000 :qreg
Step: 5     reg_a[msb] = 0
 <  reg_a:       00000011 11100010 :qreg
 -  reg_divisor: 00000111
    reg_a:       11111100 11100010 :qreg
Step: 6     reg_a[msb] = 1
 <  reg_a:       11111001 11000100 :qreg
 +  reg_divisor: 00000111
    reg_a:       00000000 11000100 :qreg
Step: 7     reg_a[msb] = 0
 <  reg_a:       00000001 10001010 :qreg
 -  reg_divisor: 00000111
    reg_a:       11111010 10001010 :qreg
Step: 8     reg_a[msb] = 1
 <  reg_a:       11110101 00010100 :qreg
 +  reg_divisor: 00000111
    reg_a:       11111100 00010100 :qreg
Remainder Correction Step: Needed     reg_a[msb] = 1
 <  reg_a:       11111100 00010100 :qreg
 +  reg_divisor: 00000111
    reg_a:       00000011 00010100 :qreg
The Computed Results are:
qreg            contains: 00010100    the quotient
reg_a           contains: 00000011    the remainder
reg_divisor contains: 00000111    the divisor
```

The subtraction operation in Step 1, after the double-precision shift of reg_a and the qreg, caused a borrow, indicated by the 1 in the msb of "reg_a" at the beginning of Step 2. The resulting quotient bit for Step 1 (0 since a borrow was generated) was shifted into the least significant end of register Q, via QIO0, in the first line of Step 2. Since the borrow was generated in the previous subtraction, the divisor is added to the shifted remainder in the second part of Step 2. This operation continues until Step 5 since the developing remainder is always too small. After the subtraction in Step 5, the remainder is once again too small so that an addition is required in Step 6. At the end of Step 8, the correct quotient is in the Q register but the remainder, in "reg_a", is too small; i.e., a borrow was generated in the previous operation. The divisor must be added back into the remainder to correct it.

As an example, we will now map the non-restoring division algorithm onto a modern RALU. The TI 74ACT8832 32-bit RALU will be used as our platform. The important feature of modern RALUs, when used for multiplication or division, is the internal control path that determines the ALU operation from the relevant data bit. We exploited this feature in the section on multiplication in this chapter to produce an algorithm that executed in the same number of clock cycles as the number of bits in the multiplier. Here, the ALU must add or subtract the divisor from the developing remainder if the previous operation resulted in a sign change. The TI RALU presents the microprogrammer with a series of "instructions" that allow the non-restoring algorithm to be implemented. The instructions are load-mq, udivis, udivi, udivit, and divrf. Each executes in one clock cycle. The first instruction, loadmq, transfers the dividend to the Q register and initializes appropriate internal flags. The next, udivis, performs the first subtraction and reports the success of the operation via the flag "ssf". The instruction, udivi, is repeated n - 1 times, where n is the number of bits in the data path. The last two operations, udivit and divrf, perform the final step of subtraction or addition, and determine if a remainder correction is needed. Notice that one clock cycle may be saved in the algorithm if the remainder is not needed in future calculations. In this case, since the quotient is already developed at the end of the udivit operation, the application of the divrf instruction may be omitted.

A pseudo-code representation of the algorithm follows:

```
loadmq();
udivis();
for (n = 1; n <= NBITS-1; n++)
{
      udivi();
}
udivit();
divrf();
```

We will now show the results of applying the algorithm by displaying the register contents and operations in the following table:

```
Compute the Quotient and Remainder of an Unsigned Divisor
 and Dividend using the Non-restoring Division Algorithm
 - Based on TI 74AS888
divisor = 7
dividend = 143
Expected results: quotient = 20_d or 14_x; remainder = 3_d or 3_x
Initial Register Configuration:
reg_a          contains:00000000     zero
reg_divisor contains:00000111     the divisor
qreg           contains:10001111     the dividend
**loadmq
    reg_divisor: 00000111    reg_a[msb] = 1
    reg_a:        00000000 10001111 :qreg
**udivis        ssf = 1
 - sub_a_divisor
    reg_divisor: 00000111    reg_a[msb] = 1
    reg_a:       11111001 10001111 :qreg  ssf = 1 mqf = 0
 < shift_aq
```

```
      reg_divisor: 00000111     reg_a[msb] = 1
      reg_a:       11110011 00011110 :qreg  ssf = 0 mqf = 0
**udivi       ssf = 0
 + add_a_divisor
      reg_divisor: 00000111     reg_a[msb] = 0
      reg_a:       11111010 00011110 :qreg  ssf = 0 mqf = 0
 < shift_aq
      reg_divisor: 00000111     reg_a[msb] = 1
      reg_a:       11110100 00111100 :qreg  ssf = 0 mqf = 0
**udivi       ssf = 0
 + add_a_divisor
      reg_divisor: 00000111     reg_a[msb] = 0
      reg_a:       11111011 00111100 :qreg  ssf = 0 mqf = 0
 < shift_aq
      reg_divisor: 00000111     reg_a[msb] = 1
      reg_a:       11110110 01111000 :qreg  ssf = 0 mqf = 0
**udivi       ssf = 0
 + add_a_divisor
      reg_divisor: 00000111     reg_a[msb] = 0
      reg_a:       11111101 01111000 :qreg  ssf = 0 mqf = 0
 < shift_aq
      reg_divisor: 00000111     reg_a[msb] = 1
      reg_a:       11111010 11110000 :qreg  ssf = 0 mqf = 0
**udivi       ssf = 0
 + add_a_divisor
      reg_divisor: 00000111     reg_a[msb] = 0
      reg_a:       00000001 11110000 :qreg  ssf = 0 mqf = 1
 < shift_aq
      reg_divisor: 00000111     reg_a[msb] = 1
      reg_a:       00000011 11100001 :qreg  ssf = 1 mqf = 1
**udivi       ssf = 1
 - sub_a_divisor
      reg_divisor: 00000111     reg_a[msb] = 1
      reg_a:       11111100 11100001 :qreg  ssf = 1 mqf = 0
 < shift_aq
      reg_divisor: 00000111     reg_a[msb] = 1
      reg_a:       11111001 11000010 :qreg  ssf = 0 mqf = 0
**udivi       ssf = 0
 + add_a_divisor
      reg_divisor: 00000111     reg_a[msb] = 0
      reg_a:       00000000 11000010 :qreg  ssf = 0 mqf = 1
 < shift_aq
      reg_divisor: 00000111     reg_a[msb] = 1
      reg_a:       00000001 10000101 :qreg  ssf = 1 mqf = 1
**udivi       ssf = 1
 - sub_a_divisor
      reg_divisor: 00000111     reg_a[msb] = 1
      reg_a:       11111010 10000101 :qreg  ssf = 1 mqf = 0
 < shift_aq
      reg_divisor: 00000111     reg_a[msb] = 1
      reg_a:       11110101 00001010 :qreg  ssf = 0 mqf = 0
**udivit      ssf = 0
 + add_a_divisor
      reg_divisor: 00000111     reg_a[msb] = 0
      reg_a:       11111100 00001010 :qreg  ssf = 0 mqf = 0
 < shift_q
      reg_divisor: 00000111     reg_a[msb] = 1
      reg_a:       11111100 00010100 :qreg  ssf = 0 mqf = 0
**divrf       ssf = 0
```

```
+ add_a_divisor
   reg_divisor: 00000111    reg_a[msb] = 0
   reg_a:       00000011 00010100 :qreg  ssf = 0 mqf = 1
Remainder Correction Required
The Computed Results are:
qreg           contains: 00010100   the quotient
reg_a          contains: 00000011   the remainder
reg_divisor    contains: 00000111   the divisor
```

Each instruction execution, flagged by "**", takes one clock cycle in the TI 74ACT8832. Only 8 of the 32 bits are shown in the data paths for clarity.

7.2.6. The Arithmetic Sum of Products

Now that we have provided ourselves with the fundamental arithmetic operations, we can build larger structures using them. One operation that is the basis of many important math functions is the arithmetic sum of products.

$$F(x) = \sum_{i=0}^{\infty} a_i \cdot x_i$$

It can be used to directly evaluate digital filters, Fourier Series, and other functions that are frequently encountered in high-speed arithmetic applications.

The algorithm for the arithmetic sum of products as expressed in PDL is shown below:

For all a's and x's: F = F + a(i) * x(i)

Since we now have an algorithmic "module" called multiply, we can avoid some of the notational complexity that would be necessary if we had to repeat our previous work. We should expect that our controller will support this modular approach.

In HDL, we will express this function as it is mapped onto the ASM shown in Figure 7-7. We will designate RALU registers to be used for indexing (counting) into the lists of numbers that are represented by a(i) and x(i). The simplest possible arrangement makes the counting and pointing part of the algorithm. The RALU Register Address inputs would be pointed by the microcode to select the operands in the register array. If this arrangement were used, the number of values of a(i) and x(i), hence the maximum value for i, would be limited by the size of the register array.

In this discussion, then, we will work with an architecture that consists of the RALU used for the multiplication algorithms extended by one or more memories connected to its DA, and/or DB, data bus input as shown in Figure 7-11. The address into the memory can be generated in one of two ways. The first way employs a field in the microword and a path to the parallel load inputs of the address registers instead of the path to DB in Figure 7-11. The microinstruction itself is used to point at the memory location desired. A larger number of data values, up to the capacity of the memory, could be included in the sum of products by pointing at the elements one after another. A value for a(i) can be loaded into the Multiplier register in the RALU and then a value for x(i) can be transferred to the Multiplicand register. This would be written as follows:

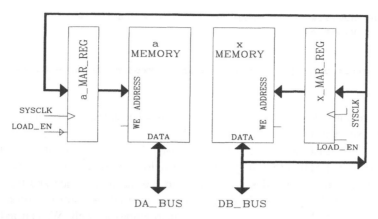

Figure 7-11: Memory added to RALU.

```
Q.Register := Memory[a[i]]
Reg[Multiplicand] := Memory[x[i]]
```

Each move will require one controller state. The actual memory location symbolized by a[i] and x[i] is contained in the field in the microword (pipeline register) attached to the address bus input of the memory. At this point in our discussion, we will assert that the elements of a(i) and x(i) have been placed in the memory by some unseen operation; i.e., maybe the memory is an EPROM. The HDL description of the entire sum of products would appear as follows:

```
Q.Register := Memory[a[0]]
Reg[Multiplicand] := Memory[x[0]]
Call Multiply Subroutine
Reg[Sum] := Reg[Sum] + Product

Q.Register := Memory[a[1]]
Reg[Multiplicand] := Memory[x[1]]
Call Multiply Subroutine
Reg[Sum] := Reg[Sum] + Product
```

Repeat until all a's and x's have been multiplied and accumulated.

Two points must be made here. The most important is that we have treated the Multiply algorithm shown earlier as a subroutine or function. The Return from Subroutine is the only addition that needs to be made to its microprogram. It is added after the final microinstruction. Since it does not involve the architecture, here the RALU, we might as well move the product accumulation into a new version of the multiply algorithm. Since this new sequence both multiplies and accumulates the product into a register, we will call it MAC for Multiply/Accumulate. This module is obviously useful in any arithmetic sum of products algorithm.

The second point involves the fact that the product is a double-width number. We have two choices as to how we handle the accumulation. We can preserve its length, in which case the HDL for the accumulation becomes:

```
        Reg[Sum.lower] := Reg[Sum.lower] + Product.lower;
        Carry.Latch := Cout;
}
{
        Reg[Sum.upper] := Reg[Sum.upper] + Product.upper + Carry.Latch;
}
```

This requires two clock cycles to perform what amounts to double-precision arithmetic with respect to the underlying data paths.

If we know that the product will never reach double-precision proportions, we can accumulate the proper half into a single-precision field, thus saving a clock cycle.

Since the data for a and x are frozen into the memory device and the addresses that access them are similarly frozen in the microprogram itself, we notice a strong lack of flexibility in this design. It occupies several microinstructions as well. We can reduce the number of microinstructions by not including the actual addresses in the microprogram and using the loop control facility provided by the controller. Both have to happen together.

First, we will use two of the registers in the RALU register array to point at the locations of the data in memory. One register will point at the "a" data points and the other at the "x" data points. We will name these registers Reg[a.Address] and Reg[x.Address], reflecting that they are located in the register array. The contents of these "address" registers may be transferred to the external address registers attached to the address inputs of the memories via the DB Bus in Figure 7-11. Another register containing the counter for the loop will be similarly set up and called Reg[Counter]. Whether this counter can be placed in the controller will be dealt with later. An accumulator register of double length will also be defined as Reg[Accumulator.upper] and Reg[Accumulator.lower]. This register is not the same as the Product, which is similarly defined by the Multiplier module. The counter and multiplicand are also already defined as above. The multiplier can be assigned directly to the Q Register since this will save a microinstruction. The HDL description of the arithmetic sum of products will now appear as follows:

```
SOP.LOOP:
{    Reg[Multiplicand] := Memory[Reg[a.Address]]; }
{    Q.Register := Memory[Reg[x.Address]]; }
{    Call Multiply/Accumulate Subroutine }
{    Reg[a.Address] := Reg[a.Address] + 1; }
{    Reg[x.Address] := Reg[x.Address] + 1; }
{    Reg[Counter] := Reg[Counter] - 1; }
{
        IF (Reg[Counter] != 0)
             THEN GOTO SOP.LOOP
             ELSE CONTINUE
}
```

The loop control construct assumes that a latch is present that contains the ALU.ZERO.FLAG loaded during the counter decrement state. If this latch is not present and the ALU.ZERO output is connected directly to the Condition Code Multiplexer of the controller, then this construct can be moved into the counter decrement state.

Control of the Sum-of-Products loop is an explicit operation involving both the register array, Reg[Counter], and the ALU. If this operation could be moved into the controller as was done for the multiplication algorithms, then the loop control construct could be performed in parallel with some part of the data manipulation section (e.g., memory address incrementation). The controller discussed previously supports only one internal counter; however, the TI 'AS890 and TI 'ACT8816 both contain two independent counters. This markedly improves the speed and implementation simplicity of nested loops.

It is assumed that the address registers as well as the counter are initialized upon entry to SOP.LOOP. The subroutine Multiply/Accumulate performs its own loop control and Product area initialization.

The last point to address occurs when this structure is integrated into a system. In other words, how does the data get into the memory in the first place? In digital filtering applications, one of the sets of numbers (a or x) is constant. One possible arrangement would place the constants in a memory that is physically separate from the variable memory and implemented using PROM or EPROM technology. The variables are placed in another memory having an entirely different implementation.

A simple situation would be to implement the variable memory as a shift register having a width equal to the number of bits in the data representation and a depth appropriate to the problem. For example, a two-pole recursive digital filter would require a two-stage shift register to support the two delayed signals. Finite impulse response (FIR) filters would generally require more stages for a similar response, but even 50 or 100 stages is reasonable for this method. The controller would be required to supply control signals for this part of the architecture as well. These would include signals to acquire data from the outside world.

One advantage of this arrangement is that the variable memory requires not addresses, but clock signals to make data available at its output. We have substituted a system with few control states for one with a large number of control states that reduced another part of the design. The output of the variable memory is connected to the DB input on the RALU. This yields two independent paths into the RALU that can reduce the number of clocks required to initialize the Multiplier/Accumulator (MAC) operation to one. No matter what the manner of implementation of the second memory, both paths can decrease the execution time of the system.

To address the external memory in the most general way will cause us to add external registers that contain the current addresses of the data to be multiplied. These registers, called *memory address registers*, stand between one of the data paths that lead out of the RALU and the address inputs of the memory containing the data. The following HDL defines address registers for both the "a" and "x" memories as shown in Figure 7-11:

```
define REGISTER a_MAR_REG[15-0]
define REGISTER x_MAR_REG[15-0]
define MEMORY a_MEM[8091-0][15-0]
define MEMORY x_MEM[8091-0][15-0]

connect RALU_DB_BUS[15-0] to a_MAR_REG[15-0]
connect RALU_DB_BUS[15-0] to x_MAR_REG[15-0]
connect a_MAR_REG_OUTPUT[15-0] to a_MEM_ADDRESS[15-0]
connect x_MAR_REG_OUTPUT[15-0] to x_MEM_ADDRESS[15-0]
connect a_MEM_DATA_OUTPUT[15-0] to RALU_DA_BUS[15-0]
connect x_MEM_DATA_OUTPUT[15-0] to RALU_DB_BUS[15-0]
```

The function, Memory[Reg[x.Address]], is actually implemented on this architecture in the following manner:

```
{

     x_MAR_REG := RALU_REG[x.Address]

}
```

```
    {
            RALU_Q := x_MEM[x_MAR_REG]
    }
```

The process requires two clock cycles since the x_MAR_REG is a clocked element that lies along the transfer path. This is required in this version since the paths used for data are the same as those used for addresses. If the "a" and "x" memories are implemented in the same memory device, then the RALU registers containing the addresses can hold each address on the DB bus while the data is transferred into the RALU by the other path, i.e., the DA bus. In the later RALUs, the DA bus is bidirectional allowing more flexibility in data routing.

A final improvement in processing speed can be made by implementing the a_MAR_REG and x_MAR_REG devices using parallel loadable, up/down counters. The data inputs to the counters are driven by a field in the microword instead of the DB bus as shown in Figure 7-11. Each register's LOAD_EN input becomes a signal that can allow the counter to be loaded, counted, or held under control of the microinstruction. These signals, along with the memory write enables, WE, can be used in a larger system context to place data in the memory from an external source such as an analog-to-digital converter.

In summary, we have explored the implementation of an important data processing algorithm using a microprogrammed state machine having an RALU and extended memory in its architecture. For very high-speed operations, the multiplication algorithm may limit the applicability of this arrangement; however, our development path has allowed us to evolve more complex structures. We have studied this algorithm without incurring any new hardware overhead. In a later section, we will explore the replacement of the multiply algorithm implemented on the RALU by a single-cycle combinatorial multiplier and accumulator.

7.2.7. The Power Series Expansion

Another important algorithm is the *power series expansion*. This algorithm supports the evaluation of Taylor and Chebyshev series as well as the Newton-Raphson method for division and root extraction. The power series expansion takes the following mathematical form:

$$F(x) = \sum_{i=0}^{\infty} a_i \cdot x^i$$

There are two important considerations in applying this algorithm to the evaluation of functions. The first is convergence: i.e., the number of terms required to evaluate the function within an acceptable error limit. The second is the minimization of the number of operations that must be performed once the first consideration is determined. The Taylor series converges quickly for small values of x but slowly for large values of x. The number of multiply/accumulation operations is dependent upon the size of the variable x. The convergence of the Chebyshev series is independent of the size of x.

A brief introduction to the Chebyshev series is appropriate at this point. An in-depth study should be directed to texts on special functions. The Chebyshev series has the following form:

The Chebyshev polynomials, T(x), are defined as follows:

$$F(x) = 0.5 \cdot C_0 + C_1 \cdot T_1(x) + C_2 \cdot T_2(x) + \dots$$

$$T_n(x) = \cos(n \cdot \text{acos}(x))$$

This form of the polynomial is not directly usable, however; by performing the algebra suggested by the defining equation we can create useful forms as follows:

$$T_0(x) = \cos(0) = 1$$

$$T_1(x) = \cos(1 \cdot \text{acos} x) = x$$

$$T_2(x) = \cos(2 \cdot \text{acos}(x)) = 2(\cos(\text{acos}(x)))^2 = 2 \cdot x^2 - 1$$

etc.

Rewriting f(x) in terms of these polynomials and collecting terms will result in an equation having the form of a power series expansion. Each of the A_i will be expressed in terms of a group of the Chebyshev coefficients, C_i. Tables of the Chebyshev coefficients are available for the common functions.

The efficiency of evaluating the power series expansion can be further increased by applying Horner's Rule. This consists of inserting parentheses to control the order of evaluation of terms in the expansion. Any truncated power series expansion takes the following form:

$$F(x) = ((((A_5 \cdot x + A_4) \cdot x + A_3) \cdot x + A_2) \cdot x + A_1) \cdot x + A_0$$

The important effect of the rearrangement is the reduction of the number of products and operand transfers. In terms of the multiplication algorithms discussed above, the contents of the accumulator at any stage is multiplied by the value of x. The summation of an A_i with the current product can take place directly into the product register in the multiplier. Then a copy of x is moved to the Q register, the upper half of the product is cleared, and the multiplication begins again. Since the Chebyshev expansion converges rapidly, the number of iterations is correspondingly small for most functions.

Mapping of the power series expansion onto the state machine shown in Figure 7-7 extended by the memory expansion in Figure 7-11 is similar to that shown for the arithmetic sum of products in the previous section. Based on Horner's Rule, the Program Design Language (PDL) description appears as follows:

```
F := 0
FOR i = n TO 1 STEP -1
      F := (F + a(i)) * x
NEXT i
```

An HDL description follows assuming the coefficients (i.e., a(i)) are in the a_MEMORY and the data value, x, is in a register in the RALU:

```
POW.LOOP:
        { Reg[F] := Memory[Reg[a.Address]] + Reg[F] + CNMUX[0]; }
        { Q.Register := Reg[x]; }
        { Reg[Multiplicand] := Reg[F]; }
        { Call Multiply Subroutine }
        { Reg[F] := Reg[Product];}
        { Reg[a.Address] := Reg[a.Address] - 1; }
        { Reg[Counter] := Reg[Counter] - 1; }
        {
                IF (Reg[Counter] != 0)
                    THEN GOTO POW.LOOP
                    ELSE CONTINUE
        }
```

There is a subtlety contained in the lines in which the accumulated function, F, is manipulated. In our simple example, the arithmetic performed by the RALU is on single precision two's complement numbers. When two such numbers are multiplied, the product is double-precision. At some point in the algorithm, the designer needs to prevent the continual build-up in the width of F. We show it here in the implied truncation that occurs when the results are returned to the F register after the multiplication. This method to reduce precision is frequently not the best in most applications. The actual meaning of the "a(i)" and "x" elements (i.e., are they integers, fixed point, or floating point numbers) will greatly affect the method used to control the precision of F.

7.3. REPLACEMENT OF ALGORITHMIC MODULES BY COMBINATORIAL MODULES

In the course of this chapter we have identified and implemented several important mathematical algorithms. These were logically separated into two groups. One contained the basic operations such as multiplication and division and the second contained "complex" algorithms that used members of the first group to reach the desired results, i.e., sum-of-products and power series expansions. We have noticed that certain functions, because of their complexity, had to be implemented using algorithms. Early in the computer era, these included the addition of multibit numbers. A single-bit full adder was implemented and then used repeatedly to sum the bits of two multibit numbers. The carry was propagated through a storage element from bit to bit in the process. As device fabrication technology improved, numbers containing several bits could be added or subtracted combinatorially.

The 4-bit wide full adder was an important step that lasted for a number of years. If one desired to add 16-bit wide numbers, then he selected either four copies of the 4-bit adder to produce a combinatorial sum or one copy that he used four times similar to the method used with the single-bit element. As data paths were added, the 4-bit wide RALU evolved. The application of four 4-bit wide copies resulted in a much faster result at the cost of more equipment, neglecting the controller's cost in both cases. As technology marched on, we have reached another plateau on which we find the 32-bit wide full adder. Several RALU's are available from various manufacturers having 32-bit wide data paths. In some cases, the width of the modern RALUs can not physically be expanded, thus indicating that the 32-bit wide plateau will not be passed by using multiple copies of that element.

All of the devices discussed above perform integer arithmetic on unsigned and two's complement binary numbers. If one wishes to use other number representations, the addition and subtraction operations become algorithms again. Binary Coded Decimal (BCD) operations are normally implemented using the binary full adder with the help of a decimal carry generator. This is accomplished by using a comparison between the sum produced in each 4-bit wide add and the maximum value represented by the notation (e.g., 10). The function can also be created using a combinatorial network similar to binary carry generation. In either case, if the carry is non-zero, then the sum must be "corrected" for the 4-bit field by another addition, e.g., 6 is added in the BCD case.

Floating point numbers can be processed using the full adder that is in the RALUs we have studied. For addition and subtraction, the binary point of the two operands must be aligned by shifting the fractional part of the smaller operand to the right and incrementing the exponent to retain the value of the operand. This "denormalization" is continued until the exponents of both operands are equal. At that point, the fractional parts of the operands are added, or subtracted, as if they were fixed point fractions with the binary point near the left end of the number. The representation of fixed point numbers was discussed in Chapter 2 where it was shown that these numbers could be manipulated using the integer arithmetic hardware. It is clear that addition and subtraction of floating point numbers requires an algorithmic approach if the architecture contains only integer arithmetic hardware.

The multiplication and division of integers and fixed point numbers has been described algorithmically in this chapter. If the numbers were expressed in floating point notation, then multiplication and division can be thought of in terms of the layers that we saw develop for the expansion algorithms above. The exponent fields are dealt with separately, while the fractional parts are treated as fixed point numbers. If these operations are performed on a single RALU of the proper width, the exponent operation is principally concerned with normalizing the fractional part after the operation. The fractional parts themselves are multiplied or divided by the same algorithms given for integers above.

The multiplication of integers using a combinatorial approach is as old as the Transistor-Transistor Logic (TTL) family. The development of these devices has been pushed by the needs of various fields that are known collectively as *digital signal processing*. As was noted above, the arithmetic sum-of-products algorithm is fundamental to this field. While addition is handled in the RALU, the multiplication takes a considerable length of time – time that was too long for most useful applications. The need for a combinatorial multiplier was quite apparent. Figure 7-12 shows the architectural portion of a microprogrammed state machine in which a combinatorial multiplier/accumulator (MAC) with memory replaces the RALU and shifters in Figure 7-7.

This structure was discussed at length in Chapter 3. Here, the MAC is combined with memories in a structure that can support either the arithmetic sum of products or the power series expansion. The multiplier-accumulator architecture is available for integers or floating point numbers. The symbol for it is the same in either case.

The flow of data through the MAC is similar to that for the RALU-based sum of products or power series expansion algorithms except that the multiplication algorithm is replaced by a combinatorial multiplier having a width appropriate to the problem. The overall control algorithm is simplified since a single clock cycle can see two numbers multiplied and then accumulated into the accumulator register. The arithmetic sum of products is accomplished by using the a_MEM_REG and x_MEM_REG to scan through a set of coefficient/data pairs in the respective memories. After initialization, there is only a single microinstruction within the sum of products loop.

The power series expansion mapping requires that the data (i.e., x) be multiplied with the developing accumulation. The order of performing the multiplications and additions is changed. This is reflected in Figure 7-13, where the dataflow through the MAC has been altered. The coefficients, a(i), are added with the product of the data and the previous terms. In order to support initialization, the adder must function as a small ALU in that it must be able add, subtract, and pass one operand, the coefficient, to the accumulator register.

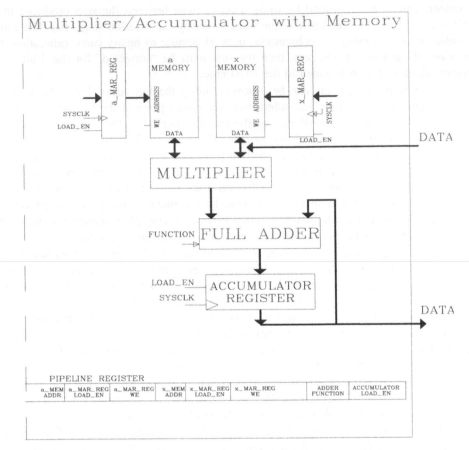

Figure 7-12: Architecture section of an ASM with a multiplier/accumulator.

7.4. COPROCESSORS

A *coprocessor* is a device that extends the instruction set and data types that a host computer can handle. As will be presented more thoroughly in Chapter 8, a central processor (CPU) in a computer is described by its instruction set, address modes, and programming model. A given CPU has instructions that can manipulate data in particular formats. A coprocessor adds additional operations (i.e., instructions) to the original CPU's instruction set. In addition, it can extend the types of data that can be manipulated by the host processor. In all cases, the additions appear as additional instructions, data types, or registers added to the description of the host CPU.

Our interest in the coprocessor at this point in our studies is in the recognition that the coprocessor can be built as a microprogrammed algorithmic state machine using a single level of control. We distinguish two versions of the coprocessor here. The first, called the memory mapped coprocessor, is constructed by adding an architecture to the external architecture of the CPU. The registers in the *memory mapped coprocessor* appear as memory elements to the host CPU. The second version is formed by attaching a complete algorithmic state machine to the host system by placing its registers at selected addresses in the external architecture of a CPU. This version is most frequently called a *coprocessor* without using any qualifying adjectives. Most math coprocessors used with microprocessors are of this type. The distinguishing feature of the two versions is involved with the location of the controller of the coprocessor ASM. In the memory-mapped coprocessor, the controller is simulated by the host CPU while in the "coprocessor", it is implemented in hardware.

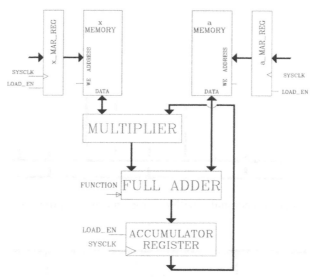

Figure 7-13: Multiplier/accumulator dataflow for power series expansion.

7.4.1. Memory-Mapped Coprocessor

Consider the table driven data transformation architecture discussed in Section 3.8. The memory that implements the table-driven architecture is also found in the external architecture of the von Neumann computer shown in Figure 7-14. Such table-driven transformations can be implemented in the computer by placing the data that is to be transformed on the part of the address bus that directly enters the memory module. The element of the memory at the "address" formed by the data contains the "transformed" data. The CPU is made to "read" the transformation table by executing its data movement instruction.

Some processors support an address mode that uses two "index" registers at the same time; see Section 8.7. The programmer can place the address of the base of the transformation table in one register and the data to be transformed in the second register. The CPU adds the two registers to create the effective address of the transformed data element in memory. For simplicity, the base address register should contain 0s in the bit positions occupied by the data in the second register; e.g., for 8-bit data, the low 8 bits of the base address register should contain 0s.

So, to begin with, any computer contains a memory-mapped coprocessor. In some cases, the CPU's address modes make the transformations easier and faster to implement.

The element labeled "Memory-Mapped Coprocessor" in Figure 7-14 contains the real reason for this section of the chapter. Registers implemented using structures similar to Figures 3-9 and 3-10 are attached to the data bus of the CPU to serve as "output" and "input" ports. The decoder is a Boolean network that creates register enable signals. These registers can be used like any other memory location addressable by the CPU, with the additional feature that data written to a coprocessor input register will be transformed when it is removed from the coprocessor's output register.

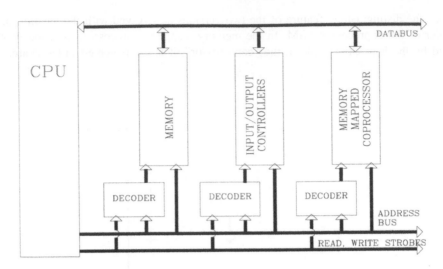

Figure 7-14: von Neumann computer with memory-mapped coprocessor.

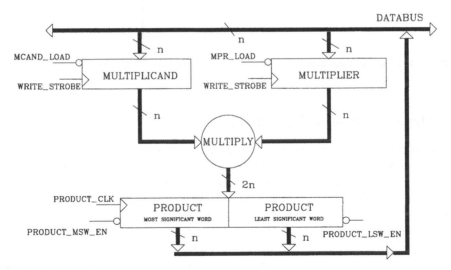

Figure 7-15: Multiplier as a memory-mapped coprocessor.

For optimal response, the propagation time of the Boolean networks that implement the coprocessor should be less than the memory cycle time of the CPU, i.e., the time required for the CPU to complete a read or write cycle with any element in the external architecture.

Figure 7-15 shows a memory-mapped coprocessor implemented using a combinatorial multiplier of the type discussed in Sections 3.9 and 7.3. The multiplier is a minimized Boolean network placed between the CPU output registers, Multiplicand and Multiplier, and the CPU input register, Product. The sense of "input" and "output" is relative to the CPU. The multiplier input registers have a common clock derived from the CPU write strobe. If the registers are implemented using the Enabled D register, then the active low MCAND_LOAD and MPR_LOAD can be generated as functions of the system address lines. These signals evaluate as true when the address bus contains a qualified address for the memory locations desired. The programmer needs to perform a data movement from the operand memory sources to the registers using the appropriate assembler language statements. For the MC68000 family, the instruction is

```
MOVE.W    0(A1),MULTIPLICAND
```

where the 0(A1) uses address register A1 to point at the source data in memory to be placed in the MULTIPLICAND register.

The double-precision product must be recovered from the multiplier. Figure 7-15 shows a PRODUCT register placed on the output of the multiplier. If this register is present, it must be clocked after the operand input registers are loaded. This clock can be implemented in one of three ways depending on the speed of the registers relative to the memory cycle time. The first method is to decode a third write strobe that is generated by a "dummy" MOVE to an address associated with the coprocessor, but in which no data is actually moved. The "decoded write strobe" causes the output of the multiplier to be loaded in the PRODUCT register.

The second method to generate the PRODUCT_CLK signal is to derive it from one of the input register load controls and the system WRITE_STROBE. If the write cycle is long enough and the data is presented by the CPU to the data bus early enough, then one of the input registers can be loaded at the beginning (falling) edge of the write strobe. The active edge of the PRODUCT_CLK can then be aligned with the rising (end) of the system WRITE_STROBE. The propagation time through the register and the multiplier must be less than the write strobe pulse width to make this method function. Obviously, the operand used for this purpose must be the second one loaded, otherwise the correct first operand will not be present.

The third method employs the beginning (falling) edge of one of the PRODUCT register output enable signals. This method works, again, if the signal (i.e., PRODUCT_MSW_EN) is the first one used to move data out of the coprocessor and if the memory cycle time is much longer than the register propagation time, as is usually the case.

Data is returned to the host computer by "MOVEing" it from the PRODUCT register to a storage location elsewhere. Since the example contains a multiplier, the number of bits in the PRODUCT is greater than in either input operand. The PRODUCT register is "broken" into two structures by partitioning the output buffers that connect the register to the CPU's data bus. The order that the parts of the product are returned is dependent upon the ordering of the data type by the host CPU. In the example, if the data were of integer format, we would want to use the same ordering supported by the host to make the transfer and further use of the data more efficient, i.e., fewer instruction fetches.

The number of registers in our example can be reduced. Assume that the PRODUCT register is replaced by tristate buffers divided as shown and controlled by the same signals, PRODUCT_MSW_EN and PRODUCT_LSW_EN. The PRODUCT_CLK signal disappears as well. The functionality of the coprocessor does not change. The control sequence is as follows if the host CPU is an MC68000:

```
MOVE.W 0(A1),MULTIPLICAND
MOVE.W 0(A2),MULTIPLIER
MOVE.L PRODUCT,0(A3)
```

The first statement moves a 16-bit word from the location pointed at by address register A1 to the MULTIPLICAND register. The second statement loads the MULTIPLIER register. The third statement moves a 32-bit quantity from the location called PRODUCT, our set of tristate buffers (input port), to the location pointed at by A3. The third statement actually performs two bus cycles over the 16-bit bus in the MC68000 system. The address of the PRODUCT (most significant word, MSW) port must be a number evenly divisible by 2 (i.e., a word boundary), while the least significant word (LSW) port must be at that address plus 2 (i.e., the next higher word boundary). Similar considerations for other CPUs make the control program compact for those machines as well.

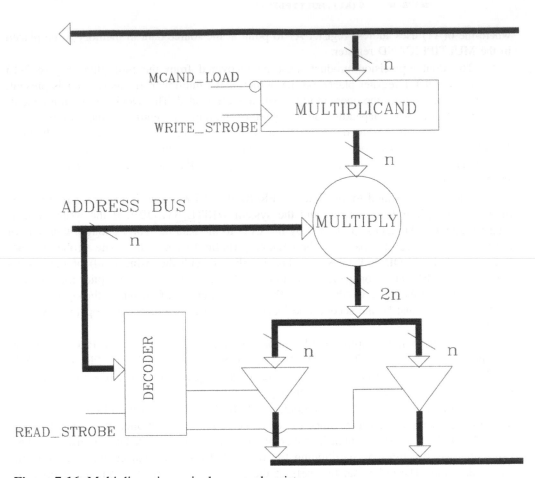

Figure 7-16: Multiplier using a single operand register.

The goal of the programmer in supporting memory-mapped coprocessors is to minimize the number of instructions in the control program. In general, the coprocessor propagation delay is much less than the instruction cycle time in most machines.

A final minimization is possible for a CPU that has wide data and address buses. For
two operand operations like our multiplier, only one of the operand input registers needs to
be retained, as shown in Figure 7-16. After loading it with one of the operands in one memory cycle, the second memory cycle can carry the second operand on part of the address bus
while returning the result on the data bus. Truncation of the product will be necessary in our
example.

For simplicity of programming, we place the output buffers at an address evenly divisible by 2n. Further, we include only the address lines more significant than bit n in the
address decoding equation, i.e., images of the output buffer completely fill the address space
for an interval of 2n above the base address of the coprocessor. For compatibility with the
host processor in our example, the least significant address bit, A0, should determine which
half of the product is gated onto the data bus. The multiplier occupies address bits A1
through An and decoding is accomplished with address bits An+1 through Am, the total
number of address bits being m + 1. The programmer "moves" the first operand to the
MULTIPLICAND register. He then moves the result from the coprocessor block using an address mode that combines the second operand, the multiplier, in a data register with the address of the base of the coprocessor block.

The final example shows how to control operations in the memory-mapped coprocessor. The simple example shown in Figure 7-17 is based on the multiplier-accumulator discussed in Chapter 3. Our example shows a unit that employs the integer data type with a single "instruction" line, FUNCTION_SELECT, that causes the adder to perform an addition or subtraction. Four registers are used – three were seen in earlier examples – and the fourth, the accumulator register, is added here. The adder and accumulator register implement a 2n+j-wide data path to accommodate the accumulation of several 2n-bit wide products. The MULTIPLIER and MULTIPLICAND registers are loaded by the CPU as in the earlier examples and the product register, if present, may be clocked by one of the methods used previously. If the product register, or pipeline, is not present, then the propagation time from the input registers to the accumulator must be less than the CPU instruction cycle time or the programmer must delay before clocking the accumulator register.

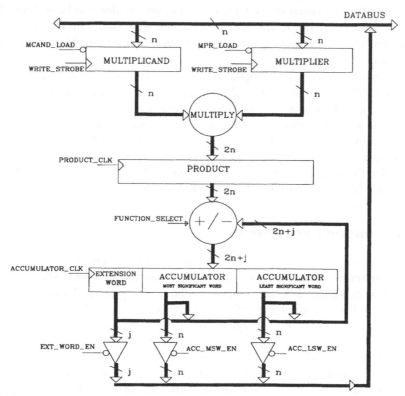

Figure 7-17: Memory-mapped multiplier/accumulator.

The additional feature added in this example involves the FUNCTION_SELECT control signal. While this signal could be placed in another output port, it is most easily driven by an address line (e.g., A0). Storing the second operand at an odd address can be used to signal an add operation, while at the adjacent even address could signal a subtract operation. The PRODUCT_CLK and ACCUMULATOR_CLK signals are generated from one of the input operand strobes using the method described earlier. If the product register is present, then the FUNCTION_SELECT signal must be delayed by the programmer to align with the data flow through the coprocessor. An "instruction" register can be placed between the address bus and the adder to accomplish the delay automatically.

The width of the accumulator register must be accommodated in its connection to the data bus. As in previous examples, the bus driver controls are "broken" into sections appropriate to the data bus width. Other partitionings are possible that place the extension word at the most significant end of the recovered number. Several modern integer multiplier-accumulators support the arrangement shown. These "microprocessor" compatible units return integer results in less than 100 nanoseconds for 32-bit input operands.

The most powerful architecture that can be used as a memory mapped coprocessor is the Floating Point Unit (FPU). As produced by several manufacturers, the FPU has the overall appearance of the multiplier-accumulator shown in Figure 7-17 except that data is in the IEEE floating point format. The FPU generally has a multi-bit FUNCTION_SELECT input that reflects the many operations that it can perform. These include multiplication with or without accumulation and various data flow functions. Presently, the FPU, under control of a program, can perform the normal group of operations supported by the standard microprocessor math coprocessor (e.g., sine, cosine, exponentiate, etc.) faster than the dedicated math coprocessor itself. Preparation of the table of control information can be performed using the microprogram assembler just as for any other microprogrammed state machine. The controller field is generally not present since the programmer can arrange loop control in the host CPU's program.

The speed of a memory-mapped coprocessor is usually limited only by the rate at which the host CPU can move data and commands to the architecture. The propagation delays for the devices discussed are less than 100 nanoseconds for all units that are appropriate for this service. Currently, this type of coprocessor is most visible as an adjunct to the IBM PC architecture where the INTEL math coprocessor is replaced by a FPU and some control tables that emulate the math coprocessor's instructions and registers. An array of FPUs makes possible the processing of several data streams at once for those problems that benefit from the Single Instruction Multiple Data (SIMD) type of architecture, e.g., processing of arrays of floating point numbers.

7.4.2. Independent Coprocessors

The coprocessor structure that is commonly used in the microcomputer field to extend the instruction set of a microprocessor is an independent algorithmic state machine of the pattern shown in Figure 1-3. In contrast to the memory-mapped coprocessor discussed above, the controller is implemented within the coprocessor instead of in the form of instruction sequences in the host CPU's program. Most common coprocessors are based on a microprogrammed controller since the sequences they run have lengths that depend on the functions they evaluate. Our interest in this version is kindled by the fact that the simplest environment to support the data needs of a state machine having a single level of control is within the external architecture of a computer. The task of supplying data and control programs can be relegated to the computer, while the high-speed data transformations can be performed in the coprocessor.

Figure 7-18 shows a von Neumann computer with an algorithmic state machine attached as a coprocessor. The attachment method is very "loose" since the output ports in the computer are input registers in the coprocessor while the input ports in the computer receive data and status from the coprocessor. This attachment method is commonly employed for normal I/O controllers such as serial ports and disk drive controllers which also are single level state machines that reformat data and control physical devices. The programmer on the computer is aware of the communications protocol with the coprocessor so the degree of integration implied by the definition of the term "coprocessor" is not reached.

The complete integration (i.e., extended instruction set) can be simulated by placing the coprocessor communication routines inside interrupt service routines that are entered as a result of an illegal instruction. This scheme is extended in the current microprocessors to make the job of the systems programmer easier. The applications programmer can be made

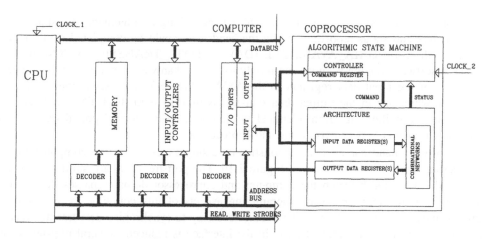

Figure 7-18: Loosely coupled coprocessor.

to think that he has a new set of instructions to invoke the operations provided by the co-processor by this simulation.

The controller in the coprocessor ASM may receive commands from the host computer that tell it which one of several sequences to run. Notice that this appears to make the coprocessor an instruction set processor; however, the coprocessor has no mechanism to fetch its own instructions. The ASM controller in turn commands its architecture using bit fields in its own pipeline register (see Figure 4-14). Information can be sent from the pipeline directly to the host computer by way of a path to the computer input port. The coprocessor controller can sense events either in its own architecture or, via an input data register, from the host computer using its *conditional multiplexer*.

The architecture of the coprocessor in Figure 7-18 is divided into input/output data registers and combinatorial networks. The data registers shown are included to support communication between the computer and its coprocessor. The combinatorial networks and other memory elements are present to allow the coprocessor to accomplish its task. It is here that the modern integer and FPUs can make their greatest contributions. The coprocessor designer can take advantage of these devices to construct networks that manipulate his chosen data format in ways that greatly increase execution speed.

One design point that should be mentioned in this connection is related to the relative speed of the coprocessor and the data communications path with the host computer. If the data transfer path is much slower than the coprocessor, then it is to the designer's advantage to place additional memory in the coprocessor's architecture and to employ long execution sequences. For many algorithms, this reduces the amount of communications needed across the interface and lets the coprocessor approach its real potential. A data type that is particularly sensitive to this approach is the vector or matrix; i.e., the array. If the designer wishes to implement a coprocessor that supports matrix operations like addition, multiplication (both types), and transposition, then it behooves him to place the entire operand(s) in the coprocessor before starting the operation or the computer interface speed will seriously limit his solution. Even slow computers can be made to do serious work on this data type using this method.

One should note that the communications protocol must support a method whereby both the computer and coprocessor know each other's completion status. The asynchronous nature of the two processes is indicated in Figure 7-18 where the CPU and coprocessor clocks are given as CLOCK_1 and CLOCK_2. The frequencies of these clocks can be, and usually are, different, thus requiring the designer to provide process synchronization. The most common method used is to implement a handshaking arrangement that supports both data transfers and status reporting. While the handshake could be completed using hardware signals of the form DATA_AVAILABLE (DA) and DATA_TRANSFER_ACKNOWLEDGE (DTA), a software version is more commonly used. The hardware version is implemented by using one of the input data registers in the ASM in Figure 7-18 to sense the DA from the computer, while one of the output data registers is used to send the DTA signal. The computer applies a polling algorithm identical to the one used for normal polled I/O operations to support transfers between itself and the coprocessor.

Software handshaking is accomplished by a subtle merging of the computer ports and coprocessor registers, as is shown in Figure 7-19. The ports/registers are similar to Figure 3-11. The dividing line between the computer and coprocessor domains passes down through the middle of the registers. The programming aspects of this coupling are essentially the same as above. A software handshake is still necessary to synchronize data transfers between systems; however, the number of parts in the interface is reduced. Several registers of each type (i.e., input and output) are mapped into the address space of the CPU by using address decoders on the computer side. The coprocessor drives the register enable signals from its controller's pipeline register using the "OTHER" field shown in Figure 4-14.

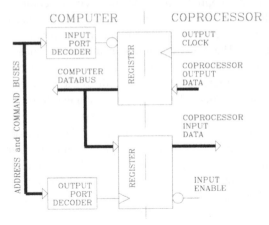

Figure 7-19: Computer-coprocessor coupling details.

The coprocessor communications algorithm that was under the systems programmer's direction in the earlier example has been made more automatic by certain CPU designers. Examples of microprocessors that include this feature are the Intel 80286 and 80386 in which the Numeric Data Processor (NDP) is supported and the Motorola 68020 and 68030 in which a general coprocessor interface is defined. Both manufacturers used the interface to extend the instruction set of their basic CPUs at a time when various large system functions could not be integrated onto a single chip. While either manufacturer's interface could be used for data other than that intended by the CPU designer, the Motorola protocol has been publicized as a general purpose coprocessor interface.

In this more tightly coupled environment, certain instructions, instead of directly accomplishing their stated purpose (e.g., a floating point addition) exercise the coprocessor interface to send control information and data from the computer to the coprocessor and return results and status information. The algorithm discussed above is run "automatically" when the host CPU fetches certain operation codes (opcodes). In other processors of the same family (e.g., the Motorola 68000), the opcode is treated as an illegal instruction allowing the system designer to supply software emulation of the desired sequence. The overall effect enjoyed by the application programmer is the speed improvement seen in the interface since the host CPU does not need to fetch "MOVE" instructions to accomplish the task. In either case, the speed of the actual coprocessor algorithm is dependent on the implementation of the coprocessor, in particular, on its internal architecture.

In the discussion given on the memory-mapped coprocessor, it was noted that a special numeric processor implemented in a memory-mapped form for the IBM PC AT was faster than the Intel 80287 Numeric Data Processor, (i.e., math coprocessor) which is a loosely coupled coprocessor of the type shown above. The speed difference is reflective of the way the Intel part actually accomplishes its operations not of the interface method. In this regard, one should notice the Texas Instruments TMS34082 which is intended as a floating point coprocessor for the TMS340 graphics processor and is based on a combinatorial floating point unit discussed in Chapter 3. It is of the form shown in Figure 7-18 with registers that map into the CPU's address space.

Brief mention will be made here of the tightly coupled coprocessor. The Intel 8087 math coprocessor used to extend the instruction set of the Intel 8086 and 8088 processors is such a unit. The tight coupling is produced by the capability of the coprocessor's controller to examine the instruction stream being fetched by the host CPU, i.e., the 8088 in the IBM PC. The 8087 was able to recognize its own operation codes in the instruction stream and execute them in parallel with operations in the 8088. This synchronization is accomplished at the cost of some duplication of control structure in the host CPU and coprocessor. Later versions of this function (i.e., the 80287 and 80387 for use with the 80286 and 80386) were built with "looser" coupling, allowing the liberated silicon to be used for other purposes. The independent designer would normally not benefit from building a tightly coupled coprocessor since the speed increase associated with the reduction of computer bus cycles can be accomplished by improvements in the architecture of the coprocessor. The loosely coupled coprocessor allows the designer free choice concerning the coprocessor clock frequency and communications protocol in a way that does not complicate the coprocessor's control algorithm.

7.5. A FINAL EXAMPLE

Our last example combines most of the machines discussed in the earlier sections of this chapter. We needed a machine that would let us test various ideas connected with "data flow" architectures. Such an architecture attempts to match the flow of data through a structure in such a way that it will appear at an operational element when needed by the algorithm being implemented. Most data flow architectures are very algorithm-sensitive, meaning that a structure that is optimum for one algorithm may actually be slower for another algorithm than a general purpose network. One algorithm and data type that has received a considerable amount of attention is the multiplication of matrices. As shown in Figure 7-20, each element of the product matrix is formed by a sum of products of the elements in a row of one matrix with those in a column of the other matrix. The "flow" of data relative to a Processing Element (PE) that forms the sum of products appears like two lines of soldiers, one line (the row from one matrix) intersects the other line (the column from the second matrix) at the PE.

Two entire matrices may be multiplied by using a single PE in a time determined by the steps needed to form a single sum of products times the number of elements in the product matrix. The speed of forming the product could be increased if the number of PEs used was increased. Many problems can be solved by exploiting the data flow inherent in an algorithm in association with the placement of the correct processing elements at the proper points in the data paths. If each element is capable of executing its own string of commands

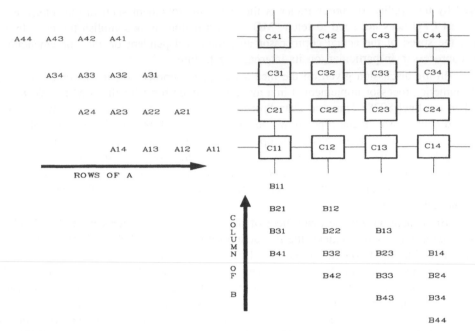

Figure 7-20: Matrix multiplication accomplished by parading elements of A and B through
an array of processing elements.

(Reprinted by permission of Ahmad Ansari.)

(i.e., program) independent of any other, then the architecture is said to involve Multiple Instruction Multiple Data (MIMD) control. While some algorithms can experience speed increases using this technique, interprocessor communications required by synchronization
problems introduces a time penalty. An alternate method in which all PEs execute the same
instruction in "lock step" removes the synchronization cost while greatly decreasing the
number of algorithms that can be handled efficiently. This method, known as Single Instruction Multiple Data (SIMD) control, is especially applicable to matrix operations if care is
taken in routing data among the available processing elements.

We chose to design a bus-connected architecture with which we could simulate the
operation of an array of PEs in several different structures. The Microprogrammed Algorithmic State Machine consists of nine FPUs (TI 74ACT8847) arranged in a three-by-three array
as depicted in Figure 7-21. Each PE is indicated as a numbered box at the intersection of a
pair of buses labeled Xi and Yj in the figure. Data from each 32-bit wide bus enters the PE
where it can undergo a floating point multiply – add operation in a single-system clock cycle. The result can be returned to the corresponding Xi bus at a time determined by the algorithm. All PEs share the same "PE instruction" field in the microword as a consequence
of the SIMD architecture.

Each of the X and Y buses is attached to a 32k × 32-bit word read-write memory. All
data operands come from these memories and results are returned to them at the appropriate
time. The system timing is such that two 32-bit operands can be placed on an appropriate
pair of buses and combined in a PE in a single bus cycle, thus determining the period of the
system clock, SYSCLK. The 32-bit results are returned in a single bus cycle, while 64-bit
results require a second cycle. The six memories and their associated bus drivers are individually controlled by way of 4-bit Y memory or 5-bit X memory subfields within the 27-
bit memory banks control field in the microword. This control philosophy allows data to
move on the appropriate X (Y) or *common bus* between one or more memories in the same

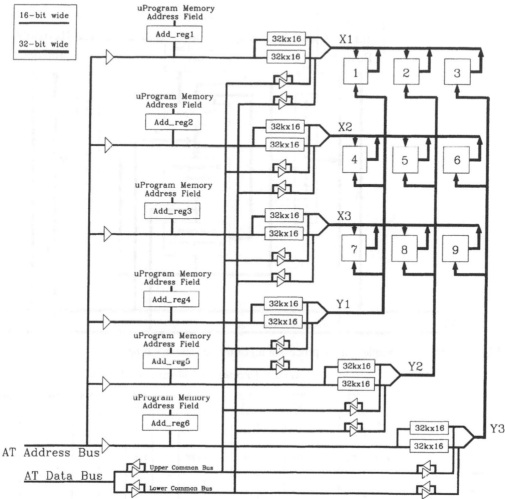

Figure 7-21: Architecture of array processor.

(Reprinted by permission of Ahmad Ansari.)

clock cycle. Clearing or parallel loading a group of memories with the same data is relatively efficient.

Data is moved from locations in the six memories determined by pointers in address registers, Add_regn. These multipurpose registers may be individually initialized from the microword, incremented, decremented, or held. Each is an independent subsystem controlled by a 2-bit subfield within the Address Registers' Control field in the microinstruction; see the micromachine pipeline format in Figure 7-24. A single 15-bit address field in the microword is used as the source of initialization address. The data in this field may be copied into from one to six address registers in a single clock cycle determined by each address register control instruction.

The entire ASM, as shown in Figure 7-22, uses a microprogrammed controller implemented with the TI 74ACT8818 Next Address Generator in a structure similar to Figure 4-14. Next Address Logic is added between the control field in the microword to reduce the width of the microprogram memory. The 4-bit controller field supports 16 specially selected control constructs appropriate to the various algorithms we wish to test. The microword width is 76 bits, requiring a total of 10 8-bit-wide memories and registers to implement.

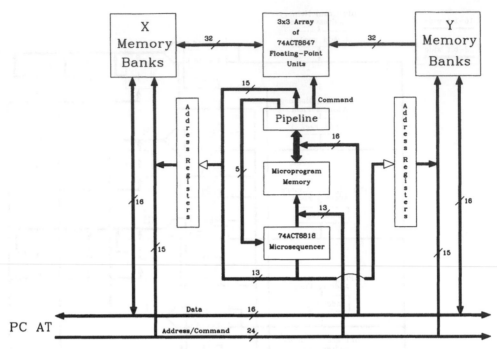

Figure 7-22: The complete array processor algorithmic state machine.

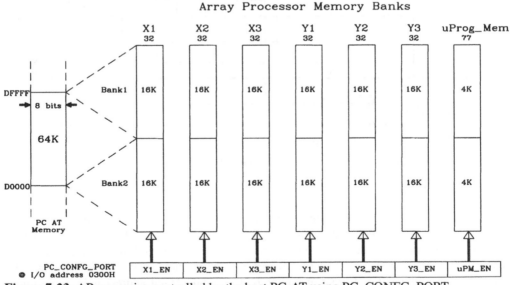

Figure 7-23: AP memories controlled by the host PC-AT using PC_CONFG_PORT.

(Reprinted by permission of Ahmad Ansari.)

The machine is placed within a host computer in order to obtain microprograms and data. The computer chosen was an IBM PC-AT that was conveniently at hand. Its 16-bit data bus was interfaced to the 76-bit wide microprogram and 32-bit wide data memories by a memory mapping scheme that reduced the amount of time required to transfer programs and data to and from the machine to a minimum. Each memory appears as a 16-bit wide structure in the unoccupied space between D0000 and DFFFF in the AT's memory space as shown in Figure 7-23.

A configuration register is placed at location 0300 in the AT's I/O space to control which 64k byte "bank" is actually visible to the AT. Program and data transfers to the machine require the setting of the desired "bank" field in the configuration register and then the execution of a repeated MOVSW instruction on the AT to copy a large block of information. The repeated MOVSW instruction can move a 16-bit quantity from the AT memory to the machine's memory in 5 AT clock cycles, each about 160 nanoseconds long on a 6-MHz AT. The "bank" field for each memory, shown labeled as EN_uPM, EN_Y3, etc. in the lower part of Figure 7-24, determines which 32k × 16-bit word block in the six data memories or microprogram memory is enabled. The host computer's address bus is coupled to each memory system's address input by tristate buffers, thus forming a two-input multiplexer with the tristate buffers on the output of the address registers. It is these buffer enable lines that are controlled by the "bank" bit fields.

The 16-bit AT data bus is connected through tristate buffers to the upper and lower halves of the *common bus*. PC-AT address bit A1 determines which half is used for a transfer between the machine and the host. This allows for blocks of data (or machine code) that are organized as a contiguous string of bytes in PC-AT memory space to be moved to the machine space without scrambling. Since the microprogram preparation and data acquisition take place in the PC-AT environment, this is a definite advantage.

The need for the host computer should be clear since it represents a simple platform for program development and data storage. The implementation of the platform in the PC-AT architecture was a matter of convenience. A host system with a wider data bus (e.g., 32 bits) would certainly double the overall speed of operation. A system with more room than that presented by the PC-AT prototyping card form factor would certainly make fabrication easier. In other words, if a viable machine could be made using the PC-AT as a host, then many other machines would be even more effective. The one simplifying feature of the AT is the simple, documented bus structure without which the architecture would have remained theoretical.

The size of the host computer's data bus also influenced the choice to add as much data memory as possible to the machine itself. The more operations that could be performed before data or results were moved out of the machine, the better its performance. Notice in Figure 7-21 the bus labeled "Common Bus". While this structure serves as the data path between the AT and the machine's memories, it also allows for data transfers between any memory and one or more other memories; i.e., simultaneous writing is supported for all data movements. In the event that the best fit that can be obtained between an algorithm and the hardware leaves a result in the wrong place to continue, then data can be moved back to its starting position via the common bus.

Control of the machine is performed by the host as master. The configuration register (lower portion of Figure 7-24) in the host processor's I/O space determines whether the host or machine has access to the machine's memories. A program in the AT has absolute control over the machine. Once the AT control program releases the machine from RESET by writing to the microprogram (μP) field in PC_CONFG_PORT, the machine performs the algorithm placed in its microprogram memory. The microinstruction to be executed last contains a 1 in the "Terminate" bit field (upper portion of Figure 7-24). All other instructions contain 0 in the field. The terminate bit field is connected to a read-only status register at the same address in AT I/O space as the write-only configuration register. The Array Processor Management program in the AT watches this bit to determine when the machine has finished its work.

Overall host ↔ machine interaction is summarized in Figure 7-25, and in more detail in Figure 7-26. A control program can be moved to the machine, the upper box in Figure 7-25, where it will remain until a new algorithm is needed. If many algorithms are to be used,

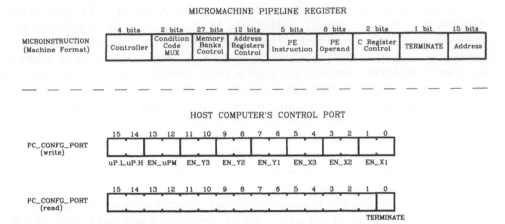

Figure 7-24: Format of microword and PC-AT configuration port.
(Reprinted by permission of Ahmad Ansari.)

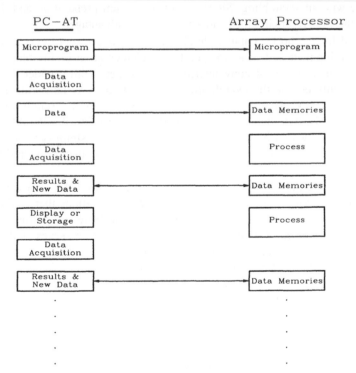

Figure 7-25: Overall host ↔ machine interaction.

(Reprinted by permission of Ahmad Ansari.)

all can be loaded while the one to be used on the current data is selected by the AT. The host then can execute a cycle, in Figure 7-26, in which it acquires data to be processed by machine, uploads the data, and then continues data acquisition while monitoring the completion status of the machine's current program. If the current results must be removed from the machine for storage, such is done before the next new data is uploaded for processing.

Several points should be made at this time concerning fabrication of the machine

- 4 AT prototyping boards were used to contain the machine
- 3 of the boards hold a portion of the architecture, i.e., 3 PEs and 2 32k × 32 memories each, while the fourth holds the controller
- Each X bus was located entirely on one board, while the 3 Y buses pass to all boards
- Interconnections are made using wrapped wire techniques. This one fact limited the overall system speed more than any other.

The design speed for SYSCLK was 5 MHz; however, the employment of faster memories, with an 80-nanosecond cycle time, has allowed some satisfactory testing at 8 MHz. Multilayer printed circuit (PC) boards would have been much more satisfactory in establishing stable operation; however, the technology was not readily available to us. Extensive power supply by-passing at each device and the creation of a power and ground "grid" partially alleviated the lack of PC boards. The SYSCLK circuit was buffered by the 8 elements in a tristate bus driver and then distributed through 47-ohm ring suppression resistors around the boards. All elements in the machine were implemented using high-speed CMOS technology except for the address registers which were built using PAL20XRP8 programmable logic devices.

Special note should be made here concerning the implementation of the address registers. As was noted above, they were fabricated using Programmable Logic Devices (PLDs) and are controllable state machines in their own right. Their function is to distribute the generation of addresses out into the architecture to increase overall speed. The alternative implementation technique would have involved placing a register array and arithmetic network at a central point in the architecture. The PLD allows the designer to perform appropriate arithmetic operations in association with providing a holding register at the desired location. While certain TTL or CMOS (complementary metal oxide semiconductor) parts could have provided some of the desired functions, the PLD approach allows the designer to group all of the functions he needs, along with the control simplicity he desires, into a single part.

In this example, the address registers were implemented as enabled D flip-flops with additional internal incrementing and decrementing paths. A parallel load path was also included, along with the tristate buffer structure on the output. A 2-bit control signal for each register determines whether the register remembers its current value, increments, decrements, or parallel loads data out of the 15-bit address field in the microword. The action of the control field is appropriate to use in the architecture of the microprogrammed algorithmic state machine since permission to perform a function is established during the present state with the action being performed on the next rising edge of the system clock. Recently introduced PLDs would have allowed us also to create a means to save the current value of the counter at each register site while introducing a new address value. Such a structure would be useful in streamlining the convolution algorithm.

The resulting system serves as an excellent test bed for many data flow algorithms. Since each PE is pipelined, allowing a floating point multiplication to take place while the previous product is accumulated, a total of 144 MFLOPS (million floating point operations per second) may be accomplished using the higher SYSCLK rate. As long as the algorithm does not require new data from the host, this rate can be sustained since addressing operations are performed within the address registers in parallel with the floating point operations. The way that a large two-dimensional matrix multiplication maps onto the 3 × 3 array of

PC−AT   Array Processor

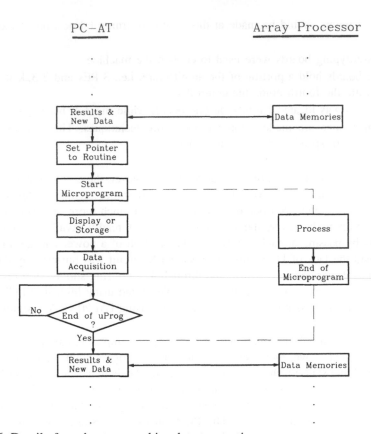

Figure 7-26: Detail of one host ↔ machine data transaction.
(Reprinted by permission of Ahmad Ansari.)

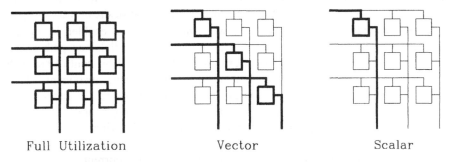

Full Utilization   Vector   Scalar

Figure 7-27: Utilization of processing elements for various data types.
(Reprinted by permission of Ahmad Ansari.)

PEs, as shown in the left part of Figure 7-27, allows each element of the product matrix to be accumulated at the rate of one per N clocks, where N is the dimension of the array. Larger arrays are mapped using a 3 x 3 subset of elements for the result matrix.

Since the memories and associated buses are individually controllable, the structure can be used for vectors and scalars as well. While not showing the spectacular speeds as for matrices, this flexibility allows the designer to embed the smaller data types without destroying the speed of the overall solution, as frequently happens in highly pipelined systems.

A description of the design of this machine was given in the industrial press (see Drafz, et al. in the Bibliography). Papers (see Dowling, et. al. and Ansari, et. al. in the Bibliography) and a thesis (see Ansari) have been written about the machine.

7.6. CONCLUSION

In this chapter, several important algorithms have been mapped onto Microprogrammed Algorithmic State Machines having a single level of control. The introductory examples based on the ROM/Latch implementation were intended to show the creation of a "programmable" control system where no explicit instruction set hardware appeared to exist. The middle group of examples showed the mapping of several important arithmetic algorithms ranging from accumulation, multiplication, and division, through the arithmetic sum of products to the power series expansion onto complex architectural elements that had been specifically evolved to be very efficient in these applications.

The coprocessor section was intended to illustrate an actual application area for these structures. Since the host machine handles data acquisition and program development, the complexity of the target first-level machine can be greatly reduced. Simulation of the microprogrammed controller, as in the memory-mapped coprocessor, further reduces the hardware complexity where the reduced speed can be accommodated. Even when the complete first-level machine is present, it benefits from operating in the host environment for reasons of algorithm development and data acquisition.

The final example shows the level of complexity that can be achieved in an educational environment for the study of algorithms on special-purpose machines. Dedicated architectures can greatly increase the speed of execution of many algorithms. The creation of a flexible, programmed environment that is appropriate for specific areas of application is presented to encourage other supercomputer "slaves" to widen the field of their choice of solution implementations.

EXERCISES

The following exercises are intended to explore complete systems using a single level of control. They build on studies done in earlier chapters involving single components and the tools used to control them. Analog-to-digital converters (A/D) and digital-to-analog converters (D/A) are used to interface systems to the "real" world. For the purpose of the exercises below that use these networks, we will define a common interface and control method so that they can be easily fitted into the resulting designs.

The A/D will perform a conversion in a fixed length of time, t_{conv}, after receiving a High-to-Low transition on its Start Conversion (SC) input. The conversion time, t_{conv}, is a characteristic of the conversion method, ranging from about 1 ms for multiple-slope, high-accuracy A/Ds used in digital voltmeters through the 20 μs–50 μs domain in 12- to 16-bit successive approximation A/Ds to the low microsecond range for 8- to 10-bit flash converters used in video applications. The first two A/D types send out an active high *End-of-Conversion* (EOC) signal that may be sampled by the controller to determine when a conversion is completed and data is available. We will assume that a *Track-and-Hold amplifier* (THA) is built into the front end of the A/D to sample and hold the incoming analog signal during the conversion. No further control signals are needed for it since SC triggers its operation.

The D/A is basically a combinatorial network consisting of switches controlling a set of weighted current sources. We will assume that two digital data buffer registers, modeled after the 74LS374, are placed in "word" serial relationship. The first buffer will be loaded with a word of data produced in our network when we assert a Low-to-High transition on its clock input, CLKBUF_1. The second buffer is loaded with data from the first buffer by a similar transition on CLKBUF_2. Double buffering reduces "glitches" in the analog output caused by irregular arrival times of data words and skew between data bits. A THA may also be used on the output of the D/A in a mode that presents a constant width pulse of a height proportional to the converted number to the following reconstruction filter. We will not assume control of this network in our problems.

In order to maintain the constant sample rate constraint of the Sampling Theorem, we will drive the SC input of the A/D and CLKBUF_2 of the D/A with a fixed frequency clock that is not derived from our state machine controller. Our controller will be used to test EOC and drive CLKBUF_1. A sequence may be run in our controller to determine if data overrun occurs anywhere in the system.

Procedure

All or part of the following procedure should be used in the execution of these problems.

1. Write a specification for the design — set sample rates, number of operations needed, and what is supposed to be accomplished in terms of the overall system.

2. Write a PDL or flowchart description of the flow of data and the algorithm used to produce the desired result. This will require that a block diagram of the system be produced so that some understanding of the control algorithm can be achieved.

3. Elaborate the block diagram in Step 2. Write an HDL or ASM description of the solution in terms of the block diagram. Here, the actual timing relationships between the blocks should be revealed.

4. Draw a logic diagram of the system in terms of real parts. This will require using the appropriate databook for each component employed. It is best to use a schematic capture program or a general purpose drawing program on a computer for this part of the task because of their superior editing capabilities. Learning at this stage involves a certain amount of "erasing" and redrawing to work out the control relationships. Even inexpensive general drawing programs available for personal computers are a great aid for this type of work relative to the paper-and-pencil method.

5. Once the parts are defined in the previous step, prepare a macro file that defines a language that can be used for writing the control algorithm. Remember to create commands that are appropriate to the solution of the problem. Do not repeat the names of the control pins on the parts in the command list. The control commands should reflect the problem – not the parts. Select a macroassembler that is available at your site and use its documentation to determine how to write macros, how to declare initialized data areas, and which host instruction names are not available for your use.

6. Write and assemble the control program. A final version of Step 3 should be produced that reflects the details that have been revealed in Step 4. Even though a top-down approach has been followed to this point, it is convenient to write a few of the sequences that control specific parts of the architecture. This method increases your familiarity with the devices under your control and allows you to take some time resolving problems caused by your lack of awareness of certain details. It conditions you to look at the problem with the "eye of your mind" as well as that of your forehead.

7. If you have been able to use a development system for producing the logic diagram in Step 4, then you can proceed to simulate the network using the control program from Step 6 and some suitable circuit stimulation information. This step should be performed

repeatedly along with Steps 6 and 4, as needed. Step 5 will be needed if the architecture is changed as a result of simulation studies.

In the event that a development system is not available along with all of its simulation models, the student may write behavioral models of his own along the lines suggested in Chapter 6 using his favorite programming language. Writing models of existing integrated circuits is a very efficient way to understand them and the "digital" abstraction that they present to the designer. Remember, the model detail needs to be sufficient only to show the correctness of the control program. The same development cycle should be used as with the development system.

Design Exercises

1. A common simple state machine exercise involves creating the controller for a "coffee" dispensing machine. Consider such a machine. It must receive coins from the user in the denominations of 5¢, 10¢, and 25¢ and dispense to the user one cup of coffee, black or creamed, or a cup of hot cocoa as he selects. Set the price of one cup of beverage as 45¢. It should also dispense the correct change. To make the problem easier, we will define three coin receivers that will signal to the control program when a coin is present. A new coin may not be entered by the user until the control program acknowledges the receipt of the current coin. There are three coin-received signals, one from each coin receiver, and three coin-acknowledge signals, one for each type of coin. Three buttons are available for the user to select his beverage. Dispensing a drink involves placing a cup in a holder and filling it with the correct amount of coffee or cocoa. A squirt of cream from a separate dispenser simultaneously with the coffee will create the "creamed coffee". Two possible solutions are available to determine the amount of material to dispense. One, the timed solution, requires no extra hardware, while the second, using a level sensor, may simplify the control program. Select a method and produce a coffee machine controller.

   a. Implement it using the ROM/Latch method.
   b. Implement it using a modern microprogrammed controller such as the Am2910 or TI 74ACT8818. Does the extra counter in the TI next address generator have an effect on the timed dispensing method?

2. Elaborate the traffic controller to include walk sign buttons and suitable improvements that you have desired in the course of dealing with intersections. How would you handle bicycles in a bike lane in the curb lane next to traffic? You have noticed that they are in a hazardous position when car traffic wishes to turn right. The same problem applies to pedestrians. Also, remember that fuel economy is destroyed at intersections and that streets are not parking lots. Consider one of the modern microprogrammed controllers for this task.

3. Design an elevator controller and implement it using a modern microprogrammed controller. Let the elevator serve four floors, each with a demand signal. Provide fire detection and override controls as are on many elevators now. Do not squeeze a passenger in the door. Use a door edge contact detector as well as photo detectors to determine motion into and out of the elevator car. Define a speed of motion for the elevator and acceleration and deceleration rates, i.e., elevators do not accelerate infinitely rapidly and they can not be two places at once despite our desires.

4. A garage door opener is found in many residences. Design one that meets the following requirements. It should respond to a single button mounted in a radio or on the wall that will raise or lower the door as appropriate. An interior light should be turned on when the door is opened or closed and left on for 10 minutes. The door should *not* crush things placed in its closing path; use an edge sensor and other methods to detect such things. Give the controller some feedback as to whether the motor is running. Keep the cost of each of the sensors to a bare minimum since price sensitivity is a fact of life. Remember, a simple sensor and a smart control program can simulate a costly sensor.

5. Let's design and build a simple waveform generator. Use a counter, an EPROM, and a D/A. Store a table of data that represents one cycle of the desired waveform in the EPROM. The data may be generated using a program written in a suitable high-level language. Make the program format the data such that it may be used by your intended EPROM programmer. The frequency of the output waveform should be determined by the D/A conversion clock, i.e., CLKBUF_2. The first buffer need not exist since it is replaced by the EPROM and counter. The counter clock should be derived from the D/A conversion clock.

6. A more capable version of the table-driven waveform generator described above may be made that keeps the D/A conversion clock running at a fixed frequency (e.g., 44100 samples per second) and creates a pointer into the waveform table that is incremented by an amount appropriate to the frequency desired.

    Consider a pointer into the waveform table. If this pointer is incremented by 1 each time it is used to look up a sample, then the output frequency will be the D/A conversion clock frequency divided by the length of the waveform table. However, if the pointer is incremented by 2, the waveform will be repeated twice as fast. Notice that the precision of the waveform is reduced. What happens if the pointer is incremented by 1.5? Design a state machine containing a 16-bit wide RALU as in Figure 7-7. Add a D/A with its double-buffered input. Control CLKBUF_1 from your controller and CLKBUF_2 from the D/A conversion clock mentioned above. Use the RALU to contain and increment a pointer into a waveform table EPROM that is suitably attached to one of its data paths. To accommodate incrementing by fractional amounts, let the pointer and its increment each occupy two registers as double-precision fixed point numbers. The binary point should be between the upper and lower 16-bit halves of each. The number passed to the EPROM should consist of the integer portion of the fixed point pointer.

    The controller should sense several switches and "play" the waveform at the frequency assigned to the one that is selected by associating a switch with an increment in the register array of the RALU. The design will serve as a one-note-at-a-time "organ".

7. Use the table-driven waveform generator in Exercise 6 as the basis of an "organ" that can play more than one note at a time. Suggested modifications to the architecture consist of providing a separate EPROM that contains the increments for the keys on a small keyboard so that the registers in the RALU can be used for other things. The waveform table's output should be made available to the RALU so that each key's waveform element can be added to that for another key. The double-buffered D/A should be attached to one of the data paths out of the RALU so that the summed waveforms for all keys for a given sample time may be converted. The CLKBUF_2 signal is still supplied by the D/A conversion clock and not the state machine controller.

8. A variation of the network in Exercise 7 can be produced by designing custom incrementing networks and eliminating the RALU. Consider a block that contains a register and a network that implements the following function table:

F_Sel	Function
0	Hold current contents unchanged
1	Add a fixed point constant to the current contents

For compactness, perform the incrementation function in serial; i.e., place a 1-bit adder on the most significant bit of the register. The remaining bits are merely shifted across the register. The right-most bit is made available to the flow control network on the adder. Notice that a single bit of carry must be saved during each clock cycle. The fixed point constant may be "hard-wired" into a MUX on the other input to the adder. The pointer may be moved to a buffer register on the address inputs of the EPROM containing the waveform table by serially shifting while it is being incremented. One of these specialized blocks is provided for each key and may be implemented using a Programmable Logic Device (PLD). A single 16-bit parallel full adder is used in an accumulator structure to receive the sample from the waveform table for each key that is currently pressed. This implies that the controller must run much faster than the D/A conversion clock. Specify the number of keys that may be played simultaneously. Notice that our controllers may *not* run at the frequency of visible light. Pay attention to the propagation times for the appropriate components as specified in their data sheets.

9. Design and implement a state machine based on Figure 7-12 to serve as a digital filter on data that is converted by an A/D at a sample rate of 44,100 Hz and applied to the input data bus. A double-buffered D/A is placed on the output data bus with CLK-BUF_2 driven by the same sample rate clock. The MAC and memories in the architecture are used by the controller to perform a sum-of-products (SOP) calculation involving the last 20 converted samples. Each time a newly converted sample replaces the "oldest" sample in the "x" memory, a new SOP calculation is performed. There are 20 coefficients stored in the "a" memory. Coefficient a_0 is always multiplied with the newest sample, while a_{19} is always used with the oldest data sample in the "x" memory. Since a new SOP must be performed on every A/D–D/A sample cycle, SYSCLK must be much greater than the frequency of the sample rate clock. Use a handshaking method to retain synchronization between the two clocks. The details of the digital filter are left to a suitable course. The nature and frequency characteristics of the filter are determined by the specific "a" vector used. The number of coefficients is appropriate for a small Finite Impulse Response (FIR) filter or a very high precision Infinite Impulse Response (IIR) filter.

10. Use the multiplier model produced in the exercises in Chapter 6 to create an architecture similar to that in Figures 7-12 and 7-13. The integer MAC with memory must be able to perform sum-of-products calculations or power series expansions. Suitable multiplexers should be added to make both possible on the same network. Be sure to create the proper control signal as well. Notice the bus width problem in the power series expansion of an integer. Show how the network should be used with fixed point numbers and truncation during such expansions, i.e., specify the location of the binary point and how truncation occurs. The result of this exercise should include a part specification,

logic diagram and simulation programs, and test data that demonstrate the part's performance.

11. Use the module produced in Exercise 10 to implement a memory-mapped coprocessor for a computer accessible to you. Eggebrecht describes the action of the IBM PC-AT bus in his book *Interfacing to the IBM Personal Computer* (see Bibliography). Implement the coprocessor in such a way that the control lines needed by the architecture are passed via the address bus of the host computer. Show assembler language code for the selected host that exercises this coprocessor in calculating the sum-of-products and power series expansion using data contained in tables in the host computer. Be sure to take advantage of the host processor's instructions and address modes to increase speed. In the case of the PC-AT, notice that the MOVSW instruction is able to move strings of words between elements in the external architecture at a very rapid pace. This implies that both the data and commands should be stored in tables in the host's memory.

    Remember that most hosts have a macroassembler available. Take advantage of this feature to create a set of instructions like SOP and POW that are invoked with pointers to the appropriate data tables in the host's registers. In other words, integrate the architecture into the machine in a meaningful way.

12. Perform Exercise 11 using the TI 74ACT8847 or the AMD Am29C327 FPUs described in Chapter 3. Be sure to obtain the appropriate data book. This exercise will benefit greatly from using a workstation-based design system so that simulation can be performed on the design. Most steps from the previous exercise can be performed without this level of hardware commitment, however. Note the wider the host's data bus, the faster the execution speed.

13. Attach the stand-alone machine designed in Exercise 9 to a host computer as an independent coprocessor. Let the host computer serve as a source and sink of data in addition to the A/D–D/A combination. This places more constraints on the handshaking between the data source and the machine's controller. Data should be able to enter the machine from the host or the A/D and leave via the D/A or the host data bus. The host should be able to poll the filter machine to determine the need for a data transfer. Once Chapters 8 and 9 have been covered, this problem should be examined again in the light of interrupts and DMA controllers.

Chapter 8

A MICROPROGRAMMED COMPUTER

8.1. INTRODUCTION

The focus of this chapter is on microprogrammed machines having a second level of control. This study will be built around the design of a hypothetical General Purpose Instruction Set Processor, GPISP, or computer. A GPISP may be realized as a Complex Instruction Set Computer (CISC) or Reduced Instruction Set Computer (RISC). Even though the basic instruction set chosen for our machine could be implemented either as a CISC or RISC, the microprogrammed controller we plan to employ is more efficiently used in a CISC. The control sequences do not affect the logic circuits in a microprogrammed controller as they do in the traditional controller. The flexibility of this type of controller makes implementing the complex sequences in the CISC easier.

The features of a basic, general purpose register-based instruction set will be discussed in the first few sections. The second-level design is roughly similar to many current machines. Methods of addressing data are discussed. While the instruction set and selected address modes are simple, the reader will find the machine to be very effective and relatively easy to program.

Once the second-level machine is specified, its implementation using a microprogrammed ASM composed of the building blocks studied in previous chapters proceeds. The overall design goal will be to employ as few blocks as necessary to provide the features required by the instruction set. Simplicity is necessary to prevent obscuring the overall structure with detail. The minimized hardware will result in more clock cycles being needed to implement certain sequences. This is a valid trade-off for both pedagogical and cost reasons.

In Chapter 9, several "improvements" will be made to the simple architecture in order to increase execution speed or make the design more flexible. In many cases, the changes will require very little hardware revision. These revisions are shown in the context of "evolving" the initial design toward a final elegant, efficient solution.

8.2. BASIC ARCHITECTURE

An Instruction Set Processor (ISP) is characterized by three elements that are visible to the programmer at the second level of control. These are the instruction set itself, the data addressing modes, and the programming model. In addition, the ISP needs other sequences of states to support the facilities visible to the programmer. The most important of these is the sequence needed to fetch the second-level instruction. This sequence allows the ISP to perform its tasks independent of operator interference and is its major distinguishing feature. Some terms used in the following discussions (e.g., CPU) look toward a particular implementation of the ISP as a computer. The definitions of these terms is given in Section 8.4. At this point, consider that a CPU is a special place and a register is a specialized storage element. In some implementations of an ISP, these items may be merely special data areas in a program that are physically implemented just as all other data areas.

8.2.1. The Instruction Set

We will provide our GPISP with a working subset of typical general purpose computer's instructions. This set, called the basic instruction set, is sufficient for implementing any control or calculation sequence possible. Missing second-level "instructions" (e.g., multiplication) may be provided using functions created from the existing basic instruction set. The set of instructions suggested here is more than sufficient for the task. Several instructions (e.g., subroutine CALL and RETURN) are provided to make the work of the second-level programmer easier. The following list shows the basic instruction set divided into several logical groupings:

Data Movement – Move data between registers and/or memory

LOAD
STORE

Arithmetic

ADC – Add using Carry In
SBC – Subtract using Borrow
ASL – Arithmetic Shift Register Left
ASR – Arithmetic Shift Register Right

Logical

AND
OR
XOR
LSL – Logical Shift Register Left
LSR – Logical Shift Register Right

Conditional Branches — Program flow control

Bcc Destination — Test various flags and branch if true

Subroutine

CALL Destination — Go to Destination and remember the return address
RETURN — Return to address remembered during the previous CALL

This instruction set will allow the implementation of all algorithms needed by a second-level program. The implementations of the instructions will be covered in more detail later.

8.2.2. The Addressing Modes

The next aspect of an ISP to be discussed is its ability to point at (i.e., address, operands). We know from the previous chapters that operands occupy registers in our earlier work. The number of registers in the parts that we have studied is far less than the number of variables in even the smallest program. Therefore, we expect that memory external to the Register Arithmetic/Logic Unit (RALU) will be used to contain these variables. Some means of conducting the data in external memory to the RALU will be needed in our implementation. The details of the addressing algorithm is not visible to the second-level programmer, but the method by which he points at a variable is. These methods are grouped under the title of "addressing modes". Each addressing mode will be implemented as a sequence on the underlying state machine that will result in the specified operand being conducted to the RALU for processing. We will identify the following address modes as a common, useful subset of all possible address modes:

- Register/Register (RR) – The operands are in the registers within the CPU.
- Register/Memory (RM) – One source operand and the destination operand are in the CPU registers, while the other operand is in external memory.
- Register/Memory Immediate (Im) – The memory operand is part of the instruction stream.

- Register/Memory Direct (ABS) – The memory operand is pointed at by an address contained entirely in the instruction.

- Register/Memory Indirect (IND) – The memory operand is pointed at by a register or a register plus an address contained in the instruction.

8.2.3. The Programming Model

The final element of an ISP specification is the *programming model*. This model is usually presented graphically. It details the resources of the CPU that can be manipulated by the second-level programmer. The main CPU resource is the set of registers whose contents the programmer manipulates using second-level assembly or high-level language instructions. Our simple computer will possess 16 registers, 14 of which are general purpose and 2 of which, while visible to the second-level programmer, are dedicated to implementing the functions of the machine. The programming model is presented graphically in Table 8-1.

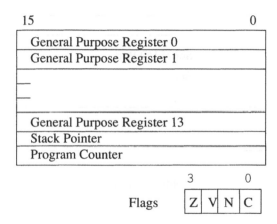

Table 8-1: Programming model for the computer example.

Frequently, the remaining resources in the complete computer are specified in an extended version of the programming model. Our simple computer allows addressing of 32k words of 16 bits. The serial I/O controller that connects to the console terminal is memory mapped. The complete memory map is presented in Table 8-2.

The details of the programming model will be discussed more thoroughly later as each element is implemented.

8.3. THE SECOND–LEVEL MACHINE INSTRUCTION FORMAT

The format of the machine instruction reflects the architecture of the system on which it runs. Each cell in memory is 16 bits wide, which upon implementation, implies that 16 bits will be transferred to the CPU in one "lump". While this may not always be true, it is in our simple design. The first word of an instruction must tell the CPU about all other parts of the instruction. Generally, the first word will contain all of the information needed to execute the instruction as well as some part of the location of the operands. In our design, the first word of the machine instruction will have the following format:

15	8	7	4	3	0
Operation Code		Reg 1		Reg 2	

Remember that the control and arithmetic elements make decisions by manipulating numbers in some representation using the digital elements that we discussed in earlier chapters. This requires that the instructions prepared by the second-level programmer must be presented to the machine in the same form as the data (i.e., as numbers). The operation code (opcode) in our machine will be an 8-bit unsigned binary number. Its interpretation will cause the machine to perform one of the sequences that will simulate the action desired by the second-level programmer.

	15	0
	Serial I/O Port	
FF02	Status	
FF00	Data	
17FF		
	Application Memory	
	"RAM"	
1000		
07FF		
	MONITOR Program	
	"EPROM"	
0000		

Table 8-2: Extended programming model of the computer example.

The Reg 1 and Reg 2 fields will allow us to point at two of the registers visible in the programming model. The programming model indicated that there were 16 registers available to the second-level programmer. Our instruction format must contain unique codes for each register. The most compact form of this code is binary, so we manage to squeeze two 4-bit codes into the remaining part of the first word of the instruction. We anticipate that our implementation will be able to fetch all of this information in one memory read operation in the external architecture. Notice that this is all of the information that is needed for the Register/Register (RR) address mode.

If more address information is needed, it must be provided in the next word of the instruction. This is the case when the second-level programmer wants to point at a location in the external architecture. The following format is appropriate for our simple computer where the range of addresses in the architecture can be handled using a 16-bit number:

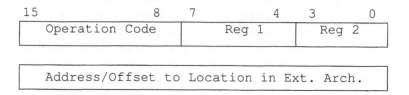

This format is appropriate for all of the Register/Memory address modes that require either an offset or complete address information.

In the address mode specification above, we notice that the second-level programmer should also be allowed to include data itself in an instruction; that is, the immediate address mode. The format of this instruction follows:

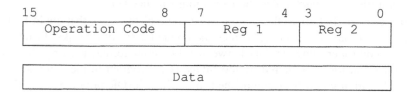

8.4. MAJOR LOGICAL SECTIONS OF THE COMPUTER

In this section, we plan to implement the ISP defined in Sections 8.2 and 8.3 as a computer. The elements of the computer that are visible to the second-level programmer begin to reveal some of the major logical sections of the computer. Referring to Figure 1-6 (reproduced below) and comparing it to Figure 1-4, we note that the computer is an Algorithmic State Machine (ASM). It is composed of the Central Processing Unit (CPU) acting as a controller and an external architecture. The external architecture contains second-level program and data memory and controllers for I/O. Comparing the CPU itself to Figure 1-4, we see it is also a state machine consisting of a controller and an internal architecture.

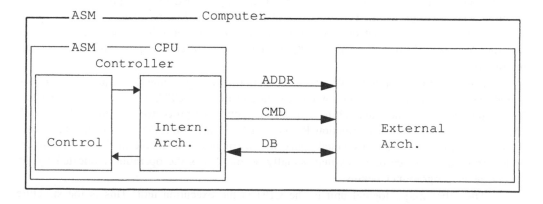

The CPU serves as a controller in the "computer" ASM, the structure seen by the second-level programmer. His programs implement state machines on a third level. The purpose of such a hierarchy is to convert the architecture of the underlying machine into one appropriate to the problem that he is trying to solve. At some point, instructions may be implemented in which a solution can be simply and efficiently written. This type of processor is known as an Application-Specific Instruction Set Processor (ASISP). Here, we have implied that the ASISP can be built on top of a GPISP (i.e., a computer). This need not be done, however, so that one could obtain exceptional speed by moving the ASISP implementation to a lower level.

While other controllers exist in the external architecture, it should be noted that they actually are outside of and invisible to the computer. The part of the I/O controller that lies inside the external architecture is an I/O port that appears to our CPU as just another memory cell. Its address is displayed in the extended programming model as for the serial port in Table 8-2. The port itself is only a place to store or load data. The I/O controller manipulates the "other" side of the port from the CPU. Other ports are provided such that the computer and I/O controller can synchronize their operations as needed.

The purpose of the I/O controller is to assume some of the tasks that could be implemented in programs on our computer. This frees the computer for other tasks. In our example, the serial I/O controller converts data between a bit serial format used by the console terminal and word serial format required by the MONITOR program in the computer. While the serialization algorithm could be accomplished in the MONITOR program using the shift instructions provided by the CPU, we find it easier to encapsulate this task in another dedicated state machine.

The logical structure of the CPU can be further divided when one considers the tasks that are necessary to implement the execution of the second-level programmer's instructions. These are the Program Control Unit (PCU) and the execution unit.

Each instruction must be moved from the external memory where it was placed by the programmer. It will then be used by the CPU to select a sequence of states that control the internal architecture to simulate the desired operation specified by the instruction. The task of pointing at the current instruction in the external architecture is accomplished by the PCU, of which the Program Counter (PC) is an important element. As is true in virtually all computers, the PC contains the pointer to the next second-level instruction to be executed. The PCU is responsible for "incrementing" the PC to point at the next instruction as well as delivering it to the external architecture to serve as an address. The actual hardware that contains the PCU will depend on the speed and flexibility needed by the applications planned for the CPU. A simple counter may suffice to increment the PC; however, we will be using second-level instructions that occupy a variable number of words in memory. This requires that the PCU be able to increment the PC by the number of cells required by the particular instruction.

Once the PCU points the PC at an instruction, it will generate the sequence of control signals necessary to move a copy of the instruction into the CPU. This action is called the instruction fetch sequence, the hallmark of all instruction set processors. In our CPU, we will place a register, called the Instruction Register (IR), in the controller to hold the first word of the instruction while it is executed. Once the instruction is available in the CPU, it is decoded and the sequence of states that actually accomplishes the operation requested by the second-level programmer begins.

The other major logical unit in the CPU is the execution unit. This is the structure that transforms application data selected by the second-level programmer. It is capable of taking the programmer's directions concerning the locations of the operands and performing the operations requested. The implementation of this logical structure will involve all of the facilities in the computer under the direction of the controller inside the CPU.

8.5. PHYSICAL IMPLEMENTATION OF THE COMPUTER EXAMPLE

Let us now provide a hardware existence for our computer example. For reasons of clarity, we will use as few components and buses as possible consistent with our goal of providing a reasonably complete machine. Our machine is divided into the Central Processing Unit and the External Architecture as suggested in Figure 1-6. The External Architecture contains the second-level program and data memory and a serial I/O controller for console communication. The CPU is further separated into the first-level controller and the internal architecture.

The first-level controller is implemented using a microprogrammed controller based on Figure 4-14. This form of the controller was selected since it allows us to efficiently use sequences of arbitrary lengths to construct a CISC instruction set. As shown in Figure 8-1, it is built using the AMD Am2910 Next Address Generator. The microprogram memory con-

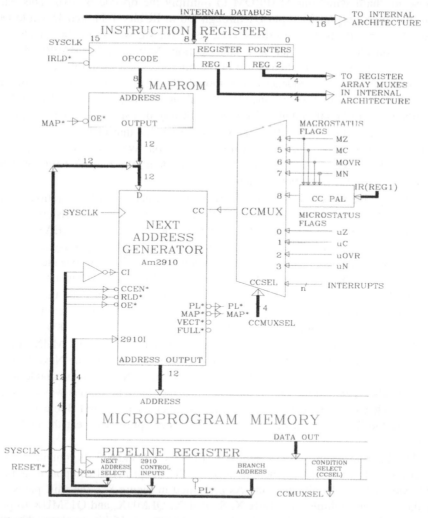

Figure 8-1: Microprogrammed controller.

tains 4096 words, each having a length of 100 bits. A pipeline register is used to hold the current microword during the course of a single microcycle. This reduces the length of the critical time path such that events in the internal and external architectures can run in parallel with accessing the next microinstruction. The first word of the second-level instruction will be retained in the IR in the controller for the duration of the decoding and execution cycle.

The opcode portion of the second-level instruction is translated to the address of the first microinstruction in an execution sequence by the MAPROM. The upper 8 bits of the IR and the MAPROM may be thought of as replacing the second address register in Figure 4-14. The translated opcode is enabled onto the input of the Am2910 Next Address Generator when the MAP signal is True and the PL signal is False. This signal combination is produced when the Am2910 executes a Jump to MAP address (JMAP) instruction. Notice that one clock cycle will be required for this program flow transition since the uninterrupted propagation path begins at the IR and ends at the pipeline register. For hardware simplicity, we chose to "hard wire" our MAPROM to multiply the opcode by two. This was accomplished by connecting Am2910 pin D1 through D8 to IR bits 8 through 15 while D0 and D9 through D11 were connected to ground (i.e., logic 0). We will show the effects this simplification has on the microprogram in a later section. Normally, the MAPROM is used in a more flexible way. For most efficient use of the microprogram memory area, no space should be left between execution sequences even though the sequences may be of arbitrary length. The MAPROM provides for this since the "data" at the address in the MAPROM specified by the opcode may be any number. This is the beauty of using a translation table.

The lower 8 bits of the IR contain two 4-bit fields that may be used to point at registers in the RALU under control of a microinstruction. While the opcode need not be retained beyond the decoding state when its work is done, the register fields may be needed at any time during the execution sequence.

The pipeline register holds the 100-bit microinstruction for the duration of one system clock cycle. It is implemented using three types of edge-triggered registers. The Next Address Select field is placed in a register similar to the 74LS273. Its CLEAR input is driven by the system reset signal RESET*. A Low value placed on this input asynchronously forces 0s into the Next Address Select field, thus causing the Am2910 to execute a Jump to Zero (JZ) instruction at the next rising edge of the clock. The result of this action is that an address of 000_x is forced onto the microprogram memory address bus and the microinstruction at that address is clocked into the pipeline register. It is the microprogrammer's responsibility to place the first microinstruction in the reset sequence at location 000_x.

A 16-input Condition Select Multiplexer is provided as the CCMUX. The four second-level status flags, MC, MZ, MOVR, and MN are presented to the Next Address Generator along with the first-level flags μC, μZ, μOVR, and μN. A programmable logic device, CC PAL, is also included to combine the second-level flags in various ways to support conditional branches on two's complement signed numbers. The desired Boolean function is selected using the Reg1 field of the IR. The functions are presented near the end of Appendix D for this chapter.

The Internal Architecture is shown in Figures 8-2 and 8-3. It contains all of the data combination and local storage functions in the CPU. Its principal component is the Am2903 RALU extended to a 16-bit data path. As presented in Figure 8-2, the shift paths are extended by the external multiplexers SLMUX, S15MUX, QLMUX, and Q15MUX to provide single- and double-precision shifts and rotate operations on 16-bit numbers. The 8-to-1 line multiplexers patterned after the 74LS251 serve as inputs to the bidirectional SIO and QIO signals on the Am2903 structure. Inputs to the MUXes are chosen to include the SIO and

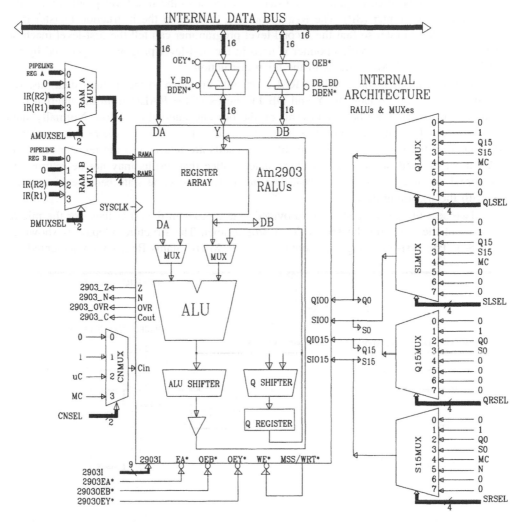

Figure 8-2: RALU, shift, and carry input multiplexers.

QIO signals for 16- and 32-bit operand shifts and the second-level carry flag, MC, to support 17- and 33-bit operand shifts. The extension shift MUXes must have outputs that can be tristated since they form a virtual multiplexer with the corresponding SIO or QIO I/O pin. The SIO or QIO pin is enabled or tristated by the selected RALU instruction, while the extension MUX is enabled by the most significant bit of its control signal (i.e., SLSEL, SR-SEL, QLSEL, or QRSEL). More information about how these paths are used is given later in the chapter.

The Carry Input to the RALU is provided by the Carry In Multiplexer (CNMUX). Values of 0, 1, or the contents of the first- or second-level carry produced by the last operation may be selected using CNSEL. The 0 and 1 values are used to provide the proper null input at the beginning a single-precision addition or subtraction operation, respectively. The status flag inputs are convenient when implementing multiprecision arithmetic operations; e.g., long-word addition of 32-bit operands using the 16-bit RALU data path provided.

Two register pointer multiplexers are provided to give the first-level programmer more flexibility in his choice of pointer sources. Two fields in the microinstruction, REG A and REG B, may be used to point at registers of interest to the first-level programmer, like the PC and Stack Pointer (SP). The Reg 1 and Reg 2 fields in the instruction register may be used for either pointer so that the second-level programmer can tell the first-level machine which register he desires. Which pointer is used in the first-level program is specified by the AMUXSEL and BMUXSEL fields in the microinstruction.

The data buses of the Am2903 RALU are all connected to the internal data bus. Bi-directional bus drivers isolate the Y and DB I/O buses of the RALU from the internal data bus to prevent interference between events on the bus and internal data flow. Normally, data from the external architecture will pass into the RALU via the DA bus where it is combined with data from the register array in the RALU. Results are returned to the register array via the Y bus at the end of the same machine state. Notice that results may not be returned to the external architecture via the internal data bus during the same state because of interference with the incoming data.

Two sets of status registers are provided, one to contain the ALU flags for the micro-program and the other for the second-level programmer. The structure adopted is shown in Figure 8-3. The first-level flags are contained in the Micro Status Register implemented us-

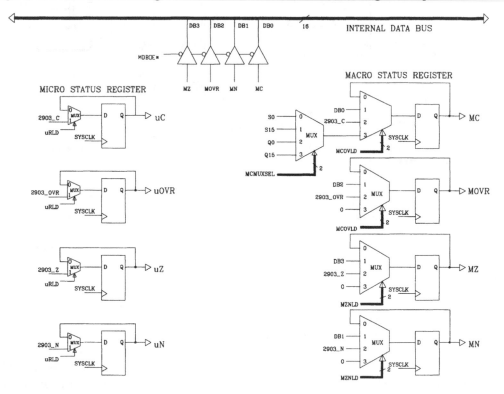

Figure 8-3: First- and second-level status registers.

ing four enabled D flip-flops. The single control input μSRLD is provided for the first-level programmer to determine whether these flags are all to be updated or not. The second-level flags are contained in the Macro Status Register. A more complex control capability is presented for this register to allow the second-level instruction set designer more freedom in the choice of flag setting strategies. The basic implementation method for each bit in the Macro Status Register is again the edge-triggered, enabled D flip-flop but with more inputs added to the multiplexer. Two control signals, MCOVLD and MZNLD, allow the first-level programmer to select the sources of the flags to be used in the register update. These independently controllable signals allow the microprogrammer to change the "logical" flags, N and Z, while retaining the arithmetic flags, C and OVR. Other combinations are also useful and will be exploited later in the chapter.

The four-input MUX on the third input to the MC multiplexer allows the choice of one of the shift output ports on the RALU as an input to the second-level carry status flag. Using MCMUXSEL, the microprogrammer can implement shift and rotate instructions that include the MC bit in the shift path.

As the capabilities of the machine are extended, there are some circumstances in which the second-level status flags must be saved between second-level instructions. While the choice of flag setting strategy can minimize the need for saving on the part of the second-level programmer, interrupts will still force flag saving. A path from the Macro Status Register to main memory is provided by the tristate buffers that connect the status register outputs to the internal data bus under control of MDBOE*. The return path is present in the connections to the enabled D flip-flops shown at input 1 on each MUX. The internal data bus bit position is labeled by DB0, DB1, etc. While more registers could have been added inside the CPU to contain copies of the status register, the nesting depth for this type of saving is drastically limited. There are more important uses for this type of expensive register.

The external architecture for our simple computer example is shown in Figure 8-4. The boundary between the CPU and the external architecture passes down through the Data Bus Buffer (DBB) and the Memory Address Register (MAR). The bidirectional DBB serves to isolate events on the external data bus from those on the internal. It also contains the necessary current driving capabilities to accommodate the many devices on the external data bus. The selection of items in the external architecture is under control of the computer designer and must satisfy loading requirements specified by the CPU designer. While these design groups may be associated as in the case of large machines, they are usually separated (i.e., in different companies) for microprocessor-based systems. The capabilities of the DBB and MAR and the external architecture bus cycle timing established by the CPU designer establish a design specification that must be followed by the computer designer if the entire system is to operate reliably.

An important trade-off in all second-level machines is that the external architecture is optimized for maximum storage space at minimum cost. Unfortunately, minimum cost implies slow response times for devices. This means that the external architecture must be operated at a lower speed than elements within the CPU. Events in the external architecture will be divided into steps called "bus cycles". The period of a bus cycle is determined by the time required to propagate a signal from the address and command buses back to the CPU by way of the external data bus. To accommodate the storage space maximization trade-off, this time will be longer than the CPU's system clock cycle. As described later in this chapter, we will allow a single transaction to be performed in the external architecture in a single bus cycle. By practicing a "sequential control philosophy", we will only require a single command, address, and data bus to connect a large number of storage locations in the external architecture. Data may be moved between any two cells in the external architecture but the operation will require two bus cycles, and the data will make a trip to a temporary register in the CPU or elsewhere. A cell in the external architecture is specified by the pointer carried by the address bus. The lower bits in the bus are used to specify a cell within a block (e.g., RAM or EPROM) and the upper bits are used by a "block" decoder to identify the block that should respond to the command. The locations in the memory map in

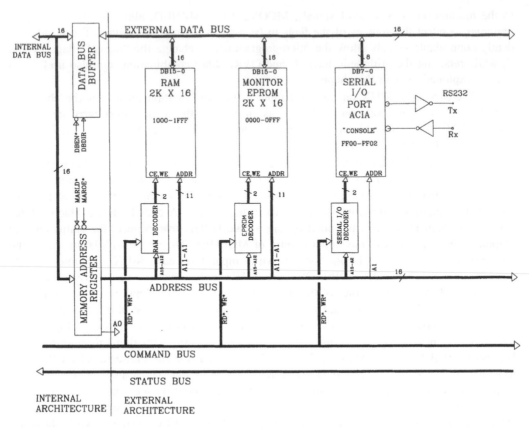

Figure 8-4: Memory address register and external architecture.

Table 8-2 are determined by the block decoders. The command bus carries information about the direction of the transfer, in- or out-bound with respect to the CPU, and its timing.

During a bus cycle, the MAR holds the pointer to the desired element in the external architecture. The data that is placed in the MAR is moved by way of the internal data bus from the RALU or other source under control of the microinstruction. Obviously, since we have only placed a single bus between the RALU, MAR, and DBB, we will be unable to transfer a pointer to the MAR while moving data from the external architecture into the CPU. Normally, this conflict should not occur since pointing at data should precede moving the data; i.e., the logical flow the our data accessing algorithms makes this a reasonable arrangement.

We have chosen to not use MAR bit 0 in the version of our machine that we implement in this chapter. This creates the appearance that our machine is able to address an entity that is only as wide as half of our data bus. The addresses 0100_x and 0101_x point at the same information. The immediate benefit to us is to allow the assemblers developed for machines that can directly address bytes (i.e., half of our data bus width) to be used directly with our machine without having to resort to the machine-word-to-byte conversion steps used in Chapter 7. We can immediately employ any such assembler as soon as the macro file is written even though we should not try to operate on bytes directly.

The blocks that we have decided to place in the external architecture of our simple computer include a Read-Only Memory (ROM), the EPROM, that contains the Monitor program. This supervisor program is normally not changed in the course of developing programs at the second-level since it contains our primary program development tool. Once the Monitor is written and debugged, it needs to be placed in a "protected" area so that an errant application program can not destroy it. Most Erasable-Programmable ROMs, EPROMs, are 8-bit wide devices that contain some multiple of 1024 bytes. We have used two devices, each containing 2048 (2k) bytes, placed side-by-side across the 16-bit data bus to build the 2k "word" EPROM memory block. The address bus drives the address inputs of the two EPROMs in parallel, i.e., A1 goes to the same pin on both devices.

A 16-bit wide read/write memory block (i.e., RAM) is also provided. It will be used by the second-level programmer to contain programs and data. A serial I/O controller is included to allow programs to communicate with the user through a serial console terminal. The terminal is connected to the RS232 drivers to the right of the port in Figure 8-4. These units serve to convert the higher signal voltage levels in the RS232 standard to the TTL thresholds used by the serial port. The serial port is modeled after the simple devices used with 8-bit microprocessors.

The computational capabilities of the computer are concentrated in the CPU. This concentration developed early in the history of computer design and was caused by the very high cost of the digital computational elements such as adders both in money and space. Our design as well as that of most modern computers follows the same general structure since computational elements that directly connect storage elements in the external architecture are not employed. Notice that the text of a program in a high-level language implies that such a connection exists; e.g.,

$$X = Y + Z$$

The storage elements X, Y, and Z used by the second-level programmer are physically implemented in the external architecture, i.e., in main memory. While the statement implies that paths directly connect storage elements Y and Z to an adder for the proper data type, we will simulate the connection by moving the contents of Y and Z into the CPU. The output of the adder appears to be connected to another storage element called "X". We will simulate this action by moving the data from the CPU, where it was generated, out to the "X" storage location in the external architecture. The following system of statements in a high-level language appear to be capable of simultaneous execution since no statement depends on the result of another:

$$A = B * C$$
$$X = Y + Z$$
$$J = I / K$$

We will simulate this apparent "parallel" operation by executing each statement in order one at a time. In order to prevent the execution time from being prohibitively long, we will invest heavily in the fastest arithmetic unit that we can afford. Further, we will interconnect it with a small collection of very fast temporary storage elements by way of fast buses. This should allow a single fast arithmetic element to replace the many operations and connections implied by the statements above. One should keep in mind, however, that this substitution has been made because one fast arithmetic unit is cheaper and the connections are "more" flexible than the many slower units and their "hard wired" data paths. Notice how quickly we work ourselves into a situation where we need ever-faster CPUs. As we require more and more equations to describe the solution to our problem, we are still trying to simulate the myriad of interconnections and operations with the same, single arithmetic unit. We can build faster CPUs by using faster basic elements, i.e., select an implementation technology for its speed. At some point we can go no faster. We can also build faster CPUs by providing an improved collection of data paths and computation elements. This is the architectural approach to solving the problem. Current CPUs represent a combination of both methods at one instance of time in the evolution of both techniques.

In some cases, a solution can be mapped very efficiently on an arithmetic network that contains many computational elements. Faster computation than ever can be achieved than with a single unit. As integrated circuit technology becomes more capable and simultaneously less expensive, this will be a more important option. For further study in this area, consider the architectures presented in the areas of distributed processing and neural networks.

8.6. IMPLEMENTATION OF THE FIRST–LEVEL MACHINE

Our CPU contains all of the computational capability that is accessible to the second-level programmer. The internal architecture consists of arithmetic elements (e.g., the ALU and shifters) as well as temporary storage elements, registers, and the data paths that make our simulation task possible. The main physical device to implement this structure will be the RALU discussed in Chapters 3 and 4. We have provided some additional registers and multiplexers to implement other functions, including status flags.

The Program Control Unit (PCU) will be simulated by using controller sequences in association with certain additional registers and data paths. Its most important input is the Instruction Register (IR). The opcode is used by the PCU to determine the sequence that simulates the second-level instruction. The PCU must contain and manipulate the second-level Program Counter (PC). It must also move it to the external architecture for addressing parts of the instruction. In our simple computer, we will not supply extra hardware for this purpose. The PC will be placed in register 15 in the RALU. The first-level controller will supply commands to the RALU to increment the PC each time it is used. It will also move it to the Memory Address Register (MAR) for use as an instruction pointer in the external architecture. The first-level controller will also manipulate the memory Read (RD*) and Write Strobes (WR*) on the command bus during the instruction fetch cycle.

In many computers, the PCU is implemented as a separate state machine with its own associated arithmetic elements. It also runs bus cycles to move data while the execution unit is processing data. Simultaneous operation of these units greatly increases the speed of execution but at the cost of more complex hardware, especially in the control area. For the sake of hardware simplicity, we have chosen to simulate the PCU on the same hardware as that used for the execution unit.

The execution unit will also be simulated using the resources in our simple CPU. The extensive arithmetic capabilities concentrated in the RALU are directed toward this task. Our execution unit is simulated by using sequences generated by the controller to orchestrate the RALU and other elements in the internal and external architecture. They must move data associated with the second-level program between the external architecture and the RALU, as well as use the RALU to perform the requested arithmetic or logical operations.

The RALU is capable of certain basic operations on data, including addition, subtraction, some logical operations, and single-bit shifting. If an operation such as multiplication is requested by the second-level programmer, then it must be implemented using the basic operations as a sequence similar to that in Section 7.2.4. Our machine can implement any instruction ever conceived for the second-level programmer. Even though first-level combinatorial networks may not be present to perform the operation, we can always string together microinstructions into a sequence that does. Our basic data type is the 16-bit unsigned or two's complement signed integer that the RALU was designed specifically to handle. We can handle any data type, including the IEEE floating point form or matrices on the same hardware without adding anything. However, the limited resources will result in long sequences and a great deal of nonproductive data movement to do the job.

8.7. SOME IMPORTANT SEQUENCES

All second-level instructions serve as requests to the first-level machine to execute a sequence of microinstructions that perform the desired operations. Other sequences are also necessary. The reset sequence initializes the second-level machine and begins instruction fetching. The instruction fetch/decode sequence, central to the Instruction Set Processor (ISP), maintains the continuous flow of sequence requests. The address mode sequences each encapsulate a particular data addressing algorithm that has been found useful by the second-level programmer. Such sequences save a great deal of "detail" programming concerned with data addressing by giving the second-level programmer many flexible ways to "point" at data.

We will now consider each of these groups of sequences individually, both as algorithms and how they interact with the hardware implementation of our simple computer example. In Chapter 9, we will extend our machine by adding two more sequences that support interrupts and bus mastership.

8.7.1. The RESET Sequence

To "start" the second-level machine, we must first start the first-level machine. In our design, resetting the first-level machine consists of supplying a 0 to the 2910I (instruction) input on the Am2910 Next Address Generator in the first-level Controller (Figure 8-1). This is accomplished by clearing the portion of the pipeline register that contains the NEXT ADDRESS SELECT field. Since these 4 bits were placed in a clearable, edge-triggered D register, resetting the first-level machine is simply a matter of supplying a *True* (Low) level on the clear input to this register. The reset signal can be generated by a circuit whose input is the RESET button or a power sensing circuit to provide "power-on" resetting. A zero on the 2910I input forces a zero on the ADDRESS OUTPUT (Y) of the Am2910 Next Address Generator, thus placing the microinstruction at location 0 in the microprogram memory on the input of the pipeline register. The next system clock tick places this instruction in the pipeline. When the RESET signal goes *False*, the first-level machine needs to begin executing the first microinstruction in the Reset Sequence.

The sequence of instructions that "RESETs" the second-level machine is called the reset sequence. This sequence must prevent the second-level machine from "hurting" itself or its users, as well as get it started. Damage prevention is absolutely necessary for machines that support second-level interrupts to be studied in the next chapter. Starting the second-level machine consists of obtaining the address of the first second-level instruction and placing it in the PC in the Program Control Unit.

There are two philosophies for obtaining the number to place in the PC. The first involves creating a number (e.g., 0000_x) in the CPU. This requires that the second-level programmer place the first instruction of his program at the specified location. In practice, many second-level programmers want to place their programs elsewhere in the external architecture. They must then place a GO TO instruction (JUMP or unconditional BRANCH) at the location decreed by the CPU designer. The second philosophy, recognizing the independence of the second-level programmer, requires him to place a pointer (address) to the entry point of his program at a specific location, also decreed by the CPU designer. The GO TO operation is then implicit in the reset sequence. This method is employed commonly for "vectoring" interrupts, but is not universally used for resetting. In our computer example, we will create a zero in the ALU, by XORing a data path with itself and use it for the address of the first second-level instruction. The 0000_x thus created is placed in the MAR and an incremented copy is placed in the PC. Notice the use of the data paths in the microinstructions shown for this sequence in Appendix H to get two operations accomplished in the same microinstruction. We expect the second-level programmer to place his first instruction at location 0000_x. The EPROM was based at 0000_x so that the entry point to the Monitor program would be used to start the second-level program.

8.7.2. The Instruction Fetch Sequence

The sequence whereby the instructions of the second-level programmer are brought, one at a time, into the CPU controller and executed is a distinguishing feature of an instruction set processor. All sequences that simulate the action requested by the second-level programmer end by returning to the instruction fetch sequence. Thus, the first-level program never ends. It continues in an infinite loop that starts with the instruction fetch sequence. An overall flowchart of this sequence is shown in Figure 8-5.

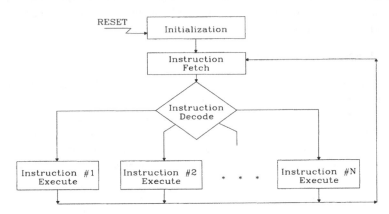

Figure 8-5: The instruction fetch, decode, and execute sequence.

Several things happen in this sequence. The first, the fetching of the second-level instruction, gives the sequence its name. A copy of the PC is placed in the MAR and the external architecture control signals, RD* and WR*, are set to cause data to be read from memory at the location specified by the MAR contents. The data bus driver that connects the external data bus to the internal data bus is enabled and directed inbound toward the CPU. After waiting for a number of states that is determined by the speed of the external devices, the data is latched into the IR in the controller. In our example, we wait for a total of four states, implying that the external architecture must respond with valid data in a period that is four times our system clock period. Since we dictate a rigid memory cycle time, our bus management protocol is a type of synchronous bus management. We share this with a large number of microprocessors. The alternative is to expect the devices in the external architecture to signal the CPU when they can respond. This is known as an asynchronous bus management protocol.

The next thing that must happen during the instruction fetch sequence is the incrementation of the PC to point at the next second-level instruction. Since the software that was to be adapted for this example required byte addressing (see Chapter 5), we will increment the PC by two and wire the MAR to the external architecture with Address bit A0 unused. The same word will be accessed whether the address is even or odd. This feature will be exploited in Chapter 9. To increment the PC by two, we will add one to it during two of the cycles that we are waiting for the external memory to get the instruction we are fetching.

The conclusion of the fetch cycle concerns decoding the second-level instruction. In a microprogrammed controller, decoding is done by "translating" the opcode of the second-level instruction into the address of the first microinstruction in the desired sequence. The MAPROM in the controller performs this task in our design by "shifting" the opcode left by 1 bit position. Therefore, each sequence starts at an address that is twice the value of the opcode. In our machine, all sequences must begin every other microword in the lower 512 microwords. If we need more space than that for our sequence, then we must jump to a large area above address 200 in micromemory.

8.7.3. The Basic Address Mode Sequences

In this section, we will discuss the basic address mode sequences given earlier in this chapter. The sequences, where appropriate, will be shown in program design language (PDL) form. Comments will be given on the manipulation of the architecture via microcode assembler fragments.

Register/Register Address Mode

The Register/Register (RR) address mode assumes that the operands are in the general purpose registers shown in the programming model. There is no explicit sequence required to access the data since it is already present in the first-level machine. The important microword fields are accessed through the microcode assembler operation mnemonics REGMUX and R2903.

REGMUX has two operands that control the RAM A and RAM B MUXes that in turn provide the register pointers to the RALU. Each MUX has three possible sources for these pointers. They include the REG 1 and REG 2 fields in the IR that allow the second-level programmer to point at his registers of choice. Each MUX can also point at fields in the pipeline register that allow the microprogrammer to directly select a register (e.g., the PC or SP) without regard for the second-level programmer.

The R2903 operation has two operands that the microprogrammer uses to specify his choice of registers. Its use is seen in the instruction fetch sequence in the microcode in Appendix H of this chapter. Note that the default for both REGMUX fields is the pipeline input (0) and R2903 has its second operand (RAM B) pointed at R15, the PC. This microinstruction increments the PC among other things.

During RR addressing, REGMUX is used to point the RAM A and RAM B MUXes at the IR REG 1 and REG 2 fields. This is visible in the execute phase of OR instruction at label ORS16 in Appendix H of this chapter where we see REGMUX R2A,R1B. The RALU uses the IR REG 2 field as a pointer to one of the input operands and the IR REG 1 field as a pointer to the other input operand and the destination of the result. Recall that our RALU uses two addresses to point at three operands with the "B" address being used twice.

Register/Memory Address Modes

In the Register/Memory (RM) address mode group, one of the source operands is in a general purpose register. The result will be returned to the same register. The other input operand is in memory. This method of addressing is quite common and will be used in this form in our computer example. In several CPU designs, the destination of the result can also be in memory. The second-level programmer specifies the direction of transfer by using the appropriate operation mnemonic or by using the order of appearance of the operands in the source code. We will examine three of the major register memory address modes for inclusion in our computer.

Immediate Address Mode

The first member of the register/memory group is the Immediate (Im) address mode. Here, one of the input operands is contained in the second word of the instruction. At the time this address mode sequence is executed, the first word of the instruction is in the IR and the PC points at the data word. The PDL for this sequence is as follows:

> MAR ← PC; PC ← PC + 2; RD* = TRUE;
> Internal Data Bus ← Mem[contents(MAR)];

The PC is incremented by two in our computer example since we are supporting byte addressing but are moving 2 bytes of data on the 16-bit data bus every memory cycle.

Once the data is on the internal data bus, it can be operated upon by the RALU based on the instruction sequence requested by the second-level programmer, i.e., specified by the opcode in the IR. Compare the second microinstruction after the label LORI with that after the label ANDI in the microcode listing in Appendix H. The microinstructions are identical in that REGMUX uses its second operand to point the RALU REG B address at the REG1 field of the IR. In both cases, data enters the RALU on the DA port from the data bus where it is combined with data in the register array. The results are returned to the same register. The difference between the two microinstructions is seen in the ALU operation. In the LORI sequence, the ALU's second operand is used to pass (PASSR) data from the DA port through the RALU unchanged (by adding zero) through the ALU shifter to the input of the register array. In the ANDI sequence, the same operand is set to AND the data from the DA port with the register contents. This comparison can be made for all of the immediate address mode instructions.

Absolute Address Mode

The second member of this group is the Register/Memory Direct (ABS) where the memory operand is pointed at by an address contained entirely in the instruction. The second word of the instruction is used to contain an address as opposed to data as in the Im case above. The PDL for this sequence is shown as follows:

> MAR ← PC; PC <- PC + 2; RD* = TRUE;
> MAR ← Mem[contents(MAR)]; RD* = TRUE;
> Internal Data Bus ← Mem[contents(MAR)]; RD* = TRUE;

In the second line, the second word of the instruction is fetched (read) and placed in the MAR. Another memory cycle is performed placing the selected data on the internal data bus. Notice that from this point on the sequence is identical with that used for the Im address mode. In the microcode for our computer example, we exploited this point to produce address mode subroutines. OPERI serves the Im address mode and OPERA serves this one. This feature makes implementing the various instructions using each address mode rather quick. For each instruction, we call the appropriate address mode subroutine and then perform a sequence that is identical for this instruction no matter what the address mode is. Our controller makes subroutine calling and branching as efficient as sequential coding so we don't lose any time with this organization.

Register Indirect Address Mode

Next in this group is the register indirect address mode in which the memory operand is pointed at by a register, or a register plus an address contained in the instruction. One version of this address mode uses a pointer to the data that has been placed in one of the CPU registers by the second-level programmer. The following PDL describes the steps taken at the first-level of control to bring the desired data onto the internal data bus.

> MAR ← Register; RD* = TRUE;
> Internal Data Bus ← Mem[contents(MAR)]; RD* = TRUE;

Comparison of this sequence with that for Im addressing shows that the only difference is the source of the address. A variation of this mode can be made by incrementing or decrementing the address register when its contents are used. The MC68000 family of microprocessors from Motorola support this variation by pre-incrementing or post-decrementing the address register, if the second-level programmer desires. The PDL for these versions follows:

Post-decrement:

> MAR ← Register; Register <- Register - 2; RD* = TRUE;
> Internal Data Bus ← Mem[contents(MAR)]; RD* = TRUE;

Pre-increment:

> Register ← Register + 2;
> MAR ← Register; RD* = TRUE;
> Internal Data Bus ← Mem[contents(MAR)]; RD* = TRUE;

The factor of two in the first line of each version is appropriate to our example computer since we are addressing words with an address register appropriate to byte addressing.

Indexed Address Mode

An extension of register/memory indirect addressing combines both register indirect and memory direct mode shown above. The contents of an address register is added to the contents of the second word of the instruction to create an effective address. This addressing mode is frequently referred to as indexed addressing or more specifically indexed with offset addressing. The "offset" field is contained in the second word of the instruction. Over the years, many variations of this address mode have been created by the fact that different designers allocated different amounts of space in the instruction for the "offset" field. If this field is too small to be used to point anywhere in the address space of the second-level machine, as was frequently the case with 8-bit microprocessors, then the second-level programmer had to rely on the register to contain the base address of any array he wanted to place at an arbitrary location in memory. The "index" in the array was then placed in the second "word", a byte in the 8-bit processor, of the instruction. If the register itself was only 8 bits wide, then arrays were confined to the lower 256 bytes of memory. All variations of this theme were performed in the 8-bit processor architectures. The shorter instructions required fewer memory cycles to fetch-making them faster to perform. The narrower registers created a simpler physical integration task.

In our design, we have both 16-bit registers and a 16-bit wide second word for our instruction. This combined with a 2^{16} byte memory space allows us to place the pointer to the head of an array in either the "offset" field or the register. For second-level programmer convenience, we probably would use the former so that the data dependent "index" was in the register. The PDL for this sequence is given next:

>MAR ← PC; PC ← PC + 2; RD* = TRUE;
>MAR ← - Register + Mem[contents(MAR)]; RD* = TRUE;
>Internal Data Bus ← Mem[contents(MAR)]; RD* = TRUE;

In our computer example, we will find that implementing the second line of this PDL will require an extra clock cycle and a temporary storage location in the RALU for the effective address. The path over which the "offset" is moved is coincident with the path needed to update the MAR with the effective address. We will examine this point in the "improvements" section in Chapter 9.

8.7.4. Additional Address Modes

Other address modes are possible. In the course of second-level program development, many algorithms for pointing at data have been produced. Any of these, if used often enough, is a candidate for inclusion at the microlevel as an address mode. Some brief examples in PDL are summarized here.

Memory Indirect

>MAR ← PC; PC ← PC + 2; RD* = TRUE;
>MAR ← Mem[contents(MAR)]; RD* = TRUE;
>MAR ← Mem[contents(MAR)]; RD* = TRUE;
>Internal Data Bus ← Mem[contents(MAR)]; RD* = TRUE;

The second word of the instruction points at a memory location which contains a pointer to data. The indexed versions of this address mode are very powerful. They are as follows.

Memory Preindexed Indirect

>MAR ← PC; PC ← PC + 2; RD* = TRUE;
>MAR ← Register + Mem[contents(MAR)]; RD* = TRUE;
>MAR ← Mem[contents(MAR)]; RD* = TRUE;
>Internal Data Bus ← Mem[contents(MAR)]; RD* = TRUE;

The preindexed indirect address mode is used to support passing of parameters with an argument list during a subroutine call. At compile or assembly time, a list of pointers to the various data fields to be passed to a subroutine is placed in memory. A pointer to this list is placed in one of the CPU registers or on the stack and the subroutine is CALLed. The subroutine uses this address mode to point at the argument list. The register points at its base and the "offset" field in the instruction points at the address of the operand in the argument list. The last memory cycle puts the data on the internal data bus. Normally, this address mode is provided for all arithmetic/logic instructions in the instruction set so that the following high-level language statement can be implemented with one instruction per symbol:

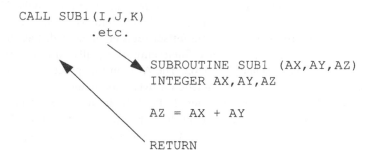

```
            CALL SUB1(I,J,K)
                  .etc.

                              SUBROUTINE SUB1 (AX,AY,AZ)
                              INTEGER AX,AY,AZ

                              AZ = AX + AY

                              RETURN
```

The subroutine assembler language might look like the following:

```
;Note: the calling program places a pointer to the argument
;list in register IX (one of the registers 0 through 13)
SUB1:
        LORXI    R0,IX,0 ;Place contents of 1st arg in reg 0
        ADDXI    R0,IX,2 ;Add contents of 2nd arg
        STRXI    R0,IX,4 ;Store results in 3rd arg
```

The "XI" extension on each operation mnemonic indicates that the preindexed indirect addressing mode is to be used. The third operand in each line specifies the offset in bytes from the head of the argument list to each argument pointer. The first-level machine uses this information in evaluating the address mode algorithm shown in the PDL above. The STORE instruction uses the effective address calculated as above; however, the last line of the PDL is reversed to show a data transfer from inside the CPU to memory (a write).

Memory Post-indexed Indirect

```
MAR ← PC; PC <- PC + 2; RD* = TRUE;
MAR ← Mem[contents(MAR)]; RD* = TRUE;
MAR ← Register + Mem[contents(MAR)]; RD* = TRUE;
Internal Data Bus ← Mem[contents(MAR)]; RD* = TRUE;
```

The Memory Post-indexed Indirect address mode may be used for specifying one component of a vector where the base address of the vector is stored in a memory location. In 8-bit microprocessors, the creation of multibyte numbers (e.g., 16 or 32 bits long) was frequently needed. Adding or subtracting two n-byte numbers could be performed in a very small loop using this address mode. Each multibyte operand was pointed to by a pointer stored in main memory. A loop similar to that shown next can efficiently add two N-word numbers in our computer example.

```
        LORI IX,COUNT-1 ;Count = Number of words in each number
        CLC             ;Clear carry flag at beginning
LOOP:
        LDIX R0,IX,A    ;Move IXth word that A points at to R0
        ADIX R0,IX,B    ;Add IXth word that B points at to R0
        STIX R0,IX,C    ;Store R0 in IXth word pointed at by C
        DECI IX,1       ;Decrement loop count/word pointer
        BPL LOOP        ;Run loop until IX underflows
```

The locations A, B, and C contain pointers to the multiword operands. They can be changed in the course of a program to point at any three operands of the same size. If the loop is called as a subroutine with the counter/offset register initialized, then it can be used to add any two numbers of arbitrary widths. Subtraction is equally simple. This address mode was available on the MOS Technology 6502. The ALU in this processor could perform binary or Binary Coded Decimal (BCD) arithmetic on byte-wide numbers. The mode of operation was signaled by a flag in the status register (D flag). If the loop was entered with COUNT = 4, then 32-bit numbers were efficiently added. The data was treated as 32-bit binary numbers if the D flag was cleared before the loop was entered, or as an 8-digit decimal number if it were set before entry.

Consider the JMP 0,IX,TABLE instruction which is a program control version of this address mode. TABLE contains a list of pointers to a group of program segments that the second-level programmer might wish to execute based on a data condition at run time. He places the table element selector in the IXth register and performs the JUMP. The final line of the PDL above would be used to place the contents of the internal data bus in the PC. This results in a very compact assembler language version of the SWITCH or computed GOTO commands in high-level languages.

One other address mode commonly encountered is based on the indexed with offset address mode described above. In our design, either the offset field or the register can be used to point anywhere in our address space. In many machines, this is not possible because the register or offset field is too narrow. In this case, the designers had the option of extending the narrow field to the width of the ALU during the ADD operation by treating it as a signed or unsigned number. The examples above treated it as an unsigned number. If the field is treated as a signed number, then the second-level programmer can consider that it represents an offset to a position above (positive) or below (negative) the place pointed at by the field to be added to it. This is called the "Relative Address Mode". In many 8-bit processor cases, the register was wide and the offset field in the instruction was narrow. If the register was the PC, the address mode was further identified as "PC Relative" and was almost always used for branching. This address mode has been used to point at data also. In this case, the data are kept in the vicinity of the program module that uses it.

8.7.5. Dealing With the Microword Symbolically

Before we deal with the execution sequences, we will introduce some detail that can be seen in the microprogram instruction. Appendix A of this chapter shows the physical format of the pipeline (microword) for the computer example. Figures 8-1 through 8-4 show the physical interconnections between the modules that comprise the physical first-level machine. While it is possible to fill the microprogram memory with suitable collections of 1s and 0s derived from the data sheets for the parts used, it is undesirable to do so. This can be avoided by using a symbolic assembler such as that described in Chapter 5.

In preparing the file containing the symbol definitions (DEF file), we assumed a viewpoint in which each logical "device" would be called an "operation". So, the operation mnemonics that follow this view in our definition file as shown in Appendix G are the following.

CCMUX – Condition Code Multiplexer in the controller
ALU – ALU in the RALU
SPF14 – special function instructions for the ALU
MSRLD – micro and macro status registers
SHFT – external ALU and Q register shift multiplexers

CNMUX – ALU Carry Input multiplexer

REGMUX – RALU RAM A and B address input multiplexers

R2903 – pipeline RALU RAM A and B address fields

DATAPATH – various bus drivers and elements on the internal
and external data buses including external architecture
control signals RD* and WR*

MCMUX – shift multiplexer input to second-level Carry register

Each of these operation mnemonics contains one or more operands. The possible values for the operands are given in EQUate statements preceding the definition of the operation mnemonic that uses them. In each case, the default value is given in the operation definition. This value is chosen to "do nothing" wherever possible.

Other operation mnemonics are related directly to their second-level counterparts to make writing easier at this level. They are as follows:

The program control group

JZ, CJS, JSB, JMAP, CJP, JMP, PUSH, PHLC, JSRP, CJV, JMPV, JRP, RFCT, RPCT, CRTN, RTN, CJPP, LDCT, LOOP, CONT, TWB

The ALU operand sources

AB, ADB, AQ, DAB, DADB, DAQ

Control signals for miscellaneous bus drivers and registers

REGWR, NOREGWR, YBDEN, NOYBDEN, DBBDEN, NODBBDEN

Simulation controls

SOPPN, PAUSE, HALT

Simulation report controls

DEBCON, DEBALU, DEBSRG, DEBMAC, DEBOUT and DEBALL

The program control group can be used in a form that is very similar to that experienced when writing second-level programs. This time, however, the program control can be performed in the same microinstruction with events that happen elsewhere in the architecture. In the microcode listing shown in Appendix H of this chapter, notice that the first field of each microinstruction is devoted to program control, most frequently the continue instruction CONT. The second line of the immediate address mode subroutine OPERI shows a return from subroutine (RTN) instruction in this field.

The last two groups, simulation controls and simulation report controls, are not present in the physical implementation of the computer example. We provided them in the simulator so that debugging information would be available to the microprogrammer from each module under control of the microword being debugged. This allowed the working part of the machine to support debugging for newly written microcode. Each section (e.g., the controller — DEBCON) can be told to report internal variables during the execution of a microinstruction by placing its corresponding operation mnemonic in the symbolic microinstruction under test.

Refer to Chapter 5 and the listing in Appendix H of this chapter for examples to assist in learning how to write in this language. The names of each element in Figures 8-1 through 8-4 are similar to those used in the definition file given in Appendix G. Writing a microprogram then consists of determining the data path needed by the operands by using the figures, followed by determining the operations needed by consulting the definition file and any ap-

328   8.7. SOME IMPORTANT SEQUENCES

propriate data sheets (e.g., for the Am2903). The symbolic microinstruction is written by combining this information in a form similar to the microinstructions given in Appendix H. Several important sequences will be discussed below and our version of them for our computer example will be pointed out in Appendix H.

8.7.6. Execution Sequences

Figure 8-6 is an overview of the hardware architecture. It should make the discussion of the execution sequences easier to follow. As outlined above, the address mode sequences are run at the appropriate position within the execution phase of the second-level instruction cycle. The actual data transformations are performed in a single clock cycle for the Arithmetic/Logic instructions in the Basic Instruction Set since the RALU was optimized for just this purpose. The point at which the data transformation is made is identical once the second operand has reached the R side of the ALU. The following table summarizes the ALU function field in the microinstruction for the members of this group regardless of the address mode of the second operand.

Instruction	ALU Function	CNMUX
LOR	PASSR	default (0)
ADC	ADD	MCF
ADD	ADD	0
SBC	SUBR	MCF
SUB	SUBR	1
CMP	SUBR	1
AND	AND	default
OR	OR	default
XOR	XOR	default
STR	PASSS	0
INC	ADD	1
DEC	SUBR	0

The Carry Input Multiplexer determines the difference between the Add with Carry Input (ADC) and Add with No Carry Input (ADD) second-level instruction. The former is implemented using the Carry Flag in Macro Status Register as an input, while the latter uses zero. The subtract instruction pair (SBC and SUB) are similarly implemented, where the Macro Carry Flag contains the borrow input. Remember that the ALU is actually performing R plus /S plus Cin to do subtraction. The logical instructions OR, AND, and XOR do not use the Cin to the ALU, so the default (0) is taken here.

The PASSR and PASSS ALU operands are actually (R plus Cin) and (S plus Cin), respectively. If the Cin is 0, then the operand is passed unchanged. Increment could be implemented using PASSS with Cin equal to 1; however, in our machine we have chosen to use a format similar to ADD using immediate addressing for the increment amount. The reason for this is given below. The Decrement instruction could have been implemented using (/S plus Cin) where Cin equals 0. This would require a second step to one's complement the result again. We chose to implement DEC as an immediate address mode instruction similar to SUB.

From the foregoing paragraph it appears that two instruction pairs, ADD/INC and SUB/DEC, are identical. We used a macro status register flag-setting strategy that distinguished between logical and arithmetic instructions in its operation. Logical instructions are allowed to change the "sign" (MN) and "zero" (MZ) flags only. Carry (MC) and overflow (MOVR) are left unchanged! The consequence of this will be shown in an example later. The arithmetic instructions change all four flags. The MN and MOVR flags are the two's complement sign and overflow, respectively. The MC flag is interpreted as unsigned overflow or carry out. The MZ flag means the result is 0, whether the number is interpreted in logical, arithmetic unsigned, or arithmetic signed form. While MN is changed during logical operations, it is not the sign flag but only a copy of the most significant bit of the number. The logical instructions are defined to be LOR, STR, AND, OR, XOR, INC, and DEC. The arithmetic instructions are defined to be ADC, ADD, SBC, and SUB.

This definition leads us to see the method in the flag setting strategy. Consider a common algorithm involving a loop shown below.

```
        LORI RX,COUNT-1 ;initialize counter to number of words
    LOOP:
        LORX R0,RX,A    ;use indexed addressing to get word of A
        ADCX R0,RX,B    ;add with carry a word of B
        STRX R0,RX,C    ;save result in word of C
        DECI RX,1       ;decrement index/loop counter by one
        BPL  LOOP       ;loop as long as counter is positive
```

This algorithm, which was shown above using indirect addressing, adds two multi-word numbers A and B each containing "COUNT" words and places the result in the multiword variable C. The carry out of the ADCX needs to be saved during each pass through the loop until it is needed in the next cycle. If the DECI had changed the carry flag (or a SUBI had been used), then the original carry would have to be saved and restored during each cycle of the loop. This costs additional instructions and memory cycles which greatly slows the program. The flag setting strategy adopted here eliminates this problem. This strategy was employed in the MOS Technology 6502 microprocessor, but the counter example was used in the INTEL 8085.

Some instruction implementations require more clock cycles than the ones shown above. These include the conditional branches, subroutine calls, and shift instructions. These will now be summarized.

The shift instructions, as we implemented them, can do logical right and left shifts as well as left and right rotations through the Macro Carry flag. While single-bit shifts could have been implemented as compactly as the arithmetic/logic instructions shown above, we chose to allow the second-level programmer to specify the number of bit positions desired in the second word of the instruction. This required more microinstructions and slower execution in cases where he needed only a single position shift. Many times, however, the second-level programmer needs to shift over one "digit" (4 bits) during formatting or an arbitrary number of bits when trying to isolate a bit for testing.

These examples also give us a chance to examine looping in the micromachine. The general implementation method consists of moving the immediate data to the Q register in two's complemented (negative) form. During shifting, we increment the Q register and compare it against zero (in the microcarry). This is more efficient than having to have a separate compare using the ALU. This strategy may be seen in the rotate implementations also. Shifting and Rotating involve the ALU shifter in the RALU and the external shift multiplexers, SLMUX and SRMUX. The ALU shifter is controlled by the ALU operation destination field function. The equates for these functions are reproduced from Appendix G in the space below:

```
; EQUATES FOR ALU DESTINATION CONTROL
;
ADR:    EQU H#0  * ; ARITHMETIC SHIFT DOWN, RESULTS INTO RAM
LDR:    EQU H#1  * ; LOGICAL SHIFT DOWN, RESULTS INTO RAM
ADRQ:   EQU H#2  * ; ARITH, SHIFT DOWN, RESULTS INTO RAM AND Q
LDRQ:   EQU H#3  * ; LOGICAL SHIFT DOWN, RESULTS INTO RAM AND Q
RPT:    EQU H#4    ; RESULTS INTO RAM, GENERATE PARITY
LDQP:   EQU H#5  * ; LOGICAL SHIFT DOWN Q, GENERATE PARITY
QPT:    EQU H#6    ; RESULTS INTO Q, ;GENERATE PARITY
RQPT:   EQU H#7    ; RESULTS INTO RAM AND Q, GENERATE PARITY
AUR:    EQU H#8  * ; ARITH, SHIFT UP, RESULTS INTO RAM
LUR:    EQU H#9  * ; LOGICAL SHIFT UP, RESULTS INTO RAM
AURQ:   EQU H#A  * ; ARITH, SHIFT UP, RESULTS INTO RAM AND Q
LURQ:   EQU H#B  * ; LOGICAL SHIFT UP, RESULTS INTO RAM AND Q
YBUS:   EQU H#C    ; RESULTS TO Y BUS ONLY
LUQ:    EQU H#D  * ; LOGICAL SHIFT UP Q
SINEX:  EQU H#E    ; SIGN EXTEND
REG:    EQU H#F    ; RESULTS TO RAM, SIGN EXTEND
```

The entries marked with an asterisk (*) indicate a shift function. The first letter of the function name shows that the shifting on the ALU shifter path is arithmetic (A) or logical (L); i.e., the number retains its sign (A) or not. The second letter of the function name shows the direction of shifting, whether to the more significant direction (U) (i.e., multiplying by 2) or the other way (D) (i.e., dividing by 2). The last letter or two indicates which path(s) get shifted; i.e., ALU (R) and/or Q (Q). The Q register always receives the QIO0 or QIO3 inputs when shifting.

Rotation is similar to shifting except that a connection is made between the bit shifted out and the shift-in position. In our case, we rotated through the Macro Carry so the resulting number was 17 bits (register and carry) long. A picture of the data flow is given below:

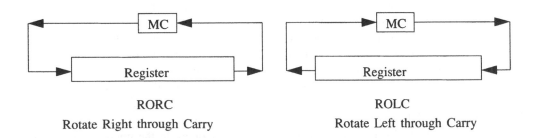

RORC   ROLC
Rotate Right through Carry   Rotate Left through Carry

This instruction is useful for performing multiprecision shifts at the second level as shown below.

```
SHRA R0,1
RORC R1,1
```

In this fragment, registers R0 through R1 form a two-word two's complement number with R0 containing the most significant end. The SHRA instruction arithmetically shifts the most significant word right, preserving its sign and sending the least significant bit into the Macro Carry. The bit is propagated into the most significant end of R1 by rotating it right through the carry. Its least significant bit is simultaneously placed in MC. A drawing of this operation is shown below:

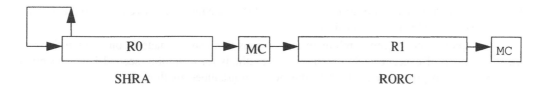

SHRA   RORC

The bit shifted out in all cases lands in the Macro Carry under control of the MSRLD and MCMUX operations.

The conditional branch instructions are relatively simple to implement since the destination of their operand is the PC. The PC will be loaded if the condition indicated by the second-level programmer is *True*; otherwise, the PC is left unchanged. This causes a GOTO if *True* or a CONTINUE if *False* in the second-level program flow. The decision-making is done at the first level using the CCMUX to select a Macro Status flag or a combination thereof and making the controller do a conditional jump instruction. This is seen at the label BNI (address 00112) in the microcode listing. The critical part of implementing the instruction comes in testing the flag, in this case, the CCMUX points at PAL a Programmable Array Logic device that implements an array of 16 Boolean functions of the Macro Status flags. These allow the second-level programmer to perform branches on conditions such as Less Than, Less Than or Equal, etc. on either signed or unsigned numbers. This function could have been implemented in several clock cycles without the network; however, it seemed like an efficient addition. The Boolean functions are listed for this instruction in Appendix E. Notice that the REG 1 field in the IR is used to select the desired function. This field is filled in by the second-level assembler when an operation mnemonic is assembled.

The remaining set of multicycle instructions that we implemented are related to subroutines. In addition to the GOTO operation implied by a subroutine CALL or RETurn, the return point must be saved. In our machine, the return address, the current value of the PC (R15), is saved on a stack addressed by the Stack Pointer (R14). Stack pointer manipulation is seen in the vicinity of address 00206/7 and PC saving is seen at 00208 for the CALL instruction. The RETurn instruction must restore the PC from the stack. Decrementing the SP is done at address 0020E/F and the PC is loaded from memory at address 0020D. Notice that conditional CALLing and RETurning are also implemented. The only difference between the return and the conditional return occurs in the first microinstruction in the execution sequence. At address 00130 (label RN), a CJP along with the CCMUX tests the condition code PAL to determine if the PC should be restored from the stack.

8.8. CONCLUSION

In this chapter, we have shown the design of a simple microprogrammed computer architecture. While exploring the possibilities of each element, we have introduced most of the important operational algorithms that any machine needs to be a state machine having a second level of control (ISP). In addition, we have explored some of the more useful data addressing algorithms that are important in making this type of processor as fast and flexible as it is. We have identified various points in the design that slow down execution with the most serious culprit being that keystone of the ISP, the instruction fetch sequence itself.

Chapter 6 showed how to create a copy of the basic computer example in software so that we are able to uncover any shortcomings at any level of design from the instruction set itself to the physical data and control paths in the architecture. It should be our goal to understand the design before any of it is committed to hardware fabrication. We can reach this level of understanding better if we have used the machine itself, even in simulated form, for the tasks for which it was designed.

Chapter 9 shows some architectural additions that can be made to our design that improve its speed of execution. Some alternative methods for designing second-level machines are also given to provide a contrast to the microprogrammed method.

EXERCISES

1. Create a *second-level* subroutine that multiplies the unsigned binary numbers in R0 and R1 and returns the double-precision product in those registers (most significant part in R0). Use the assembler language for the computer example defined in Appendix D of this chapter. Develop the product by using the SHIFT and CONDITIONAL ADD operation presented in Chapter 7, except use it at the *second level*. You will need to use two temporary registers, one to contain the multiplicand during the operation and the other to serve as a loop counter. Use R2 and R3 for these purposes. To reduce the amount of register movement in getting started, define the contents of R1 as the multiplier and develop the partial product in R0 and R1 concatenated together. The RORC instruction will be quite useful.

2. Write a macro in terms of *second-level* instructions for the computer example that, when expanded, calculates the product of two unsigned binary numbers in the registers specified in its operands. The macro should be invoked by the following:

    UMUL Rx,Ry

    The registers, Rx and Ry, may be any of the general purpose registers in the example computer. The best starting point for this problem is to consider the subroutine that you wrote in answer to Exercise 1 above. Suitable modification should result in producing the body of the macro. Is there a problem with the label on the loop if the macro is invoked more than once? How do you plan to keep the temporary register assignments from interfering with the registers used by the programmer in other parts of his program?

3. Write the unsigned multiplication algorithm specified in Exercise 1 as a *first-level* execution sequence. Review the corresponding part of Chapter 7. In this case, any pair of general purpose registers may be used for operands. Take advantage of the additional architecture available to the first-level programmer; e.g., the advanced single cycle conditional add and shift instruction for the RALU and the loop counter in the controller. You will also need to write a macro that helps the assembler create the *second-level* machine code. The macro should be invoked by an instruction similar to that in Exercise 2.

4. Redo Exercise 3 to create a signed multiply instruction. Exploit the special signed operation in the RALU in the computer example. The bulk of the writing at the *first* or *second level* is done by editing the code produced in answer to Exercise 3.

5. Create an unsigned division instruction for the computer example in this chapter. You will need to make a *second-level* macro that invokes the correct *first-level* sequence. The first-level sequence may be implemented with the aid of the data book for the Am2903. Take advantage of the special RALU instructions in writing the *first-level* sequence.

6. Write *first-level* subroutines that implement each of the address modes described in this chapter. Models for two of the subroutines may be found in the microcode in Appendix H beginning with the label OPERR for the register/register sequence and OPERA for the register/absolute sequence. Begin with the PDL for each sequence shown in the text and elaborate it into an HDL for the computer example; then write the actual microprogram after selecting a suitable assembler.

7. In the computer example, the designers had to place some constants in the RALU registers for use by the first-level program. This obviously tied up a valuable resource and was an unreasonable thing to expect of the second-level programmer. One solution to this problem is to supply a small memory for constants near the RALU. Design such a memory and its control structure. Provide a logic diagram that shows how it should be attached to the RALU. Consider means to control the memory that does not add greatly to the number of bits in the microword; i.e., use some of the bits already provided without impacting other applications.

8. Provide a means to generate the constants described in Exercise 7 by using a field in the microword. Show a logic diagram of how this is accomplished. Include suitable field definitions for the assembler and a sequence of statements that use the field. Are there any fields that could be borrowed for this purpose?

9. Create simulation modules for the components in Figure 8-6 and build an integrated circuit level, behavioral simulator for the basic computer example. The structure of the modules and of the overall simulator are suggested in Chapter 6. This program, written in your favorite high-level language, will allow you to write and debug first- and second-level programs by using the facilities of the second-level Monitor program presented in Appendix E2 of this chapter.

   The second-level Monitor's machine code, in Appendix F, may be loaded into the external architecture EPROM area beginning at address 0000_x. Test programs written for the second level may be located at 1000_x. They may be executed by using the Monitor's Go command. Be sure to place an unconditional branch at the end of the test program to take program flow back to the Monitor at the 0004_x entry point. Registers may be examined by typing Rn where "n" is a four-digit hexadecimal number specifying one of the 16 registers. Memory may be examined by typing Ma where "a" is a four-digit hexadecimal number specifying a memory address in the external architecture. If you wish to modify the memory or register location, answer yes, "y", to the question and immediately type a four-digit hexadecimal number for the new value. Note: the GO command is typed Ga where "a" is a four-digit hexadecimal number as before.

   The first-level machine code generated by the program in Appendix H and displayed in Appendix C must be loaded into the microprogram memory before the second level can execute. The debug switches shown at the extreme right end of the parsed microcode display in Appendix C have been set to zero, i.e., no display. This allows the Monitor program to run at full speed. If you desire to test code at the first level, then set the appropriate variables (i.e., DEBCON, DEBALU, and DEBSRG) to the values specified in Appendix A so that the internal variables will be printed when the test code executes. Be sure that you have sensed these switches in the reporting sections of the appropriate modules in your simulator.

While the simulator program takes some time to write, it is well worth the effort. A level of understanding of the individual integrated circuits is achieved that is even difficult using the physical devices. The overall machine allows one to write and debug programs at both the first and second level, a feat that is impossible with most real computers. The author's version of the program, adapted from one written by Bruce Kleinman and Brian Miller, runs acceptably fast on an IBM PC, but is spectacular on the current round of personal computers. With the advent of windowing systems, one is tempted to present a graphical interface derived from Figure 8-6 so that the register and bus contents are continuously updated. The speed penalty for such an interface is partially offset in current machines by their much higher computation speed. Notice that the earlier exercises may actually be performed on the simulator. Exercises 7 and 8 can also be accomplished by modifying the simulator program. This level of experimentation is much more efficient than implementing a prototype in hardware. While no timing information is provided, the designer will establish the logical consistency and effectiveness of his design.

10. Design the basic computer example using similar trade-offs; i.e., minimal parts and buses, except incorporate recent parts sets by TI (74ACT8818 and 8832) and AMD (Am29C331, 332 and 334). It is also instructive to include the appropriate FPU and support the floating point data type at the second level. Be sure to include status register sets for the first- and second-level machine or at least a means to remember the second-level flags while performing first-level sequences. Follow the procedure shown for the first-level exercises in Chapter 7. Simplified simulation modules may be written for these parts or complete ones available on some development systems may be used. Since a second level is present, adapt the procedure shown in the previous chapter's exercises to accommodate it. Do a top-down design procedure for the second-level machine. By specifying the operations, data types and overall architecture of the second-level machine, the designer will have a better understanding of how to use the capabilities of the architectural components in the first-level machine. To crown this virtuoso project, the designer should provide two second-level instructions that embody the CISC ideal, i.e., the sum-of-products (SOP) and power series expansions (POW).

11. Use the computer example's integer-based hardware to implement a Floating Point Multiply instruction, FMUL. Let the data be represented as an IEEE single-precision floating point number. The procedure followed in Exercises 1 through 3 is appropriate here. Design the algorithm in PDL. Write a second-level subroutine that performs the algorithm. This will help you identify any temporary registers and other details needed. Write a second-level macro that creates machine code that can be used by the first-level machine to invoke the first-level sequence. Prepare an HDL description of the algorithm based on the second-level work. Take advantage of any appropriate first-level architecture. Notice that the integer multiply routine needs only to be expanded to 24 bits to adapt it for multiplying the fractional parts. Since the IEEE single-precision number is 32 bits long, it is more efficient to point at it in memory (e.g., using register indirect addressing) and not keep it in a "register" in our 16-bit machine. Bring pieces into the CPU as needed. This is a common practice when CISC machines deal with data types that occupy more space than is available inside the CPU.

   a. Many of the registers within the RALU will be used by the first-level algorithm in a temporary capacity. How should you inform the second-level programmer; i.e., how should this affect the programming model?

   b. Add another register bank to the RALU using the DA and DB buses. Implement the algorithm using these registers.

12. Create CISC instructions for the computer example that move and compare strings. A means to specify or determine the length of the string is needed. Three ways are commonly used. A character count is placed in the first byte of the one-dimensional array that contains the string. The string is terminated with a special character; i.e., one that is *not* a character in the code used. The length of the string or strings are specified in the second-level machine code for the instruction. Select one of the string-length specification methods and implement the MOVSTR and CMPSTR instructions. Note that the strings are too long to bring into the CPU all at once. Use an address mode that will make it easy to perform the comparison. This exercise will also benefit from using the top-down procedure shown in Exercises 1 through 3.

13. Refer to Chapter 8, Appendix I for a demonstration of the appearance and execution of a high-level language program on a microcomputer. The demonstration shows a short FORTRAN program as it is compiled, linked, and executed using the Microsoft FORTRAN compiler and linker and the Digital Research CP/M 80 operating system and debugger program. In the first part of the demonstration, you are given a tour of the appearance of the code produced by the language processors as the high-level language is compiled and linked with the run-time library to create an executable program. Notice that the compiler places the data area for the program at a lower address than the executable code. Use the MAP produced by the compiler to locate the variables. Notice that all addresses used by the compiler are relative to the front of the program module. The linker is responsible for converting these addresses to an absolute form taking into account the physical location of the program in memory, here 0103_x. A memory image of the machine code is presented next. Notice that all of the relative address have been replaced with physical addresses. The program is then simulated using the debugger program. The programmer can look at any element visible in the programming model (e.g., registers, flags, memory, and the next instruction) as the program is single-stepped through it instructions. The next step performed is the execution of the program in a Z-80-based microcomputer. The operating system was used to load the program and cause it to go into execution. A logic analyzer was attached to the address, data, and control buses of the computer. It collected one line of data, as shown, each time the processor's SYSCLK underwent a Low–High transition. A memory-read bus cycle is signaled for those clock cycles when RD = 0. A memory-write bus cycle occurs when WR = 0. An operation code is being fetched when M1 = 0. Bus cycles get data from the memory space when MEMREQ = 0 and the Input/Output address space when IOREQ = 0. Bus cycles are explained and linked to FORTRAN and Assembler instructions in the program by comments attached to each line.

    Answer the questions attached to the bus cycles given in the exercise at the end of the appendix. If available, a Zilog Z-80 data sheet may shed extra light on the workings of the signals in the external architecture. Note that the logic analyzer reveals the maximum amount of detail that can be seen without entering the CPU itself. The "illusion" of the computer's executing the high-level FORTRAN instructions is seen to be created by many small state machine cycles that can be partially seen on the system buses. If an internal architectural figure of the Z-80 is consulted (e.g., as in [Anceau 86]), then the details of the instructions witnessed above can be understood.

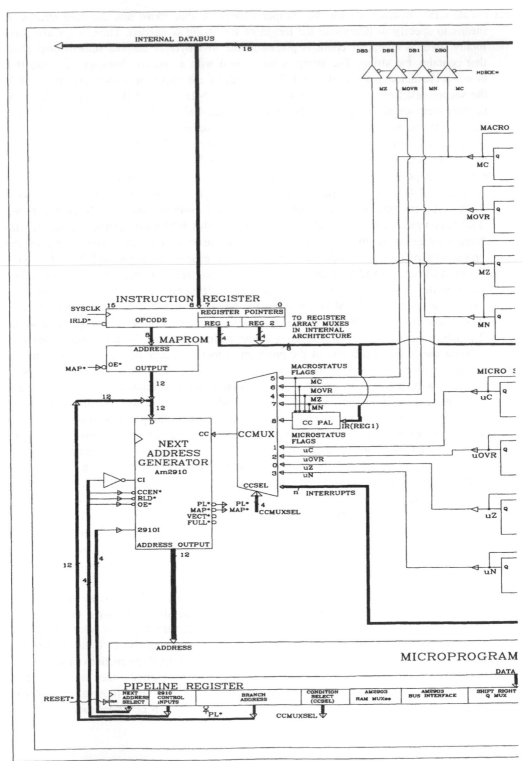

Figure 8-6: The complete architecture of the computer example.

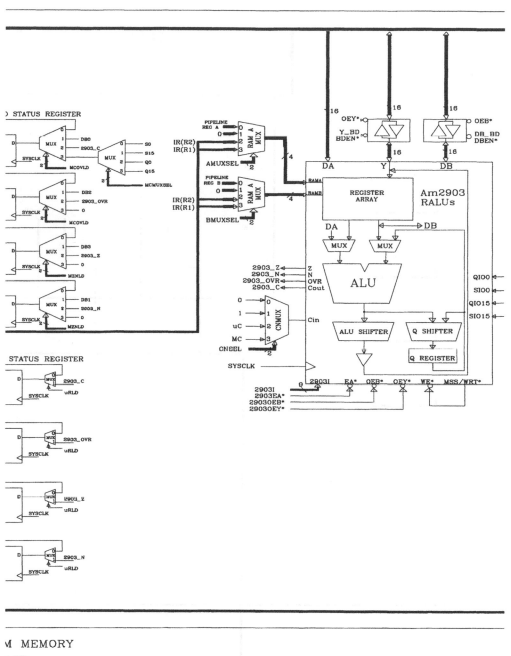

Figure 8-6: (continued)

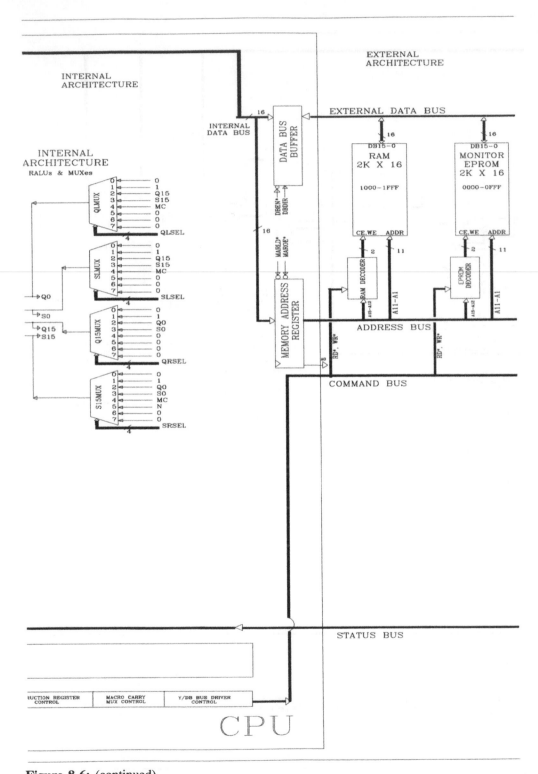

Figure 8-6: (continued)

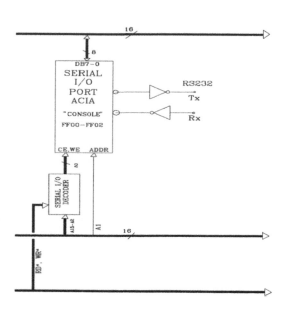

COMPUTER

Figure 8-6: (continued)

Appendices to Chapter 8

Detailed Description of the Basic Computer Example

The appendices to this chapter give detailed information about the basic computer example. The subject matter of each appendix is given in the following table of contents:

Appendices A and B give a thorough description of each bit field in the microword and the connections of all multiplexers. Figures 8-1 through 8-4 are graphic examples of this same area. Appendix C describes a utility program that can be used to print the microprogram object code in a form that is easier to use during the debugging process for programs on the first-level machine. Appendices G and H complete the description of the first-level machine by showing the definition and assembler listing files for the microprogram.

The second-level machine is described in the remaining appendices. Appendix D contains an instruction set reference manual given in two forms. The first part is a compact summary of the assembler syntax, machine code, and operations. The second part contains a more detailed description of the instructions including the flag setting strategy. Appendices E1 and E2 complete the second-level machine information by giving the macro definition file and a Monitor program listing produced by a cross assembler created from the VAX MACRO assembler. Note: not all instructions given in Appendix D are actually implemented in microcode. The instruction set summary serves as a template for a large subset of a complete instruction set for this machine. Consult the microprogram listing in Appendix H to determine the current set of implemented instructions.

Appendix I is intended to give the reader an overall view of a program in various forms ranging from high-level language, through assembler and machine code, to being in execution. Most computer users never see a program both in its high-level form and in execution. This example shows the large number of states required to execute even the simplest of high-level programs.

Appendix A. Computer Example—Microcode Bit Field Definitions

NOTE 1: Most Significant BIT is at LOWEST Numbered BIT Position.

NOTE 2: Signal Names followed by an asterisk (*) are active low.

EXAMPLE: MICROPROGRAM MEMORY U6 (Q7 - MS BIT IN EPROM) corresponds to PIPELINE U4 (8D & 8Q IN LS273) which in turn corresponds to 2910 Instruction bit I3 on U9 and PIPELINE bit position 0.

PIPELINE BIT POSITION	NAME OF FIELD
......Am2910 Instruction...	
0	2910 I3 – MS BIT OF MICRO OPCODE
1	2910 I2
2	2910 I1
3	2910 I0 – LS BIT OF MICRO OPCODE
......Am2910 Control..	
4	OE* – AM2910 Y OUTPUT ENABLE
5	RLD* – AM2910 REG/CNT LOAD CONTROL
6	CCEN* – AM2910 CONDITION CODE ENABLE
7	CI* – COMPLEMENT OF AM2910 CARRY IN
..... Am2910 Branch Address..	
8	BA11 – BRANCH ADDRESS FIELD – MS BIT
9	BA10
10	BA9
11	BA8
12	BA7
13	BA6
14	BA5
15	BA4
16	BA3
17	BA2
18	BA1
19	BA0 – LS BIT (TO 2910 D0)
........CCMUX Control...	
20	CCMUXSEL3 – CCMUX ENABLE (Active low on 74LS251)
21	CCMUXSEL2 – 74LS251 A2
22	CCMUXSEL1 – 74LS251 A1
23	CCMUXSEL0 – 74LS251 A0 – LS BIT
......Am2903 RAM MUXes...	
24	2903 RAMA3 – ADDRESS MS BIT
25	2903 RAMA2
26	2903 RAMA1
27	2903 RAMA0 – ADDRESS LS BIT
28	AMUXSEL1 – ADDRESS MUX SELECT MS BIT
29	AMUXSEL0 – LS BIT
30	2903 RAMB3 – ADDRESS MS BIT
31	2903 RAMB2
32	2903 RAMB1
33	2903 RAMB0 – ADDRESS LS BIT

34	BUXSEL1 – ADDRESS MUX SELECT MS BIT
35	BMUXSEL0 – LS BIT

.......Am2903 Bus Interface...

36	2903 EA* – ENABLE A
37	2903 OEB* – ENABLE DB OUTPUT
38	2903 OEY* – OUTPUT ENABLE Y
39	2903 WE* – WRITE ENABLE 2903

.......Shift Right Q MUX...

40	QRSEL3 – 2903 QIO3 INPUT MUX SEL (MS BIT)
	Serves as 74LS251 Enable*
41	QRSEL2
42	QRSEL1
43	QRSEL0 – SELECT LS BIT

.......Shift Right SIO MUX..

44	SRSEL3 – 2903 SIO3 INPUT MUX SEL (MS BIT)
	Serves as 74LS251 Enable*
45	SRSEL2
46	SRSEL1
47	SRSEL0 – SELECT LS BIT

........Shift Left Q MUX...

48	QLSEL3 – 2903 QIO0 INPUT MUX SEL (MS BIT)
	Serves as 74LS251 Enable*
49	QLSEL2
50	QLSEL1
51	QLSEL0 – SELECT LS BIT

........Shift Left SIO MUX..

52	SLSEL3 – 2903 SIO0 INPUT MUX SEL (MS BIT)
	Serves as 74LS251 Enable*
53	SLSEL2
54	SLSEL1
55	SLSEL0 – SELECT LS BIT

.........Carry Input MUX...

56	CNSEL – MS BIT
57	CNSEL – LS BIT

.........Micro Status Register Control...

58	USRLD* – MICROSTATUS REGISTER LOAD

........Macro Status Register Control...

59	MZNLD – MACRO SR ZERO/SIGN FLAG LOAD (MSB)
60	MZNLD – LEAST SIGNIFICANT BIT OF A TWO BIT FIELD

MZNLD Effect
0 No Change
1 ALU Z & N Flags
2 Data Bus (DB3 & DB1)
3 Unused

61	MCOVLD – MACRO SR CARRY/OVERFLOW FLAG LOAD (MSB)
62	MCOVLD – LSB OF A TWO BIT FIELD

MCOVLD Effect
0 No Change
1 ALU C & OVR Flags
2 Data Bus (DB0 & DB2)
3 Unused

63	MDBOE* – Connect MACRO SR to DATABUS

........Am2903 Instruction..

64	2903I8 – 2903 INSTRUCTION (MS BIT)
65	2903I7
66	2903I6
67	2903I5
68	2903I4
69	2903I3
70	2903I2
71	2903I1
72	2903I0 – LS BIT

.......CPU/External Architecture Control..............................

73	DBEN* – External Databus <–> Internal Databus Driver Enable
74	DBDIR – Databus Driver Direction (0 : Internal –> External)
75	MARLD* – MEMORY ADDRESS REGISTER LOAD
76	MAROE* – MAR OUTPUT ENABLE
77	RD* – EXTERNAL READ ENABLE STROBE
78	WR* – EXTERNAL WRITE ENABLE STROBE

.......Instruction Register Control....................................

79	IRLD* – INSTRUCTION REGISTER LOAD

.......Macro Carry MUX Control...

80	CMUX – Macro Carry MUX Control
81	LS bit of a 2 bit field

MCMUX	Input Selected
0	ALU Right Shift S0
1	ALU Left Shift S15
2	ALU Right Shift Q0
3	ALU Left Shift Q15

.......Y Bidirectional Bus Driver on the Am2903........................

82	Y_BBDEN* – Enable 2903 Y output to/from Data Bus Driver
	Direction is controlled by 2903OEY*

Y_BBDEN*	2903OEY*	Function
0	0	2903 Y output -> Internal Data Bus
0	1	Internal Data Bus -> 2903 Y input
1	X	Driver is high impedance

.......DB Bidirectional Bus Driver on the Am2903........................

83	DB_BDDEN* – Enable 2903 DB to/from Data Bus Driver
	Direction is controlled by 2903OEB*

DB_BDBDEN*	2903OEB*	Function
0	0	2903 DB output -> Internal Data Bus
0	1	Internal Data Bus -> 2903 DB input
1	X	Drive is high impedance

Simulator Debug Facilities

......Halt and Process Message...

| 84 | HALT(1) |
| 85 | HALT(0) |

HALT(1)	(0)	Function
0	0	Continue processing (RUN)
0	1	Print message on each cycle
1	0	Unused
1	1	Halt operation and return to operating system

Print Control Fields: Priority – 0 Print Nothing, N Highest Priority

......CCU Print Control..

86	DEBCON(3) – If 1 Then Print 2910 outputs
87	DEBCON(2) – If 1 Then Print 2910 inputs
88	DEBCON(1) – Not used
89	DEBCON(0) – If 1 Then Print contents of 2910 registers

.....RALU Print Control...

90	DEBALU(3) – If 1 Then Print 2903 outputs
91	DEBALU(2) – If 1 Then Print 2903 inputs
92	DEBALU(1) – Not used
93	DEBALU(0) – If 1 Then Print contents of 2903 registers

.......Miscellaneous Print Control....................................

| 94 | DEBSRG – Status Register Print Control – MS bit |
| 95 | – LS bit |

| 96 | DEBMAC – Macro Level Machine Print Control – MS bit |
| 97 | – LS bit |

| 98 | DEBOUT – Data Bus Print Control – MS bit |
| 99 | – LS bit |

....

Pipeline Format – Each Microword

Controller

0　　　3	4	5	6　　　7	8	19　20　23
I2910 mOPCODE	OE*	RLD*	CCEN*　CI*	BRANCH ADDR	CCMXSL

2903 RAM Address MUXes and Miscellaneous

24　　27	28　29	30　　　　33	34　　35	36	37	38	39
RAMA	AMXSEL	RAMB	BMXSEL	EA*	YBEN*	OEY*	YDIR

2903 Shift MUXes and Carry In MUX Controls

40　43	44　47	48　51	52　55	56　57
QRSEL	SRSEL	QLSEL	SLSEL	CNSEL

Status Register Load Controls

58	59	60　61	62　　63
USRLD*	MZNLD*	MCOVLD*	MDBOE*

2903 Instruction Code

64　65　66　67　68　69　70　71　72	73	74
I2903 I8　I7　I6　I5　I4　I3　I2　I1　I0	DBEN*	DBDIR

External Bus Control

75	76	77	78	79
MARLD*	MAROE*	RD*	WR*	IRLD*

Internal Bus Control for the RALU

80　81	82	83
MCMUX	YBDEN*	DBDEN*

Print Control Fields

84　85　86　　89	90　　　93	94　　　95	96　　　97	98　　　99
HALT　DEBCON	DEBALU	DEBSRG	DEBMAC	DEBOUT

HALT CONTROL FIELD:　　　　　　　0　　　RUN
　　　　　　　　　　　　　　　　　　15　　　STOP SIMULATION

PRINT CONTROL FIELD PRIORITY:

　　　　　　　　　　　　　　　　　　0　　　PRINT NOTHING
　　　　　　　　　　　　　　　　　　N　　　HIGHEST PRIORITY

Appendix B. Computer Example – Multiplexer Select Definitions

Instruction Register Bit Positions – Internal Data Bus Connection

```
DATABUS   15  14  13  12  11  10  9   8   7   6   5   4   3   2   1   0
IR        15  14  13  12  11  10  9   8   7   6   5   4   3   2   1   0
OPCODE    7   6   5   4   3   2   1   0
R1                                        3   2   1   0
R2                                                        3   2   1   0
```

(Most Significant Bit of Each Field Has Highest Number)

Memory Address Register (MAR)

External Address Bus Connection (with Internal Data Bus)

Note: Byte ADDRESSING is implemented by discarding bit 0 of address (MAR).

```
ADDRESS BUS 15  14  13  12  11  10  9   8   7   6   5   4   3   2   1   0
INTERNAL DB 15  14  13  12  11  10  9   8   7   6   5   4   3   2   1   0
MAR (U12-PIN) 19 16  15  12  9   6   5   2
MAR (U23-PIN)                           19  16  15  12  9   6   5   2
RAM (U17-A)MSB         A10 A9  A8  A7  A6  A5  A4  A3  A2  A1  A0  -
RAM (U25-A)LSB         A10 A9  A8  A7  A6  A5 A4 A3 A2 A1 A0 -
EPROM (U28-A)MSB       A10 A9  A8  A7  A6  A5 A4 A3 A2 A1 A0 -
EPROM (U20-A)LSB       A10 A9  A8  A7  A6  A5  A4  A3  A2  A1  A0  -
8251 (U27-USART)                                                  C/D
```

UART (8251) BASE ADDRESS – FF00

COMMAND:SENSE REGISTER – FF02

DATA REGISTER (READ OR WRITE) – FF00

Am2903 Data Bus connections:

```
DATABUS    15  14  13  12  11  10  9   8   7   6   5   4   3   2   1   0
U18 (MSD) DA 3 2   1   0
U18 (MSD) DBI/O 3 2 1 0
U18 (MSD) YI/O 3 2  1   0
U30 DA              3   2   1   0
U30 DBI/O           3   2   1   0
U30 YI/O            3   2   1   0
U22 DA                              3   2   1   0
U22 DBI/O                           3   2   1   0
U22 YI/O                            3   2   1   0
U24 (LSD) DA                                        3   2   1   0
U24 (LSD) DBI/O                                     3   2   1   0
U24 (LSD) YI/O                                      3   2   1   0
```

RAM A and RAM B address multiplexers – Am2903 A3–A0, B3–B0 inputs

```
AMUXSEL            Source
(PL28/PL29)
00                 PIPELINE          PL24 -> 2903 A3
                                     PL25 -> 2903 A2
                                     PL26 -> 2903 A1
                                     PL27 -> 2903 A0
01                 UNUSED
10                 INSTRUCTION REGISTER - R2 FIELD
                                     IR3 -> 2903 A3
                                     IR2 -> 2903 A2
                                     IR1 -> 2903 A1
                                     IR0 -> 2903 A0
11                 INSTRUCTION REGISTER - R1 FIELD
                                     IR7 -> 2903 A3
                                     IR6 -> 2903 A2
                                     IR5 -> 2903 A1
                                     IR4 -> 2903 A0
```

```
BMUXSEL              Source
(PL34/PL35)
00                   PIPELINE              PL30 -> 2903 B3
                                           PL31 -> 2903 B2
                                           PL32 -> 2903 B1
                                           PL33 -> 2903 B0
01                   UNUSED
10                   INSTRUCTION REGISTER - R2 FIELD
                                           IR3 -> 2903 B3
                                           IR2 -> 2903 B2
                                           IR1 -> 2903 B1
                                           IR0 -> 2903 B0
11                   INSTRUCTION REGISTER - R1 FIELD
                                           IR7 -> 2903 B3
                                           IR6 -> 2903 B2
                                           IR5 -> 2903 B1
                                           IR4 -> 2903 B0
```

Microstatus Register Input MUXes

```
USRLD (PL58)         SOURCE
   0                 Current contents of Microstatus Register
   1                 2903 RALU flags (C, N, OVR, Z)
```

Macrostatus Register Input MUXes

```
MZNLD (PL59:PL60) SOURCE
   0                 Current value of MZ and MN
   1                 Data Bus (DB1 -> N, DB3 -> Z)
   2                 2903 RALU flags (N and Z)
   3                 unused
MCOVLD (PL61:PL62) SOURCE
   0                 Current value of MC and MOVR
   1                 Data Bus (DB0 -> C, DB2 MOVR)
   2                 2903 RALU flags (C and OVR)
   3                 unused
```

Macrostatus Register –> Databus

```
MDBOE* (PL63)        Function
   0                 Connect Macrostatus Register to Databus
                     (MZ -> DB3, MOVR -> DB2, MN -> DB1, MC -> DB0)
   1                 Disconnect Macrostatus Register from Databus
```

Carry Input MUX (Carry Input Selection for 2903 RALU)

```
CNSEL (PL56:PL57) Source
   0                 Carry Input = 0
   1                 Carry Input = 1
   2                 Microstatus Register Carry Flag (UC)
   3                 Macrostatus Register Carry Flag (MC)
```

2903 RALU (LS Digit) SIO0 Input MUX

```
SLSEL (PL52:PL55) Source (SLSEL3 {SLEN} = 0 {TRUE})
   0                 Shift In 0
   1                 Shift In 1
   2                 Shift In 2903 RALU (MS Digit QIO3) - Q15
   3                 Shift In 2903 RALU (MS Digit SIO3) - S15
   4-7               Unused
   8-15              MUX output tristated - disabled
```

2903 RALU (LS Digit) QIO0 Input MUX

```
QLSEL (PL48:PL51) Source (QLSEL3 {QLSEL} = 0 {TRUE})
   0                 Shift in 0
   1                 Shift in 1
   2                 Shift in 2903 RALU (MS Digit QIO3) - Q15
   3                 Shift in 2903 RALU (MS Digit SIO3) - S15
   4-7               Unused
   8-15              MUX output tristated - disabled
```

2903 RALU (MS Digit) SIO3 Input MUX

```
SRSEL (PL44:PL47) Source (SRSEL3 = 0 {TRUE})
0               Shift in 0 (opposite schematic ***)
1               Shift in 1 ***
2               Shift in 2903 RALU (LS Digit QIO0) - Q0
3               Shift in 2903 RALU (LS Digit SIO0) - S0
4-7             Unused
8-15            MUX output tristated - disabled
```

2903 RALU (MS Digit) QIO3 Input MUX

```
QRSEL (PL40:PL43) Source (QRSEL3 = 0 {TRUE})
0               Shift in 0 (opposite to schematic****)
1               Shift in 1 ***
2               Shift in 2903 RALU (LS Digit QIO0) - Q0 **
3               Shift in 2903 RALU (LS Digit SIO0) - S0 **
4-7             Unused
8-15            MUX output tristated - disabled
```

CCMUX Selections (CCU Section)

```
CCMUXSEL        FLAG OR INPUT
0               MICRO Z
1               MICRO C
2               MICRO OVR
3               MICRO N
4               MACRO Z
5               MACRO C
6               MACRO OVR
7               MACRO N
8-15            NOT IMPLEMENTED (FUTURE EXPANSION)
```

Some Macroinstruction Formats Supported by this Hardware Architecture

```
REGISTER - REGISTER
15                    8 7           4 3           0
+--------------------+-------------+-------------+
|    OP   CODE       |     R1      |     R2      |
+--------------------+-------------+-------------+


IMMEDIATE: Data is contained in second word of Instruction
15                    8 7           4 3           0
+--------------------+-------------+-------------+
|    OP   CODE       |     R1      |     R2      |
+--------------------+-------------+-------------+
+-----------------------------------------------+
|                 IMMEDIATE DATA                |
+-----------------------------------------------+


ABSOLUTE OR INDIRECT ADDRESSING:
Effective Address = c(Operand Address Field)
15                    8 7           4 3           0
+--------------------+-------------+-------------+
|    OP   CODE       |     R1      |     R2      |
+--------------------+-------------+-------------+
+-----------------------------------------------+
|          ABSOLUTE or INDIRECT ADDRESS         |
+-----------------------------------------------+


INDEXED ADDRESSING:
Effective Address = c(Operand Address Field) + c(X2)
15                    8 7           4 3           0
+--------------------+-------------+-------------+
|    OP   CODE       |     R1      |     R2      |
+--------------------+-------------+-------------+
+-----------------------------------------------+
|          DISPLACEMENT of ABSOLUTE ADDRESS     |
+-----------------------------------------------+
```

Appendix C. Parsed Microcode Display

The MCH microcode parser produces a formatted microcode dump in hex
from Microtec object files which is very useful for debugging purposes.
To use this program on the IBM PC type:

> MCH micro.dat

The output is sent to the default display device. To have this program
send the formatted output to a file called micro.txt type:

> MCH micro.dat >micro.txt

On the VAX prepare a .COM file in the following form:

$assign/user_mode micro.txt sys$output
$mch :== $vlsi$dja1:[user.mal.sw16]mch.exe
$mch 'p1'

Then type the following command at the VAX prompt:

$ @MCH micro.dat

The formatted file will be placed in micro.txt as above. If screen output
is desired then merely type:

$ mch micro.dat

Note: a similar program called MCL will produce the same output as MCH except that
it uses the Motorola S-Record format produced by UPASM as input. See Chapter 5 Appendix A
for more information on this suite of programs.

Below is a dump of the current MICRO.DAT file as produced by MCH:

```
_____HARDWARE BITS_____SIMULATOR
                 C                                                 M
               A   C   A   B                                     C   D      SMO
       2       BD  M   M   M                    MM          MM   M   YB    2222TAU
       9   C   RDF U   U   U       Q S Q S C UMCD      DD AA    I   U  BB   9999ACT
       1  RC   ARI X R X R X   OO  R R L L N SZOB 2 I BB RR    R X  DD H   1100THS
       0 OLEC  NEE S A S A S  EEEW S S S S S RNVO 9AN ED LO RW L S  EE A   0033RII
         EDNI  CSL E M E M E  ABYE E E E E E LLLE OLS NI DE DR D E  NN L   IIIIEND
       I ****  HSD L A L B L  **** L L L L L DDD* 3UT *R ** ** * L  ** T   ONOOGEE
       +-+----+---+-+-+-+-+-+----+-+-+-+-+-+----+---+--+--+--+-+-+--+-+------+
     0 |e|0110|  0|a|0|0|0|0|0001|8|8|8|8|0|1001|190|10|00|01|1|0|01|0|0000000
     1 |e|0110|  0|a|0|0|f|0|0000|8|8|8|8|1|1001| 88|01|10|11|1|0|11|0|0000000
     2 |e|0110|  0|a|0|0|0|0|0001|8|8|8|8|0|1001|190|01|10|11|0|0|11|0|0000000
     3 |2|0110|  0|a|0|0|0|0|0001|8|8|8|8|0|1001|190|10|10|11|1|0|11|0|0000000
```

```
 4 |e|0110|   0|a|0|0|f|0|0000|8|8|8|8|1|1001| 88|10|00|01|1|0|01|0|0000000
 5 |e|0110|   0|a|0|0|f|0|0000|8|8|8|8|1|1001| 88|01|10|11|1|0|11|0|0000000
 6 |e|0110|   0|a|0|0|0|0|0001|8|8|8|8|0|1001|190|01|10|11|0|0|11|0|0000000
 7 |2|0110|   0|a|0|0|0|0|0001|8|8|8|8|0|1001|190|10|10|11|1|0|11|0|0000000
 8 |e|0110|   0|a|0|0|f|0|0000|8|8|8|8|1|1001| 88|10|00|01|1|0|01|0|0000000
 9 |a|0110|   0|a|0|0|f|0|0000|8|8|8|8|1|1001| 88|01|10|11|1|0|11|0|0000000
 a |e|0110|   0|a|0|0|0|2|0001|8|8|8|8|0|1001|188|10|00|01|1|0|01|0|0000000
 b |a|0110|   0|a|0|0|0|0|0001|8|8|8|8|0|1001|190|01|10|11|1|0|11|0|0000000
 c |e|0110|   0|a|0|0|f|0|0000|8|8|8|8|1|1001| 88|10|00|01|1|0|01|0|0000000
 d |e|0110|   0|a|0|0|f|0|0000|8|8|8|8|1|1001| 88|01|10|11|1|0|11|0|0000000
 e |e|0110|   0|a|0|0|0|0|0001|8|8|8|8|0|1001|190|01|00|01|1|0|11|0|0000000
 f |a|0110|   0|a|0|0|0|0|0001|8|8|8|8|0|1001|190|01|10|11|1|0|11|0|0000000
20 |3|0110|   4|a|0|2|0|3|0000|8|8|8|8|0|1001| 8c|10|10|11|1|0|11|0|0000000
22 |1|0110|   8|a|0|0|0|0|0001|8|8|8|8|0|1001|190|10|10|11|1|0|11|0|0000000
23 |3|0110|   4|a|0|0|0|3|1000|8|8|8|8|0|1001| 8c|01|10|11|1|0|11|0|0000000
24 |1|0110|   c|a|0|0|0|0|0001|8|8|8|8|0|1001|190|10|10|11|1|0|11|0|0000000
25 |3|0110|   4|a|0|0|0|3|1000|8|8|8|8|0|1001| 8c|01|10|11|1|0|11|0|0000000
26 |1|0110|   a|a|0|0|0|0|0001|8|8|8|8|0|1001|190|10|10|11|1|0|11|0|0000000
27 |3|0110|   4|a|0|0|0|3|1000|8|8|8|8|0|1001| 8c|01|10|11|1|0|11|0|0000000
32 |3|0110|200|a|0|0|0|0|0001|8|8|8|8|0|1001|190|10|10|11|1|0|11|0|0000000
36 |e|0110|   0|a|0|0|0|3|0001|8|8|8|8|0|1001|188|10|00|10|1|0|01|0|0000000
37 |3|0110|   4|a|0|0|0|2|0001|8|8|8|8|0|1001|188|00|10|11|1|0|01|0|0000000
62 |1|0110|   8|a|0|0|0|0|0001|8|8|8|8|0|1001|190|10|10|11|1|0|11|0|0000000
63 |3|0110|   4|a|0|0|0|3|1000|8|8|8|8|0|1001| 98|01|10|11|1|0|11|0|0000000
92 |1|0110|   8|a|0|0|0|0|0001|8|8|8|8|0|1001|190|10|10|11|1|0|11|0|0000000
93 |3|0110|   4|a|0|0|0|3|1001|8|8|8|8|1|1221| 82|01|10|11|1|0|11|0|0000000
c2 |3|0110|210|a|0|0|0|0|0001|8|8|8|8|0|1001|190|10|10|11|1|0|11|0|0000000
c6 |3|0110|215|a|0|0|0|0|0001|8|8|8|8|0|1001|190|10|10|11|1|0|11|0|0000000
e0 |e|0110|   0|a|0|3|f|0|0000|8|8|8|8|0|1001| 8c|10|10|11|1|0|11|0|0000000
e1 |3|0110|   4|a|4|0|f|0|0000|8|8|8|8|0|1001| 82|10|10|11|1|0|11|0|0000000
e2 |1|0110|   8|a|0|0|0|0|0001|8|8|8|8|0|1001|190|10|10|11|1|0|11|0|0000000
e3 |3|0110|   4|a|4|0|f|0|0100|8|8|8|8|0|1001| 82|01|10|11|1|0|10|0|0000000
f2 |3|0110|204|a|0|0|0|0|0001|8|8|8|8|0|1001|190|10|10|11|1|0|11|0|0000000
100|3|0110|20b|a|0|0|0|0|0001|8|8|8|8|0|1001|190|10|10|11|1|0|11|0|0000000
110|3|0100| e0|8|0|0|0|0|0001|8|8|8|8|0|1001|190|10|10|11|1|0|11|0|0000000
112|3|0100| e2|8|0|0|0|0|0001|8|8|8|8|0|1001|190|10|10|11|1|0|11|0|0000000
113|3|0110|21a|a|0|0|0|0|0001|8|8|8|8|0|1001|190|10|10|11|1|0|11|0|0000000
130|3|0100|20b|8|0|0|0|0|0001|8|8|8|8|0|1001|190|10|10|11|1|0|11|0|0000000
131|3|0110|   4|a|0|0|0|0|0001|8|8|8|8|0|1001|190|10|10|11|1|0|11|0|0000000
140|3|0110|   4|a|0|2|0|3|0000|8|8|8|8|0|1221| 86|10|10|11|1|0|11|0|0000000
142|1|0110|   8|a|0|0|0|0|0001|8|8|8|8|0|1001|190|10|10|11|1|0|11|0|0000000
143|3|0110|   4|a|0|0|0|3|1000|8|8|8|8|0|1001| 86|01|10|11|1|0|11|0|0000000
152|1|0110|   8|a|0|0|0|0|0001|8|8|8|8|0|1001|190|10|10|11|1|0|11|0|0000000
153|3|0110|   4|a|0|0|0|3|1000|8|8|8|8|1|1221| 82|01|10|11|1|0|11|0|0000000
1fe|e|0110|   0|a|0|0|0|0|0001|8|8|8|8|0|1001|190|10|10|11|1|0|11|3|0000000
200|e|0110|   0|a|0|0|f|0|0000|8|8|8|8|1|1001| 88|10|00|01|1|0|01|0|0000000
201|e|0110|   0|a|0|0|f|0|0000|8|8|8|8|1|1001| 88|01|10|11|1|0|11|0|0000000
202|e|0110|   0|a|0|0|0|0|0001|8|8|8|8|0|1001|190|01|00|10|1|0|11|0|0000000
203|3|0110|   4|a|0|0|0|3|0001|8|8|8|8|0|1001|188|00|10|11|1|0|01|0|0000000
204|1|0110|   8|a|0|0|0|0|0001|8|8|8|8|0|1001|190|10|10|11|1|0|11|0|0000000
205|e|0110|   0|a|0|0|4|0|1000|8|8|8|8|0|1001| 84|01|10|11|1|0|11|0|0000000
206|e|0110|   0|a|0|0|e|0|0000|8|8|8|8|1|1001| 88|10|10|11|1|0|11|0|0000000
207|e|0110|   0|a|0|0|e|0|0000|8|8|8|8|1|1001| 88|10|00|10|1|0|01|0|0000000
208|e|0110|   0|a|0|0|f|0|0001|8|8|8|8|0|1001| 88|00|10|11|1|0|01|0|0000000
209|e|0110|   0|a|4|0|f|0|0000|8|8|8|8|0|1001| 8c|10|10|11|1|0|11|0|0000000
20a|3|0110|   4|a|0|0|4|0|0000|8|8|8|8|0|1001| 90|10|10|11|1|0|11|0|0000000
20b|e|0110|   0|a|0|0|e|0|0001|8|8|8|8|0|1001| 88|10|00|01|1|0|01|0|0000000
20c|e|0110|   0|a|0|0|0|0|0001|8|8|8|8|0|1001|190|01|10|11|1|0|11|0|0000000
20d|e|0110|   0|a|0|0|f|0|1000|8|8|8|8|0|1001| 8c|01|10|11|1|0|11|0|0000000
```

```
20e |e|0110|   0|a|4|0|e|0|0000|8|8|8|8|0|1001| 82|10|10|11|1|0|10|0|0000000
20f |3|0110|   4|a|4|0|e|0|0000|8|8|8|8|0|1001| 82|10|10|11|1|0|10|0|0000000
210 |1|0110|   8|a|0|0|0|0|0001|8|8|8|8|0|1001|190|10|10|11|1|0|11|0|0000000
211 |e|0110|   0|a|0|0|0|0|1001|8|8|8|8|0|1301| cc|01|10|11|1|0|11|0|0000000
212 |3|0100|   4|4|0|0|0|0|0001|8|8|8|8|0|1001|190|10|10|11|1|0|11|0|0000000
213 |e|0110|   0|a|0|0|0|3|0000|8|8|8|0|0|1001|128|10|10|11|1|0|11|0|0000000
214 |3|0110|212|a|4|0|0|0|0001|8|8|8|8|0|1201| c3|10|10|11|1|0|11|0|0000000
215 |1|0110|   8|a|0|0|0|0|0001|8|8|8|8|0|1001|190|10|10|11|1|0|11|0|0000000
216 |e|0110|   0|a|0|0|0|0|1001|8|8|8|8|0|1301| cc|01|10|11|1|0|11|0|0000000
217 |3|0100|   4|4|0|0|0|0|0001|8|8|8|8|0|1001|190|10|10|11|1|0|11|0|0000000
218 |e|0110|   0|a|0|0|0|3|0000|8|0|8|8|0|1001| 28|10|10|11|1|0|11|0|0000000
219 |3|0110|217|a|4|0|0|0|0001|8|8|8|8|0|1201| c3|10|10|11|1|0|11|0|0000000
21a |e|0110|   0|a|0|0|f|0|0000|8|8|8|8|1|1001| 88|10|10|11|1|0|11|0|0000000
21b |3|0110|   4|a|0|0|f|0|0000|8|8|8|8|1|1001| 88|10|10|11|1|0|11|0|0000000
d0  |3|0110|220|a|0|0|0|0|0001|8|8|8|8|0|1001|190|10|10|11|1|0|11|0|0000000
d4  |3|0110|225|a|0|0|0|0|0001|8|8|8|8|0|1001|190|10|10|11|1|0|11|0|0000000
220 |1|0110|   8|a|0|0|0|0|0001|8|8|8|8|0|1001|190|10|10|11|1|0|11|0|0000000
221 |e|0110|   0|a|0|0|0|0|1001|8|8|8|8|0|1001| cc|01|10|11|1|0|11|0|0000000
222 |e|0110|   0|a|4|0|0|0|0001|8|8|8|8|0|1001| c3|10|10|11|1|0|11|0|0000000
223 |3|0100|   4|0|0|0|0|3|0000|8|8|8|4|0|1231|128|10|10|11|1|0|11|0|0000000
224 |3|0110|223|a|4|0|0|0|0001|8|8|8|8|0|1001| c3|10|10|11|1|0|11|0|3333300
225 |1|0110|   8|a|0|0|0|0|0001|8|8|8|8|0|1001|190|10|10|11|1|0|11|0|0000000
226 |e|0110|   0|a|0|0|0|0|1001|8|8|8|8|0|1001| cc|01|10|11|1|0|11|0|0000000
227 |e|0110|   0|a|4|0|0|0|0001|8|8|8|8|0|1001| c3|10|10|11|1|0|11|0|0000000
228 |3|0100|   4|0|0|0|0|3|0000|8|4|8|8|0|1231| 28|10|10|11|1|0|11|0|0000000
229 |3|0110|228|a|4|0|0|0|0001|8|8|8|8|0|1001| c3|10|10|11|1|0|11|0|0000000
```

Appendix D. Instruction Set Reference

Please refer to the "Sweet/16 Instruction Set User's Manual" in the next section for complete implementation details.

Key to notation: (Contents of) [Address]

** Note — Commands with optional suffix "R" accept relocatable **
** labels as Data operands. Such is the default in STR instructions **

Load

```
10          LOR R1,R2 (R2) --> (R1)
11          LORI(R) R1,D2 D2 --> (R1)
12          LORA(R) R1,D2 ([D2]) --> (R1)
13          LORR R1,R2 ([(R2)]) --> (R1)
14          LORX(R) R1,R2,D3 ([D1 + (R2)]) --> (R1)
```

Store

```
19          STRI R1,D2 (R1) --> [D2]
1A          STRA R1,D2 (R1) --> [([D2])]
1B          STRR R1,R2 (R2) --> [(R1)]
1C          STRX R1,R2,D3 (R2) --> [(R1) + D3]
```

Add w/Carry

```
20          ADC R1,R2 (R1) + (R2) --> (R1)
21          ADIC R1,D2 D2 + (R1) --> (R1)
22          ADAC R1,D2 ([D2]) + (R1) --> (R1)
23          ADRC R1,R2 ([(R2)]) + (R1) --> (R1)
24          ADXC R1,R2,D3 ([D3 + (R2)]) + (R1) --> (R1)
```

Add

```
A0          ADD R1,R2 (R1) + (R2) --> (R1)
A1          ADI R1,D2 D2 +(R1) --> (R1)
A2          ADA R1,D2 ([D2]) + (R1) --> (R1)
A3          ADR R1,R2 ([(R2)]) + (R1) --> (R1)
A4          ADX R1,R2,D3 ([D3 + (R2)]) + (R1) --> (R1)
A5          INCI R1,D2 Logical Flags D2 +(R1) --> (R1)
```

Subtract w/Carry

```
28          SBC R1,R2 (R1) - (R2) --> (R1)
29          SBIC R1,D2 (R1) - D2 --> (R1)
2A          SBAC R1,D2 (R1) - ([D2]) --> (R1)
2B          SBRC R1,R2 (R1) - ([(R2)]) --> (R1)
2C          SBXC R1,R2,D3 (R1) - ([D3 + (R2)] --> (R1)
```

Subtract

```
A8          SB R1,R2 (R1) - (R2) --> (R1)
A9          SBI R1,D2 (R1) - D2 --> (R1)
AA          SBA R1,D2 (R1) - ([D2]) --> (R1)
AB          SBR R1,R2 (R1) - ([(R2)]) --> (R1)
AC          SBX R1,R2,D3 (R1) - ([D3 + (R2)]) --> (R1)
AD          DECI R1,D2 Logical Flags (R1) - D2 --> (R1)
```

And

```
30          AND R1,R2 (R1) . (R2) --> (R1)
31          ANDI R1,D2 D2 . (R1) --> (R1)
32          ANDA R1,D2 ([D2]) . (R1) --> (R1)
33          ANDR R1,R2 ([(R2)]) . (R1) --> (R1)
34          ANDX R1,R2,D3 ([D3 + (R2)]) . (R2) --> (R1)
```

Or

38	OR R1,R2 (R1) + (R2) --> (R1)
39	ORI R1,D2 D2 + (R1) --> (R1)
3A	ORA R1,D2 ([D2]) + (R1) --> (R1)
3B	ORR R1,R2 ([(R2)]) + (R1) --> (R1)
3C	ORX R1,R2,D3 ([D3 + (R2)]) + (R1) --> (R1)

Exclusive OR

40	XOR R1,R2 (R1) XOR (R2) --> (R1)
41	XORI R1,D2 D2 XOR (R1) --> (R1)
42	XORA R1,D2 ([D2]) XOR (R1) --> (R1)
43	XORR R1,R2 ([(R2)]) XOR (R1) --> (R1)
44	XORX R1,R2,D3 ([D3 + (R2)]) XOR (R1) --> (R1)

Compare

48	CMP R1,R2 (R1) - (R2)
49	CMPI R1,D2 (R1) - D2
4A	CMPA R1,D2 (R1) - ([D2])
4B	CMPR R1,R2 (R1) - ([(R2)])
4C	CMPX R1,R2,D3 (R1) - ([D3 + (R)])

Set

50	SETN Set Sign Flag
51	SETV Set Overflow Flag
52	SETC Set Carry Flag
53	SETZ Set Zero Flag

Clear

58	CEN Clear Sign Flag
59	CEV Clear Overflow Flag
5A	CEC Clear Carry Flag
5B	CEZ Clear Zero Flag

Shift

60	SHLA R1,D2 Shift left D2 times, over sign bit, inject 0's
61	SHLL R1,D2 Shift left D2 times, through sign bit, inject 0's
62	SHRA R1,D2 Shift right D2 times, over sign bit, inject 0's
63	SHRL R1,D2 Shift right D2 times, through sign bit, inject 0's

Rotate

68	ROLC R1,D2 Rotate left D2 times, through carry flag
69	ROL R1,D2 Rotate left D2 times, over carry flag
6A	RORC R1,D2 Rotate right D2 times, through carry flag
6B	ROR R1,D2 Rotate right D2 times, over carry flag

Branch

70	B R1 (R1) --> PC
71	BI(R) D1 D1 --> PC
72	BA(R) D1 ([D1]) --> PC
73	BARR R1 ([(R1)]) --> PC
74	BX(R) R1,D2 ([D1 + (R1)]) --> PC

Call Subroutine

78	CA R1 PC+1 --> Stack, (R1) --> PC
79	CAI(R) D1 PC+1 --> Stack, D1 --> PC
7A	CAA(R) D1 PC+1 --> Stack, ([D1]) --> PC
7B	CAR R1 PC+1 --> Stack, ([(R1)]) --> PC
7C	CAX(R) R1,D2 PC+1 --> Stack, ([D1 + (R1)]) --> PC

Return

80	R Stack --> PC

The Flags:
```
N --> Negative (Sign)
V --> Overflow
C --> Carry
Z --> Zero
```

Test Conditions Based on the Flags:
```
X1   Test Name Function
0000 CC   Carry Clear /C
0001 CS   Carry Set C
0010 EQ   Equal Z
0011 GE   Greater or Equal N*V + /N*/V
0100 GT   Greater N*V*/Z + /N*/V*/Z
0101 HI   High (for unsigned) /C*/Z
0110 LE   Less or Equal Z + N*/V + /N*V
0111 LS   Less (for unsigned) C + Z
1000 LT   Less N*/V + /N*V
1001 MI   Minus N
1010 NE   Not Equal /Z
1011 PL   Plus /N
1100 VC   Overflow Clear /V
1101 VS   Overflow Set V
```

```
**************************************************************************
* For the following conditional instructions, substitute the above *
* conditionals for 'N'. *
**************************************************************************
```

Conditional Branch
```
88        BN R1 if F{N} then (R1) --> PC
89        BNI(R) D1 if f{N} then D1 --> PC
8A        BNA(R) D1 if f{N} then ([D1]) --> PC
8B        BNRR R1 if f{N} then ([[(R1)]]) --> PC
8C        BNX(R) R1,D2 if f{N} then ([D2 + (R1)]) --> PC
```

Conditional Call Subroutine
```
90        CN R1 if f{N} then PC+1 --> Stack, (R1) --> PC
91        CNI(R) D1 if f{N} then PC+1 --> Stack, D1 --> PC
92        CNA(R) D1 if f{N} then PC+1 --> Stack, ([D1]) --> PC
93        CNR R1 if f{N} then PC+1 --> Stack, ([[(R1)]]) --> PC
94        CNX(R) R1,D2 if f{N} then PC+1 --> Stack, ([D3 + (R2)]) --> PC
```

Conditional Return
```
98        RN if f{N} then Stack --> PC
```

SWEET/16 INSTRUCTION SET USER'S MANUAL

Created by Brian Miller and Bruce Kleinman — Version 1/13/89
Documentation Format:

NAME OF INSTRUCTION
 Format: --> ASSEMBLER MNEMONIC [OPERAND], [OPERAND], [OPERAND]
Register Transfer Language Function Description:
Machine Code Format:
Flags Modified:
Remarks:
Notes:
1. There may be any number of operands up to three.
2. The following shorthand notation is used in the Function
Description:
(xxxx) — Signifies "The contents of"
[xxxx] — Signifies "The address specified by"

Data Movement Operations

***** LOAD REGISTER TO REGISTER
Format: LOR R1,R2
Function Description: (R2) --> (R1)
Machine Code Format:
 Word 1: [10] [R1] [R2]
 Word 2: NONE
Flags Modified: NONE
Remarks: LOR copies the contents of one Register into another.

***** LOAD REGISTER IMMEDIATE
Format: LORI(R) R1,D2
Function Description: D2 --> (R1)
Machine Code Format:
 Word 1: [11] [R1]
 Word 2: D2
Flags Modified: NONE
Remarks: LORI(R) loads the designated Register with the data specified in the second word of the instructions. The LORI form is used when the data is a numeric constant defined to the assembler. The LORIR for is used when the data is an address.

***** STORE REGISTER INDEXED INDIRECT
Format: STRX R1,R2,D3
Function Description: (R2) --> [(R1) + D3]
Machine Code Format:
 Word 1: [1C] [R1] [R2]
 Word 2: D3
Flags Modified: NONE
Remarks: STRX places the data in R1 in the memory address specified by the sum of the contents of R2 and the second word of the instruction.

Arithmetic Operations

***** ADD WITH CARRY REGISTER TO REGISTER
Format: ADC R1,R2

Function Description: (R1) + (R2) + C --> (R1)
Machine Code Format:
 Word 1: [20] [R1] [R2]
 Word 2: NONE
Flags Modified: Carry, Overflow, Sign and Zero
Remarks: ADC sums the contents of R1 with R2 and the Carry Flag and places the result in R1. The Carry produced is placed in the Carry Flag.

***** ADD WITH CARRY REGISTER TO REGISTER
Format: ADD R1,R2
Function Description: (R1) + (R2) + 0 --> (R1)
Machine Code Format:
 Word 1: [A0] [R1] [R2]
 Word 2: NONE
Flags Modified: Carry, Overflow, Sign and Zero
Remarks: ADD sums the contents of R1 with R2 without the Carry Flag and places the result in R1. The Carry produced is placed in the Carry Flag.

***** ADD WITH CARRY REGISTER IMMEDIATE
Format: ADIC R1,D2
Function Description: D2 + (R1) + C --> (R1)
Machine Code Format:
 Word 1: [21] [R1]
 Word 2: D2
Flags Modified: Carry, Overflow, Sign and Zero
Remarks: The ADIC instruction sums the contents of the register specified by the R1 field with the immediate data and the Carry Flag. The sum is placed in the source register and all flags are updated.

***** ADD REGISTER IMMEDIATE

Format: ADI R1,D2
Function Description: D2 + (R1) + 0 --> (R1)
Machine Code Format:
 Word 1: [A1] [R1]
 Word 2: D2
Flags Modified: Carry, Overflow, Sign and Zero
Remarks: The ADI instruction sums the contents of the register specified by the R1 field with the immediate data. The sum is placed in the source register and all flags are updated.

***** ADD WITH CARRY REGISTER INDIRECT
Format: ADAC R1,D2
Function Description: ([D2]) + (R1) + C --> (R1)
Machine Code Format:
 Word 1: [22] [R1]
 Word 2: D2
Flags Modified: Carry, Overflow, Sign and Zero
Remarks: The ADAC instruction sums the contents of the register specified by the R1 field with the data pointed at by the second word of the instruction, D2 and the Carry Flag. The sum is placed in R1 and all flags are updated.

***** ADD REGISTER INDIRECT
Format: ADA R1,D2
Function Description: ([D2]) + (R1) --> (R1)
Machine Code Format:

Word 1: [A2] [R1]
Word 2: D2
Flags Modified: Carry, Overflow, Sign and Zero
Remarks: The ADA instruction sums the contents of the register specified by the R1 field with the data pointed at by the second word of the instruction, D2. The sum is placed in R1 and all flags are updated.

***** ADD WITH CARRY REGISTER INDIRECT REGISTER

Format: ADRC R1,R2
Function Description: ([(R2)]) + (R1) + C --> (R1)
Machine Code Format:
 Word 1: [23] [R1] [R2]
 Word 2: NONE
Flags Modified: Carry, Overflow, Sign and Zero
Remarks: The ADRC instruction sums R1 with the contents of the memory location specified by the contents of R2 and includes the Carry Flag. The result is placed in R1 and all flags are updated.

***** ADD REGISTER INDIRECT REGISTER

Format: ADR R1,R2
Function Description: ([(R2)]) + (R1) --> (R1)
Machine Code Format:
 Word 1: [A3] [R1] [R2]
 Word 2: NONE
Flags Modified: Carry, Overflow, Sign and Zero
Remarks: The ADR instruction sums R1 with the contents of the memory location specified by the contents of R2. The result is placed in R1 and all flags are updated.

***** ADD WITH CARRY REGISTER INDEXED INDIRECT

Format: ADXC R1,R2,D3
Function Description: ([D3 + (R2)]) + (R1) + C --> (R1)
Machine Code Format:
 Word 1: [24] [R1] [R2]
 Word 2: D3
Flags Modified: Carry, Overflow, Sign and Zero
Remarks: The contents of R1 are summed with the contents of the location pointed to by the sum of the immediate data and R2. The Carry Flag is included. The result is placed in R1 and all flags are updated.

***** ADD REGISTER INDEXED INDIRECT

Format: ADX R1,R2,D3
Function Description: ([D3 + (R2)]) + (R1) --> (R1)
Machine Code Format:
 Word 1: [A4] [R1] [R2]
 Word 2: D3
Flags Modified: Carry, Overflow, Sign and Zero
Remarks: The contents of R1 are summed with the contents of the location pointed to by the sum of the immediate data and R2. The result is placed in R1 and all flags are updated.

***** INCREMENT REGISTER IMMEDIATE

Format: INCI R1,D2
Function Description: D2 + (R1) + 0 --> (R1)
Machine Code Format:
 Word 1: [A1] [R1]
 Word 2: D2
Flags Modified: Sign and Zero

Remarks: The INCI instruction sums the contents of the register specified by the R1 field with the immediate data. The sum is placed in the source register and only the logical flags are updated. This is a convenient instruction to use for counter operations during loop control.

***** SUBTRACT WITH CARRY REGISTER TO REGISTER
Format: SBC R1,R2
Function Description: (R1) - (R2) - B --> (R1)
Machine Code Format:
 Word 1: [28] [R1] [R2]
 Word 2: NONE
Flags Modified: Carry, Overflow, Sign and Zero
Remarks: The contents of R2 is subtracted from the contents of R1 along with the Borrow (.NOT. Carry) Flag. The results are placed in R1 and all flags are updated.

***** SUBTRACT REGISTER TO REGISTER
Format: SB R1,R2
Function Description: (R1) - (R2) --> (R1)
Machine Code Format:
 Word 1: [A8] [R1] [R2]
 Word 2: NONE
Flags Modified: Carry, Overflow, Sign and Zero
Remarks: The contents of R2 is subtracted from the contents of R1. The results are placed in R1 and all flags are updated.

***** SUBTRACT WITH CARRY REGISTER IMMEDIATE
Format: SBIC R1,D2
Function Description: (R1) - D2 - B --> (R1)
 Machine Code Format:
 Word 1: [29] [R1]
 Word 2: D2
Flags Modified: Carry, Overflow, Sign and Zero
Remarks: The SBIC instruction subtracts the immediate data, along with the Borrow (.NOT. Carry) flag, from the register specified by the R1 field. The result is returned to R1 and all flags are updated.

***** SUBTRACT REGISTER IMMEDIATE
Format: SBI R1,D2
Function Description: (R1) - D2 --> (R1)
 Machine Code Format:
 Word 1: [A9] [R1]
 Word 2: D2
Flags Modified: Carry, Overflow, Sign and Zero
Remarks: The SBI instruction subtracts the immediate data from the register specified by the R1 field. The result is returned to R1 and all flags are updated.

***** SUBTRACT WITH CARRY REGISTER INDIRECT
Format: SBAC R1,D2
Function Description: (R1) - ([D2]) - B --> (R1)
Machine Code Format:
 Word 1: [2A] [R1]
 Word 2: D2
Flags Modified: Carry, Overflow, Sign and Zero
Remarks: The SBAC instruction subtract the contents of the address specified by the immediate data from R1 along with the Borrow (.NOT. Carry) Flag. The results are returned to the register and all flags are updated.

***** SUBTRACT REGISTER INDIRECT
Format: SBA R1,D2
Function Description: (R1) - ([D2]) --> (R1)
Machine Code Format:
 Word 1: [AA] [R1]
 Word 2: D2
Flags Modified: Carry, Overflow, Sign and Zero
Remarks: The SBA instruction subtract the contents of the address specified by the immediate data
from R1. The results are returned to the register and all flags are updated.

***** SUBTRACT WITH CARRY REGISTER INDIRECT REGISTER
Format: SBRC R1,R2
Function Description: (R1) - ([(R2)]) - B --> (R1)
Machine Code Format:
 Word 1: [2B] [R1] [R2
 Word 2: NONE
Flags Modified: Carry, Overflow, Sign and Zero
Remarks: The SBRC instruction subtracts the contents of the location specified by R2 from the contents
of R1 along with the Borrow (.NOT. Carry) Flag. The result is returned to R1 and all flags are updated.

***** SUBTRACT REGISTER INDIRECT REGISTER
Format: SBR R1,R2
Function Description: (R1) - ([(R2)]) --> (R1)
Machine Code Format:
 Word 1: [AB] [R1] [R2
 Word 2: NONE
Flags Modified: Carry, Overflow, Sign and Zero
Remarks: The SBR instruction subtracts the contents of the location specified by R2 from the contents
of R1. The result is returned to R1 and all flags are updated.

***** SUBTRACT WITH CARRY REGISTER INDEXED INDIRECT
Format: SBXC R1,R2,D3
Function Description: (R1) - ([D3 + (R2)]) - B --> (R1)
Machine Code Format:
 Word 1: [2C] [R1] [R2]
 Word 2: D3
Flags Modified: Carry, Overflow, Sign and Zero
Remarks: The contents of the address specified by the sum of the immediate data field and R2 are sub-
tracted from R1 along with the Borrow (.NOT. Carry) Flag. The result is returned to R1 and all flags
are updated.

***** SUBTRACT REGISTER INDEXED INDIRECT
Format: SBX R1,R2,D3
Function Description: (R1) - ([D3 + (R2)]) --> (R1)
Machine Code Format:
 Word 1: [AC] [R1] [R2]
 Word 2: D3
Flags Modified: Carry, Overflow, Sign and Zero
Remarks: The contents of the address specified by the sum of the immediate data field and R2 are sub-
tracted from R1. The result is returned to R1 and all flags are updated.

***** DECREMENT REGISTER IMMEDIATE
Format: DECI R1,D2
Function Description: (R1) - D2 --> (R1)

Machine Code Format:
Word 1: [AD] [R1]
Word 2: D2
Flags Modified: Sign and Zero
Remarks: The DECI instruction subtracts the immediate data from the register specified by the R1 field. The result is returned to R1 and only the logical flags are updated. Consider this instruction for use in loop control operations.

Logical Operations (Register/Register)
***** AND REGISTER TO REGISTER
***** OR REGISTER TO REGISTER
***** XOR REGISTER TO REGISTER
***** COMPARE REGISTER TO REGISTER

Assembler Syntax	Machine Code Format: Word 1
AND R1,R2	30] [R1] [R2]
OR R1,R2	[38] [R1] [R2]
XOR R1,R2	[40] [R1] [R2]
CMP R1,R2	[48] [R1] [R2]

Function Description:
(R1) . (R2) --> (R1)
(R1) + (R2) --> (R1)
(R1) XOR (R2) --> (R1)
(R1) - (R2)
Flags Modified: ZERO SIGN
Remarks: These instructions perform the specified logical operation between the two registers specified. The result is returned to R1. Only the logical flags are set.

Logical Operations (Register/Immediate)
***** AND REGISTER IMMEDIATE
***** OR REGISTER IMMEDIATE
***** XOR REGISTER IMMEDIATE
***** COMPARE REGISTER IMMEDIATE

Assembler Syntax	Machine Code Format: Word 1 Word 2
ANDI R1,D2	[31] [R1] D2
ORI R1,D2	[39] [R1] D2
XORI R1,D2	[41] [R1] D2
CMPI R1,D2	[49] [R1] D2

Function Description:
(R1) . D2 --> (R1)
(R1) + D2 --> (R1)
(R1) XOR D2 --> (R1)
(R1) - D2
Flags Modified: ZERO SIGN
Remarks: These instructions perform the specified logical operation between the register specified by the R1 field and the immediate data. The result is returned to R1 and the logical flags are updated.

Logical Operations (Register Absolute)
***** AND REGISTER ABSOLUTE
***** OR REGISTER ABSOLUTE
***** XOR REGISTER ABSOLUTE
***** COMPARE REGISTER ABSOLUTE

Assembler Syntax   Machine Code Format:
                                          Word 1Word 2
ANDA R1,D2   [32] [R1] D2
ORA R1,D2   [3A] [R1] D2
XORA R1,D2   [42] [R1] D2
CMPA R1,D2   [4A] [R1] D2
Function Description:
(R1) . ([D2]) --> (R1)
(R1) + ([D2]) --> (R1)
(R1) XOR ([D2]) --> (R1)
(R1) - ([D2])
Flags Modified: ZERO SIGN
Remarks: These instructions perform the specified logical operation between the contents of R1 and
that of the location specified by the second word of the instruction. The result is placed in R1 and the
logical flags are updated.

Logical Operations (Register/Indirect)
 ***** AND REGISTER INDIRECT REGISTER
 ***** OR REGISTER INDIRECT REGISTER
 ***** XOR REGISTER INDIRECT REGISTER
 ***** COMPARE REGISTER INDIRECT REGISTER
Assembler Syntax   Machine Code Format:
                                          Word 1
ANDR R1,R2   [33] [R1] [R2]
ORR R1,R2   [3B] [R1] [R2]
XORR R1,R2   [43] [R1] [R2]
CMPR R1,R2   [4B] [R1] [R2]
Function Description:
(R1) . ([(R2)]) --> (R1)
(R1) + ([(R2)]) --> (R1)
(R1) XOR ([(R2)]) --> (R1)
(R1) - ([(R2)])
Flags Modified: ZERO SIGN
Remarks: These instructions perform the specified logical operation between the contents of the reg-
ister specified by the R1 field and the memory location specified by the R2 field in the instruction. The
result is returned to R1 and the logical flags are updated.

Logical Operations (Register/Indexed Indirect)
 ***** AND REGISTER INDEXED INDIRECT
 ***** OR REGISTER INDEXED INDIRECT
 ***** XOR REGISTER INDEXED INDIRECT
 ***** COMPARE REGISTER INDEXED INDIRECT
Assembler Syntax   Machine Code Format:
                                          Word 1   Word 2
ANDX R1,R2,D3   [34] [R1] [R2] D3
ORX R1,R2,D3   [3C] [R1] [R2]D3
XORX R1,R2,D3   [44] [R1] [R2]D3
CMPX R1,R2,D3   [4C] [R1] [R2]D3
Function Description:
([D3 + (R2)]) . (R1) --> (R1)
([D3 + (R2)]) + (R1) --> (R1)
([D3 + (R2)]) XOR (R1) --> (R1)
([D3 + (R2)]) - (R1) --> (R1)
Flags Modified: ZERO SIGN

Remarks: These instructions perform the specified logical operation between the data in the register specified by the R1 field and that in the location specified by the sum of the contents of R2 and the second word of the instruction. The result is placed in R1 and the logical flags are updated.

Mode Setting Operations

Operation	Assembler Syntax	Machine Code Format
SET THE SIGN FLAG	SETN	[50]
SET THE OVERFLOW FLAG	SETV	[51]
SET THE CARRY FLAG	SETC	[52]
SET THE ZERO FLAG	SETZ	[53]

Flags Modified: The Flag Specified is SET.

Operation	Assembler Syntax	Machine Code Format
CLEAR THE SIGN FLAG	CEN	[58]
CLEAR THE OVERFLOW FLAG	CEV	[59]
CLEAR THE CARRY FLAG	CEC	[5A]
CLEAR THE ZERO FLAG	CEZ	[5B]

Flags Modified: The Flag Specified is CLEARED.

Shift Operations

Operation	Assembler Syntax	Machine Code Format
SHIFT LEFT ARITHMETIC	SHLA R1,D2	[60]
SHIFT LEFT LOGICAL	SHLL R1,D2	[61]
SHIFT RIGHT ARITHMETIC	SHRA R1,D2	[62]
SHIFT RIGHT LOGICAL	SHRL R1,D2	[63]

Flags Modified: NONE

Remarks: These instructions shift the contents of the specified register to the left or right as specified. In the arithmetic shift, the sign is extended. All bits are treated equally in the logical shift. The number of bit positions to shift is specified in the D2 field.

Operation	Assembler Syntax	Machine Code Format
ROTATE LEFT with Carry	ROLC R1,D2	[68]
ROTATE LEFT	ROL R1,D2	[69]
ROTATE RIGHT with Carry	RORC R1,D2	[6A]
ROTATE RIGHT	ROR R1,D2	[6B]

Flags Modified: Carry for ROLC and RORC
                    NONE for ROL and ROR

Remarks: These instructions rotate the contents of the specified register to the left or right as specified. In the arithmetic shift the Carry increases the shift path to 17 bits. Only the 16 bits in the specified register are rotated in the logical shift. The number of bit positions to shift is specified in the D2 field.

Decision Making Instructions

The Flags:
N --> Negative (Sign)
V --> Overflow
C --> Carry
Z --> Zero

Test Conditions Based on the Flags:

X1	Test	Name	Function
0000	CC	Carry Clear	/C
0001	CS	Carry Set	C
0010	EQ	Equal	Z

0011	GE	Greater or Equal	N*V + /N*/V
0100	GT	Greater	N*V*/Z + /N*/V*/Z
0101	HI	High (for unsigned)	/C*/Z
0110	LE	Less or Equal	Z + N*/V + /N*V
0111	LS	Less (for unsigned)	C + Z
1000	LT	Less	N*/V + /N*V
1001	MI	Minus	N
1010	NE	Not Equal	Z
1011	PL	Plus	/N
1100	VC	Overflow Clear	/V
1101	VS	Overflow Set	V

```
*************************************************************************
* For the following conditional instructions, substitute the above *
* conditionals for 'N'. *
*************************************************************************
```

***** BRANCH TO REGISTER
***** BRANCH TO REGISTER

Assembler Syntax	Machine Code Format:
	Word 1
B R1	[70] [R1]
BN R1	[88] [R1]

Function Description:
(R1) --> PC B
if f{N} then (R1) --> PC BN
Flags Modified: NONE
Remarks: The Branch and Branch Conditional instructions cause the PC to be loaded with the contents of the specified register. The transfer is conditional in the BC case upon satisfying the condition specified by "N".

***** BRANCH TO REGISTER IMMEDIATE
***** BRANCH TO REGISTER IMMEDIATE CONDITIONAL

Assembler Syntax	Machine Code Format:	
	Word 1	Word 2
BI(R) D1	[71] [R1]	D1
BNI(R) D1	[89] [R1]	D1

Function Description:
D1 --> PC BI(R)
if f{N} then D1 --> PC BNI(R)
Flags Modified: NONE
Remarks: These instructions transfer the contents of the immediate data field, D1, in the instruction to the PC. In the BNI and BNIR instructions, the transfer is dependent on the condition "N" being true else the PC is incremented. The BIR and BNIR mnemonics are used when the symbol for D1 is a relocatable address.

***** BRANCH TO REGISTER INDIRECT
***** BRANCH TO REGISTER INDIRECT CONDITIONAL

Assembler Syntax	Machine Code Format:	
	Word 1	Word 2
BA(R) D1	[72]	D1
BNA(R) D1	[8A]	D1

Function Description:
([D1]) --> PC BA(R)

if f{N} then ([D1]) --> PC BNA(R)
Flags Modified: NONE
Remarks:
***** BRANCH TO REGISTER INDIRECT REGISTER
***** BRANCH TO REGISTER INDIRECT CONDITIONAL REGISTER

Assembler Syntax	Machine Code Format:
	Word 1
BARR R1	[73]
BNRR R1	[8B]

Function Description:
([(R1)]) --> PC BARR
if f{N} then ([(R1)]) --> PC BNRR
Flags Modified: NONE
Remarks:

***** BRANCH TO REGISTER INDIRECT INDEXED
***** BRANCH TO REGISTER INDIRECT CONDITIONAL INDEXED

Assembler Syntax	Machine Code Format:
	Word 1 Word 2
BX(R) R1,D2	[74] [R1] D2
BNX(R) R1,D2	[8C] [R1]D2

Function Description:
([D2 + (R1)]) --> PC
if f{N} then ([D2 + (R1)]) --> PC BNX(R)
Flags Modified: NONE
Remarks:

***** CALL TO REGISTER
***** CALL TO REGISTER CONDITIONAL

Assembler Syntax	Machine Code Format:
	Word 1
CA R1	[78] [R1]
CN R1	[90] [R1]

Function Description:
PC + 1 --> STACK   (R1) --> PC CA
if f{N} then (R1) --> STACK (R1) ---> PC CN
Flags Modified: NONE
Remarks:

***** CALL TO REGISTER IMMEDIATE
***** CALL TO REGISTER IMMEDIATE CONDITIONAL

Assembler Syntax	Machine Code Format:
	Word 1 Word 2
CAI(R) R1	[79] D1
CNI(R) R1	[91] D1

Function Description:
PC+1 --> Stack, D1 --> PC CAI(R)
if f{N} then PC+1 --> Stack, D1 --> PC CNI(R)
Flags Modified: NONE
Remarks:

***** CALL TO REGISTER INDIRECT
***** CALL TO REGISTER INDIRECT CONDITIONAL

Assembler Syntax	Machine Code Format:

	Word 1	Word 2
CAA(R) D1	[7A]	D1
CNA(R) D1	[92]	D1

Function Description:

PC+1 --> Stack, ([D1]) --> PC CAA(R)

if f{N} then PC+1 --> Stack, ([D1]) --> PC CNA(R)

Flags Modified: NONE

Remarks:

***** CALL TO REGISTER INDIRECT REGISTER

***** CALL TO REGISTER INDIRECT CONDITIONAL REGISTER

Assembler Syntax	Machine Code Format:
	Word 1
CAR R1	[7B] [R1]
CNR R1	[93] [R1]

Function Description:

PC+1 --> Stack, ([(R1)]) --> PC CNR

if f{N} then PC+1 --> Stack, ([(R1)]) --> PC CAR

Flags Modified: NONE

Remarks:

***** CALL TO REGISTER INDIRECT INDEXED

***** CALL TO REGISTER INDIRECT CONDITIONAL INDEXED

Assembler Syntax	Machine Code Format:
	Word 1 Word 2
CAX(R) R1,D2	[7C] [R1]D2
CNX(R) R1,D2	[94] [R1] D2

Function Description:

PC+1 --> Stack, ([D3 + (R2)]) --> PC CAX(R)

if f{N} then PC+1 --> Stack, ([D3 + (R2)]) --> PC CNX(R)

Flags Modified: NONE

Remarks:

***** RETURN

***** RETURN CONDITIONAL

Assembler Syntax	Machine Code Format:
	Word 1
R	[80]
RN	[98]

Function Description:

Stack --> PC R

if f{N} then Stack --> PC RN

Flags Modified: NONE

Remarks: The Return instructions cause the PC to be loaded with the address on top of the system stack. The RN instruction will cause a return if the condition N is satisfied.

Appendix E1. Sweet16 Macro Definition File

```
   (SW16MACRO.MAR - Source File)
   (SW16MACRO.MLB - Macro Library File)
   0000 1
;************************************************************************
   0000 2 ;        Address mode macros
   0000 3
;************************************************************************
   0000 4
   0000 5 ;        Register/Register address mode
   0000 6 ;        & Register Indirect
   0000 7          .macro moderr opnum,reg1,reg2
   0000 8          .word <<opnum>*<^x100>>!<<reg1>*<^x10>>!<reg2>
   0000 9          .endm
   0000 10
   0000 11 ; ******************************************************
   0000 12 ;        Immediate address mode
   0000 13 ;        & Absolute address mode
   0000 14 ;        & Indexed address mode
   0000 15
   0000 16          .macro moderd opnum,reg1,reg2,data
   0000 17          .word <<opnum>*<^x100>>!<<reg1>*<^x10>>!<reg2>
   0000 18          .word data
   0000 19          .endm
   0000 20
   0000 21
;************************************************************************
   0000 22 ;        Register/Register address mode instructions
   0000 23
;************************************************************************
   0000 24
   0000 25 ; Data Movement
   0000 26
   0000 27          .macro LOR reg1,reg2
   0000 28          moderr <^x10>,<reg1>,<reg2>
   0000 29          .endm
   0000 30
   0000 31 ; Arithmetic
   0000 32
   0000 33          .macro ADC reg1,reg2
   0000 34          moderr <^x20>,<reg1>,<reg2>
   0000 35          .endm
   0000 36
   0000 37          .macro ADD reg1,reg2
   0000 38          moderr <^xA0>,<reg1>,<reg2>
   0000 39          .endm
   0000 40
   0000 41          .macro SBC reg1,reg2
   0000 42          moderr <^x28>,<reg1>,<reg2>
   0000 43          .endm
   0000 44
   0000 45          .macro SB reg1,reg2
   0000 46          moderr <^xA8>,<reg1>,<reg2>
   0000 47          .endm
   0000 48
   0000 49          .macro CMP reg1,reg2
   0000 50          moderr <^x48>,<reg1>,<reg2>
   0000 51          .endm
   0000 52
   0000 53 ; Logical
   0000 54
   0000 55          .macro OR reg1,reg2
   0000 56          moderr <^x38>,<reg1>,<reg2>
   0000 57          .endm
```

```
0000 58
0000 59              .macro AND reg1,reg2
0000 60                moderr <^x30>,<reg1>,<reg2>
0000 61              .endm
0000 62
0000 63              .macro XOR reg1,reg2
0000 64                moderr <^x40>,<reg1>,<reg2>
0000 65              .endm
0000 66
0000 67
;*************************************************************************
0000 68 ;          Immediate address mode instructions
0000 69
;*************************************************************************
0000 70
0000 71 ; Data Movement
0000 72
0000 73              .macro LORI reg1,data
0000 74                moderd <^x11>,<reg1>,0,<data>
0000 75              .endm
0000 76
0000 77              .macro LORIR reg1,data
0000 78                moderd <^x11>,<reg1>,0,<data-begin>
0000 79              .endm
0000 80
0000 81              .macro STRI reg1,data
0000 82                moderd <^x19>,<reg1>,0,<data-begin>
0000 83              .endm
0000 84
0000 85 ; Arithmetic
0000 86
0000 87              .macro ADIC reg1,data
0000 88                moderd <^x21>,<reg1>,0,<data>
0000 89              .endm
0000 90
0000 91              .macro ADI reg1,data
0000 92                moderd <^xA1>,<reg1>,0,<data>
0000 93              .endm
0000 94
0000 95              .macro SBCI reg1,data
0000 96                moderd <^x29>,<reg1>,0,<data>
0000 97              .endm
0000 98
0000 99              .macro SBI reg1,data
0000 100               moderd <^xA9>,<reg1>,0,<data>
0000 101             .endm
0000 102
0000 103             .macro CMPI reg1,data
0000 104               moderd <^x49>,<reg1>,0,<data>
0000 105             .endm
0000 106
0000 107 ;Arithmetic Shifts
0000 108
0000 109             .macro SHLA reg1,data
0000 110               moderd <^x60>,<reg1>,0,<data>
0000 111             .endm
0000 112
0000 113             .macro SHRA reg1,data
0000 114               moderd <^x62>,<reg1>,0,<data>

0000 115             .endm
0000 116
0000 117 ; Logical
0000 118
0000 119             .macro ORI reg1,data
```

```
0000 120            moderd <^x39>,<reg1>,0,<data>
0000 121            .endm
0000 122
0000 123            .macro ANDI reg1,data
0000 124            moderd <^x31>,<reg1>,0,<data>
0000 125            .endm
0000 126
0000 127            .macro XORI reg1,data
0000 128            moderd <^x41>,<reg1>,0,<data>
0000 129            .endm
0000 130
0000 131 ;Special
0000 132            .macro INCI reg1,data
0000 133            moderd <^xA5>,<reg1>,0,<data>
0000 134            .endm
0000 135
0000 136            .macro DECI reg1,data
0000 137            moderd <^xAD>,<reg1>,0,<data>
0000 138            .endm
0000 139
0000 140 ;Logical Shifts
0000 141
0000 142            .macro SHLL reg1,data
0000 143            moderd <^x61>,<reg1>,0,<data>
0000 144            .endm
0000 145
0000 146            .macro SHRL reg1,data
0000 147            moderd <^x63>,<reg1>,0,<data>
0000 148            .endm
0000 149
0000 150 ;Rotates
0000 151            .macro ROLC reg1,data
0000 152            moderd <^x68>,<reg1>,0,<data>
0000 153            .endm
0000 154
0000 155            .macro ROL reg1,data
0000 156            moderd <^x69>,<reg1>,0,<data>
0000 157            .endm
0000 158
0000 159            .macro RORC reg1,data
0000 160            moderd <^x6A>,<reg1>,0,<data>
0000 161            .endm
0000 162
0000 163            .macro ROR reg1,data
0000 164            moderd <^x6B>,<reg1>,0,<data>
0000 165            .endm
0000 166
0000 167
;********************************************************************************
0000 168 ;     Absolute address mode instructions
0000 169
;********************************************************************************
0000 170
0000 171 ; Data Movement
0000 172
0000 173            .macro LORA reg1,data
0000 174            moderd <^x12>,<reg1>,0,<data>
0000 175            .endm
0000 176
0000 177            .macro LORAR reg1,data
0000 178            moderd <^x12>,<reg1>,0,<data-begin>
0000 179            .endm
0000 180
0000 181            .macro STRA reg1,data
0000 182            moderd <^x1A>,<reg1>,0,<data-begin>
0000 183            .endm
```

```
0000 184
0000 185 ; Arithmetic
0000 186
0000 187         .macro ADAC reg1,data
0000 188          moderd <^x22>,<reg1>,0,<data>
0000 189         .endm
0000 190
0000 191         .macro ADA reg1,data
0000 192          moderd <^xA2>,<reg1>,0,<data>
0000 193         .endm
0000 194
0000 195         .macro SBAC reg1,data
0000 196          moderd <^x2A>,<reg1>,0,<data>
0000 197         .endm
0000 198
0000 199         .macro SBA reg1,data
0000 200          moderd <^xAA>,<reg1>,0,<data>
0000 201         .endm
0000 202
0000 203         .macro CMPA reg1,data
0000 204          moderd <^x4A>,<reg1>,0,<data>
0000 205         .endm
0000 206
0000 207 ; Logical
0000 208
0000 209         .macro ORA reg1,data
0000 210          moderd <^x3A>,<reg1>,0,<data>
0000 211         .endm
0000 212
0000 213         .macro ANDA reg1,data
0000 214          moderd <^x32>,<reg1>,<data>
0000 215         .endm
0000 216
0000 217         .macro XORA reg1,data
0000 218          moderr <^x42>,<reg1>,<data>
0000 219         .endm
0000 220
0000 221
;*****************************************************************************
0000 222 ;        Register Indirect address mode instructions
0000 223
;*****************************************************************************
0000 224
0000 225 ; Data Movement
0000 226
0000 227         .macro LORR reg1,reg2
0000 228          moderr <^x13>,<reg1>,<reg2>
0000 229         .endm
0000 230
0000 231         .macro STRR reg1,reg2
0000 232          moderr <^x1B>,<reg1>,<reg2>
0000 233         .endm
0000 234
0000 235 ; Arithmetic
0000 236
0000 237         .macro ADRC reg1,reg2
0000 238          moderr <^x23>,<reg1>,<reg2>
0000 239         .endm
0000 240
0000 241         .macro ADR reg1,reg2
0000 242          moderr <^xA3>,<reg1>,<reg2>
0000 243         .endm
0000 244
0000 245         .macro SBRC reg1,reg2
0000 246          moderr <^x2B>,<reg1>,<reg2>
0000 247         .endm
```

```
0000 248
0000 249          .macro SBR reg1,reg2
0000 250           moderr <^xAB>,<reg1>,<reg2>
0000 251          .endm
0000 252
0000 253          .macro CMPR reg1,reg2
0000 254           moderr <^x4B>,<reg1>,<reg2>
0000 255          .endm
0000 256
0000 257 ; Logical
0000 258
0000 259          .macro ORR reg1,reg2
0000 260           moderr <^x3B>,<reg1>,<reg2>
0000 261          .endm
0000 262
0000 263          .macro ANDR reg1,reg2
0000 264           moderr <^x33>,<reg1>,<reg2>
0000 265          .endm
0000 266
0000 267          .macro XORB reg1,reg2
0000 268           moderr <^x43>,<reg1>,<reg2>
0000 269          .endm
0000 270
0000 271
0000 272
;******************************************************************************
0000 273 ;        Indexed address mode instructions
0000 274
;******************************************************************************
0000 275
0000 276 ; Data Movement
0000 277
0000 278          .macro LORX reg1,reg2,data
0000 279           moderd <^x14>,<reg1>,<reg2>,<data>
0000 280          .endm
0000 281
0000 282          .macro LORXR reg1,reg2,data
0000 283           moderd <^x14>,<reg1>,<reg2>,<data-begin>
0000 284          .endm
0000 285
0000 286          .macro STRX reg1,reg2,data
0000 287           moderd <^x1C>,<reg1>,<reg2>,<data-begin>
0000 288          .endm
0000 289
0000 290 ; Arithmetic
0000 291
0000 292          .macro ADXC reg1,reg2,data
0000 293           moderd <^x24>,<reg1>,<reg2>,<data>
0000 294          .endm
0000 295
0000 296          .macro ADX reg1,reg2,data
0000 297           moderd <^xA4>,<reg1>,<reg2>,<data>
0000 298          .endm
0000 299
0000 300          .macro SBXC reg1,reg2,data
0000 301           moderd <^x2C>,<reg1>,<reg2>,<data>
0000 302          .endm
0000 303
0000 304          .macro SBX reg1,reg2,data
0000 305           moderd <^xAC>,<reg1>,<reg2>,<data>
0000 306          .endm
0000 307
0000 308          .macro CMPX reg1,reg2,data
0000 309           moderd <^x4C>,<reg1>,<reg2>,<data>
0000 310          .endm
0000 311
```

```
0000 312 ; Logical
0000 313
0000 314         .macro ORX reg1,reg2,data
0000 315          moderd <^x3C>,<reg1>,<reg2>,<data>
0000 316         .endm
0000 317
0000 318         .macro ANDX reg1,reg2,data
0000 319          moderd <^x34>,<reg1>,<reg2>,<data>
0000 320         .endm
0000 321
0000 322         .macro XORX reg1,reg2,data
0000 323          moderr <^x44>,<reg1>,<reg2>,<data>
0000 324         .endm
0000 325
0000 326
;*********************************************************************
0000 327 ;       Implied address mode instructions
0000 328
;*********************************************************************
0000 329
0000 330 ;       Set flags in macro status register
0000 331         .macro SETN
0000 332          moderr <^x50>,0,1
0000 333         .endm
0000 334
0000 335         .macro SETV
0000 336          moderr <^x51>,0,1
0000 337         .endm
0000 338
0000 339         .macro SETC
0000 340          moderr <^x52>,0,1
0000 341         .endm
0000 342
0000 343         .macro SETZ
0000 344          moderr <^x53>,0,1
0000 345         .endm
0000 346
0000 347 ;       Clear flags in macro status register
0000 348         .macro CEN
0000 349          moderr <^x58>,0,0
0000 350         .endm
0000 351
0000 352         .macro CEV
0000 353          moderr <^x59>,0,0
0000 354         .endm
0000 355
0000 356         .macro CEC
0000 357          moderr <^x5A>,0,0
0000 358         .endm
0000 359
0000 360         .macro CEZ
0000 361          moderr <^x5B>,0,0
0000 362         .endm
0000 363
0000 364
;*********************************************************************
0000 365 ;       Decision making instructions
0000 366
;*********************************************************************
0000 367
0000 368         .macro BREG reg1
0000 369          moderr <^x70>,<reg1>,0
0000 370         .endm
0000 371
0000 372         .macro BI data
0000 373          moderd <^x71>,0,0,<data>
```

```
0000 374            .endm
0000 375
0000 376            .macro BIR data
0000 377             moderd <^x71>,0,0,<data-begin>
0000 378            .endm
0000 379
0000 380            .macro BA data
0000 381             moderd <^x72>,0,0,<data>
0000 382            .endm
0000 383
0000 384            .macro BAR data
0000 385             moderd <^x72>,0,0,<data-begin>
0000 386            .endm
0000 387
0000 388            .macro BARR reg1
0000 389             moderr <^x73>,<reg1>,0
0000 390            .endm
0000 391
0000 392            .macro BX reg1,data
0000 393             moderd <^x74>,<reg1>,0,<data>
0000 394            .endm
0000 395
0000 396            .macro BXR reg1,data
0000 397             moderd <^x74>,<reg1>,0,<data-begin>
0000 398            .endm
0000 399
0000 400
;***************************************************************************
0000 401 ; Conditional BRANCH Instructions
0000 402
;***************************************************************************
0000 403
0000 404            .macroBCCIRdata
0000 405             moderd <^x89>,0,0,<data-begin>
0000 406            .endm
0000 407
0000 408            .macroBCSIRdata
0000 409             moderd <^x89>,1,0,<data-begin>
0000 410            .endm
0000 411
0000 412            .macroBEQIRdata
0000 413             moderd <^x89>,2,0,<data-begin>
0000 414            .endm
0000 415
0000 416            .macroBGEIRdata
0000 417             moderd <^x89>,3,0,<data-begin>
0000 418            .endm
0000 419
0000 420            .macroBGTIRdata
0000 421             moderd <^x89>,4,0,<data-begin>
0000 422            .endm
0000 423
0000 424            .macroBHIIRdata
0000 425             moderd <^x89>,5,0,<data-begin>
0000 426            .endm
0000 427
0000 428            .macroBLEIRdata
0000 429             moderd <^x89>,6,0,<data-begin>
0000 430            .endm
0000 431
0000 432            .macroBLSIRdata
0000 433             moderd <^x89>,7,0,<data-begin>
0000 434            .endm
0000 435
0000 436            .macroBLTIRdata
0000 437             moderd <^x89>,8,0,<data-begin>
```

```
0000 438            .endm
0000 439
0000 440            .macroBMIIRdata
0000 441             moderd <^x89>,9,0,<data-begin>
0000 442            .endm
0000 443
0000 444            .macroBNEIRdata
0000 445             moderd <^x89>,10,0,<data-begin>
0000 446            .endm
0000 447
0000 448            .macroBPLIRdata
0000 449             moderd <^x89>,11,0,<data-begin>
0000 450            .endm
0000 451
0000 452            .macroBVCIRdata
0000 453             moderd <^x89>,12,0,<data-begin>
0000 454            .endm
0000 455
0000 456            .macroBVSIRdata
0000 457             moderd <^x89>,13,0,<data-begin>
0000 458            .endm
0000 459
0000 460
0000 461
;**************************************************************************
0000 462 ;       Subroutine CALL instructions
0000 463
;**************************************************************************
0000 464
0000 465            .macro CA reg1
0000 466             moderr <^x78>,<reg1>,0
0000 467            .endm
0000 468
0000 469            .macro CAI data
0000 470             moderd <^x79>,0,0,<data>
0000 471            .endm
0000 472
0000 473            .macro CAIR data
0000 474             moderd <^x79>,0,0,<data-begin>
0000 475            .endm
0000 476
0000 477            .macro CAA data
0000 478             moderd <^x7A>,0,0,<data>
0000 479            .endm
0000 480
0000 481            .macro CAAR data
0000 482             moderd <^x7A>,0,0,<data-begin>
0000 483            .endm
0000 484
0000 485            .macro CAR reg1
0000 486             moderr <^x7B>,<reg1>,0
0000 487            .endm
0000 488
0000 489            .macro CAX reg1,data
0000 490             moderd <^x7C>,<reg1>,0,<data>
0000 491            .endm
0000 492
0000 493            .macro CAXR reg1,data
0000 494             moderd <^x7C>,<reg1>,0,<data-begin>
0000 495            .endm
0000 496
0000 497
;**************************************************************************
0000 498 ;       Subroutine RETURN instructions
0000 499
;**************************************************************************
```

```
0000 500
0000 501 ;Unconditional RETURN
0000 502
0000 503         .macro R
0000 504          moderr <^x80>,0,0
0000 505         .endm
0000 506
0000 507 ;Unimplemented opcode to force simulator termination
0000 508
0000 509         .macro GFO
0000 510          moderr <^xFF>,0,0
0000 511         .endm
0000 512
0000 513 ;Conditional RETURN
0000 514
0000 515         .macro RCC
0000 516          moderr <^x98>,0,0
0000 517         .endm
0000 518
0000 519         .macro RCS
0000 520          moderr <^x98>,1,0
0000 521         .endm
0000 522
0000 523         .macro REQ
0000 524          moderr <^x98>,2,0
0000 525         .endm
0000 526
0000 527         .macro RGE
0000 528          moderr <^x98>,3,0
0000 529         .endm
0000 530
0000 531         .macro RGT
0000 532          moderr <^x98>,4,0
0000 533         .endm
0000 534
0000 535         .macro RHI
0000 536          moderr <^x98>,5,0
0000 537         .endm
0000 538
0000 539         .macro RLE
0000 540          moderr <^x98>,6,0
0000 541         .endm
0000 542
0000 543         .macro RLS
0000 544          moderr <^x98>,7,0
0000 545         .endm
0000 546
0000 547         .macro RLT
0000 548          moderr <^x98>,8,0
0000 549         .endm
0000 550
0000 551         .macro RMI
0000 552          moderr <^x98>,9,0
0000 553         .endm
0000 554
0000 555         .macro RNE
0000 556          moderr <^x98>,10,0
0000 557         .endm
0000 558
0000 559         .macro RPL
0000 560          moderr <^x98>,11,0
0000 561         .endm
0000 562
0000 563         .macro RVC
0000 564          moderr <^x98>,12,0
0000 565         .endm
```

```
0000 566
0000 567          .macro RVS
0000 568           moderr <^x98>,13,0
0000 569          .endm
0000 570 ;
0000 571 ; End of macro definition file for SWEET16
0000 572 ; In order to use this file with the VAX MACRO assembler, a SWEET16
0000 573 ; assembler source code file will be concatenated to this file at this
0000 574 ; point.
0000 575 ;
0000 576 ; eg. MACRO/LIST SW16MACRO+SW16MON
```

Appendix E2. Sweet16 Monitor Program

Assembled Using the Sweet16 Macro Library (SW16MACRO.MLB)

```
0000  1 ;**************************************************
0000  2 ;
0000  3 ; MONITOR program
0000  4 ;
0000  5 ; logic by : Bruce Kleinman and Brian Miller
0000  6 ; code by : Brian Miller
0000  7 ; date : 12/11/85
0000  8 ; adapted for VAX MACRO by M. Lynch 1/88
0000  9 ;
0000 10 ; Initialize assembler for this target system space
0000 11        .show expansions
00000000 12    .psectPROG,pic,ovr,rel,gbl,shr,noexe,rd,wrt,long
0000 13 ; Tell assembler about register names
0000000E 0000 14 msp = 14
0000000F 0000 15 mpc = 15
0000 16 ;
0000FF00 0000 17 SER = ^xFF00
0000FF02 0000 18 SENSE = ^xFF02
00000044 0000 19 ASCD = ^x0044
00000047 0000 20 ASCG = ^x0047
0000004D 0000 21 ASCM = ^x004D
0000004E 0000 22 ASCN = ^x004E
00000052 0000 23 ASCR = ^x0052
0000 24 ;
0000 25 ; Get boundary alignment
00000000 0000 26 .blkw0
0000 27 ;
0000 28 ;
0000 29 begin: LORI 14,<^x1FC0>
0000            moderd <^x11>,<14>,0,<^x1FC0>
11E0 0000       .word<<^x11>*<^x100>>!<<14>*<^x10>>!<0>
1FC0 0002       .word^x1FC0
0004
0004
0004 30 MAIN: LORI 4,<^x0>
0004            moderd <^x11>,<4>,0,<^x0>
1140 0004       .word<<^x11>*<^x100>>!<<4>*<^x10>>!<0>
0000 0006       .word^x0
0008
0008
0008 31 LORIR 13,RDY
0008            moderd <^x11>,<13>,0,<RDY-begin>
11D0 0008       .word<<^x11>*<^x100>>!<<13>*<^x10>>!<0>
0242' 000A      .wordRDY-begin
000C
000C
000C 32 CAIR PRINT
000C            moderd <^x79>,0,0,<PRINT-begin>
7900 000C       .word<<^x79>*<^x100>>!<<0>*<^x10>>!<0>
01FA' 000E      .wordPRINT-begin
0010
0010
0010 33 MLOP: LORA 9,SENSE
0010            moderd <^x12>,<9>,0,<SENSE>
1290 0010       .word<<^x12>*<^x100>>!<<9>*<^x10>>!<0>
FF02 0012       .wordSENSE
0014
0014
0014 34 ANDI 9,<^x06>
0014            moderd <^x31>,<9>,0,<^x06>
3190 0014       .word<<^x31>*<^x100>>!<<9>*<^x10>>!<0>
```

```
0006 0016          .word^x06
0018
0018
0018 35 CMPI 9,<^x06>
0018               moderd <^x49>,<9>,0,<^x06>
4990 0018          .word<<^x49>*<^x100>>!<<9>*<^x10>>!<0>
0006 001A          .word^x06
001C
001C
001C 36 BNEIR MLOP
001C               moderd <^x89>,10,0,<MLOP-begin>
89A0 001C          .word<<^x89>*<^x100>>!<<10>*<^x10>>!<0>
0010 001E          .wordMLOP-begin
0020
0020
0020 37 LORA 9,SER
0020               moderd <^x12>,<9>,0,<SER>
1290 0020          .word<<^x12>*<^x100>>!<<9>*<^x10>>!<0>
FF00 0022          .wordSER
0024
0024
0024 38 CMPI 9,ASCD
0024               moderd <^x49>,<9>,0,<ASCD>
4990 0024          .word<<^x49>*<^x100>>!<<9>*<^x10>>!<0>
0044 0026          .wordASCD
0028
0028
0028 39 BNEIR CHREG
0028               moderd <^x89>,10,0,<CHREG-begin>
89A0 0028          .word<<^x89>*<^x100>>!<<10>*<^x10>>!<0>
0034' 002A         .wordCHREG-begin
002C
002C
002C 40 CAIR DISP
002C               moderd <^x79>,0,0,<DISP-begin>
7900 002C          .word<<^x79>*<^x100>>!<<0>*<^x10>>!<0>
006C' 002E         .wordDISP-begin
0030
0030
0030 41 BIR MAIN
0030               moderd <^x71>,0,0,<MAIN-begin>
7100 0030          .word<<^x71>*<^x100>>!<<0>*<^x10>>!<0>
0004 0032          .wordMAIN-begin
0034
0034
0034 42 CHREG: CMPI 9,ASCR
0034               moderd <^x49>,<9>,0,<ASCR>
4990 0034          .word<<^x49>*<^x100>>!<<9>*<^x10>>!<0>
0052 0036          .wordASCR
0038
0038
0038 43 BNEIR CHMEM
0038               moderd <^x89>,10,0,<CHMEM-begin>
89A0 0038          .word<<^x89>*<^x100>>!<<10>*<^x10>>!<0>
0044' 003A         .wordCHMEM-begin
003C
003C
003C 44 CAIR REG
003C               moderd <^x79>,0,0,<REG-begin>
7900 003C          .word<<^x79>*<^x100>>!<<0>*<^x10>>!<0>
009C' 003E         .wordREG-begin
0040
0040
0040 45 BIR MAIN
0040               moderd <^x71>,0,0,<MAIN-begin>
7100 0040          .word<<^x71>*<^x100>>!<<0>*<^x10>>!<0>
```

```
0004 0042          .wordMAIN-begin
0044
0044
0044 46 CHMEM: CMPI 9,ASCM
0044               moderd <^x49>,<9>,0,<ASCM>
4990 0044          .word<<^x49>*<^x100>>!<<9>*<^x10>>!<0>
004D 0046          .wordASCM
0048
0048
0048 47 BNEIR CHGT
0048               moderd <^x89>,10,0,<CHGT-begin>
89A0 0048          .word<<^x89>*<^x100>>!<<10>*<^x10>>!<0>
0054' 004A         .wordCHGT-begin
004C
004C
004C 48 CAIR MEM
004C               moderd <^x79>,0,0,<MEM-begin>
7900 004C          .word<<^x79>*<^x100>>!<<0>*<^x10>>!<0>
0172' 004E         .wordMEM-begin
0050
0050
0050 49 BIR MAIN
0050               moderd <^x71>,0,0,<MAIN-begin>
7100 0050          .word<<^x71>*<^x100>>!<<0>*<^x10>>!<0>
0004 0052          .wordMAIN-begin
0054
0054
0054 50 CHGT: CMPI 9,ASCG
0054               moderd <^x49>,<9>,0,<ASCG>
4990 0054          .word<<^x49>*<^x100>>!<<9>*<^x10>>!<0>
0047 0056          .wordASCG
0058
0058
0058 51 BNEIR FINISH
0058               moderd <^x89>,10,0,<FINISH-begin>
89A0 0058          .word<<^x89>*<^x100>>!<<10>*<^x10>>!<0>
006A' 005A         .wordFINISH-begin
005C
005C
005C 52 LORI 5,<^x00>
005C               moderd <^x11>,<5>,0,<^x00>
1150 005C          .word<<^x11>*<^x100>>!<<5>*<^x10>>!<0>
0000 005E          .word^x00
0060
0060
0060 53 CAIR RED
0060               moderd <^x79>,0,0,<RED-begin>
7900 0060          .word<<^x79>*<^x100>>!<<0>*<^x10>>!<0>
018C' 0062         .wordRED-begin
0064
0064
0064 54 BREG 6
0064               moderr <^x70>,<6>,0
7060 0064 .word <<^x70>*<^x100>>!<<6>*<^x10>>!<0>
0066
0066
0066 55 BIR MAIN
0066               moderd <^x71>,0,0,<MAIN-begin>
7100 0066          .word<<^x71>*<^x100>>!<<0>*<^x10>>!<0>
0004 0068          .wordMAIN-begin
006A
006A
006A 56 FINISH: GFO
006A               moderr <^xFF>,0,0
FF00 006A .word <<^xFF>*<^x100>>!<<0>*<^x10>>!<0>
006C
```

```
006C
006C 57 ;
006C 58 ;
006C 59 ;
006C 60 DISP: LORI 5,<^x0>
006C            moderd <^x11>,<5>,0,<^x0>
1150 006C      .word<<^x11>*<^x100>>!<<5>*<^x10>>!<0>
0000 006E      .word^x0
0070
0070
0070 61 CAIR RED
0070            moderd <^x79>,0,0,<RED-begin>
7900 0070      .word<<^x79>*<^x100>>!<<0>*<^x10>>!<0>
018C' 0072     .wordRED-begin
0074
0074
0074 62 LORI 8,<^x10>
0074            moderd <^x11>,<8>,0,<^x10>
1180 0074      .word<<^x11>*<^x100>>!<<8>*<^x10>>!<0>
0010 0076      .word^x10
0078
0078
0078 63 DLOP: LORR 11,6
0078            moderr <^x13>,<11>,<6>
13B6 0078 .word <<^x13>*<^x100>>!<<11>*<^x10>>!<6>
007A
007A
007A 64 CAIR RPR
007A            moderd <^x79>,0,0,<RPR-begin>
7900 007A      .word<<^x79>*<^x100>>!<<0>*<^x10>>!<0>
0212' 007C     .wordRPR-begin
007E
007E
007E 65 LORIR 13,ONESP
007E            moderd <^x11>,<13>,0,<ONESP-begin>
11D0 007E      .word<<^x11>*<^x100>>!<<13>*<^x10>>!<0>
0266' 0080     .wordONESP-begin
0082
0082
0082 66 CAIR PRINT
0082            moderd <^x79>,0,0,<PRINT-begin>
7900 0082      .word<<^x79>*<^x100>>!<<0>*<^x10>>!<0>
01FA' 0084     .wordPRINT-begin
0086
0086
0086 67 ADI 6,<^x02>
0086            moderd <^xA1>,<6>,0,<^x02>
A160 0086      .word<<^xA1>*<^x100>>!<<6>*<^x10>>!<0>
0002 0088      .word^x02
008A
008A
008A 68 SBI 8,<^x01>
008A            moderd <^xA9>,<8>,0,<^x01>
A980 008A      .word<<^xA9>*<^x100>>!<<8>*<^x10>>!<0>
0001 008C      .word^x01
008E
008E
008E 69 BNEIR DLOP
008E            moderd <^x89>,10,0,<DLOP-begin>
89A0 008E      .word<<^x89>*<^x100>>!<<10>*<^x10>>!<0>
0078 0090      .wordDLOP-begin
0092
0092
0092 70 LORIR 13,CRLF
0092            moderd <^x11>,<13>,0,<CRLF-begin>
11D0 0092      .word<<^x11>*<^x100>>!<<13>*<^x10>>!<0>
```

```
026A' 0094          .wordCRLF-begin
0096
0096
0096 71 CAIR PRINT
0096                moderd <^x79>,0,0,<PRINT-begin>
7900 0096          .word<<^x79>*<^x100>>!<<0>*<^x10>>!<0>
01FA' 0098          .wordPRINT-begin
009A
009A
009A 72 R
009A                moderr <^x80>,0,0
8000 009A .word <<^x80>*<^x100>>!<<0>*<^x10>>!<0>
009C
009C
009C 73 ;
009C 74 ;
009C 75 ;
009C 76 REG: LORI 5,<^x00>
009C                moderd <^x11>,<5>,0,<^x00>
1150 009C          .word<<^x11>*<^x100>>!<<5>*<^x10>>!<0>
0000 009E          .word^x00
00A0
00A0
00A0 77 CAIR RED
00A0                moderd <^x79>,0,0,<RED-begin>
7900 00A0          .word<<^x79>*<^x100>>!<<0>*<^x10>>!<0>
018C' 00A2          .wordRED-begin
00A4
00A4
00A4 78 NR0: CMPI 6,<^x00>
00A4                moderd <^x49>,<6>,0,<^x00>
4960 00A4          .word<<^x49>*<^x100>>!<<6>*<^x10>>!<0>
0000 00A6          .word^x00
00A8
00A8
00A8 79 BNEIR NR1
00A8                moderd <^x89>,10,0,<NR1-begin>
89A0 00A8          .word<<^x89>*<^x100>>!<<10>*<^x10>>!<0>
00C0' 00AA          .wordNR1-begin
00AC
00AC
00AC 80 LOR 11,0
00AC                moderr <^x10>,<11>,<0>
10B0 00AC .word <<^x10>*<^x100>>!<<11>*<^x10>>!<0>
00AE
00AE
00AE 81 CAIR LODX
00AE                moderd <^x79>,0,0,<LODX-begin>
7900 00AE          .word<<^x79>*<^x100>>!<<0>*<^x10>>!<0>
0116' 00B0          .wordLODX-begin
00B2
00B2
00B2 82 CMPI 5,<^x01>
00B2                moderd <^x49>,<5>,0,<^x01>
4950 00B2          .word<<^x49>*<^x100>>!<<5>*<^x10>>!<0>
0001 00B4          .word^x01
00B6
00B6
00B6 83 BNEIR REGEND
00B6                moderd <^x89>,10,0,<REGEND-begin>
89A0 00B6          .word<<^x89>*<^x100>>!<<10>*<^x10>>!<0>
0114' 00B8          .wordREGEND-begin
00BA
00BA
00BA 84 LOR 0,7
00BA                moderr <^x10>,<0>,<7>
```

```
1007 00BA   .word <<<^x10>*<^x100>>!<<0>*<^x10>>!<7>
     00BC
     00BC
     00BC 85 BIR REGEND
     00BC        moderd <^x71>,0,0,<REGEND-begin>
7100 00BC        .word<<<^x71>*<^x100>>!<<0>*<^x10>>!<0>
0114' 00BE       .wordREGEND-begin
     00C0
     00C0
     00C0 86 NR1: CMPI 6,<^x01>
     00C0        moderd <^x49>,<6>,0,<^x01>
4960 00C0        .word<<<^x49>*<^x100>>!<<6>*<^x10>>!<0>
0001 00C2        .word^x01
     00C4
     00C4
     00C4 87 BNEIR NR2
     00C4        moderd <^x89>,10,0,<NR2-begin>
89A0 00C4        .word<<<^x89>*<^x100>>!<<10>*<^x10>>!<0>
00DC' 00C6       .wordNR2-begin
     00C8
     00C8
     00C8 88 LOR 11,1
     00C8        moderr <^x10>,<11>,<1>
10B1 00C8 .word <<<^x10>*<^x100>>!<<11>*<^x10>>!<1>
     00CA
     00CA
     00CA 89 CAIR LODX
     00CA        moderd <^x79>,0,0,<LODX-begin>
7900 00CA        .word<<<^x79>*<^x100>>!<<0>*<^x10>>!<0>
0116' 00CC       .wordLODX-begin
     00CE
     00CE
     00CE 90 CMPI 5,<^x01>
     00CE        moderd <^x49>,<5>,0,<^x01>
4950 00CE        .word<<<^x49>*<^x100>>!<<5>*<^x10>>!<0>
0001 00D0        .word^x01
     00D2
     00D2
     00D2 91 BNEIR REGEND
     00D2        moderd <^x89>,10,0,<REGEND-begin>
89A0 00D2        .word<<<^x89>*<^x100>>!<<10>*<^x10>>!<0>
0114' 00D4       .wordREGEND-begin
     00D6
     00D6
     00D6 92 LOR 1,7
     00D6        moderr <^x10>,<1>,<7>
1017 00D6 .word <<<^x10>*<^x100>>!<<1>*<^x10>>!<7>
     00D8
     00D8
     00D8 93 BIR REGEND
     00D8        moderd <^x71>,0,0,<REGEND-begin>
7100 00D8        .word<<<^x71>*<^x100>>!<<0>*<^x10>>!<0>
0114' 00DA       .wordREGEND-begin
     00DC
     00DC
     00DC 94 NR2: CMPI 6,<^x02>
     00DC        moderd <^x49>,<6>,0,<^x02>
4960 00DC        .word<<<^x49>*<^x100>>!<<6>*<^x10>>!<0>
0002 00DE        .word^x02
     00E0
     00E0
     00E0 95 BNEIR NR3
     00E0        moderd <^x89>,10,0,<NR3-begin>
89A0 00E0        .word<<<^x89>*<^x100>>!<<10>*<^x10>>!<0>
00F8' 00E2       .wordNR3-begin
     00E4
```

```
00E4
00E4 96 LOR 11,2
00E4            moderr <^x10>,<11>,<2>
10B2 00E4 .word <<^x10>*<^x100>>!<<11>*<^x10>>!<2>
00E6
00E6
00E6 97 CAIR LODX
00E6            moderd <^x79>,0,0,<LODX-begin>
7900 00E6      .word<<^x79>*<^x100>>!<<0>*<^x10>>!<0>
0116' 00E8     .wordLODX-begin
00EA
00EA
00EA 98 CMPI 5,<^x01>
00EA            moderd <^x49>,<5>,0,<^x01>
4950 00EA      .word<<^x49>*<^x100>>!<<5>*<^x10>>!<0>
0001 00EC      .word^x01
00EE
00EE
00EE 99 BNEIR REGEND
00EE            moderd <^x89>,10,0,<REGEND-begin>
89A0 00EE      .word<<^x89>*<^x100>>!<<10>*<^x10>>!<0>
0114' 00F0     .wordREGEND-begin
00F2
00F2
00F2 100 LOR 2,7
00F2            moderr <^x10>,<2>,<7>
1027 00F2 .word <<^x10>*<^x100>>!<<2>*<^x10>>!<7>
00F4
00F4
00F4 101 BIR REGEND
00F4            moderd <^x71>,0,0,<REGEND-begin>
7100 00F4      .word<<^x71>*<^x100>>!<<0>*<^x10>>!<0>
0114' 00F6     .wordREGEND-begin
00F8
00F8
00F8 102 NR3: CMPI 6,<^x03>
00F8            moderd <^x49>,<6>,0,<^x03>
4960 00F8      .word<<^x49>*<^x100>>!<<6>*<^x10>>!<0>
0003 00FA      .word^x03
00FC
00FC
00FC 103 BNEIR REGEND
00FC            moderd <^x89>,10,0,<REGEND-begin>
89A0 00FC      .word<<^x89>*<^x100>>!<<10>*<^x10>>!<0>
0114' 00FE     .wordREGEND-begin
0100
0100
0100 104 LOR 11,3
0100            moderr <^x10>,<11>,<3>
10B3 0100 .word <<^x10>*<^x100>>!<<11>*<^x10>>!<3>
0102
0102
0102 105 CAIR LODX
0102            moderd <^x79>,0,0,<LODX-begin>
7900 0102      .word<<^x79>*<^x100>>!<<0>*<^x10>>!<0>
0116' 0104     .wordLODX-begin
0106
0106
0106 106 CMPI 5,<^x01>
0106            moderd <^x49>,<5>,0,<^x01>
4950 0106      .word<<^x49>*<^x100>>!<<5>*<^x10>>!<0>
0001 0108      .word^x01
010A
010A
010A 107 BNEIR REGEND
010A            moderd <^x89>,10,0,<REGEND-begin>
```

```
89A0 010A          .word<<<^x89>*<^x100>>!<<10>*<^x10>>!<0>
0114' 010C         .wordREGEND-begin
010E
010E
010E 108 LOR 3,7
010E               moderr <^x10>,<3>,<7>
1037 010E .word <<<^x10>*<^x100>>!<<3>*<^x10>>!<7>
0110
0110
0110 109 BIR REGEND
0110               moderd <^x71>,0,0,<REGEND-begin>
7100 0110          .word<<<^x71>*<^x100>>!<<0>*<^x10>>!<0>
0114' 0112         .wordREGEND-begin
0114
0114
0114 110 REGEND: R
0114               moderr <^x80>,0,0
8000 0114 .word <<<^x80>*<^x100>>!<<0>*<^x10>>!<0>
0116
0116
0116 111 ;
0116 112 ;
0116 113 ;
0116 114 LODX: CAIR RPR
0116               moderd <^x79>,0,0,<RPR-begin>
7900 0116          .word<<<^x79>*<^x100>>!<<0>*<^x10>>!<0>
0212' 0118         .wordRPR-begin
011A
011A
011A 115 LORIR 13,CNG
011A               moderd <^x11>,<13>,0,<CNG-begin>
11D0 011A          .word<<<^x11>*<^x100>>!<<13>*<^x10>>!<0>
0252' 011C         .wordCNG-begin
011E
011E
011E 116 CAIR PRINT
011E               moderd <^x79>,0,0,<PRINT-begin>
7900 011E          .word<<<^x79>*<^x100>>!<<0>*<^x10>>!<0>
01FA' 0120         .wordPRINT-begin
0122
0122
0122 117 WAIT1: LORA 11,SENSE
0122               moderd <^x12>,<11>,0,<SENSE>
12B0 0122          .word<<<^x12>*<^x100>>!<<11>*<^x10>>!<0>
FF02 0124          .wordSENSE
0126
0126
0126 118 ANDI 11,<^x06>
0126               moderd <^x31>,<11>,0,<^x06>
31B0 0126          .word<<<^x31>*<^x100>>!<<11>*<^x10>>!<0>
0006 0128          .word^x06
012A
012A
012A 119 CMPI 11,<^x06>
012A               moderd <^x49>,<11>,0,<^x06>
49B0 012A          .word<<<^x49>*<^x100>>!<<11>*<^x10>>!<0>
0006 012C          .word^x06
012E
012E
012E 120 BNEIR WAIT1
012E               moderd <^x89>,10,0,<WAIT1-begin>
89A0 012E          .word<<<^x89>*<^x100>>!<<10>*<^x10>>!<0>
0122 0130          .wordWAIT1-begin
0132
0132
0132 121 LORA 11,SER
```

```
0132              moderd <^x12>,<11>,0,<SER>
12B0 0132         .word<<^x12>*<^x100>>!<<11>*<^x10>>!<0>
FF00 0134         .wordSER
0136
0136
0136 122 WAIT2: LORA 5,SENSE
0136              moderd <^x12>,<5>,0,<SENSE>
1250 0136         .word<<^x12>*<^x100>>!<<5>*<^x10>>!<0>
FF02 0138         .wordSENSE
013A
013A
013A 123 ANDI 5,<^x06>
013A              moderd <^x31>,<5>,0,<^x06>
3150 013A         .word<<^x31>*<^x100>>!<<5>*<^x10>>!<0>
0006 013C         .word^x06
013E
013E
013E 124 CMPI 5,<^x06>
013E              moderd <^x49>,<5>,0,<^x06>
4950 013E         .word<<^x49>*<^x100>>!<<5>*<^x10>>!<0>
0006 0140         .word^x06
0142
0142
0142 125 BNEIR WAIT2
0142              moderd <^x89>,10,0,<WAIT2-begin>
89A0 0142         .word<<^x89>*<^x100>>!<<10>*<^x10>>!<0>
0136 0144         .wordWAIT2-begin
0146
0146
0146 126 LORA 5,SER
0146              moderd <^x12>,<5>,0,<SER>
1250 0146         .word<<^x12>*<^x100>>!<<5>*<^x10>>!<0>
FF00 0148         .wordSER
014A
014A
014A 127 CMPI 11,ASCN
014A              moderd <^x49>,<11>,0,<ASCN>
49B0 014A         .word<<^x49>*<^x100>>!<<11>*<^x10>>!<0>
004E 014C         .wordASCN
014E
014E
014E 128 BNEIR YES
014E              moderd <^x89>,10,0,<YES-begin>
89A0 014E         .word<<^x89>*<^x100>>!<<10>*<^x10>>!<0>
0160' 0150        .wordYES-begin
0152
0152
0152 129 LORI 5,<^x02>
0152              moderd <^x11>,<5>,0,<^x02>
1150 0152         .word<<^x11>*<^x100>>!<<5>*<^x10>>!<0>
0002 0154         .word^x02
0156
0156
0156 130 LORIR 13,CRLF
0156              moderd <^x11>,<13>,0,<CRLF-begin>
11D0 0156         .word<<^x11>*<^x100>>!<<13>*<^x10>>!<0>
026A' 0158        .wordCRLF-begin
015A
015A
015A 131 CAIR PRINT
015A              moderd <^x79>,0,0,<PRINT-begin>
7900 015A         .word<<^x79>*<^x100>>!<<0>*<^x10>>!<0>
01FA' 015C        .wordPRINT-begin
015E
015E
015E 132 R
```

```
015E              moderr <^x80>,0,0
8000 015E .word <<<^x80>*<^x100>>!<<0>*<^x10>>!<0>
0160
0160
0160 133 YES: LORI 5,<^x01>
0160              moderd <^x11>,<5>,0,<^x01>
1150 0160         .word<<<^x11>*<^x100>>!<<5>*<^x10>>!<0>
0001 0162         .word^x01
0164
0164
0164 134 CAIR RED
0164              moderd <^x79>,0,0,<RED-begin>
7900 0164         .word<<<^x79>*<^x100>>!<<0>*<^x10>>!<0>
018C' 0166        .wordRED-begin
0168
0168
0168 135 LORIR 13,CRLF
0168              moderd <^x11>,<13>,0,<CRLF-begin>
11D0 0168         .word<<<^x11>*<^x100>>!<<13>*<^x10>>!<0>
026A' 016A        .wordCRLF-begin
016C
016C
016C 136 CAIR PRINT
016C              moderd <^x79>,0,0,<PRINT-begin>
7900 016C         .word<<<^x79>*<^x100>>!<<0>*<^x10>>!<0>
01FA' 016E        .wordPRINT-begin
0170
0170
0170 137 R
0170              moderr <^x80>,0,0
8000 0170 .word <<<^x80>*<^x100>>!<<0>*<^x10>>!<0>
0172
0172
0172 138 ;
0172 139 ;
0172 140 ;
0172 141 MEM: LORI 5,<^x00>
0172              moderd <^x11>,<5>,0,<^x00>
1150 0172         .word<<<^x11>*<^x100>>!<<5>*<^x10>>!<0>
0000 0174         .word^x00
0176
0176
0176 142 CAIR RED
0176              moderd <^x79>,0,0,<RED-begin>
7900 0176         .word<<<^x79>*<^x100>>!<<0>*<^x10>>!<0>
018C' 0178        .wordRED-begin
017A
017A
017A 143 LORR 11,6
017A              moderr <^x13>,<11>,<6>
13B6 017A .word <<<^x13>*<^x100>>!<<11>*<^x10>>!<6>
017C
017C
017C 144 CAIR LODX
017C              moderd <^x79>,0,0,<LODX-begin>
7900 017C         .word<<<^x79>*<^x100>>!<<0>*<^x10>>!<0>
0116 017E         .wordLODX-begin
0180
0180
0180 145 CMPI 5,<^x01>
0180              moderd <^x49>,<5>,0,<^x01>
4950 0180         .word<<<^x49>*<^x100>>!<<5>*<^x10>>!<0>
0001 0182         .word^x01
0184
0184
0184 146 BNEIR MEMEND
```

```
0184               moderd <^x89>,10,0,<MEMEND-begin>
89A0 0184          .word<<^x89>*<^x100>>!<<10>*<^x10>>!<0>
018A' 0186         .wordMEMEND-begin
0188
0188
0188 147 STRR 6,7
0188               moderr <^x1B>,<6>,<7>
1B67 0188 .word <<^x1B>*<^x100>>!<<6>*<^x10>>!<7>
018A
018A
018A 148 MEMEND: R
018A               moderr <^x80>,0,0
8000 018A .word <<^x80>*<^x100>>!<<0>*<^x10>>!<0>
018C
018C
018C 149 ;
018C 150 ;
018C 151 ;
018C 152 RED: LORI 12,<^x1FA0>
018C               moderd <^x11>,<12>,0,<^x1FA0>
11C0 018C          .word<<^x11>*<^x100>>!<<12>*<^x10>>!<0>
1FA0 018E          .word^x1FA0
0190
0190
0190 153 RLOOP: LORA 11,SENSE
0190               moderd <^x12>,<11>,0,<SENSE>
12B0 0190          .word<<^x12>*<^x100>>!<<11>*<^x10>>!<0>
FF02 0192          .wordSENSE
0194
0194
0194 154 ANDI 11,<^x06>
0194               moderd <^x31>,<11>,0,<^x06>
31B0 0194          .word<<^x31>*<^x100>>!<<11>*<^x10>>!<0>
0006 0196          .word^x06
0198
0198
0198 155 CMPI 11,<^x06>
0198               moderd <^x49>,<11>,0,<^x06>
49B0 0198          .word<<^x49>*<^x100>>!<<11>*<^x10>>!<0>
0006 019A          .word^x06
019C
019C
019C 156 BNEIR RLOOP
019C               moderd <^x89>,10,0,<RLOOP-begin>
89A0 019C          .word<<^x89>*<^x100>>!<<10>*<^x10>>!<0>
0190 019E          .wordRLOOP-begin
01A0
01A0
01A0 157 LORA 11,SER
01A0               moderd <^x12>,<11>,0,<SER>
12B0 01A0          .word<<^x12>*<^x100>>!<<11>*<^x10>>!<0>
FF00 01A2          .wordSER
01A4
01A4
01A4 158 CMPI 11,<^x0>
01A4               moderd <^x49>,<11>,0,<^x0>
49B0 01A4          .word<<^x49>*<^x100>>!<<11>*<^x10>>!<0>
0000 01A6          .word^x0
01A8
01A8
01A8 159 BEQIR INPT
01A8               moderd <^x89>,2,0,<INPT-begin>
8920 01A8          .word<<^x89>*<^x100>>!<<2>*<^x10>>!<0>
01B6' 01AA         .wordINPT-begin
01AC
01AC
```

```
01AC 160 STRR 12,11
01AC              moderr <^x1B>,<12>,<11>
1BCB 01AC .word <<<^x1B>*<^x100>>!<<12>*<^x10>>!<11>
01AE
01AE
01AE 161 ADI 12,<^x02>
01AE              moderd <^xA1>,<12>,0,<^x02>
A1C0 01AE        .word<<^xA1>*<^x100>>!<<12>*<^x10>>!<0>
0002 01B0        .word^x02
01B2
01B2
01B2 162 BIR RLOOP
01B2              moderd <^x71>,0,0,<RLOOP-begin>
7100 01B2        .word<<^x71>*<^x100>>!<<0>*<^x10>>!<0>
0190 01B4        .wordRLOOP-begin
01B6
01B6
01B6 163 INPT: SBI 12,<^x08>
01B6              moderd <^xA9>,<12>,0,<^x08>
A9C0 01B6        .word<<^xA9>*<^x100>>!<<12>*<^x10>>!<0>
0008 01B8        .word^x08
01BA
01BA
01BA 164 LORI 13,<^x00>
01BA              moderd <^x11>,<13>,0,<^x00>
11D0 01BA        .word<<^x11>*<^x100>>!<<13>*<^x10>>!<0>
0000 01BC        .word^x00
01BE
01BE
01BE 165 LORI 10,<^x04>
01BE              moderd <^x11>,<10>,0,<^x04>
11A0 01BE        .word<<^x11>*<^x100>>!<<10>*<^x10>>!<0>
0004 01C0        .word^x04
01C2
01C2
01C2 166 LOOP: SHLL 13,<^x04>
01C2              moderd <^x61>,<13>,0,<^x04>
61D0 01C2        .word<<^x61>*<^x100>>!<<13>*<^x10>>!<0>
0004 01C4        .word^x04
01C6
01C6
01C6 167 LORR 11,12
01C6              moderr <^x13>,<11>,<12>
13BC 01C6 .word <<<^x13>*<^x100>>!<<11>*<^x10>>!<12>
01C8
01C8
01C8 168 CMPI 11,<^x3D>
01C8              moderd <^x49>,<11>,0,<^x3D>
49B0 01C8        .word<<^x49>*<^x100>>!<<11>*<^x10>>!<0>
003D 01CA        .word^x3D
01CC
01CC
01CC 169 BGTIR CHTR
01CC              moderd <^x89>,4,0,<CHTR-begin>
8940 01CC        .word<<^x89>*<^x100>>!<<4>*<^x10>>!<0>
01D8' 01CE       .wordCHTR-begin
01D0
01D0
01D0 170 SBI 11,<^x30>
01D0              moderd <^xA9>,<11>,0,<^x30>
A9B0 01D0        .word<<^xA9>*<^x100>>!<<11>*<^x10>>!<0>
0030 01D2        .word^x30
01D4
01D4
01D4 171 BIR NX
01D4              moderd <^x71>,0,0,<NX-begin>
```

```
7100 01D4        .word<<^x71>*<^x100>>!<<0>*<^x10>>!<0>
01DC' 01D6       .wordNX-begin
01D8
01D8
01D8 172 CHTR: SBI 11,<^x37>
01D8             moderd <^xA9>,<11>,0,<^x37>
A9B0 01D8        .word<<^xA9>*<^x100>>!<<11>*<^x10>>!<0>
0037 01DA        .word^x37
01DC
01DC
01DC 173 NX: ADD 13,11
01DC             moderr <^xA0>,<13>,<11>
A0DB 01DC .word <<^xA0>*<^x100>>!<<13>*<^x10>>!<11>
01DE
01DE
01DE 174 ADI 12,<^x02>
01DE             moderd <^xA1>,<12>,0,<^x02>
A1C0 01DE        .word<<^xA1>*<^x100>>!<<12>*<^x10>>!<0>
0002 01E0        .word^x02
01E2
01E2
01E2 175 SBI 10,<^x01>
01E2             moderd <^xA9>,<10>,0,<^x01>
A9A0 01E2        .word<<^xA9>*<^x100>>!<<10>*<^x10>>!<0>
0001 01E4        .word^x01
01E6
01E6
01E6 176 BNEIR LOOP
01E6             moderd <^x89>,10,0,<LOOP-begin>
89A0 01E6        .word<<^x89>*<^x100>>!<<10>*<^x10>>!<0>
01C2 01E8        .wordLOOP-begin
01EA
01EA
01EA 177 CMPI 5,<^x00>
01EA             moderd <^x49>,<5>,0,<^x00>
4950 01EA        .word<<^x49>*<^x100>>!<<5>*<^x10>>!<0>
0000 01EC        .word^x00
01EE
01EE
01EE 178 BNEIR DAT
01EE             moderd <^x89>,10,0,<DAT-begin>
89A0 01EE        .word<<^x89>*<^x100>>!<<10>*<^x10>>!<0>
01F6' 01F0       .wordDAT-begin
01F2
01F2
01F2 179 LOR 6,13
01F2             moderr <^x10>,<6>,<13>
106D 01F2 .word <<^x10>*<^x100>>!<<6>*<^x10>>!<13>
01F4
01F4
01F4 180 R
01F4             moderr <^x80>,0,0
8000 01F4 .word <<^x80>*<^x100>>!<<0>*<^x10>>!<0>
01F6
01F6
01F6 181 DAT: LOR 7,13
01F6             moderr <^x10>,<7>,<13>
107D 01F6 .word <<^x10>*<^x100>>!<<7>*<^x10>>!<13>
01F8
01F8
01F8 182 R
01F8             moderr <^x80>,0,0
8000 01F8 .word <<^x80>*<^x100>>!<<0>*<^x10>>!<0>
01FA
01FA
01FA 183 ;
```

```
01FA 184 ;
01FA 185 ;
01FA 186 PRINT: LORR 12,13
01FA              moderr <^x13>,<12>,<13>
13CD 01FA .word <<<^x13>*<^x100>>!<<12>*<^x10>>!<13>
01FC
01FC
01FC 187 ADD 13,12
01FC              moderr <^xA0>,<13>,<12>
A0DC 01FC .word <<<^xA0>*<^x100>>!<<13>*<^x10>>!<12>
01FE
01FE
01FE 188 PLOOP: LORR 11,13
01FE              moderr <^x13>,<11>,<13>
13BD 01FE .word <<<^x13>*<^x100>>!<<11>*<^x10>>!<13>
0200
0200
0200 189 STRI 11,SER
0200              moderd <^x19>,<11>,0,<SER-begin>
19B0 0200        .word<<<^x19>*<^x100>>!<<11>*<^x10>>!<0>
FF00' 0202       .wordSER-begin
0204
0204
0204 190 SBI 13,<^x02>
0204              moderd <^xA9>,<13>,0,<^x02>
A9D0 0204        .word<<<^xA9>*<^x100>>!<<13>*<^x10>>!<0>
0002 0206        .word^x02
0208
0208
0208 191 SBI 12,<^x02>
0208              moderd <^xA9>,<12>,0,<^x02>
A9C0 0208        .word<<<^xA9>*<^x100>>!<<12>*<^x10>>!<0>
0002 020A        .word^x02
020C
020C
020C 192 REQ
020C              moderr <^x98>,2,0
9820 020C .word <<<^x98>*<^x100>>!<<2>*<^x10>>!<0>
020E
020E
020E 193 BIR PLOOP
020E              moderd <^x71>,0,0,<PLOOP-begin>
7100 020E        .word<<<^x71>*<^x100>>!<<0>*<^x10>>!<0>
01FE 0210        .wordPLOOP-begin
0212
0212
0212 194 ;
0212 195 ;
0212 196 ;
0212 197 RPR: LORI 10,<^x04>
0212              moderd <^x11>,<10>,0,<^x04>
11A0 0212        .word<<<^x11>*<^x100>>!<<10>*<^x10>>!<0>
0004 0214        .word^x04
0216
0216
0216 198 RCT: LOR 12,11
0216              moderr <^x10>,<12>,<11>
10CB 0216 .word <<<^x10>*<^x100>>!<<12>*<^x10>>!<11>
0218
0218
0218 199 ANDI 12,<^xF000>
0218              moderd <^x31>,<12>,0,<^xF000>
31C0 0218        .word<<<^x31>*<^x100>>!<<12>*<^x10>>!<0>
F000 021A        .word^xF000
021C
021C
```

```
021C 200 SHRL 12,<^x0C>
021C            moderd <^x63>,<12>,0,<^x0C>
63C0 021C       .word<<^x63>*<^x100>>!<<12>*<^x10>>!<0>
000C 021E       .word^x0C
0220
0220
0220 201 CMPI 12,<^x0A>
0220            moderd <^x49>,<12>,0,<^x0A>
49C0 0220       .word<<^x49>*<^x100>>!<<12>*<^x10>>!<0>
000A 0222       .word^x0A
0224
0224
0224 202 BLTIR RNUM
0224            moderd <^x89>,8,0,<RNUM-begin>
8980 0224       .word<<^x89>*<^x100>>!<<8>*<^x10>>!<0>
022C' 0226      .wordRNUM-begin
0228
0228
0228 203 ADI 12,<^x07>
0228            moderd <^xA1>,<12>,0,<^x07>
A1C0 0228       .word<<^xA1>*<^x100>>!<<12>*<^x10>>!<0>
0007 022A       .word^x07
022C
022C
022C 204 RNUM: ADI 12,<^x30>
022C            moderd <^xA1>,<12>,0,<^x30>
A1C0 022C       .word<<^xA1>*<^x100>>!<<12>*<^x10>>!<0>
0030 022E       .word^x30
0230
0230
0230 205 SEND: STRI 12,SER
0230            moderd <^x19>,<12>,0,<SER-begin>
19C0 0230       .word<<^x19>*<^x100>>!<<12>*<^x10>>!<0>
FF00' 0232      .wordSER-begin
0234
0234
0234 206 SHLL 11,<^x04>
0234            moderd <^x61>,<11>,0,<^x04>
61B0 0234       .word<<^x61>*<^x100>>!<<11>*<^x10>>!<0>
0004 0236       .word^x04
0238
0238
0238 207 SBI 10,<^x01>
0238            moderd <^xA9>,<10>,0,<^x01>
A9A0 0238       .word<<^xA9>*<^x100>>!<<10>*<^x10>>!<0>
0001 023A       .word^x01
023C
023C
023C 208 BNEIR RCT
023C            moderd <^x89>,10,0,<RCT-begin>

89A0 023C       .word<<^x89>*<^x100>>!<<10>*<^x10>>!<0>
0216 023E       .wordRCT-begin
0240
0240
0240 209 R
0240            moderr <^x80>,0,0
8000 0240 .word <<^x80>*<^x100>>!<<0>*<^x10>>!<0>
0242
0242
0242 210 ;
0242 211 ;
0242 212 ;
0242 213 ;************************************************
0242 214 ;
0242 215 ; WE'LL PUT THE DATA TO BE PRINTED RIGHT AFTER THE PROGRAM.
```

```
0242 216 ;
0242 217 ;
000E 0242 218 RDY: .word ^x000E
000A 0244 219 .word ^x000A
000D 0246 220 .word ^x000D
0059 0248 221 .word ^x0059
0044 024A 222 .word ^x0044
0041 024C 223 .word ^x0041
0045 024E 224 .word ^x0045
0052 0250 225 .word ^x0052
0252 226 ;
0012 0252 227 CNG: .word ^x0012
0020 0254 228 .word ^x0020
003F 0256 229 .word ^x003F
0045 0258 230 .word ^x0045
0047 025A 231 .word ^x0047
004E 025C 232 .word ^x004E
0041 025E 233 .word ^x0041
0048 0260 234 .word ^x0048
0043 0262 235 .word ^x0043
0020 0264 236 .word ^x0020
0266 237 ;
0002 0266 238 ONESP: .word ^x0002
0020 0268 239 .word ^x0020
026A 240 ;
0004 026A 241 CRLF: .word ^x0004
000A 026C 242 .word ^x000A
000D 026E 243 .word ^x000D
0270 244 ;
0270 245 ;
0000 0270 246 MONEND: .word0
0272 247 ;
0272 248 ; This psect places the length of the above program in a common
0272 249 ; area named LENGTH in an integer*2 variable
0272 250 ;
00000000 251 .psectLENGTH,pic,ovr,rel,gbl,shr,noexe,rd,wrt,long
0270 0000 252   .word<MONEND-begin>
0002 253        .end


Symbol table
ASCD = 00000044
ASCG = 00000047
ASCM = 0000004D
ASCN = 0000004E
ASCR = 00000052
BEGIN 00000000 R 01
CHGT 00000054 R 01
CHMEM 00000044 R 01
CHREG 00000034 R 01
CHTR 000001D8 R 01
CNG 00000252 R 01
CRLF 0000026A R 01
DAT 000001F6 R 01
DISP 0000006C R 01
DLOP 00000078 R 01
FINISH 0000006A R 01
INPT 000001B6 R 01
LODX 00000116 R 01
LOOP 000001C2 R 01
MAIN 00000004 R 01
MEM 00000172 R 01
MEMEND 0000018A R 01
MLOP 00000010 R 01
MONEND 00000270 R 01
MPC = 0000000F
MSP = 0000000E
```

```
NR0 000000A4 R 01
NR1 000000C0 R 01
NR2 000000DC R 01
NR3 000000F8 R 01
NX 000001DC R 01
ONESP 00000266 R 01
PLOOP 000001FE R 01
PRINT 000001FA R 01
RCT 00000216 R 01
RDY 00000242 R 01
RED 0000018C R 01
REG 0000009C R 01
REGEND 00000114 R 01
RLOOP 00000190 R 01
RNUM 0000022C R 01
RPR 00000212 R 01
SEND 00000230 R 01
SENSE = 0000FF02
SER = 0000FF00
WAIT1 00000122 R 01
WAIT2 00000136 R 01
YES 00000160 R 01


    +----------------+
    ! Psect synopsis !
    +----------------+

PSECT name Allocation PSECT No. Attributes
---------- ---------- --------- ----------
  . ABS . 00000000 ( 0.) 00 ( 0.) NOPIC USR CON ABS LCL NOSHR NOEXE NORD NOWRT
NOVEC BYTE
  PROG 00000272 ( 626.) 01 ( 1.) PIC USR OVR REL GBL SHR NOEXE RD WRT NOVEC LONG
  LENGTH 00000002 ( 2.) 02 ( 2.) PIC USR OVR REL GBL SHR NOEXE RD WRT NOVEC LONG
```

Appendix F. Sweet16 Monitor Object Code in Intel HEX Format

MON.H — Upper byte of each word with checksum

```
:10000000111F11001102790112FF3100490089000E
:1000100012FF49008900790071004900890079000C8
:1000200071004900890079017100490089001100BF
:10003000790170710 0FF11007901110 0137902112B
:10004000027901A100A9008900110279018001100 43
:1000500079014900890010790149008900890110710175
:100060004900890010790149008901107101490096
:100070008900107901490089011071014900890145
:10008000107901490089011071018079021102790A
:100090000112FF3100490089 0112FF12FF310049AE
:1000A00000089 0112FF490089011001102790180C4
:1000B000110079011102790180110 0790113790190
:1000C000049 0089011B80111F12FF310049 0089017D
:1000D00012FF49 0089011BA1007101A90011001143
:1000E0000061001349008901A9007101A900A0A1C4
:1000F00000A9008901490089011080108013A01314
:1001000019FFA900A90098710111001031F06300D6
:1001100049008902A100A10019FF6100A90089021C
:100120000800000000000000000000000000000004F
:10013000000000000000000000000000000000000BF
:00000001FF
```

MON.L — Lower byte of each word

```
:10000000E0C04000D04200FA900290069006A01096
:10001000900090044A034006C00049052A044009CD6
:100020000004904DA054007200049047A06A500054
:100030000 08C600004005000008C8010B60012D0CC
:100040006600FA60028001A078D06A00FA005000D1
:100050000 08C6000A0C0B000165001A0140700146E
:1000600006001A0DCB100165001A01417001460025A
:10007000A0F8B200165001A0142700146003A014C9
:10008000B300165001A014370014000012D0520023
:10009000FAB002B006B006A022B00050025 00650DE
:1000A0006A0365000B04EA0605002D06A00FA00A0
:1000B0005001008CD06A00FA005000 08CB6001687
:1000C0005001A08A6700C0A0B002B006B006A090A0
:1000D000B000B00020B6CBC0020090C008D000A095
:1000E0004D004BCB03D40D8B03000DCB037DBC039
:1000F0002A001A0C25000A0F66D007D00CDDCBDC5
:10010000B000D002C0022000FEA004CBC000C00C92
:10011000C00A802CC007C030C000B004A001A016E7
:10012000000E0A0D594441455212203F45474E41A9
:100130004843200220040A0D0000000000000000D7
:00000001FF
```

SW16TEST — Example test program machine code in Intel HEX Format

```
MEM.H Upper byte record
:1000000011001100A67100000000000000000000C7
:00000001FF
MEM.L Lower byte record
:10000000001910210100040000000000000000000B1
:00000001FF
```

Appendix G. First-Level Definition File for Basic Computer Example

```
; REVISION: 12/5/85
;
; AMDASM DEFINITION FILE FOR 16-BIT COMPUTER
; USING AM2903 RALU AND AM2910 CONTROLLER
; FILE CREATED BY MICHEL LYNCH 11/14/85
;
; REVISIONS: 12/7/85
;
; MODIFIED FOR THE 'SWEET 16' 16 BIT COMPUTER SIMULATOR
; CREATED BY BRIAN MILLER AND BRUCE KLEINMAN
;
;
 WORD 100
; 84 BIT MICROWORD TO CONTROL HARDWARE
; 16 BIT EXTENSION TO CONTROL SIMULATOR
;
;
; EQUATES FOR CCU - AM2910 SEQUENCER
;
RLD: EQU B#0
;
;
; DEFINITIONS FOR AM2910 SEQUENCER
;
JZ: DEF H#0,B#0,1V:%B#1,B#1,B#0,92X ; JUMP ZERO
CJS: DEF H#1,B#0,1V:%B#1,B#0,B#0,12V:%H#000,80X ; COND JSB PL
JSB: DEF H#1,B#0,1V:%B#1,B#1,B#0,12V:%H#000,80X ; UNCOND JSB PL
JMAP: DEF H#2,B#0,1V:%B#1,B#1,B#0,92X ; JUMP MAP
CJP: DEF H#3,B#0,1V:%B#1,B#0,B#0,12V:%H#000,80X ; COND JUMP PL
JMP: DEF H#3,B#0,1V:%B#1,B#1,B#0,12V:%H#000,80X ; UNCOND JUMP PL
PUSH: DEF H#4,B#0,1V:%B#1,B#0,B#0,12V:%H#000,80X ; PUSH/COND LD CNTR
PHLC: DEF H#4,B#1,1V:%B#1,B#0,B#0,12V:%H#000,80X ; PUSH AND LD CNTR
JSRP: DEF H#5,B#0,1V:%B#1,B#0,B#0,12V:%H#000,80X ; COND JSB R/PL
CJV: DEF H#6,B#0,1V:%B#1,B#0,B#0,92X ; COND JUMP VECTOR
JMPV: DEF H#6,B#0,1V:%B#1,B#1,B#0,92X ; UNCOND JUMP VECTOR
JRP: DEF H#7,B#0,1V:%B#1,B#0,B#0,12V:%H#000,80X ; CONDI JUMP R/PL
RFCT: DEF H#8,B#0,1V:%B#1,B#1,B#0,12V:%H#000,80X ; RPT LOOP, CNTR <> O
RPCT: DEF H#9,B#0,1V:%B#1,B#1,B#0,12V:%H#000,80X ; RPT PL, CNTR <> O
CRTN: DEF H#A,B#0,1V:%B#1,B#0,B#0,92X ; COND RTN
RTN: DEF H#A,B#0,1V:%B#1,B#1,B#0,92X ; UNCOND RETURN
CJPP: DEF H#B,B#0,1V:%B#1,B#0,B#0,12V:%H#000,80X ; COND JUMP PL & POP
LDCT: DEF H#C,B#0,1V:%B#1,B#1,B#0,12V:%H#000,80X ; LD CNTR & CONT
LOOP: DEF H#D,B#0,1V:%B#1,B#0,B#0,92X ; TEST END LOOP
CONT: DEF H#E,B#0,1V:%B#1,B#1,B#0,92X ; CONTINUE
TWB: DEF H#F,B#0,1V:%B#1,B#0,B#0,12V:%H#000,80X ; THREE-WAY BRANCH
;
;
; EQUATES FOR CCMUX SELECT FIELD
;
UZ: EQU H#0 ;MICROSTATUS REGISTER ZERO FLAG
UC: EQU H#1 ;MICROSTATUS REGISTER CARRY FLAG
UOVR: EQU H#2 ;MICROSTATUS REGISTER OVERFLOW FLAG
UN: EQU H#3 ;MICROSTATUS REGISTER SIGN FLAG
;
MZ: EQU H#4 ;MACROSTATUS REGISTER ZERO FLAG
MC: EQU H#5 ;MACROSTATUS REGISTER CARRY FLAG
MOVR: EQU H#6 ;MACROSTATUS REGISTER OVERFLOW FLAG
MN: EQU H#7 ;MACROSTATUS REGISTER SIGN FLAG
;
PAL: EQU H#8 ;CONDITION CODE PAL
;
```

```
; REMAINING 7 INPUTS ARE UNUSED
;
;
; DEFINITION FOR CCMUX SELECT FIELD
;
CCMUX: DEF 20X,4V:%H#A,76X ;CCMUX - DEFAULT OUTPUT IS AN UNUSED MUX INPUT
;
;
; DEFINITIONS FOR AM2903 ALU
;
; THE ALU DEFINITION IS OF THE FOLLOWING FORMAT
; ALU DESTINATION CONTROL, FUNCTION
;
; EQUATES FOR ALU DESTINATION CONTROL
;
ADR: EQU H#0 ; ARITHMETIC SHIFT DOWN, RESULTS INTO RAM
LDR: EQU H#1 ; LOGICAL SHIFT DOWN, RESULTS INTO RAM
ADRQ: EQU H#2 ; ARITH, SHIFT DOWN, RESULTS INTO RAM AND Q
LDRQ: EQU H#3 ; LOGICAL SHIFT DOWN, RESULTS INTO RAM AND Q
RPT: EQU H#4 ; RESULTS INTO RAM, GENERATE PARITY
LDQP: EQU H#5 ; LOGICAL SHIFT DOWN Q, GENERATE PARITY
QPT: EQU H#6 ; RESULTS INTO Q, ;GENERATE PARITY
RQPT: EQU H#7 ; RESULTS INTO RAM AND Q, GENERATE PARITY
AUR: EQU H#8 ; ARITH, SHIFT UP, RESULTS INTO RAM
LUR: EQU H#9 ; LOGICAL SHIFT UP, RESULTS INTO RAM
AURQ: EQU H#A ; ARITH, SHIFT UP, RESULTS INTO RAM AND Q
LURQ: EQU H#B ; LOGICAL SHIFT UP, RESULTS INTO RAM AND Q
YBUS: EQU H#C ; RESULTS TO Y BUS ONLY
LUQ: EQU H#D ; LOGICAL SHIFT UP Q
SINEX: EQU H#E ; SIGN EXTEND
REG: EQU H#F ; RESULTS TO RAM, SIGN EXTEND
;
; EQUATES FOR ALU FUNCTIONS
;
HIGH: EQU H#0 ; FI=1
SUBR: EQU H#1 ; SUBTRACT R FROM S
SUBS: EQU H#2 ; SUBTRACT S FROM R
ADD: EQU H#3 ; ADD R AND S
PASSS: EQU H#4 ; PASS S
COMPLS: EQU H#5 ; 2'S COMPLEMENT OF S
PASSR: EQU H#6 ; PASS R
COMPLR: EQU H#7 ; 2'S COMPLEMENT OF R
LOW: EQU H#8 ; FI = 0
NOTRS: EQU H#9 ; COMPLEMENT R AND WITH S
EXNOR: EQU H#A ; EXCLUSIVE NOR R WITH S
EXOR: EQU H#B ; EXCLUSIVE OR R WITH S
AND: EQU H#C ; AND R WITH S
NOR: EQU H#D ; NOR R WITH S
NAND: EQU H#E ; NAND R WITH S
OR: EQU H#F ; OR R WITH S
;
; ALU DEFINITION
;
ALU: DEF 64X,4VH#C,4VH#8,28X
;
; ALU OPERAND SOURCES
;
AB: DEF 36X,B#0,B#0,34X,B#0,27X ; R = RAM A, S = RAM B
ADB: DEF 36X,B#0,B#1,34X,B#0,27X ; R = RAM A, S = DB
AQ: DEF 36X,B#0,35X,B#1,27X ; R = RAM A, S = Q
DAB: DEF 36X,B#1,B#0,34X,B#0,27X ; R = DA, S = RAM B
DADB: DEF 36X,B#1,B#1,34X,B#0,27X ; R = DA, S = DB
DAQ: DEF 36X,B#1,35X,B#1,27X ; R = DA, S = Q
;
; OUTPUT Y ENABLE IS CONTROLLED IN DATAPATH
;
```

```
;
; SPECIAL FUNCTIONS FOR AM2903
;
; TO USE THE SPECIAL FUNCTIONS, THE DESTINATION
; CONTROL MUST NOT BE AQ OR DAQ
;
; SPECIAL FUNCTION EQUATES
;
USMUL: EQU H#00 ; UNSIGNED MULTIPLY
TCMUL: EQU H#20 ; TWO'S COMPLEMENT MULTIPLY
INCTWO: EQU H#40 ; INCREMENT BY ONE OR TWO
SMTC: EQU H#50 ; SIGN-MAGNITUDE/TWO'S COMPLEMENT
TCMLS: EQU H#60 ; TWO'S COMPLEMENT MULT. LAST STEP
SLN: EQU H#80 ; SINGLE LENGTH NORMALIZE
DLN: EQU H#A0 ; DOUBLE LENGTH NORMALIZE AND 1ST DIVIDE OP.
TCDIV: EQU H#C0 ; TWO'S COMPLEMENT DIVIDE
TCDC: EQU H#E0 ; TWO'S COMPLEMENT DIVISION CORRECTION
;
; SPECIAL FUNCTION DEFINITION
;
SPF14: DEF 64X,8VH#,28X
;
;
; DEFINITIONS FOR STATUS REGISTER RELATED CONTROL BITS
;
; STATUS REGISTER BIT DEFINITIONS ARE AS FOLLOWS:
;
; BIT 58 - MICROSTATUS REGISTER LOAD ENABLE
; BITS 59, 60 - MACRO SR ZERO/SIGN FLAG LOAD ENABLE
; BITS 61, 62 - MACRO SR CARRY/OVERFLOW FLAG LOAD ENABLE
;
; BIT 63 - MACRO SR TO DATABUS ENABLE
;
NOUSLD: EQU B#0 ; HOLD MICROSTATUS REGISTER
;
MDBZN: EQU B#01 ; LOAD MACRO SR Z & N FROM DB3 & DB1
MZNLD: EQU B#10 ; LOAD MACRO SR Z & N FROM RALU
MZNUL: EQU B#11 ; CLEAR MACRO SR Z & N
;
MDBCOV: EQU B#01 ; LOAD MACRO SR C & OVR FROM DB0 & DB2
MCOVLD: EQU B#10 ; LOAD MACRO SR C & OVR FROM RALU
MCSHFT: EQU B#11 ; LOAD MACRO SR C FROM 2903 SHIFT MUXES
;
MDBOE: EQU B#0
;
; STATUS REGISTER CONTROL FIELDS DEFINITION
;
; ORDER OF FIELDS: MICROSTATUS REGISTER LOAD
; : MACROSTATUS REGISTER Z & N
; : MACROSTATUS REGISTER C & OVR
; : MACROSTATUS REGISTER --> DATABUS
;
MSRLD: DEF 58X,1VB#1,2VB#00,2VB#00,1VB#1,36X
;
; SHIFT MUX CONTROL FIELDS
;
; SHIFT MUX EQUATES
;
ZERO: EQU H#0 ;SHIFT IN LOGICAL ZERO
ONE: EQU H#1 ;SHIFT IN LOGICAL ONE
Q0: EQU H#2 ;SHIFT IN QIO0
S0: EQU H#3 ;SHIFT IN SIO0
Q15: EQU H#2 ;SHIFT IN Q15
S15: EQU H#3 ;SHIFT IN SIO3 FROM MSD 2903 (S15)
MACC: EQU H#4 ;MACRO CARRY REGISTER
NEG: EQU H#5 ;NEGATIVE OUTPUT FROM 2903
```

```
;
; SHIFT MUX DEFINITIONS
; MS BIT OF EACH 4 BIT FIELD ENABLES MUX OUTPUT IF 0
;
SHFT: DEF 40X,4V:%H#8,4V:%H#8,4V:%H#8,4V:%H#8,44X
;
;
; CARRY INPUT MUX EQUATES
; ZERO AND ONE ARE DEFINED ABOVE
;
UCF: EQU B#10 ;MICRO STATUS REGISTER CARRY BIT
MCF: EQU B#11 ;MACRO STATUS REGISTER CARRY BIT
;
; CARRY INPUT MUX DEFINITION
;
CNMUX: DEF 56X,2V:%B#00,42X
;
;
; REGISTER MUX SELECT
;
; REGISTER MUX SELECT EQUATES
;
; REGISTER-TO-2903 RAM A INPUT SELECT EQUATES
;
PIPAA: EQU B#00 ;POINT AT REGISTER FIELD A IN PIPELINE
R2A: EQU B#10 ;POINT AT R2 FIELD IN INSTRUCTION REG
R1A: EQU B#11 ;POINT AT R1 FIELD IN INSTRUCTION REG
;
; REGISTER-TO-2903 RAM B INPUT SELECT EQUATES
;
PIPBB: EQU B#00 ;POINT AT REGISTER FIELD B IN PIPELINE
R2B: EQU B#10 ;POINT AT R2 FIELD IN INSTRUCTION REG
R1B: EQU B#11 ;POINT AT R1 FIELD IN INSTRUCTION REG
;
; REGISTER MUX SELECT DEFINITION
;
REGMUX: DEF 28X,2VB#00,4X,2VB#00,64X
;
; 2903 REGISTER FIELD EQUATES
;
R0: EQU H#0
R1: EQU H#1
R2: EQU H#2
R3: EQU H#3
R4: EQU H#4
R5: EQU H#5
R6: EQU H#6
R7: EQU H#7
R8: EQU H#8
R9: EQU H#9
R10: EQU H#A
R11: EQU H#B
R12: EQU H#C
R13: EQU H#D
R14: EQU H#E
R15: EQU H#F
;
PC: EQU H#F
;
; PIPELINE - 2903 REGISTER FIELD DEFINITION
;
R2903: DEF 24X,4V:%H#0,2X,4V:%H#0,66X
;
;
; EQUATES FOR DATAPATH DEFINITION
;
```

```
OEY: EQU B#0 ;CONNECT 2903 FUNCTION TO DATABUS
;
DBOUT: EQU B#00 ;SET DATABUS OUTBOUND AND ENABLE DRIVER
DBIN: EQU B#01 ;SET DATABUS INBOUND AND ENABLE DRIVER
;
MARLD: EQU B#0 ;LOAD MAR FROM DATABUS
;
READ: EQU B#01 ;READ EXTERNAL DEVICES
WRITE: EQU B#10 ;WRITE TO EXTERNAL DEVICES
;
IRLD: EQU B#0 ;LOAD INSTRUCTION REGISTER FROM DATABUS
;
; DATAPATH DEFINITION
;
DATAPATH: DEF 38X,1V:%B#0,34X,2V:%B#10,1V:%B#1,
/ B#0,2V:%B#11,1V:%B#1,20X
;
;
; 2903 RAM WRITE ENABLE DEFINITION
;
REGWR: DEF 39X,B#0,60X
NOREGWR: DEF 39X,B#1,60X
;
;
; MCMUX SELECT DEFINITION
;
SIO0IN: EQU B#00 ;2903 SHIFT PORTS
SIO3IN: EQU B#01
QIO0IN: EQU B#10
QIO3IN: EQU B#11
;
MCMUX: DEF 80X,2V:%B#00,18X
;
;
; Y BI-DIRECTIONAL BUS DRIVER FOR THE 2903
;
YBDEN: DEF 82X,B#0,17X
NOYBDEN: DEF 82X,B#1,17X
;
;
; DB BI-DIRECTIONAL BUS DRIVER FOR THE 2903
;
DBBDEN: DEF 83X,B#0,16X
NODBBDEN: DEF 83X,B#1,16X
;
;
; HALT AND PROCESS MESSAGE DEFINITIONS
;
SOPPN: DEF 84X,B#01,14X
PAUSE: DEF 84X,B#10,14X
HALT: DEF 84X,B#11,14X
;
;
; DEBUG DEFINTIONS
DEBCON: DEF 86X,B#1111,10X
DEBALU: DEF 90X,B#1111,6X
DEBSRG: DEF 94X,B#11,4X
DEBMAC: DEF 96X,B#11,2X
DEBOUT: DEF 98X,B#11
DEBALL: DEF 86X,B#11111111111111
 ;
  END
```

Appendix H. First-Level Source Program for the Basic Computer Example

```
Microtec Microcode Assembler Source for Sweet16

LINE ADDR
0001 00000 ;
0002 00000 ; MICROCODE ASSEMBLER SOURCE PROGRAM FOR 4713 COMPUTER
0003 00000 ;
0004 00000 LIST B,L,O
0005 00000 ;
0006 00000 ;
*************************************************************************
0007 00000 ; * YES, HERE WE GO. THE ACTUAL MICROCODE FOR THE 'SWEET 16' SIMULATOR
0008 00000 ; * A MONUMENTAL EFFORT BY BRIAN MILLER AND BRUCE KLEINMAN
0009 00000 ; * ADAPTED BY M. LYNCH
*************************************************************************
0010 00000 ;
0011 00000 ; Version 1/24/89
0012 00000 ;
0013 00000 ;
0014 00000 ; Reset sequence - Fetches op-code at location 0000H.
0015 00000 ;
0016 00000 START: CONT & CCMUX & ALU & AB & MSRLD & SHFT & CNMUX &
0017 00000 / REGMUX & R2903 & DATAPATH ,,MARLD,READ, &
0018 00000 / NOREGWR & MCMUX & YBDEN & NODBBDEN
 11100110 XXXXXXXX XXXX1010 00000000 00000001 10001000
 10001000 00100001 11001000 01000011 0001XXXX XXXXXXXX
 XXXX
0019 00001 ;/ & DEBCON & DEBALU & DEBSRG
0020 00001 ;
0021 00001 CONT & CCMUX & ALU RPT,PASSS & AB & MSRLD & SHFT & CNMUX ONE &
0022 00001 / REGMUX & R2903 ,R15 & DATAPATH ,DBIN,,READ, &
0023 00001 / REGWR & MCMUX & NOYBDEN & NODBBDEN
 11100110 XXXXXXXX XXXX1010 00000011 11000000 10001000
 10001000 01100001 01000100 00110011 0011XXXX XXXXXXXX
 XXXX
0024 00002 ;/ & DEBCON & DEBALU & DEBSRG
0025 00002 ;
0026 00002 CONT & CCMUX & ALU & AB & MSRLD & SHFT & CNMUX &
0027 00002 / REGMUX & R2903 & DATAPATH ,DBIN,,READ,IRLD &
0028 00002 / NOREGWR & MCMUX & NOYBDEN & NODBBDEN
 11100110 XXXXXXXX XXXX1010 00000000 00000001 10001000
 10001000 00100001 11001000 00110010 0011XXXX XXXXXXXX
 XXXX
0029 00003 ;/ & DEBCON & DEBALU & DEBSRG
0030 00003 ;
0031 00003 JMAP & CCMUX & ALU & AB & MSRLD & SHFT & CNMUX &
0032 00003 / REGMUX & R2903 & DATAPATH &
0033 00003 / NOREGWR & MCMUX & NOYBDEN & NODBBDEN
 00100110 XXXXXXXX XXXX1010 00000000 00000001 10001000
 10001000 00100001 11001000 01010111 0011XXXX XXXXXXXX
 XXXX
0034 00004 ;/ & DEBCON & DEBALU & DEBSRG
0035 00004 ;
0036 00004 ; Op-code fetch sequence. Note that the program counter 'rests'
0037 00004 ; at an odd number, and is incremented to the next even number before
0038 00004 ; it is used to load up the MAR.
0039 00004 ;
0040 00004 FETCH: CONT & CCMUX & ALU RPT,PASSS & AB & MSRLD & SHFT & CNMUX ONE &
0041 00004 / REGMUX & R2903 ,R15 & DATAPATH ,,MARLD,READ, &
0042 00004 / REGWR & MCMUX & YBDEN & NODBBDEN
 11100110 XXXXXXXX XXXX1010 00000011 11000000 10001000
```

```
     10001000 01100001 01000100 01000011 0001XXXX XXXXXXXX
     XXXX
0043 00005 ;/ & DEBCON & DEBALU & DEBSRG
0044 00005 ;
0045 00005 CONT & CCMUX & ALU RPT,PASSS & AB & MSRLD & SHFT & CNMUX ONE &
0046 00005 / REGMUX & R2903 ,R15 & DATAPATH ,DBIN,,READ, &
0047 00005 / REGWR & MCMUX & NOYBDEN & NODBBDEN
     11100110 XXXXXXXX XXXX1010 00000011 11000000 10001000
     10001000 01100001 01000100 00110011 0011XXXX XXXXXXXX
     XXXX
0048 00006 ;/ & DEBCON & DEBALU & DEBSRG
0049 00006 ;
0050 00006 CONT & CCMUX & ALU & AB & MSRLD & SHFT & CNMUX &
0051 00006 / REGMUX & R2903 & DATAPATH ,DBIN,,READ,IRLD &
0052 00006 / NOREGWR & MCMUX & NOYBDEN & NODBBDEN
     11100110 XXXXXXXX XXXX1010 00000000 00000001 10001000
     10001000 00100001 11001000 00110010 0011XXXX XXXXXXXX
     XXXX
0053 00007 ;/ & DEBCON & DEBALU & DEBSRG
0054 00007 ;
0055 00007 JMAP & CCMUX & ALU & AB & MSRLD & SHFT & CNMUX &
0056 00007 / REGMUX & R2903 & DATAPATH &
0057 00007 / NOREGWR & MCMUX & NOYBDEN & NODBBDEN
     00100110 XXXXXXXX XXXX1010 00000000 00000001 10001000
     10001000 00100001 11001000 01010111 0011XXXX XXXXXXXX
     XXXX
0058 00008 ;/ & DEBCON & DEBALU & DEBSRG
0059 00008 ;
0060 00008 ; Get operand for IMMEDIATE address mode. Data will end up on the
0061 00008 ; internal data bus.
0062 00008 ;
0063 00008 OPERI: CONT & CCMUX & ALU RPT,PASSS & AB & MSRLD & SHFT & CNMUX ONE &
0064 00008 / REGMUX & R2903 ,R15 & DATAPATH ,,MARLD,READ, &
0065 00008 / REGWR & MCMUX & YBDEN & NODBBDEN
     11100110 XXXXXXXX XXXX1010 00000011 11000000 10001000
     10001000 01100001 01000100 01000011 0001XXXX XXXXXXXX
     XXXX
0066 00009 ;
0067 00009 RTN & CCMUX & ALU RPT,PASSS & AB & MSRLD & SHFT & CNMUX ONE &
0068 00009 / REGMUX & R2903 ,R15 & DATAPATH ,DBIN,,READ, &
0069 00009 / REGWR & MCMUX & NOYBDEN & NODBBDEN
     10100110 XXXXXXXX XXXX1010 00000011 11000000 10001000
     10001000 01100001 01000100 00110011 0011XXXX XXXXXXXX
     XXXX
0070 0000A ;
0071 0000A ; Get operand for REGISTER INDIRECT mode. Data will end up on the
0072 0000A ; internal data bus.
0073 0000A ;
0074 0000A OPERR: CONT & CCMUX & ALU ,PASSS & AB & MSRLD & SHFT & CNMUX &
0075 0000A / REGMUX ,R2B & R2903 & DATAPATH ,,MARLD,READ, &
0076 0000A / NOREGWR & MCMUX & YBDEN & NODBBDEN
     11100110 XXXXXXXX XXXX1010 00000000 00100001 10001000
     10001000 00100001 11000100 01000011 0001XXXX XXXXXXXX
     XXXX
0077 0000B ;
0078 0000B RTN & CCMUX & ALU & AB & MSRLD & SHFT & CNMUX &
0079 0000B / REGMUX & R2903 & DATAPATH ,DBIN,,READ, &
0080 0000B / NOREGWR & MCMUX & NOYBDEN & NODBBDEN
     10100110 XXXXXXXX XXXX1010 00000000 00000001 10001000
     10001000 00100001 11001000 00110011 0011XXXX XXXXXXXX
     XXXX
0081 0000C ;
0082 0000C ; Get operand for DATA INDIRECT mode. Data will, of course, end
0083 0000C ; up on the internal data bus.
0084 0000C ;
0085 0000C OPERA: CONT & CCMUX & ALU RPT,PASSS & AB & MSRLD & SHFT & CNMUX ONE &
```

```
0086 0000C / REGMUX & R2903 ,R15 & DATAPATH ,,MARLD,READ, &
0087 0000C / REGWR & MCMUX & YBDEN & NODBBDEN
 11100110 XXXXXXXX XXXX1010 00000011 11000000 10001000
 10001000 01100001 01000100 01000011 0001XXXX XXXXXXXX
 XXXX
0088 0000D ;
0089 0000D CONT & CCMUX & ALU RPT,PASSS & AB & MSRLD & SHFT & CNMUX ONE &
0090 0000D / REGMUX & R2903 ,R15 & DATAPATH ,DBIN,,READ, &
0091 0000D / REGWR & MCMUX & NOYBDEN & NODBBDEN
 11100110 XXXXXXXX XXXX1010 00000011 11000000 10001000
 10001000 01100001 01000100 00110011 0011XXXX XXXXXXXX
 XXXX
0092 0000E ;
0093 0000E CONT & CCMUX & ALU & AB & MSRLD & SHFT & CNMUX &
0094 0000E / REGMUX & R2903 & DATAPATH ,DBIN,MARLD,READ, &
0095 0000E / NOREGWR & MCMUX & NOYBDEN & NODBBDEN
 11100110 XXXXXXXX XXXX1010 00000000 00000001 10001000
 10001000 00100001 11001000 00100011 0011XXXX XXXXXXXX
 XXXX
0096 0000F ;
0097 0000F RTN & CCMUX & ALU & AB & MSRLD & SHFT & CNMUX &
0098 0000F / REGMUX & R2903 & DATAPATH ,DBIN,,READ, &
0099 0000F / NOREGWR & MCMUX & NOYBDEN & NODBBDEN
 10100110 XXXXXXXX XXXX1010 00000000 00000001 10001000
 10001000 00100001 11001000 00110011 0011XXXX XXXXXXXX
 XXXX
0100 00010 ;
0101 00010 ; Load register from register.
0102 00020 ORG H#20
0103 00020 ;
0104 00020 LOR: JMP ,FETCH & CCMUX & ALU RPT,PASSR & AB &
0105 00020 / MSRLD ,MZNLD,, & SHFT &
0106 00020 / CNMUX & REGMUX R2A,R1B & R2903 & DATAPATH &
0107 00020 / REGWR & MCMUX & NOYBDEN & NODBBDEN
 00110110 00000000 01001010 00001000 00110000 10001000
 10001000 00110001 01000110 01010111 0011XXXX XXXXXXXX
 XXXX
0108 00021 ;
0109 00021 ; load register from immediate.
0110 00022 ORG H#22
0111 00022 ;
0112 00022 LORI: JSB ,OPERI & CCMUX & ALU & AB &
0113 00022 / MSRLD ,MZNLD,, & SHFT & CNMUX &
0114 00022 / REGMUX & R2903 & DATAPATH &
0115 00022 / NOREGWR & MCMUX & NOYBDEN & NODBBDEN
 00010110 00000000 10001010 00000000 00000001 10001000
 10001000 00110001 11001000 01010111 0011XXXX XXXXXXXX
 XXXX
0116 00023 ;
0117 00023 JMP ,FETCH & CCMUX & ALU RPT,PASSR & DAB &
0118 00023 / MSRLD ,MZNLD,, & SHFT &
0119 00023 / CNMUX & REGMUX ,R1B & R2903 & DATAPATH ,DBIN,,, &
0120 00023 / REGWR & MCMUX & NOYBDEN & NODBBDEN
 00110110 00000000 01001010 00000000 00111000 10001000
 10001000 00110001 01000110 00110111 0011XXXX XXXXXXXX
 XXXX
0121 00024 ;
0122 00024 ; load register from data indirect.
0123 00024 ORG H#24
0124 00024 ;
0125 00024 LORA: JSB ,OPERA & CCMUX & ALU & AB & MSRLD & SHFT & CNMUX &
0126 00024 / REGMUX & R2903 & DATAPATH &
0127 00024 / NOREGWR & MCMUX & NOYBDEN & NODBBDEN
 00010110 00000000 11001010 00000000 00000001 10001000
 10001000 00100001 11001000 01010111 0011XXXX XXXXXXXX
 XXXX
```

```
0128 00025 ;
0129 00025 JMP ,FETCH & CCMUX & ALU RPT,PASSR & DAB &
0130 00025 / MSRLD ,MZNLD,, & SHFT &
0131 00025 / CNMUX & REGMUX ,R1B & R2903 & DATAPATH ,DBIN,,, &
0132 00025 / REGWR & MCMUX & NOYBDEN & NODBBDEN
 00110110 00000000 01001010 00000000 00111000 10001000
 10001000 00110001 01000110 00110111 0011XXXX XXXXXXXX
 XXXX
0133 00026 ;
0134 00026 ; load register from register indirect.
0135 00026 ORG H#26
0136 00026 ;
0137 00026 LORR: JSB ,OPERR & CCMUX & ALU & AB & MSRLD & SHFT & CNMUX &
0138 00026 / REGMUX & R2903 & DATAPATH &
0139 00026 / NOREGWR & MCMUX & NOYBDEN & NODBBDEN
 00010110 00000000 10101010 00000000 00000001 10001000
 10001000 00100001 11001000 01010111 0011XXXX XXXXXXXX
 XXXX
0140 00027 ;
0141 00027 JMP ,FETCH & CCMUX & ALU RPT,PASSR & DAB &
0142 00027 / MSRLD ,MZNLD,, & SHFT &
0143 00027 / CNMUX & REGMUX ,R1B & R2903 & DATAPATH ,DBIN,,, &
0144 00027 / REGWR & MCMUX & NOYBDEN & NODBBDEN
 00110110 00000000 01001010 00000000 00111000 10001000
 10001000 00110001 01000110 00110111 0011XXXX XXXXXXXX
 XXXX
0145 00028 ;
0146 00028 ; store register from immediate.
0147 00032 ORG H#32
0148 00032 ;
0149 00032 STRI: JMP ,STRI2 & CCMUX & ALU & AB & MSRLD & SHFT & CNMUX &
0150 00032 / REGMUX & R2903 & DATAPATH &
0151 00032 / NOREGWR & MCMUX & NOYBDEN & NODBBDEN
 00110110 00100000 00001010 00000000 00000001 10001000
 10001000 00100001 11001000 01010111 0011XXXX XXXXXXXX
 XXXX
0152 00033 ;
0153 00033 ; store register from register indirect.
0154 00036 ORG H#36
0155 00036 ;
0156 00036 STRR: CONT & CCMUX & ALU ,PASSS & AB & MSRLD & SHFT & CNMUX &
0157 00036 / REGMUX ,R1B & R2903 & DATAPATH ,,MARLD,WRITE, &
0158 00036 / NOREGWR & MCMUX & YBDEN & NODBBDEN
 11100110 XXXXXXXX XXXX1010 00000000 00110001 10001000
 10001000 00100001 11000100 01000101 0001XXXX XXXXXXXX
 XXXX
0159 00037 ;
0160 00037 JMP ,FETCH & CCMUX & ALU ,PASSS & AB & MSRLD & SHFT & CNMUX &
0161 00037 / REGMUX ,R2B & R2903 & DATAPATH ,DBOUT,,, &
0162 00037 / NOREGWR & MCMUX & YBDEN & NODBBDEN
 00110110 00000000 01001010 00000000 00100001 10001000
 10001000 00100001 11000100 00010111 0001XXXX XXXXXXXX
 XXXX
0163 00038 ;
0164 00038 ; AND register to register.
0165 00060 ORG H#60
0166 00060 ;
0167 00060 ANDS16: JMP ,FETCH & CCMUX & ALU RPT,AND & AB & MSRLD ,MZNLD,, &
0168 00060 / SHFT & CNMUX & REGMUX R2A,R1B & R2903 & DATAPATH &
0169 00060 / REGWR & MCMUX & NOYBDEN & NODBBDEN
 00110110 00000000 01001010 00001000 00110000 10001000
 10001000 00110001 01001100 01010111 0011XXXX XXXXXXXX
 XXXX
0170 00061 ;/ & DEBCON & DEBALU & DEBSRG
0171 00061 ;
0172 00061 ; AND register from immediate.
```

```
0173 00062 ORG H#62
0174 00062 ;
0175 00062 ANDI: JSB ,OPERI & CCMUX & ALU & AB & MSRLD & SHFT & CNMUX &
0176 00062 / REGMUX & R2903 & DATAPATH &
0177 00062 / NOREGWR & MCMUX & NOYBDEN & NODBBDEN
 00010110 00000000 10001010 00000000 00000001 10001000
 10001000 00100001 11001000 01010111 0011XXXX XXXXXXXX
 XXXX
0178 00063 ;/ & DEBCON & DEBALU & DEBSRG
0179 00063 ;
0180 00063 JMP ,FETCH & CCMUX & ALU RPT,AND & DAB &
0181 00063 / MSRLD ,MZNLD,, & SHFT &
0182 00063 / CNMUX & REGMUX ,R1B & R2903 & DATAPATH ,DBIN,,, &
0183 00063 / REGWR & MCMUX & NOYBDEN & NODBBDEN
 00110110 00000000 01001010 00000000 00111000 10001000
 10001000 00110001 01001100 00110111 0011XXXX XXXXXXXX
 XXXX
0184 00064 ;/ & DEBCON & DEBALU & DEBSRG
0185 00064 ;
0186 00064 ; OR register to register.
0187 00070 ORG H#70
0188 00070 ;
0189 00070 ORS16: JMP ,FETCH & CCMUX & ALU RPT,OR & AB & MSRLD ,MZNLD,, &
0190 00070 / SHFT & CNMUX & REGMUX R2A,R1B & R2903 & DATAPATH &
0191 00070 / REGWR & MCMUX & NOYBDEN & NODBBDEN
 00110110 00000000 01001010 00001000 00110000 10001000
 10001000 00110001 01001111 01010111 0011XXXX XXXXXXXX
 XXXX
0192 00071 ;/ & DEBCON & DEBALU & DEBSRG
0193 00071 ;
0194 00071 ; OR register from immediate.
0195 00072 ORG H#72
0196 00072 ;
0197 00072 ORI: JSB ,OPERI & CCMUX & ALU & AB & MSRLD & SHFT & CNMUX &
0198 00072 / REGMUX & R2903 & DATAPATH &
0199 00072 / NOREGWR & MCMUX & NOYBDEN & NODBBDEN
 00010110 00000000 10001010 00000000 00000001 10001000
 10001000 00100001 11001000 01010111 0011XXXX XXXXXXXX
 XXXX
0200 00073 ;/ & DEBCON & DEBALU & DEBSRG
0201 00073 ;
0202 00073 JMP ,FETCH & CCMUX & ALU RPT,OR & DAB &
0203 00073 / MSRLD ,MZNLD,, & SHFT &
0204 00073 / CNMUX & REGMUX ,R1B & R2903 & DATAPATH ,DBIN,,, &
0205 00073 / REGWR & MCMUX & NOYBDEN & NODBBDEN
 00110110 00000000 01001010 00000000 00111000 10001000
 10001000 00110001 01001111 00110111 0011XXXX XXXXXXXX
 XXXX
0206 00074 ;/ & DEBCON & DEBALU & DEBSRG
0207 00074 ;
0208 00074 ; XOR register to register.
0209 00080 ORG H#80
0210 00080 ;
0211 00080 XOR: JMP ,FETCH & CCMUX & ALU RPT,EXOR & AB &
0212 00080 / MSRLD ,MZNLD,, &
0213 00080 / SHFT & CNMUX & REGMUX R2A,R1B & R2903 & DATAPATH &
0214 00080 / REGWR & MCMUX & NOYBDEN & NODBBDEN
 00110110 00000000 01001010 00001000 00110000 10001000
 10001000 00110001 01001011 01010111 0011XXXX XXXXXXXX
 XXXX
0215 00081 ;/ & DEBCON & DEBALU & DEBSRG
0216 00081 ;
0217 00081 ; XOR register from immediate.
0218 00082 ORG H#82
0219 00082 ;
0220 00082 XORI: JSB ,OPERI & CCMUX & ALU & AB & MSRLD & SHFT & CNMUX &
```

```
0221 00082 / REGMUX & R2903 & DATAPATH &
0222 00082 / NOREGWR & MCMUX & NOYBDEN & NODBBDEN
  00010110 00000000 10001010 00000000 00000001 10001000
  10001000 00100001 11001000 01010111 0011XXXX XXXXXXXX
  XXXX
0223 00083 ;/ & DEBCON & DEBALU & DEBSRG
0224 00083 ;
0225 00083 JMP ,FETCH & CCMUX & ALU RPT,EXOR & DAB &
0226 00083 / MSRLD ,MZNLD,, & SHFT &
0227 00083 / CNMUX & REGMUX ,R1B & R2903 & DATAPATH ,DBIN,,, &
0228 00083 / REGWR & MCMUX & NOYBDEN & NODBBDEN
  00110110 00000000 01001010 00000000 00111000 10001000
  10001000 00110001 01001011 00110111 0011XXXX XXXXXXXX
  XXXX
0229 00084 ;/ & DEBCON & DEBALU & DEBSRG
0230 00084 ;
0231 00084 ;
0232 00084 ; Compare immediate. same as SUBI but do not store result.
0233 00092 ORG H#92
0234 00092 ;
0235 00092 CMPI: JSB ,OPERI & CCMUX & ALU & AB & MSRLD & SHFT & CNMUX &
0236 00092 / REGMUX & R2903 & DATAPATH &
0237 00092 / NOREGWR & MCMUX & NOYBDEN & NODBBDEN
  00010110 00000000 10001010 00000000 00000001 10001000
  10001000 00100001 11001000 01010111 0011XXXX XXXXXXXX
  XXXX
0238 00093 ;
0239 00093 JMP ,FETCH & CCMUX & ALU RPT,SUBR & DAB &
0240 00093 / MSRLD ,MZNLD,MCOVLD & SHFT & CNMUX ONE & REGMUX ,R1B &
0241 00093 / R2903 & DATAPATH ,DBIN,,, & NOREGWR & MCMUX &
0242 00093 / NOYBDEN & NODBBDEN
  00110110 00000000 01001010 00000000 00111001 10001000
  10001000 01110101 01000001 00110111 0011XXXX XXXXXXXX
  XXXX
0243 00094 ;
0244 00094 ; shift register left logical.
0245 000C2 ORG H#C2
0246 000C2 ;
0247 000C2 SHLL: JMP ,SHLL2 & CCMUX & ALU & AB & MSRLD & SHFT & CNMUX &
0248 000C2 / REGMUX & R2903 & DATAPATH &
0249 000C2 / NOREGWR & MCMUX & NOYBDEN & NODBBDEN
  00110110 00100001 00001010 00000000 00000001 10001000
  10001000 00100001 11001000 01010111 0011XXXX XXXXXXXX
  XXXX
0250 000C3 ;
0251 000C3 ;
0252 000C3 ; shift register right logical.
0253 000C6 ORG H#C6
0254 000C6 ;
0255 000C6 SHRL: JMP ,SHRL2 & CCMUX & ALU & AB & MSRLD & SHFT & CNMUX &
0256 000C6 / REGMUX & R2903 & DATAPATH &
0257 000C6 / NOREGWR & MCMUX & NOYBDEN & NODBBDEN
  00110110 00100001 01011010 00000000 00000001 10001000
  10001000 00100001 11001000 01010111 0011XXXX XXXXXXXX
  XXXX
0258 000C7 ;
0259 000C7 ;
0260 000C7 ; Rotate register left through Macro Carry.
0261 000D0 ORG H#D0
0262 000D0 ;
0263 000D0 ROLC: JSB ,OPERI & CCMUX & ALU & AB & MSRLD & SHFT & CNMUX &
0264 000D0 / REGMUX & R2903 & DATAPATH &
0265 000D0 / NOREGWR & MCMUX & NOYBDEN & NODBBDEN
  00010110 00000000 10001010 00000000 00000001 10001000
  10001000 00100001 11001000 01010111 0011XXXX XXXXXXXX
  XXXX
```

```
0266 000D1 ;/ & DEBCON & DEBALU & DEBSRG
0267 000D1 ;
0268 000D1 JMP ,ROLC2 & CCMUX & ALU QPT,COMPLR & DAB & MSRLD &
0269 000D1 / SHFT & CNMUX ONE & REGMUX & R2903 & DATAPATH ,DBIN,,, &
0270 000D1 / NOREGWR & MCMUX & NOYBDEN & NODBBDEN
 00110110 00100010 00001010 00000000 00001001 10001000
 10001000 01100001 01100111 00110111 0011XXXX XXXXXXXX
 XXXX
0271 000D2 ;/ & DEBCON & DEBALU & DEBSRG
0272 000D2 ;
0273 000D2 ; Rotate register right through Macro Carry.
0274 000D4 ORG H#D4
0275 000D4 ;
0276 000D4 RORC: JSB ,OPERI & CCMUX & ALU & AB & MSRLD & SHFT & CNMUX &
0277 000D4 / REGMUX & R2903 & DATAPATH &
0278 000D4 / NOREGWR & MCMUX & NOYBDEN & NODBBDEN
 00010110 00000000 10001010 00000000 00000001 10001000
 10001000 00100001 11001000 01010111 0011XXXX XXXXXXXX
 XXXX
0279 000D5 ;/ & DEBCON & DEBALU & DEBSRG
0280 000D5 ;
0281 000D5 ; 2's complement data bus and place in Q - negate
0282 000D5 JMP ,RORC2 & CCMUX & ALU QPT,COMPLR & DAB & MSRLD &
0283 000D5 / SHFT & CNMUX ONE & REGMUX & R2903 & DATAPATH ,DBIN,,, &
0284 000D5 / NOREGWR & MCMUX & NOYBDEN & NODBBDEN
 00110110 00100010 00111010 00000000 00001001 10001000
 10001000 01100001 01100111 00110111 0011XXXX XXXXXXXX
 XXXX
0285 000D6 ;/ & DEBCON & DEBALU & DEBSRG
0286 000D6 ;
0287 000D6 ;
0288 000D6 ; branch register.
0289 000E0 ORG H#E0
0290 000E0 ;
0291 000E0 B: CONT & CCMUX & ALU RPT,PASSR & AB & MSRLD & SHFT &
0292 000E0 / CNMUX & REGMUX R1A, & R2903 ,R15 & DATAPATH &
0293 000E0 / REGWR & MCMUX & NOYBDEN & NODBBDEN
 11100110 XXXXXXXX XXXX1010 00001111 11000000 10001000
 10001000 00100001 01000110 01010111 0011XXXX XXXXXXXX
 XXXX
0294 000E1 ;
0295 000E1 JMP ,FETCH & CCMUX & ALU RPT,SUBR & AB & MSRLD & SHFT &
0296 000E1 / CNMUX & REGMUX & R2903 R4,R15 & DATAPATH &
0297 000E1 / REGWR & MCMUX & NOYBDEN & NODBBDEN
 00110110 00000000 01001010 01000011 11000000 10001000
 10001000 00100001 01000001 01010111 0011XXXX XXXXXXXX
 XXXX
0298 000E2 ;
0299 000E2 ; branch immediate. note the need to decrement the immediate value
0300 000E2 ; due in order to produce an odd number for the program counter.
0301 000E2 ORG H#E2
0302 000E2 ;
0303 000E2 BI: JSB ,OPERI & CCMUX & ALU & AB & MSRLD & SHFT & CNMUX &
0304 000E2 / REGMUX & R2903 & DATAPATH &
0305 000E2 / NOREGWR & MCMUX & NOYBDEN & NODBBDEN
 00010110 00000000 10001010 00000000 00000001 10001000
 10001000 00100001 11001000 01010111 0011XXXX XXXXXXXX
 XXXX
0306 000E3 ;
0307 000E3 JMP ,FETCH & CCMUX & ALU RPT,SUBR & ADB & MSRLD & SHFT & CNMUX &
0308 000E3 / REGMUX & R2903 R4,R15 & DATAPATH ,DBIN,,, &
0309 000E3 / REGWR & MCMUX & NOYBDEN & DBBDEN
 00110110 00000000 01001010 01000011 11000100 10001000
 10001000 00100001 01000001 00110111 0010XXXX XXXXXXXX
 XXXX
0310 000E4 ;
```

```
0311 000E4 ; call subroutine immediate. note the need to increment the stack
0312 000E4 ; pointer twice, before pushing the program counter.
0313 000F2 ORG H#F2
0314 000F2 ;
0315 000F2 CAI: JMP ,CAI2 & CCMUX & ALU & AB & MSRLD & SHFT & CNMUX &
0316 000F2 / REGMUX & R2903 & DATAPATH &
0317 000F2 / NOREGWR & MCMUX & NOYBDEN & NODBBDEN
 00110110 00100000 01001010 00000000 00000001 10001000
 10001000 00100001 11001000 01010111 0011XXXX XXXXXXXX
 XXXX
0318 000F3 ;
0319 000F3 ; return.
0320 00100 ORG H#100
0321 00100 ;
0322 00100 RE: JMP ,RE2 & CCMUX & ALU & AB & MSRLD & SHFT & CNMUX &
0323 00100 / REGMUX & R2903 & DATAPATH &
0324 00100 / NOREGWR & MCMUX & NOYBDEN & NODBBDEN
 00110110 00100000 10111010 00000000 00000001 10001000
 10001000 00100001 11001000 01010111 0011XXXX XXXXXXXX
 XXXX
0325 00101 ;
0326 00101 ; branch conditional register.
0327 00110 ORG H#110
0328 00110 ;
0329 00110 BN: CJP ,B & CCMUX PAL & ALU & AB & MSRLD & SHFT & CNMUX &
0330 00110 / REGMUX & R2903 & DATAPATH &
0331 00110 / NOREGWR & MCMUX & NOYBDEN & NODBBDEN
 00110100 00001110 00001000 00000000 00000001 10001000
 10001000 00100001 11001000 01010111 0011XXXX XXXXXXXX
 XXXX
0332 00111 ;
0333 00111 ; branch conditional immediate. if condition is TRUE, goto BI above.
0334 00111 ; if FALSE, goto fetch.
0335 00112 ORG H#112
0336 00112 ;
0337 00112 BNI: CJP ,BI & CCMUX PAL & ALU & AB & MSRLD & SHFT & CNMUX &
0338 00112 / REGMUX & R2903 & DATAPATH &
0339 00112 / NOREGWR & MCMUX & NOYBDEN & NODBBDEN
 00110100 00001110 00101000 00000000 00000001 10001000
 10001000 00100001 11001000 01010111 0011XXXX XXXXXXXX
 XXXX
0340 00113 ;
0341 00113 JMP ,DUMINC & CCMUX & ALU & AB & MSRLD & SHFT & CNMUX &
0342 00113 / REGMUX & R2903 & DATAPATH &
0343 00113 / NOREGWR & MCMUX & NOYBDEN & NODBBDEN
 00110110 00100001 10101010 00000000 00000001 10001000
 10001000 00100001 11001000 01010111 0011XXXX XXXXXXXX
 XXXX
0344 00114 ;
0345 00114 ; return conditional.
0346 00130 ORG H#130
0347 00130 ;
0348 00130 RN: CJP ,RE2 & CCMUX PAL & ALU & AB & MSRLD & SHFT & CNMUX &
0349 00130 / REGMUX & R2903 & DATAPATH &
0350 00130 / NOREGWR & MCMUX & NOYBDEN & NODBBDEN
 00110100 00100000 10111000 00000000 00000001 10001000
 10001000 00100001 11001000 01010111 0011XXXX XXXXXXXX
 XXXX
0351 00131 ;
0352 00131 JMP ,FETCH & CCMUX & ALU & AB & MSRLD & SHFT & CNMUX &
0353 00131 / REGMUX & R2903 & DATAPATH &
0354 00131 / NOREGWR & MCMUX & NOYBDEN & NODBBDEN
 00110110 00000000 01001010 00000000 00000001 10001000
 10001000 00100001 11001000 01010111 0011XXXX XXXXXXXX
 XXXX
0355 00132 ;
```

```
0356 00132 ; add register to register - Flags C, OVR, Z, N
0357 00140 ORG H#140
0358 00140 ;
0359 00140 ADD1: JMP ,FETCH & CCMUX & ALU RPT,ADD & AB &
0360 00140 / MSRLD ,MZNLD,MCOVLD, &
0361 00140 / SHFT & CNMUX & REGMUX R2A,R1B & R2903 & DATAPATH &
0362 00140 / REGWR & MCMUX & NOYBDEN & NODBBDEN
 00110110 00000000 01001010 00001000 00110000 10001000
 10001000 00110101 01000011 01010111 0011XXXX XXXXXXXX
 XXXX
0363 00141 ;
0364 00141 ; Add immediate - Flags C, OVR, Z, N
0365 00142 ORG H#142
0366 00142 ;
0367 00142 ADI: JSB ,OPERI & CCMUX & ALU & AB & MSRLD & SHFT & CNMUX &
0368 00142 / REGMUX & R2903 & DATAPATH &
0369 00142 / NOREGWR & MCMUX & NOYBDEN & NODBBDEN
 00010110 00000000 10001010 00000000 00000001 10001000
 10001000 00100001 11001000 01010111 0011XXXX XXXXXXXX
 XXXX
0370 00143 ;
0371 00143 JMP ,FETCH & CCMUX & ALU RPT,ADD & DAB &
0372 00143 / MSRLD ,MZNLD,MCOVLD, & SHFT &
0373 00143 / CNMUX & REGMUX ,R1B & R2903 & DATAPATH ,DBIN,,, &
0374 00143 / REGWR & MCMUX & NOYBDEN & NODBBDEN
 00110110 00000000 01001010 00000000 00111000 10001000
 10001000 00110101 01000011 00110111 0011XXXX XXXXXXXX
 XXXX
0375 00144 ;
0376 00144 ; INCREMENT immediate - Flags N, Z
0377 0014A ORG H#14A
0378 0014A ;
0379 0014A INCI: JSB ,OPERI & CCMUX & ALU & AB &
0380 0014A / MSRLD & SHFT & CNMUX &
0381 0014A / REGMUX & R2903 & DATAPATH &
0382 0014A / NOREGWR & MCMUX & NOYBDEN & NODBBDEN
 00010110 00000000 10001010 00000000 00000001 10001000
 10001000 00100001 11001000 01010111 0011XXXX XXXXXXXX
 XXXX
0383 0014B ;
0384 0014B JMP ,FETCH & CCMUX & ALU RPT,ADD & DAB &
0385 0014B / MSRLD ,MZNLD,, & SHFT &
0386 0014B / CNMUX & REGMUX ,R1B & R2903 & DATAPATH ,DBIN,,, &
0387 0014B / REGWR & MCMUX & NOYBDEN & NODBBDEN
0388 0014B / & DEBALU & DEBSRG
 00110110 00000000 01001010 00000000 00111000 10001000
 10001000 00110001 01000011 00110111 0011XXXX XX111111
 XXXX
0389 0014C ;
0390 0014C ; Subtract immediate - Flags C, OVR, Z, N
0391 00152 ORG H#152
0392 00152 ;
0393 00152 SBI: JSB ,OPERI & CCMUX & ALU & AB & MSRLD & SHFT & CNMUX &
0394 00152 / REGMUX & R2903 & DATAPATH &
0395 00152 / NOREGWR & MCMUX & NOYBDEN & NODBBDEN
 00010110 00000000 10001010 00000000 00000001 10001000
 10001000 00100001 11001000 01010111 0011XXXX XXXXXXXX
 XXXX
0396 00153 ;
0397 00153 JMP ,FETCH & CCMUX & ALU RPT,SUBR & DAB &
0398 00153 / MSRLD ,MZNLD,MCOVLD & SHFT & CNMUX ONE & REGMUX ,R1B &
0399 00153 / R2903 & DATAPATH ,DBIN,,, & REGWR & MCMUX & NOYBDEN & NODBBDEN
 00110110 00000000 01001010 00000000 00111000 10001000
 10001000 01110101 01000001 00110111 0011XXXX XXXXXXXX
 XXXX
0400 00154 ;
```

```
0401 00154 ; DECREMENT immediate - Flags Z, N
0402 0015A ORG H#15A
0403 0015A ;
0404 0015A DECI: JSB ,OPERI & CCMUX & ALU & AB & MSRLD & SHFT & CNMUX &
0405 0015A / REGMUX & R2903 & DATAPATH &
0406 0015A / NOREGWR & MCMUX & NOYBDEN & NODBBDEN
 00010110 00000000 10001010 00000000 00000001 10001000
 10001000 00100001 11001000 01010111 0011XXXX XXXXXXXX
 XXXX
0407 0015B ;
0408 0015B JMP ,FETCH & CCMUX & ALU RPT,SUBR & DAB &
0409 0015B / MSRLD ,MZNLD,, & SHFT & CNMUX ONE & REGMUX ,R1B &
0410 0015B / R2903 & DATAPATH ,DBIN,,, & REGWR & MCMUX & NOYBDEN & NODBBDEN
 00110110 00000000 01001010 00000000 00111000 10001000
 10001000 01110001 01000001 00110111 0011XXXX XXXXXXXX
 XXXX
0411 0015C ;/ & DEBALU & DEBSRG
0412 0015C ;
0413 0015C ; halt execution.
0414 001FE ORG H#1FE
0415 001FE ;
0416 001FE CONT & CCMUX & ALU & AB & MSRLD & SHFT & CNMUX &
0417 001FE / REGMUX & R2903 & DATAPATH &
0418 001FE / NOREGWR & MCMUX & NOYBDEN & NODBBDEN & HALT
0419 001FE / & DEBCON & DEBALU & DEBSRG
 11100110 XXXXXXXX XXXX1010 00000000 00000001 10001000
 10001000 00100001 11001000 01010111 00111111 11111111
 XXXX
0420 001FF ;
0421 001FF ;
0422 001FF ;
0423 001FF ; ****************************************************
0424 001FF ; * instructions requiring extra space jump to here. *
0425 001FF ; ****************************************************
0426 00200 ORG H#200
0427 00200 ;
0428 00200 ; store register indirect continued.
0429 00200 ;
0430 00200 STRI2: CONT & CCMUX & ALU RPT,PASSS & AB & MSRLD & SHFT & CNMUX ONE &
0431 00200 / REGMUX & R2903 ,R15 & DATAPATH ,,MARLD,READ, &
0432 00200 / REGWR & MCMUX & YBDEN & NODBBDEN
 11100110 XXXXXXXX XXXX1010 00000011 11000000 10001000
 10001000 01100001 01000100 01000011 0001XXXX XXXXXXXX
 XXXX
0433 00201 ;
0434 00201 CONT & CCMUX & ALU RPT,PASSS & AB & MSRLD & SHFT & CNMUX ONE &
0435 00201 / REGMUX & R2903 ,R15 & DATAPATH ,DBIN,,, &
0436 00201 / REGWR & MCMUX & NOYBDEN & NODBBDEN
 11100110 XXXXXXXX XXXX1010 00000011 11000000 10001000
 10001000 01100001 01000100 00110111 0011XXXX XXXXXXXX
 XXXX
0437 00202 ;
0438 00202 CONT & CCMUX & ALU & AB & MSRLD & SHFT & CNMUX &
0439 00202 / REGMUX & R2903 & DATAPATH ,DBIN,MARLD,WRITE, &
0440 00202 / NOREGWR & MCMUX & NOYBDEN & NODBBDEN
 11100110 XXXXXXXX XXXX1010 00000000 00000001 10001000
 10001000 00100001 11001000 00100101 0011XXXX XXXXXXXX
 XXXX
0441 00203 ;
0442 00203 JMP ,FETCH & CCMUX & ALU ,PASSS & AB & MSRLD & SHFT & CNMUX &
0443 00203 / REGMUX ,R1B & R2903 & DATAPATH ,DBOUT,,, &
0444 00203 / NOREGWR & MCMUX & YBDEN & NODBBDEN
 00110110 00000000 01001010 00000000 00110001 10001000
 10001000 00100001 11000100 00010111 0001XXXX XXXXXXXX
 XXXX
0445 00204 ;
```

```
0446 00204 ; call subroutine indirect continued.
0447 00204 ;
0448 00204 CAI2: JSB ,OPERI & CCMUX & ALU & AB & MSRLD & SHFT & CNMUX &
0449 00204 / REGMUX & R2903 & DATAPATH &
0450 00204 / NOREGWR & MCMUX & NOYBDEN & NODBBDEN
 00010110 00000000 10001010 00000000 00000001 10001000
 10001000 00100001 11001000 01010111 0011XXXX XXXXXXXX
 XXXX
0451 00205 ;
0452 00205 CONT & CCMUX & ALU RPT,SUBS & DAB & MSRLD & SHFT & CNMUX &
0453 00205 / REGMUX & R2903 ,R4 & DATAPATH ,DBIN,,, &
0454 00205 / REGWR & MCMUX & NOYBDEN & NODBBDEN
 11100110 XXXXXXXX XXXX1010 00000001 00001000 10001000
 10001000 00100001 01000010 00110111 0011XXXX XXXXXXXX
 XXXX
0455 00206 ;
0456 00206 CONT & CCMUX & ALU RPT,PASSS & AB & MSRLD & SHFT & CNMUX ONE &
0457 00206 / REGMUX & R2903 ,R14 & DATAPATH &
0458 00206 / REGWR & MCMUX & NOYBDEN & NODBBDEN
 11100110 XXXXXXXX XXXX1010 00000011 10000000 10001000
 10001000 01100001 01000100 01010111 0011XXXX XXXXXXXX
 XXXX
0459 00207 ;
0460 00207 CONT & CCMUX & ALU RPT,PASSS & AB & MSRLD & SHFT & CNMUX ONE &
0461 00207 / REGMUX & R2903 ,R14 & DATAPATH ,,MARLD,WRITE, &
0462 00207 / REGWR & MCMUX & YBDEN & NODBBDEN
 11100110 XXXXXXXX XXXX1010 00000011 10000000 10001000
 10001000 01100001 01000100 01000101 0001XXXX XXXXXXXX
 XXXX
0463 00208 ;
0464 00208 CONT & CCMUX & ALU RPT,PASSS & AB & MSRLD & SHFT & CNMUX &
0465 00208 / REGMUX & R2903 ,R15 & DATAPATH ,DBOUT,,, &
0466 00208 / NOREGWR & MCMUX & YBDEN & NODBBDEN
 11100110 XXXXXXXX XXXX1010 00000011 11000001 10001000
 10001000 00100001 01000100 00010111 0001XXXX XXXXXXXX
 XXXX
0467 00209 ;
0468 00209 CONT & CCMUX & ALU RPT,PASSR & AB & MSRLD & SHFT & CNMUX &
0469 00209 / REGMUX & R2903 R4,R15 & DATAPATH &
0470 00209 / REGWR & MCMUX & NOYBDEN & NODBBDEN
 11100110 XXXXXXXX XXXX1010 01000011 11000000 10001000
 10001000 00100001 01000110 01010111 0011XXXX XXXXXXXX
 XXXX
0471 0020A ;
0472 0020A JMP ,FETCH & CCMUX & ALU RPT,LOW & AB & MSRLD & SHFT & CNMUX &
0473 0020A / REGMUX & R2903 ,R4 & DATAPATH &
0474 0020A / REGWR & MCMUX & NOYBDEN & NODBBDEN
 00110110 00000000 01001010 00000001 00000000 10001000
 10001000 00100001 01001000 01010111 0011XXXX XXXXXXXX
 XXXX
0475 0020B ;
0476 0020B ; return conditional continued.
0477 0020B ;
0478 0020B RE2: CONT & CCMUX & ALU RPT,PASSS & AB & MSRLD & SHFT & CNMUX &
0479 0020B / REGMUX & R2903 ,R14 & DATAPATH ,,MARLD,READ, &
0480 0020B / NOREGWR & MCMUX & YBDEN & NODBBDEN
 11100110 XXXXXXXX XXXX1010 00000011 10000001 10001000
 10001000 00100001 01000100 01000011 0001XXXX XXXXXXXX
 XXXX
0481 0020C ;
0482 0020C CONT & CCMUX & ALU & AB & MSRLD & SHFT & CNMUX &
0483 0020C / REGMUX & R2903 & DATAPATH ,DBIN,,, &
0484 0020C / NOREGWR & MCMUX & NOYBDEN & NODBBDEN
 11100110 XXXXXXXX XXXX1010 00000000 00000001 10001000
 10001000 00100001 11001000 00110111 0011XXXX XXXXXXXX
 XXXX
```

```
0485 0020D ;
0486 0020D CONT & CCMUX & ALU RPT,PASSR & DAB & MSRLD & SHFT & CNMUX &
0487 0020D / REGMUX & R2903 ,R15 & DATAPATH ,DBIN,,, &
0488 0020D / REGWR & MCMUX & NOYBDEN & NODBBDEN
 11100110 XXXXXXXX XXXX1010 00000011 11001000 10001000
 10001000 00100001 01000110 00110111 0011XXXX XXXXXXXX
 XXXX
0489 0020E ;
0490 0020E CONT & CCMUX & ALU RPT,SUBR & AB & MSRLD & SHFT & CNMUX &
0491 0020E / REGMUX & R2903 R4,R14 & DATAPATH &
0492 0020E / REGWR & MCMUX & NOYBDEN & DBBDEN
 11100110 XXXXXXXX XXXX1010 01000011 10000000 10001000
 10001000 00100001 01000001 01010111 0010XXXX XXXXXXXX
 XXXX
0493 0020F ;
0494 0020F JMP ,FETCH & CCMUX & ALU RPT,SUBR & AB & MSRLD & SHFT & CNMUX &
0495 0020F / REGMUX & R2903 R4,R14 & DATAPATH &
0496 0020F / REGWR & MCMUX & NOYBDEN & DBBDEN
 00110110 00000000 01001010 01000011 10000000 10001000
 10001000 00100001 01000001 01010111 0010XXXX XXXXXXXX
 XXXX
0497 00210 ;
0498 00210 ; shift register left logical continued.
0499 00210 ;
0500 00210 SHLL2: JSB ,OPERI & CCMUX & ALU & AB & MSRLD & SHFT & CNMUX &
0501 00210 / REGMUX & R2903 & DATAPATH &
0502 00210 / NOREGWR & MCMUX & NOYBDEN & NODBBDEN
 00010110 00000000 10001010 00000000 00000001 10001000
 10001000 00100001 11001000 01010111 0011XXXX XXXXXXXX
 XXXX
0503 00211 ;
0504 00211 CONT & CCMUX & ALU QPT,PASSR & DAB & MSRLD ,MZNUL,, &
0505 00211 / SHFT & CNMUX & REGMUX & R2903 & DATAPATH ,DBIN,,, &
0506 00211 / NOREGWR & MCMUX & NOYBDEN & NODBBDEN
 11100110 XXXXXXXX XXXX1010 00000000 00001001 10001000
 10001000 00111001 01100110 00110111 0011XXXX XXXXXXXX
 XXXX
0507 00212 ;
0508 00212 SLLOP: CJP ,FETCH & CCMUX UZ & ALU & AB & MSRLD & SHFT & CNMUX &
0509 00212 / REGMUX & R2903 & DATAPATH &
0510 00212 / NOREGWR & MCMUX & NOYBDEN & NODBBDEN
 00110100 00000000 01000000 00000000 00000001 10001000
 10001000 00100001 11001000 01010111 0011XXXX XXXXXXXX
 XXXX
0511 00213 ;
0512 00213 CONT & CCMUX & ALU LUR,PASSS & AB &
0513 00213 / MSRLD ,MZNLD,, & SHFT ,,,ZERO &
0514 00213 / CNMUX & REGMUX ,R1B & R2903 & DATAPATH &
0515 00213 / REGWR & MCMUX & NOYBDEN & NODBBDEN
 11100110 XXXXXXXX XXXX1010 00000000 00110000 10001000
 10000000 00110001 10010100 01010111 0011XXXX XXXXXXXX
 XXXX
0516 00214 ;
0517 00214 JMP ,SLLOP & CCMUX & ALU QPT,SUBR & AQ & MSRLD &
0518 00214 / SHFT & CNMUX & REGMUX & R2903 R4, & DATAPATH &
0519 00214 / NOREGWR & MCMUX & NOYBDEN & NODBBDEN
 00110110 00100001 00101010 01000000 00000X01 10001000
 10001000 00100001 01100001 11010111 0011XXXX XXXXXXXX
 XXXX
0520 00215 ;
0521 00215 ; shift register right logical continued.
0522 00215 ;
0523 00215 SHRL2: JSB ,OPERI & CCMUX & ALU & AB & MSRLD & SHFT & CNMUX &
0524 00215 / REGMUX & R2903 & DATAPATH &
0525 00215 / NOREGWR & MCMUX & NOYBDEN & NODBBDEN
 00010110 00000000 10001010 00000000 00000001 10001000
```

```
     10001000 00100001 11001000 01010111 0011XXXX XXXXXXXX
     XXXX
0526 00216 ;
0527 00216 CONT & CCMUX & ALU QPT,PASSR & DAB & MSRLD ,MZNUL,, &
0528 00216 / SHFT & CNMUX & REGMUX & R2903 & DATAPATH ,DBIN,,, &
0529 00216 / NOREGWR & MCMUX & NOYBDEN & NODBBDEN
     11100110 XXXXXXXX XXXX1010 00000000 00001001 10001000
     10001000 00111001 01100110 00110111 0011XXXX XXXXXXXX
     XXXX
0530 00217 ;
0531 00217 SRLOP: CJP ,FETCH & CCMUX UZ & ALU & AB & MSRLD & SHFT & CNMUX &
0532 00217 / REGMUX & R2903 & DATAPATH &
0533 00217 / NOREGWR & MCMUX & NOYBDEN & NODBBDEN
     00110100 00000000 01000000 00000000 00000001 10001000
     10001000 00100001 11001000 01010111 0011XXXX XXXXXXXX
     XXXX
0534 00218 ;
0535 00218 CONT & CCMUX & ALU LDR,PASSS & AB &
0536 00218 / MSRLD ,MZNLD,, & SHFT ,ZERO,, &
0537 00218 / CNMUX & REGMUX ,R1B & R2903 & DATAPATH &
0538 00218 / REGWR & MCMUX & NOYBDEN & NODBBDEN
     11100110 XXXXXXXX XXXX1010 00000000 00110000 10000000
     10001000 00110001 00010100 01010111 0011XXXX XXXXXXXX
     XXXX
0539 00219 ;
0540 00219 JMP ,SRLOP & CCMUX & ALU QPT,SUBR & AQ & MSRLD &
0541 00219 / SHFT & CNMUX & REGMUX & R2903 R4, & DATAPATH &
0542 00219 / NOREGWR & MCMUX & NOYBDEN & NODBBDEN
     00110110 00100001 01111010 01000000 00000X01 10001000
     10001000 00100001 01100001 11010111 0011XXXX XXXXXXXX
     XXXX
0543 0021A ;
0544 0021A ; dummy increment of the program counter. jump to fetch when done.
0545 0021A ;
0546 0021A DUMINC: CONT & CCMUX & ALU RPT,PASSS & AB & MSRLD & SHFT & CNMUX ONE &
0547 0021A / REGMUX & R2903 ,R15 & DATAPATH &
0548 0021A / REGWR & MCMUX & NOYBDEN & NODBBDEN
     11100110 XXXXXXXX XXXX1010 00000011 11000000 10001000
     10001000 01100001 01000100 01010111 0011XXXX XXXXXXXX
     XXXX
0549 0021B ;
0550 0021B JMP ,FETCH & CCMUX & ALU RPT,PASSS & AB & MSRLD & SHFT &
0551 0021B / CNMUX ONE & REGMUX & R2903 ,R15 & DATAPATH &
0552 0021B / REGWR & MCMUX & NOYBDEN & NODBBDEN
     00110110 00000000 01001010 00000011 11000000 10001000
     10001000 01100001 01000100 01010111 0011XXXX XXXXXXXX
     XXXX
0553 0021C ;
0554 00220 ORG H#220
0555 00220 ;
0556 00220 ; Rotate register left through Macro Carry continued.
0557 00220 ; Preincrement the negative shift count in Q - UZ is ready to test
0558 00220 ;
0559 00220 ROLC2: CONT & CCMUX & ALU QPT,PASSS & AQ & MSRLD &
0560 00220 / SHFT & CNMUX ONE & REGMUX & R2903 & DATAPATH &
0561 00220 / NOREGWR & MCMUX & NOYBDEN & NODBBDEN
     11100110 XXXXXXXX XXXX1010 00000000 00000X01 10001000
     10001000 01100001 01100100 11010111 0011XXXX XXXXXXXX
     XXXX
0562 00221 ;/ & DEBCON & DEBALU & DEBSRG
0563 00221 ;
0564 00221 ;
0565 00221 ROLOP: CJP ,FETCH & CCMUX UZ & ALU LUR,PASSS &
0566 00221 / AB & MSRLD ,MZNLD,MCSHFT, &
0567 00221 / SHFT ,,,MACC & CNMUX &
0568 00221 / REGMUX ,R1B & R2903 & DATAPATH &
```

```
0569 00221 / REGWR & MCMUX & NOYBDEN & NODBBDEN
 00110100 00000000 01000000 00000000 00110000 10001000
 10000100 00110111 10010100 01010111 0011XXXX XXXXXXXX
 XXXX
0570 00222 ;/ & DEBCON & DEBALU & DEBSRG
0571 00222 ;
0572 00222 JMP ,ROLOP & CCMUX & ALU QPT,PASSS & AQ & MSRLD &
0573 00222 / SHFT & CNMUX ONE & REGMUX & R2903 & DATAPATH &
0574 00222 / NOREGWR & MCMUX & NOYBDEN & NODBBDEN
 00110110 00100010 00011010 00000000 00000X01 10001000
 10001000 01100001 01100100 11010111 0011XXXX XXXXXXXX
 XXXX
0575 00223 ;/ & DEBCON & DEBALU & DEBSRG
0576 00223 ;
0577 00223 ; Rotate register right through Macro Carry continued.
0578 00223 ; Preincrement Q - count register
0579 00223 RORC2: CONT & CCMUX & ALU QPT,PASSS & AQ & MSRLD &
0580 00223 / SHFT & CNMUX ONE & REGMUX & R2903 & DATAPATH &
0581 00223 / NOREGWR & MCMUX & NOYBDEN & NODBBDEN
 11100110 XXXXXXXX XXXX1010 00000000 00000X01 10001000
 10001000 01100001 01100100 11010111 0011XXXX XXXXXXXX
 XXXX
0582 00224 ;/ & DEBCON & DEBALU & DEBSRG
0583 00224 ;
0584 00224 ROROP: CJP ,FETCH & CCMUX UZ & ALU LDR,PASSS &
0585 00224 / AB & MSRLD ,MZNLD,MCSHFT, &
0586 00224 / SHFT ,MACC,, & CNMUX &
0587 00224 / REGMUX ,R1B & R2903 & DATAPATH &
0588 00224 / REGWR & MCMUX & NOYBDEN & NODBBDEN
 00110100 00000000 01000000 00000000 00110000 10000100
 10001000 00110111 00010100 01010111 0011XXXX XXXXXXXX
 XXXX
0589 00225 ;/ & DEBCON & DEBALU & DEBSRG
0590 00225 ;
0591 00225 ; Increment Q - count register
0592 00225 JMP ,ROROP & CCMUX & ALU QPT,PASSS & AQ & MSRLD &
0593 00225 / SHFT & CNMUX ONE & REGMUX & R2903 & DATAPATH &
0594 00225 / NOREGWR & MCMUX & NOYBDEN & NODBBDEN
 00110110 00100010 01001010 00000000 00000X01 10001000
 10001000 01100001 01100100 11010111 0011XXXX XXXXXXXX
 XXXX
0595 00226 ;/ & DEBCON & DEBALU & DEBSRG
0596 00226 ;
0597 00226 ;
0598 00226 ;
0599 00226 END

 TOTAL ASSEMBLY ERRORS = 0
```

Appendix I. Demonstration

The following file is TEST.FOR which is a FORTRAN program that will be used to demonstrate the execution of a program in a CP/M-80 operating system environment on an Z80 microprocessor based computer.

```
C
C Demonstration program to show 8080 machine code for a DO loop
C
  INTEGER*2 A(10),I,ONE,TEN
C
  ONE=1
  TEN=10
C
  DO 5 I=ONE,TEN
  A(I)=ONE
5 CONTINUE
C
  DO 20 I=ONE,TEN
  A(I)=A(I)+I
20 CONTINUE
C
  STOP
  END
```

This is the assembler listing file TEST.PRN produced by the FORTRAN compiler using TEST.FOR (above) as input. The compiler also produced a relocatable object file (TEST.OBJ) for the linker to combine with the runtime library to produce the executable load module TEST.CPM shown below. Note the 8080 assembler language segments produced by the FORTRAN compiler.

```
1         C
2         C Demonstration program to show 8080 machine code for a DO loop
3         C
4           INTEGER*2 A(10),I,ONE,TEN
5         C
*****     0000'   LXI   B,$$L         ;Initialization routine
*****     0003'   JMP   $INIT

6         ONE=1
*****     0006'   LXI   H,0001        ;First problem solving code
*****     0009'   SHLD  ONE           ;Get a 16 bit 0001 and place in location ONE

7         TEN=10
*****     000C'   LXI   H,000A        ;Get a 16 bit 000A (10 decimal)
*****     000F'   SHLD  TEN           ;Place it in location TEN

  Beginning of first DO loop
8         C
9           DO 5 I=ONE,TEN
*****     0012'   LHLD  ONE           ;Get contents of ONE (beginning index value)
*****     0015'   SHLD  I             ;Place into loop counter I

10        A(I)=ONE
*****     0018'   LHLD  I             ;Turn index I into an address into Array A
*****     001B'   DAD   H             ; I*2 -> HL
*****     001C'   LXI   D,A-0002      ; address of LSB of A(1) - 2 into DE
*****     001F'   DAD   D             ; (ADDR of A(1) -2) + I*2 -> HL
*****     0020'   SHLD  T:000000      ;store temporarily in T:000000
*****     0023'   LHLD  ONE           ;Get contents of ONE into HL
```

```
*****     0026'   XCHG                      ;Swap HL (ONE) with DE
*****     0027'   LHLD   T:000000           ;Get actual address of A(I) into HL
*****     002A'   MOV    M,E ;Store LSB of ONE here (HL points at LSB of A(I)
*****     002B'   INX    H                  ;Point at following byte
*****     002C'   MOV    M,D                ;Store MSB of ONE here

11        5 CONTINUE
*****     002D'   LHLD   I                  ;Get loop counter into HL
*****     0030'   INX    H                  ;Increment loop counter copy in HL
*****     0031'   XCHG                      ;Swap HL and DE
*****     0032'   LHLD   TEN                ;Place contents of TEN in HL (loop end count)
*****     0035'   XCHG                      ;Swap HL and DE
*****     0036'   MOV    A,E                ;These four instructions perform a 16 bit
*****     0037'   SUB    L                  ; subtraction of the contents of TEN (in DE)
*****     0038'   MOV    A,D                ; and the loop counter (in HL) that sets the
*****     0039'   SBB    H                  ; sign flag
*****     003A'   JP     0015'  ;Go around again if sign is positive (TEN => I)
          End of first DO loop - incremented loop counter is stored at line 0015'
          Beginning of second DO loop


12        C
13          DO 20 I=ONE,TEN

*****     003D'   LHLD   ONE                ;Loop initialization
*****     0040'   SHLD   I                  ;Set I to the contents of ONE (which is 0001)

14          A(I)=A(I)+I

*****     0043'   LHLD   I                  ;Convert array index into address as above
*****     0046'   DAD    H
*****     0047'   LXI    D,A-0002
*****     004A'   DAD    D
*****     004B'   SHLD   T:000000           ;Save address of element A(I)
*****     004E'   MOV    A,M            ;Copy location pointed to by HL into Acc, A(I)
*****     004F'   INX    H                  ;Point at next by in memory, MSB of A(I)
*****     0050'   MOV    H,M                ;Copy it into H using address in HL
*****     0051'   MOV    L,A                ;Copy LSB of A(I) from Acc to L
*****     0052'   XCHG                      ;Swap HL and DE - 16 bit A(I) is now in DE
*****     0053'   LHLD   I                  ;Get I into HL
*****     0056'   DAD    D                  ;A(I)+I, DE has 16 bit number from A(I)
*****     0057'   XCHG                      ;Swap DE and HL, A(I)+I -> DE
*****     0058'   LHLD   T:000000           ;Address of A(I) -> HL
*****     005B'   MOV    M,E                ;A(I)+I -> A(I) Note: LSB first
*****     005C'   INX    H
*****     005D'   MOV    M,D                ;A(I)+I -> A(I) MSB next

15        20 CONTINUE

*****     005E'   LHLD   I                  ;Get loop index, I -> HL
*****     0061'   INX    H                  ;Increment loop index in HL
*****     0062'   XCHG   ;Swap HL and DE, incremented loop index now in DE
*****     0063'   LHLD   TEN                ;Get contents of TEN in HL (loop end count)
*****     0066'   XCHG                      ; TEN -> DE and I -> HL
*****     0067'   MOV    A,E ;Compare TEN and incremented I and set sign flag
*****     0068'   SUB    L
*****     0069'   MOV    A,D
*****     006A'   SBB    H ;Incremented loop counter is stored in line 0040'
*****     006B'   JP     0040';Continue running loop if I <= TEN
```

```
      End of second DO loop


            End of program
  16        C
  17          STOP


  *****     006E'  CALL  $ST           ;Return to operating system


  18        END
  *****     0071'  202020202020
```

The compiler also produces a MAP of the variable locations to aid in looking
at the program in execution. The addresses given are relative to the beginning
of the program at line 0000' which is the LXI B,$$L above.

```
Program Unit Length=0077 (119) Bytes
Data Area Length=001D (29) Bytes

Subroutines Referenced:

$INIT                    $ST

Variables:

A         0001"       I            0015"ONE0017"
TEN       0019"       T:000000     001B"

LABELS:

$$L       0006'       5L           002D'20L005E'
```

Now run the debugging utility DDT with the load module TEST.CPM in order that
the actual execution of the program TEST can be followed.

```
DDT VERS 2.2

        Load TEST.CPM
NEXT PC
0400 0100End of module is at 0400 and beginning address is 0100
```

-D0100,0200 Dump the area containing the FORTRAN program. Address of the first
byte is the 4 digit number on the left. The data at location 0100 is C3, the data
at 0101 is 20, etc. with the data at 010F being the 00 at the right end of the
line. The characters, mostly periods, to the right of the numbers are the ASCII
character equivalents of the data just given. They are of little use here.

```
0100 C3 20 01 00 00 00 00 00 00 00 00 00 00 00 00 00  . . . . . . . . . . . . . . . .
0110 00 00 00 00 00 00 00 00 00 00 00 00 00 00 00 00  . . . . . . . . . . . . . . . .
0120 01 26 01 C3 0B 02 21 01 00 22 1A 01 21 0A 00 22  .&....!.."..!.."
0130 1C 01 2A 1A 01 22 18 01 2A 18 01 29 11 02 01 19  ..*.."..*..).....
0140 22 1E 01 2A 1A 01 EB 2A 1E 01 73 23 72 2A 18 01  "..*...*..s#r*..
0150 23 EB 2A 1C 01 EB 7B 95 7A 9C F2 35 01 2A 1A 01  #.*...{.z..5.*..
0160 22 18 01 2A 18 01 29 11 02 01 19 22 1E 01 7E 23  "..*..)....."..~#
0170 66 6F EB 2A 18 01 19 EB 2A 1E 01 73 23 72 2A 18  fo.*....*..s#r*.
0180 01 23 EB 2A 1C 01 EB 7B 95 7A 9C F2 60 01 CD E9  .#.*...{.z..`...
0190 01 20 20 20 20 20 20 CD C7 01 3E 20 32 FD 02 21  . ...> 2..!
01A0 50 41 22 FE 02 21 55 53 22 00 03 21 45 20 22 02  PA"..!US"..!E ".
01B0 03 CD D8 01 2A 8D 02 36 0A CD 9E 02 2A 8D 02 36  ....*..6....*..6
```

```
01C0 0A FE 54 CA 23 02 C9 E1 E3 06 06 11 04 03 7E 12   ..T.#.........~.
01D0 23 13 05 C2 CE 01 E3 E9 21 FD 02 06 0D C5 7E CD   #.......!.....~.
01E0 8F 02 C1 05 23 C2 DD 01 C9 CD C7 01 3E 20 32 FD   ....#.......> 2.
```

-L0100,0106 Disassemble the area starting at 0100 which is given as the point
to which the operating system (CP/M-80) will transfer execution to start things
off.

```
  0100 JMP 0120
```

-L0120,0191 Disassemble the main program. Compare this listing with that produced by the FORTRAN compiler above (TEST.PRN). The instructions are the same but
the symbolic variables and relative addresses have been replaced with actual addresses in memory so that the CPU can execute the program. Annotations have been
placed to emphasize the relationship between this listing and the one in TEST.PRN.
NOTE: The variables ONE, TEN, I, and elements of array A are 16-bit two's complement binary numbers stored least significant byte first (LSB at lower address).

```
  0120 LXI B,0126 ;Initialization
  0123 JMP 020B

  0126 LXI H,0001 ;ONE = 1
  0129 SHLD 011A  ;ONE is at address 011A (LSB) and 011B (MSB)
  012C LXI H,000A ;TEN = 10
  012F SHLD 011C  ;TEN is at address 011C (LSB) and 011D (MSB)

  0132 LHLD 011A  ;DO 5 I=ONE,TEN
  0135 SHLD 0118  ;I is at address 0118 (LSB) and 0119 (MSB)
  0138 LHLD 0118  ;A(I) = ONE
  013B DAD H
  013C LXI D,0102 ;The 0102 is the address of A(1) lowered by 2
  013F DAD D
  0140 SHLD 011E  ;Temporary storage T:000000 is at 011E and 011F
  0143 LHLD 011A
  0146 XCHG
  0147 LHLD 011E
  014A MOV M,E
  014B INX H
  014C MOV M,D
  014D LHLD 0118  ;5 CONTINUE
  0150 INX H
  0151 XCHG
  0152 LHLD 011C
  0155 XCHG
  0156 MOV A,E
  0157 SUB L
  0158 MOV A,D
  0159 SBB H
  015A JP 0135     ;If I<=TEN go back to start of loop body at 0135

  015D LHLD 011A  ;DO 20 I=ONE,TEN
  0160 SHLD 0118
  0163 LHLD 0118  ;calculate address of A(I)
  0166 DAD H
  0167 LXI D,0102
  016A DAD D
  016B SHLD 011E
```

```
016E MOV A,M      ;get contents of A(I)
016F INX H
0170 MOV H,M
0171 MOV L,A
0172 XCHG         ;place contents of A(I) in DE
0173 LHLD 0118    ;get contents of I in HL
0176 DAD D        ;calculate A(I)+I and place in HL
0177 XCHG         ;place A(I)+I in DE
0178 LHLD 011E    ;get address of A(I) from temporary T:000000
017B MOV M,E      ;place 2 byte sum at location using HL as pointer, A(I)
017C INX H
017D MOV M,D
017E LHLD 0118    ;20 CONTINUE
0181 INX H        ;HL contains I
0182 XCHG
0183 LHLD 011C    ;compare loop counter with TEN and set sign flag
0186 XCHG
0187 MOV A,E
0188 SUB L
0189 MOV A,D
018A SBB H
018B JP 0160 ;test sign flag for positive I<=TEN and go again if true
018E CALL 01E9  ;STOP return to operating system
```

The remaining pages show the results of executing the program by giving
dumps of the FORTRAN program's data areas and the CPU registers and then
causing small segments of the program to be executed.

-D0100,0120 Dump of the FORTRAN program's data area – area of interest starts
 at 0104 (the LSB of A(1)) and ends with 011F (the MSB of T:000000). All
 contain zero before the program starts.

```
0100 C3 20 01 00 00 00 00 00 00 00 00 00 00 00 00 00 . . . . . . . . . . . . . . .
0110 00 00 00 00 00 00 00 00 00 00 00 00 00 00 00 00 . . . . . . . . . . . . . . . .
```

-X Examine the contents of the CPU registers before the program starts.

```
Flags        Acc    BC     DE     HL    Stk Ptr  PC    Next Instruction
----------+----+------+------+------+------+------+------------------
C0Z0M0E0I0 A=00 B=0000 D=0000 H=0000 S=0100 P=0100 JMP 0120
```

-G0100,12C Go to 0100 and continue until 012C. This executes the initialization
*012C   and the ONE = 1 part of the code. The *12C says the execution
            stopped at 012C.

-X   Examine the registers. Note: HL contains 0001

```
C0Z1M0E1I0 A=00 B=0126 D=0000 H=0001 S=DBFF P=012C LXI H,000A
```

-D0100,120 Look in the data area. Note: 011A (LSB of I) contains 01 and
 011B (MSB of I) contains 00, i.e., I contains 0001.

```
0100 C3 20 01 00 00 00 00 00 00 00 00 00 00 00 00 00 . . . . . . . . . . . . . . .
0110 00 00 00 00 00 00 00 00 00 00 01 00 00 00 00 00 . . . . . . . . . . . . . . . .
```

-G,12F   Execute to the instruction at 012F.
*012F

```
-X        Examine the registers. Note: HL contains 000A now.

C0Z1M0E1I0 A=00 B=0126 D=0000 H=000A S=DBFF P=012F SHLD 011C

-G,132    Execute to the instruction at 0132.
*0132

-X        Examine the registers. Note: HL still contains 000A.

C0Z1M0E1I0 A=00 B=0126 D=0000 H=000A S=DBFF P=0132 LHLD 011A

-D0100,0120 Look in the program's data area. Note: 011C (LSB of TEN) contains
            0A and 011D (MSB of TEN) contains 00, i.e., TEN contains 000A.

0100 C3 20 01 00 00 00 00 00 00 00 00 00 00 00 00 00 . ..............
0110 00 00 00 00 00 00 00 00 00 00 01 00 0A 00 00 00 ...............

-G,138    Execute until the instruction at 0138.
*0138

-X        Examine the registers. Note: the program initialized the loop counter
          with the contents on ONE (at 011A and 011B). HL contains 0001.

C0Z1M0E1I0 A=00 B=0126 D=0000 H=0001 S=DBFF P=0138 LHLD 0118

-D100,120 Look in the data area. Loop counter I is at 0118 and 0119. It
 contains 0001 in reverse byte order.

0100 C3 20 01 00 00 00 00 00 00 00 00 00 00 00 00 00 . ..............
0110 00 00 00 00 00 00 00 00 01 00 01 00 0A 00 00 00 ...............

-T21      Use the Trace mode to display the registers and then run the
          instruction shown at the right end of each line below:

C0Z1M0E1I0 A=00 B=0126 D=0000 H=0001 S=DBFF P=0138 LHLD 0118
C0Z1M0E1I0 A=00 B=0126 D=0000 H=0001 S=DBFF P=013B DAD H
C0Z1M0E1I0 A=00 B=0126 D=0000 H=0002 S=DBFF P=013C LXI D,0102
C0Z1M0E1I0 A=00 B=0126 D=0102 H=0002 S=DBFF P=013F DAD D
C0Z1M0E1I0 A=00 B=0126 D=0102 H=0104 S=DBFF P=0140 SHLD 011E
C0Z1M0E1I0 A=00 B=0126 D=0102 H=0104 S=DBFF P=0143 LHLD 011A
C0Z1M0E1I0 A=00 B=0126 D=0102 H=0001 S=DBFF P=0146 XCHG
C0Z1M0E1I0 A=00 B=0126 D=0001 H=0102 S=DBFF P=0147 LHLD 011E
C0Z1M0E1I0 A=00 B=0126 D=0001 H=0104 S=DBFF P=014A MOV M,E
C0Z1M0E1I0 A=00 B=0126 D=0001 H=0104 S=DBFF P=014B INX H
C0Z1M0E1I0 A=00 B=0126 D=0001 H=0105 S=DBFF P=014C MOV M,D
C0Z1M0E1I0 A=00 B=0126 D=0001 H=0105 S=DBFF P=014D LHLD 0118
C0Z1M0E1I0 A=00 B=0126 D=0001 H=0001 S=DBFF P=0150 INX H
C0Z1M0E1I0 A=00 B=0126 D=0001 H=0002 S=DBFF P=0151 XCHG
C0Z1M0E1I0 A=00 B=0126 D=0002 H=0001 S=DBFF P=0152 LHLD 011C
C0Z1M0E1I0 A=00 B=0126 D=0002 H=000A S=DBFF P=0155 XCHG
C0Z1M0E1I0 A=00 B=0126 D=000A H=0002 S=DBFF P=0156 MOV A,E
C0Z1M0E1I0 A=0A B=0126 D=000A H=0002 S=DBFF P=0157 SUB L
C0Z0M0E0I0 A=08 B=0126 D=000A H=0002 S=DBFF P=0158 MOV A,D
C0Z0M0E0I0 A=00 B=0126 D=000A H=0002 S=DBFF P=0159 SBB H
C0Z1M0E1I0 A=00 B=0126 D=000A H=0002 S=DBFF P=015A JP 0135
```

At this point, the first DO loop has been executed once, i.e., A(1) = ONE. The
trace that follows starts at the first instruction in the loop body.

```
C0Z1M0E1I0 A=00 B=0126 D=000A H=0002 S=DBFF P=0135 SHLD 0118 Save incremented I
C0Z1M0E1I0 A=00 B=0126 D=000A H=0002 S=DBFF P=0138 LHLD 0118
C0Z1M0E1I0 A=00 B=0126 D=000A H=0002 S=DBFF P=013B DAD H
C0Z1M0E1I0 A=00 B=0126 D=000A H=0004 S=DBFF P=013C LXI D,0102
C0Z1M0E1I0 A=00 B=0126 D=0102 H=0004 S=DBFF P=013F DAD D
C0Z1M0E1I0 A=00 B=0126 D=0102 H=0106 S=DBFF P=0140 SHLD 011E
C0Z1M0E1I0 A=00 B=0126 D=0102 H=0106 S=DBFF P=0143 LHLD 011A
C0Z1M0E1I0 A=00 B=0126 D=0102 H=0001 S=DBFF P=0146 XCHG
C0Z1M0E1I0 A=00 B=0126 D=0001 H=0102 S=DBFF P=0147 LHLD 011E
C0Z1M0E1I0 A=00 B=0126 D=0001 H=0106 S=DBFF P=014A MOV M,E
C0Z1M0E1I0 A=00 B=0126 D=0001 H=0106 S=DBFF P=014B INX H
C0Z1M0E1I0 A=00 B=0126 D=0001 H=0107 S=DBFF P=014C MOV M,D
*014D    Stop at 014D in the second pass

-D0100,0120 Look in the data area. Note: A(1) 0104/0105 and A(2) 0106/0107
            contain 0001's in reverse order. T:000000 (011E/011F) contains
            0106 the address of A(2)'s LSB. I (0118/0119) contains 02 since the
            second pass has started.
0100 C3 20 01 00 01 00 01 00 00 00 00 00 00 00 00 00 ...............
0110 00 00 00 00 00 00 00 00 02 00 01 00 0A 00 06 01 ...............

-G,15D Execute from this point until the first DO loop is completed at 015D.
*015D

-D0100,120 Look at the data area. Note: all elements of array A contain 0001's
           The loop counter I contains 000A (10 decimal) and T:000000 contains
           0116 the address of A(10).
0100 C3 20 01 00 01 00 01 00 01 00 01 00 01 00 01 00 ...............
0110 01 00 01 00 01 00 01 00 0A 00 01 00 0A 00 16 01 ...............

-T29    Trace the second DO loop during its first pass.

C1Z0M1E1I1 A=FF B=0126 D=000A H=000B S=DBFF P=015D LHLD 011A
C1Z0M1E1I1 A=FF B=0126 D=000A H=0001 S=DBFF P=0160 SHLD 0118
C1Z0M1E1I1 A=FF B=0126 D=000A H=0001 S=DBFF P=0163 LHLD 0118
C1Z0M1E1I1 A=FF B=0126 D=000A H=0001 S=DBFF P=0166 DAD H
C0Z0M1E1I1 A=FF B=0126 D=000A H=0002 S=DBFF P=0167 LXI D,0102
C0Z0M1E1I1 A=FF B=0126 D=0102 H=0002 S=DBFF P=016A DAD D
C0Z0M1E1I1 A=FF B=0126 D=0102 H=0104 S=DBFF P=016B SHLD 011E
C0Z0M1E1I1 A=FF B=0126 D=0102 H=0104 S=DBFF P=016E MOV A,M
C0Z0M1E1I1 A=01 B=0126 D=0102 H=0104 S=DBFF P=016F INX H
C0Z0M1E1I1 A=01 B=0126 D=0102 H=0105 S=DBFF P=0170 MOV H,M
C0Z0M1E1I1 A=01 B=0126 D=0102 H=0005 S=DBFF P=0171 MOV L,A
C0Z0M1E1I1 A=01 B=0126 D=0102 H=0001 S=DBFF P=0172 XCHG
C0Z0M1E1I1 A=01 B=0126 D=0001 H=0102 S=DBFF P=0173 LHLD 0118
C0Z0M1E1I1 A=01 B=0126 D=0001 H=0001 S=DBFF P=0176 DAD D
C0Z0M1E1I1 A=01 B=0126 D=0001 H=0002 S=DBFF P=0177 XCHG
C0Z0M1E1I1 A=01 B=0126 D=0002 H=0001 S=DBFF P=0178 LHLD 011E
C0Z0M1E1I1 A=01 B=0126 D=0002 H=0104 S=DBFF P=017B MOV M,E
C0Z0M1E1I1 A=01 B=0126 D=0002 H=0104 S=DBFF P=017C INX H
C0Z0M1E1I1 A=01 B=0126 D=0002 H=0105 S=DBFF P=017D MOV M,D
C0Z0M1E1I1 A=01 B=0126 D=0002 H=0105 S=DBFF P=017E LHLD 0118
C0Z0M1E1I1 A=01 B=0126 D=0002 H=0001 S=DBFF P=0181 INX H
C0Z0M1E1I1 A=01 B=0126 D=0002 H=0002 S=DBFF P=0182 XCHG
C0Z0M1E1I1 A=01 B=0126 D=0002 H=0002 S=DBFF P=0183 LHLD 011C
C0Z0M1E1I1 A=01 B=0126 D=0002 H=000A S=DBFF P=0186 XCHG
C0Z0M1E1I1 A=01 B=0126 D=000A H=0002 S=DBFF P=0187 MOV A,E
C0Z0M1E1I1 A=0A B=0126 D=000A H=0002 S=DBFF P=0188 SUB L
C0Z0M0E0I0 A=08 B=0126 D=000A H=0002 S=DBFF P=0189 MOV A,D
```

```
C0Z0M0E0I0 A=00 B=0126 D=000A H=0002 S=DBFF P=018A SBB H
C0Z1M0E1I0 A=00 B=0126 D=000A H=0002 S=DBFF P=018B JP 0160
 End of execution of first pass through the second DO loop
 Beginning of second pass through the second DO loop
C0Z1M0E1I0 A=00 B=0126 D=000A H=0002 S=DBFF P=0160 SHLD 0118 save loop counter
C0Z1M0E1I0 A=00 B=0126 D=000A H=0002 S=DBFF P=0163 LHLD 0118
C0Z1M0E1I0 A=00 B=0126 D=000A H=0002 S=DBFF P=0166 DAD H
C0Z1M0E1I0 A=00 B=0126 D=000A H=0004 S=DBFF P=0167 LXI D,0102
C0Z1M0E1I0 A=00 B=0126 D=0102 H=0004 S=DBFF P=016A DAD D
C0Z1M0E1I0 A=00 B=0126 D=0102 H=0106 S=DBFF P=016B SHLD 011E
C0Z1M0E1I0 A=00 B=0126 D=0102 H=0106 S=DBFF P=016E MOV A,M
C0Z1M0E1I0 A=01 B=0126 D=0102 H=0106 S=DBFF P=016F INX H
C0Z1M0E1I0 A=01 B=0126 D=0102 H=0107 S=DBFF P=0170 MOV H,M
C0Z1M0E1I0 A=01 B=0126 D=0102 H=0007 S=DBFF P=0171 MOV L,A
C0Z1M0E1I0 A=01 B=0126 D=0102 H=0001 S=DBFF P=0172 XCHG
C0Z1M0E1I0 A=01 B=0126 D=0001 H=0102 S=DBFF P=0173 LHLD 0118
*0176    Stop execution during second pass through the second DO loop.

-D0100,0120 Dump the data area. Note: A(1) is now 0002 from A(1)=A(1)+1 and
            the loop counter I contains a 0002 since the second pass has been
            started.
0100 C3 20 01 00 02 00 01 00 01 00 01 00 01 00 01 00 . ..............
0110 01 00 01 00 01 00 01 00 02 00 01 00 0A 00 06 01 ................

-G,18E   Execute the remaining passes through the second DO loop and stop just
*018E    beyond the exit at 018E.

-D0100,120 Look at the data area. Note: the array A contains 0002 through
           000B (stored in byte reverse order) which was produced by A(I)=A(I)+I.
0100 C3 20 01 00 02 00 03 00 04 00 05 00 06 00 07 00 . ..............
0110 08 00 09 00 0A 00 0B 00 0A 00 01 00 0A 00 16 01 ................
```

The listing below comes from the Dolch 40C50 logic analyzer during the execution of the FORTRAN statement ONE=1 in the file TEST.FOR. The program was run on a Z80 that is a compatible with the 8080 processor at the machine code level. The microprocessor's clock signal input is being used by the logic analyzer to sample the value of the signals on the microprocessor's address, data and command buses when the clock signal changes from low to high (rising edge). The leftmost column of 4 digits shows the position of the data in the logic analyzer's memory and is of no consequence here. The next two 2-digit columns are the contents of the processor's address bus at the time the sample was taken. The third 2-digit column is the contents of the data bus at the rising edge of the processor's clock. The 8-digit column shows the values of several signals in the command bus in binary. The signals, from left to right, are RD.L, WR.L, M1.L, MREQ.L, and IOREQ.L, where the .L means that the signal is *True* when 0 and *False* when 1. The remaining 3 bits are unused.

The sequence starts (i.e., the logic analyzer was triggered) when the address bus contained the address 0126 and continues until the address bus contains 012C the beginning of the next FORTRAN statement. Remember that the assembly language representation on page 5 of the file t.run is equivalent to the instructions that the processor actually sees. The dump on the same page, in particular the line that starts at address 0120, contains the machine code that the processor is seeing as a result of the INSTRUCTION FETCH SEQUENCE.

Note: M1 indicates the processor is fetching an opcode of an instruction and MREQ indicates the address space referred to is memory.

```
Raw Logic Analyzer Data     Translation & Comments
----------------------+------------------------------------------------------
      A   A   D   RWMMI   |
      1   7   7   DR1RO   |
      5   -   -     ER    |
      -   0   0     QE    |
      8             Q     |
                          | Will now do LXI H,0001 or 0126 21 01 00
0010 01 26 21 01001000    | Addr=0126; Data=21; RD, M1, MREQ - Fetch opcode "21"
0011 FF 39 21 11101000    |
0012 FF 39 21 11111000    |
0013 01 27 21 01101000    |
0014 01 27 01 01101000    | Fetch Least Significant Byte of immediate data "01"
0015 01 27 01 11111000    |
0016 01 28 01 01101000    |
0017 01 28 00 01101000    | Fetch Most Significant Byte of immediate data "00"
0018 01 28 00 11111000    |
0019 01 29 00 01001000    | Will now do SHLD 011A or 0129 22 1A 01
0020 01 29 22 01001000    | Fetch opcode "22"
0021 FF 3A 22 11101000    |
0022 FF 3A 22 11111000    |
0023 01 2A 22 01101000    |
0024 01 2A 1A 01101000    | Fetch Least Significant Byte of address "1A" of dest.
0025 01 2A 1A 11111000    |
0026 01 2B 1A 01101000    |
0027 01 2B 01 01101000    | Fetch Most Significant Byte of address "01" of dest.
0028 01 2B 01 11111000    |
0029 01 1A 01 11101000    | Perform transfer of HL contents to memory at 011A
0030 01 1A 01 10101000    | Addr=011A; Data=01; WR; Write LSB of data to 011A
0031 01 1A 01 11111000    |
0032 01 1B 00 11101000    |
0033 01 1B 00 10101000    | Addr=011B; Data=00; WR; Write MSB of data to 011B
0034 01 1B 00 11111000    |
```

This demonstration shows the operation of a Z80 microprocessor executing the machine code necessary to initialize the loop counter I in the first DO loop in the FORTRAN program TEST.FOR. An assembly language equivalent of the machine code can be seen in the compiled program starting at address 0132 and ending at address 0138. The instructions to be done are as follows:

```
0132     LHLD 011A ;Place the contents of location TEN in register HL
0135     SHLD 0118 ;Place the contents of register pair HL in TEN.
```

Note: the number being moved is 16 bits long and will require two memory cycles each time it is moved, one for each byte.

The table below shows the Logic Analyzer sample address in the first 4 digits, the most significant byte of the processor's address bus in the next 2-digit field, the least significant byte of the processor's address bus in the next 2-digit field, the processor's data bus, and then the one bit RD.L, WR.L, M1.L, MREQ.L, and IOREQ.L command bus signals.

```
0000SFF FB 8A 11111000
0001 F1 80 32 01001000
0002 F1 80 DD 01001000
0003 FF 2A DD 11101000
0004 FF 2A DD 11111000
0005 F1 81 DD 01001000
```

```
0006 F1 81 E9 01001000
0007 FF 2B E9 11101000
0008 FF 2B E9 11111000

TRIG 01 32 E9 01001000 ;Begin fetching LHLD ONE
0010 01 32 2A 01001000 ;Addr=0132; Data=2A; RD, M1, MREQ; Get opcode "2A" from
0011 FF 2C 2A 11101000 ; address 0132 in memory.
0012 FF 2C 2A 11111000
0013 01 33 2A 01101000
0014 01 33 1A 01101000 ;Get LSB of address of ONE "1A" from memory at address
0015 01 33 1A 11111000 ; 0133 - next byte after opcode.
0016 01 34 1A 01101000
0017 01 34 01 01101000;Get MSB of address of ONE "01" from memory at address
0018 01 34 01 11111000 ; 0134 - sencond byte after opcode.
0019 01 1A 01 01101000
0020 01 1A 01 01101000 ;Read the data at memory address 011A into the CPU -
0021 01 1A 01 11111000; should go into reg. L of the HL pair but this can't
0022 01 1B 01 01101000 ; be seen from outside the CPU.
0023 01 1B 00 01101000 ;Read the data at memory address 011B into the CPU -
0024 01 1B 00 11111000 ; should go into reg. H of the HL pair.
0025 01 35 00 01001000 ;Begin fetching SHLD I
0026 01 35 22 01001000 ;Get opcode "22" from memory at 0135
0027 FF 2D 22 11101000
0028 FF 2D 22 11111000
0029 01 36 22 01101000
0030 01 36 18 01101000 ;Get LSB of address of I "18" from memory at 0136
0031 01 36 18 11111000
0032 01 37 18 01101000
0033 01 37 01 01101000 ;Get MSB of address of I "01" from memory at 0137
0034 01 37 01 11111000
0035 01 18 01 01101000
0036 01 18 01 10101000 ;Write LSB of data "01" to address 0118 - LSB of I
0037 01 18 01 11111000
0038 01 19 00 11101000
0039 01 19 00 10101000 ;Write MSB of data "00" to address 0119 - MSB of I
0040 01 19 00 11111000
0041 01 38 00 01001000 ;Begin fetching the next opcode - a RET instruction
0042 01 38 C9 01001000 ; was substituted here for this demonstration.
0043 FF 2E C9 11101000
0044 FF 2E C9 11111000
0045 FF FB C9 01101000
0046 FF FB 8A 01101000
0047 FF FB 8A 11111000
0048 FF FC 8A 01101000
0049 FF FC F2 01101000
0050 FF FC F2 11111000
0051 F2 8A F2 01001000
0052 F2 8A B7 01001000
```

Exercise - The Z80 based computer is executing part of TEST.CPM. Its starting
address is 0138 (the Trigger Point).

```
Raw Logic Analyzer Data            Translation & Comments
-----------------------+------------------------------------------------------
    -
      A   A   D   RWMMI         |
      1   7   7   DR1RO         |
      5   -   -     ER          |
```

```
              -   0   0    QE      |
              8            Q       |
0000SFF FB 8A 11111000
0001 F1 80 38 01001000
0002 F1 80 DD 01001000
0003 FF 30 DD 11101000
0004 FF 30 DD 11111000
0005 F1 81 DD 01001000
0006 F1 81 E9 01001000
0007 FF 31 E9 11101000
0008 FF 31 E9 11111000
TRIG 01 38 E9 01001000
0010 01 38 2A 01001000 ;What instruction is being fetched here?
0011 FF 32 2A 11101000 ; Give the machine code for the complete instruction
0012 FF 32 2A 11111000 ; and then give the symbolic assembly code and show
0013 01 39 2A 01101000 ; what part of the FORTRAN code that this instruction
0014 01 39 18 01101000 ; implements.
0015 01 39 18 11111000
0016 01 3A 18 01101000
0017 01 3A 01 01101000
0018 01 3A 01 11111000
0019 01 18 01 01101000
0020 01 18 01 01101000
0021 01 18 01 11111000
0022 01 19 01 01101000
0023 01 19 00 01101000
0024 01 19 00 11111000
0025 01 3B 00 01001000
0026 01 3B 29 01001000 ;What instruction is being fetched here? Give same
0027 FF 33 29 11101000 ; treatment as above.
0028 FF 33 29 11111000
0029 FF 33 29 11111000
0030 FF 33 29 11111000
0031 FF 33 2F 11111000
0032 FF 33 2F 11111000
0033 FF 33 2F 11111000
0034 FF 33 2F 11111000
0035 FF 33 0F 11111000
0036 01 3C 29 01001000 ;What is going on from 0027 to here.
0037 01 3C 11 01001000 ;What instruction is being fetched here? Give same
0038 FF 34 11 11101000 ; treatment as above.
0039 FF 34 11 11111000
0040 01 3D 11 01101000
0041 01 3D 02 01101000
0042 01 3D 02 11111000
0043 01 3E 02 01101000
0044 01 3E 01 01101000
0045 01 3E 01 11111000
0046 01 3F 01 01001000
0047 01 3F 19 01001000 ;What instruction is being fetched here? Give same
0048 FF 35 19 11101000 ; treatment as above.
0049 FF 35 19 11111000
0050 FF 35 19 11111000
0051 FF 35 1D 11111000
0052 FF 35 1F 11111000
0053 FF 35 1F 11111000
0054 FF 35 1F 11111000
0055 FF 35 1F 11111000
0056 FF 35 0F 11111000 ;What is going on from 0048 to here.
```

```
0057 01 40 19 01001000 ;What instruction is being fetched here? Give same
0058 01 40 22 01001000 ; treatment as above.
0059 FF 36 22 11101000
0060 FF 36 22 11111000
0061 01 41 22 01101000
0062 01 41 1E 01101000 ;What is happening here in terms of the FORTRAN program?
0063 01 41 1E 11111000
0064 01 42 1E 01101000
0065 01 42 01 01101000 ;Is this some more of the operation that started in 0062?
0066 01 42 01 11111000
0067 01 1E 04 11101000 ;What is happening here in terms of the FORTRAN program?
0068 01 1E 04 10101000
0069 01 1E 04 11111000
0070 01 1F 01 11101000
0071 01 1F 01 10101000
0072 01 1F 01 11111000
0073 01 43 00 01001000
0074 01 43 2A 01001000 ;What instruction is being fetched here? Give same
0075 FF 37 2A 11101000 ; treatment as above.
0076 FF 37 2A 11111000
0077 01 44 2A 01101000
0078 01 44 1A 01101000
0079 01 44 1A 11111000
0080 01 45 1A 01101000
0081 01 45 01 01101000
0082 01 45 01 11111000
0083 01 1A 01 01101000
0084 01 1A 01 01101000
0085 01 1A 01 11111000
0086 01 1B 01 01101000
0087 01 1B 00 01101000
0088 01 1B 00 11111000 ;What should data memory look like at this time?
0089 01 46 00 01001000 ;How does one determine from the CPU command bus that
0090 01 46 C9 01001000 ; this is an opcode fetch?
0091 FF 38 C9 11101000
0092 FF 38 C9 11111000
```

Chapter 9

IMPROVEMENTS, VARIATIONS, AND CONCLUSION

9.1. INTRODUCTION

Our concluding chapter is intended to put the finishing touches on the discussions started in earlier chapters. The first few sections deal with some optimizations that are possible for the computer example discussed in Chapter 8. The need for the optimizations was discovered in the course of writing microprograms to implement instructions and address modes for the example. Since a simulator had been written for the example based on the principles discussed in Chapter 6, we were able to write first- and second-level programs for the machine and test them in a running environment.

Section 9.2.1 describes a simple wiring change that was detected in the course of implementing the register indirect with offset address mode in the first-level machine. If the simulator had not been written, this change would have been detected after the hardware had been built. The change could not have been made and tested as easily as it was on the simulator. Therefore, it serves as a convincing argument for the simulation of solutions before hardware commitment is made.

Section 9.2.2 shows an improvement made to the overall performance of the computer example by introducing hardware support for a new data type, the byte. While the byte data type could have been added by using additions to the microprogram alone, the hardware changes greatly increase the speed of execution. Again, these changes were explored using the simulator to gather relative timing information. Notice that if the speed increase obtained was not very large, the hardware modifications would not be justified. It is much easier to get this information from the simulator than from a hardware prototype.

Sections 9.2.3 through 9.2.5 are intended to show how the important features found in most computer systems are implemented in the Central Processing Unit (CPU). Interrupts are important since the CPU can perform other tasks while "simultaneously" paying attention to asynchronous external events. A reader with microprocessor experience should wonder how the instruction flow can be diverted to another task by an agency outside the CPU. Section 9.2.3 should make this point clear.

The CPU can do many types of instructions in the performance of an application program. None of the instructions is performed as efficiently as a machine having a single level of control since so much time is spent fetching second-level instructions. Some algorithms, such as data movement, can be moved from the CPU to other optimized controllers. To do this, the CPU must know how to relinquish its external command, address, and data buses. This addition to the computer example is made in Section 9.2.4.

We implemented a Complex Instruction Set Computer (CISC) for our example in Chapter 8. We chose to implement complex instruction sequences since the overhead to fetch an instruction was great in our machine. Also, the microprogrammed controller made this very easy to do since the control code was contained in a memory, not in wiring paths as in the traditional controller. There is a major alternative to this approach. What would a CPU architecture look like in which the designers directly addressed the instruction fetching overhead? Could the overhead be minimized to the point that fetching and execution occur at the same time? A possible answer is the Reduced Instruction Set Computer (RISC), an architecture that implements only the instructions that can be performed in one "cycle". This important architecture based on the traditional controller is discussed in Section 9.3.2. The purpose of including it in a book about microprogrammed state machines is to demonstrate some of the alternative solutions to our "engineering problem" with which we began.

The last two sections deal with Application-Specific Instruction Set Processors (ASISP). These machines support instructions that are dedicated to the problems areas to which they are applied. Both the Digital Signal Processor (DSP) and the Video Display Processor (VDP) have a complete set of general purpose instructions qualifying them to be General Purpose Instruction Set Processors (GPISPs). Added to this set are instructions and internal data paths that decrease the execution time for certain algorithms specific to their application areas. While both machines may be implemented as CISC or RISC processors, it is the application-specific nature of their instruction sets that interests us here.

9.2. IMPROVEMENTS

The original specification for the hardware implementation of the computer example in Chapter 8 was that it must contain the smallest number of devices and the simplest bus structure possible for pedagogical reasons. While several small changes were made to this in the beginning, the hardware still meets these restrictions. At this point, we would like to point out several possible "improvements" that require relatively minor changes. These "case studies" provide a scenario in which the student can study how a design may evolve toward a practical, but elegant, solution. The original specification was made intentionally simple because added complexity would have obscured the understanding needed on the basic level. Now that we have reached this level of competence, let us see about refining several points of our design and extending its flexibility within the framework of the original.

The following sections show some of the improvements that are possible. They range from simple hardware changes that reduce the number of clock cycles needed to implement a sequence (e.g., the MAR bus) to changes that extend the flexibility of the machine (e.g., interrupts). The addition of byte manipulation makes handling this common data type much more efficient even though the operations could have been performed in the current system.

9.2.1. The Memory Address Register Bus

In the course of implementing the indexed address mode, we discovered that we needed to store the effective address in a temporary location in the Register Arithmetic/Logic Unit (RALU) because the path to the Memory Address Register (MAR) was blocked by the "offset" operand. This blockage is caused by the fact that the internal data bus represents the only path between all elements in the internal architecture as well as the path to the external architecture. Referring to Figures 8-2 and 8-4, one can see that the MAR input is taken from a stub of the internal data bus. What if the MAR input were disconnected from the internal data bus and reconnected at a place where the incoming offset from memory would not interfere with it?

In placing the Am2903 RALU in the internal architecture, we discovered that we had to protect its internal Y bus from data on the data bus; otherwise, we could not return the results to the register array if the operation used external data. This resulted in the introduction of the bidirectional Y-to-Databus driver shown above the Am2903 in Figure 8-2. Incidentally, the same consideration applies to the bidirectional DB-to-Databus driver shown immediately to the right of the Y bus driver in the same figure.

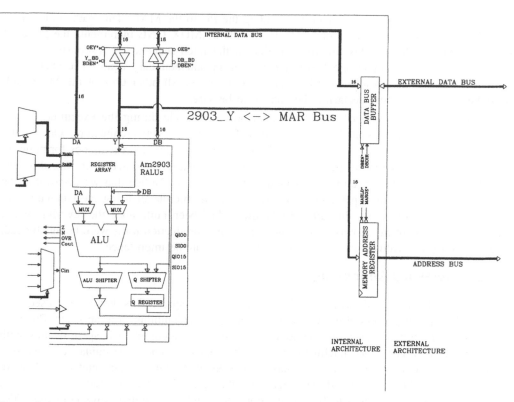

Figure 9-1: Am2903 Y bus to MAR data path.

Our "improvement" then will be to connect the input of the MAR directly to the Y bus output of the Am2903 where it is protected from the internal data bus by the Y-to-Databus driver as is shown in Figure 9-1. Tracing the flow of data in the internal architecture for the indexed with offset address mode shows the following high-level Hardware Description Language (HDL) description.

```
{
External Databus = Memory[MAR];
Internal Databus = External Databus;
2903_DA = Internal Databus;
2903_ALU_R = 2903_DA;
2903_ALU_S = 2903_Reg[RAM_A_Address];
2903_ALU_F = 2903_ALU_R + 2903_ALU_S + zero;
2903_Y_bus = 2903_ALU_F;
MAR_bus = 2903_Y_bus;
MAR := MAR_bus;
}
```

The curly brackets indicate that everything in the paragraph happens in the same clock cycle. The controller instruction is not shown.

After a change like this is made, the designer should review all places this path is used in the earlier implementation. Our immediate concern centers around the program counter (PC). We notice in the microcode for the PC in the instruction fetch sequence that the PC was left pointing at an odd byte at the end of the fetch sequence. This was caused by our choice of the Y bus as a path for sending the PC to the MAR. This path comes from the ALU, and the ALU PASSS operand along with CNMUX ONE increments the PC as it is sent to the MAR. The first microinstruction in the fetch sequence is responsible for the incrementation. The second one leaves the PC on an odd boundary again. The change we have just made does not affect the PC manipulation at all. All other uses of the MAR path must be examined before declaring the change to be safe.

What have we gained? We have saved one microcycle during the execution of any effective address calculation that involves data from the external architecture. One of the ways a CISC like this one gets its speed is by providing complex address modes for pointing at data. These modes tend to be used frequently if all instructions have them. So, this represents an often used change that costs us no additional parts. Notice that we discovered the change in the course of implementing algorithms at the microlevel using the simulator. We would not want to discover this after the hardware had been built since the rewiring would have been tedious in the prototype and impossible once production units were being shipped. The capability to evaluate and evolve a design is of supreme importance.

9.2.2. Byte/Word Addressing

Most character codes require only 8 bits of storage space. This data type is used continually during the process of editing text, compiling, or assembling programs, etc. Possibly, the greatest amount of execution time and storage space on a computer is spent dealing with the character data type. Our machine can deal with characters under the guidance of the second-level programmer. One character may be placed in the left or right end of each word and then manipulated as a word. The character, thus placed, can be treated as an unsigned or signed number allowing it to be added, subtracted, and, most importantly, compared. The problem with this method is that the programmer is wasting half of the available memory space along with extra processing to deal with these short numbers. Could we help him by creating a "byte" or "character" data type having its own set of instructions? We do not need at this time to discuss the actual character codings, but only establish a method for manipulating them. The one property of these codings we need to be aware of is that the position of a character in the alphabet is mapped onto the counting sequence in an increasing order, i.e., the character for "Z" has a numerically greater value than that for "A". One important operation then consists of comparing characters in order to alphabetize them. Another important operation is the capability to manipulate strings of characters. These normally look like one-dimensional arrays in many high-level languages so we should certainly consider the use of the various indexed address modes in instructions for this data type.

The hardware modifications that we need to make are all addressed to increasing the efficiency of storage and transportation of the byte between the RALU and the external architecture. Our changes will be concentrated in the external memory, the data bus driver between the internal and external architecture, and in the control of the RALU. In each case, change will consist of dividing the element under consideration into half; i.e., separating a 16-bit wide entity into two 8-bit wide entities. The upper and lower halves of each entity are so named because of their relation to the numerical value of the original 16-bit entity. We will create two 8-bit memory halves named MEM_H and MEM_L. The RALU will become RALU_H and RALU_L and the data bus driver will become DBDRV_H and DBDRV_L, respectively. The "_H" part will encompass the old bits 15 through 8 and the "_L" part will contain the old bits 7 through 0. Notice that RALU_H is still physically connected to the status registers while the carry out of RALU_L logically drives the carry input of RALU_H. We will examine each one of these changes individually.

Memory Partitioning

There are two modifications in the external architecture that must be made to support byte addressing and manipulation. The first was made when we originally designed the computer example and was caused by our need to use an assembler that supported byte addressing. We wired the MAR to the external memories in such a way that MAR bit 1 connected to memory address pin 0. MAR bit 0 is unused and the remaining bits are offset similar to MAR bit 1; i.e., MAR bit 2 to memory address pin 1, etc. A memory cell is 16 bits wide so it appears to the second-level programmer that addresses 0 and 1 point at different bytes in the same word. A similar statement could be made about all odd/even address pairs. As currently wired, this machine can address a byte but *not* manipulate it directly since the same pair of bytes is manipulated whether an odd or even address of an odd/even pair is used.

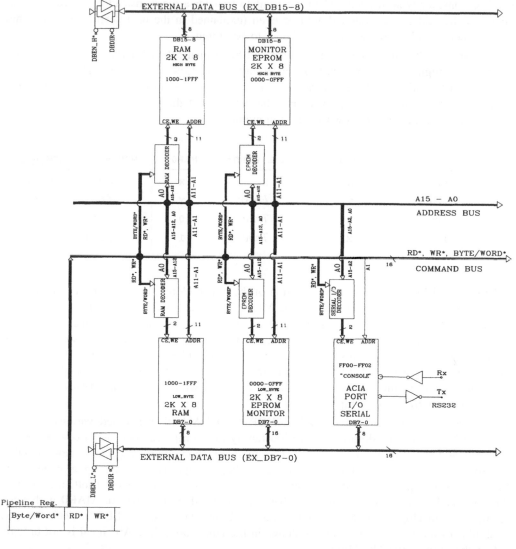

Figure 9-2: External architecture modifications needed to support byte manipulation.

The second modification is needed now to give the second-level programmer direct byte manipulation capability. The following discussion is based on Figure 9-2. The only part of this work that will be done in the memory is associated with writing the data. When a data transfer into the CPU is requested, our memory will function just as it does now. It will place the even byte on the upper half of the data bus and the odd byte on the lower; i.e., MEM_H and EXT_DB_H in the external architecture contain even addressed bytes. This is an arbitrary assignment but it requires fewer changes to the existing design. When the CPU transfers byte data to the external memory, the data on the undesired half of the data bus should not be written into the memory. This is prevented by using the least significant bit of the MAR (the bit 0 that was unconnected until now) to determine whether MEM_H or MEM_L responds to the write strobe WR*. In our example, WR* will not be gated but MAR0 will be used as a term in the chip select equations for MEM_H and MEM_L; i.e., MEM_H_CS will contain /MAR0 and MEM_L_CS will contain MAR0. In doing this, we need to consider what happens when we want to transfer a word instead of a byte. We expect that the address generated should be even (explained in the next section) but our chip select scheme will cause the upper half only to be stored.

We will therefore create a new command bus signal called BYTE/WORD*. This signal is *True* (high) when a byte is transferred and *False* (Low) for a word. The signal itself is placed in the microword, i.e., in the pipeline. It will be used in the other parts of the CPU architecture to support this improvement. The behavior of this signal along with MAR0 in the external architecture is shown in the following table.

MAR15-m+1	MARm-1	BYTE/WORD*	MAR0	MEM_H_CS	MEM_L_CS
False	x	x	x	*False*	*False*
True	x	*False*	x	*True*	*True*
True	x	*True*	0	*True*	*False*
True	x	*True*	1	*False*	*True*

"MAR15-m+1" signifies a Boolean function of the upper address lines that forms the address part of the memory chip select equations MEM_H_CS and MEM_L_CS. The symbol "m" is the top address line needed to address all of the cells within the physical memory blocks. We have made each of our blocks using memory chips containing 2k bytes, so each block contains 2k words or 4k bytes. The upper address line MAR bit 11 (i.e., m = 11). The address lines MAR15 through MAR12 determine the memory block to be activated, while address lines MAR11 through MAR1 determine the word within the block. MAR0 determines which byte (or byte-wide) chip responds.

The function provided by BYTE/WORD* and MAR0 is handled various ways by other designers. In the Motorola MC68000 family of microprocessors, the strobe portion of our read and write strobes is separated from their direction control aspects. A new direction control signal R/W* is created that is not a strobe. A single new timing signal called Data Strobe (DS*) is provided that times all data transfers. In the MC68000 itself, the external data bus is 16 bits wide. Two data strobes are present that combine our MAR0 with DS*, resulting in an Upper Data Strobe (UDS*) and a Lower Data Strobe (LDS*). Their function is similar to the memory chip selects shown in the table above. The A0 signal is not brought off of the chip in the MC68000.

Data Bus Partitioning

At this point, we have only prevented undesired byte-sized data from overwriting the wrong half of the memory word. Our next problem is concerned with moving a byte on the "wrong" half of the bus to the "right" half. What determines the correct half of the bus? We hope to make our change with a minimum amount of hardware modification. Currently RALU_H is wired to the status registers. If we perform arithmetic on bytes in RALU_H, we can avoid supplying an additional connection on the status register multiplexers to get the status from RALU_L. Our first thought then is that RALU_H is the place to perform byte arithmetic. Words will still map alright, as will bytes in MEM_H. We will need to provide an 8-bit bidirectional crossover, in Figure 9-3, between the lower half of the data bus and the upper half to move data from MEM_L to RALU_H. Notice that this scheme works well for combining byte data with byte data, and word data with word data.

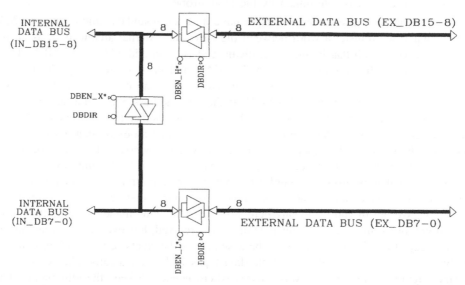

Figure 9-3: Data bus partitioning.

What happens if we want to combine byte data with word data? Small constants could efficiently be stored in a byte-sized field and then turned into words before they are combined with words. Let's examine this possibility. Most of the work involved in this implementation choice will be handled in the next section. Here, all we need to do is decide that byte data will be provided to the RALU on the lower half of the data bus instead of to RALU_H. It will be up to our RALU design to make it combine words and bytes. The crossover we indicated above will be reversed so that bytes in MEM_H passed on EXT_D-B_H will be moved to INT_DB_L. Bytes in MEM_L will be passed directly to INT_DB_L and words will be passed as before, i.e., MEM_H to INT_DB_H and MEM_L to INT_DB_L.

Now examine the 16-bit data bus driver DBDVR that lies between the internal and external data buses. These are implemented using two devices similar to the 74LS245 Octal Bidirectional Bus Driver. One device handles the DB_H and the other DB_L. Each has two control signals, Enable and Direction. Thus far, these signals have been paralleled on the two devices creating a 16-bit bidirectional bus driver with the control signals DBEN and DBDIR. At this point, we will separate these signals and add a third octal bus driver connecting

DB_H to DB_L. There are a total of six control signals and their function is given in the following table.

BYTE/WORD*	MAR0	DBEN_H	DBEN_L	DBEN_X
False	X	True	True	False
True	0	True	False	True
True	1	False	True	False

```
DBDIR_H = DBDIR_L = DBDIR_X = RD*
```

Direction control is determined by the read strobe.

Notice that we have forced words to have an even boundary alignment, i.e., MAR0 must be 0 when we transfer a word between the CPU and external architecture. What if we try to place a word such that its most significant end is at an odd address (in MEM_L)? We start the transfer with the odd address in the MAR and pass the most significant end on EXT_DB_L to INT_DB_H so that it will enter RALU_H where it belongs. But where is the lower end of the number? It is in MEM_H at the next address above the current one. We will need to increment the contents of the MAR and run another memory cycle. This is actually done in all major machines now; however, notice that an extra memory cycle cannot be avoided. The moral of this story is that if you allow odd boundary alignment for the longer data type, then the second-level programmer will pay for it by having slower executing programs. Since it unnecessarily complicates our control algorithm, we will not implement the extra memory cycle and tell the second-level programmer that he must use even addresses for words. We are in good company here since the IBM /360 made the same requirement. The requirement in the /370 and later machines was relaxed; however, the programmer continues to accept the alignment efforts of the assembler if he wants fast execution. Notice that virtually all assemblers that allow multiple data types will automatically take account of the boundary alignments needed and waste bytes where needed to keep the alignment. Our instructions required such alignment from the beginning.

RALU Partitioning

We have now positioned a byte operand on the lower half of the internal data bus regardless of its origin in the external architecture. Also, a byte on INT_DB_L will be written in the external architecture into a byte on the correct half under control of whether its address is odd or even (MAR0). Words will be transferred as before. It is up to our implementation of the RALU controls to properly manipulate the bytes. This effort will take advantage of various features of the Am2903 architecture that were included to deal with this problem. There are several details that can be added that make flag handling more complete. These will be indicated but not performed here.

The first step is to logically separate the control functions of the two RALU halves, RALU_H_2903I and RALU_L_2903I, as shown in the complete picture of our modifications given in Figure 9-4. Our first job will be to create a word-length value from a byte-length value appearing on INT_DB_L. The Am2903 ALU shifter under control of the ALU destination field of 2903I will pass the ALU F bus unchanged (F -> Y) for I8765 = 1111 or copy the Shift I/O bit 0 across the Y bus (SIO0 -> Y0, Y1, Y2, Y3) for I8765 = 1110. This means that all of the 9 instruction bits (2903I) can be paralleled for RALU_H and RALU_L except 2903I5. For all other operations RALU_H_2903I5 = RALU_L_2903I5. The following table summarizes the operation:

BYTE/WORD*	RALU_H_2903I5	RALU_L_2903I5	RALU_H_SIO0
False	RALU_L_2903I5	RALU_H_2903I5	RALU_L_SIO3
True	0	1	RALU_L_SIO3

Notice that the sign bit position in the byte located in RALU_L bit 7 is copied across the upper half of the 16-bit Y bus by RALU_H if BYTE/WORD* is *True*. This is the sign extend function. The following HDL describes the operation:

```
{
    EXT_DB_H = MEM_H[MAR]; EXT_DB_L = MEM_L[MAR];
    INT_DB_L = EXT_DB[MAR]; RALU_DA_L = INT_DB_L;
    RALU_ALU_F_BUS_L = RALU_DA_L + zero;
    RALU_Y_BUS_L = RALU_ALU_F_BUS_L;
    RALU_SIO3_L = RALU_F_BUS_bit7_L;
    RALU_Y_BUS_H = RALU_SIO3_L;
    RALU_REG_H[Reg B Addr] := RALU_Y_BUS_H;
    RALU_REG_L[Reg B Addr] := RALU_Y_BUS_L;
}
```

The events in the HDL fragment above occur during a single clock cycle and end with the sign-extended byte being placed in one of the elements of the RALU register array. This scheme requires that the byte be sign extended into a temporary location and then used as dictated by the instruction in the next cycle.

Handling the flags for a byte operation is another matter. All arithmetic/logic operations so far have assumed that words were being used or that bytes were sign extended before they were used. If one wants to preserve a byte overflow, sign, or carry out, then added positions on the corresponding macrostatus register multiplexers will be needed to collect the flag outputs of RALU_L. The zero flag results from wire-ORing RALU_Z_H and RALU_Z_L when using this part. If a byte zero flag is desired, this connection must be broken and created under control of BYTE/WORD* using a Boolean function.

This concludes our discussion of hardware modifications needed to support byte manipulation on the computer example. The microprogrammer now must supply new sequences that implement byte arithmetic and logic instructions for the second-level programmer. The sequences are initiated by the use of different opcodes in the second-level program for byte and word operations. The correct opcode is inserted by the second-level assembler in response to a symbolic operation (e.g., ADDB or ADDW).

9.2.3. Interrupts to the Second Level of Control

As the computer example described in Chapter 8 is currently configured, a second-level program cannot respond to events that are asynchronous to its own flow of execution. Input/Output (I/O) controllers must be polled to determine if data is available or the controller is available for data output. The act of polling can waste a large number of potentially useful CPU cycles if the I/O controller events occur infrequently. The interrupt allows the CPU to concentrate on the second-level application program until asynchronous events require attention. This modification will eliminate a large part of the wasted time.

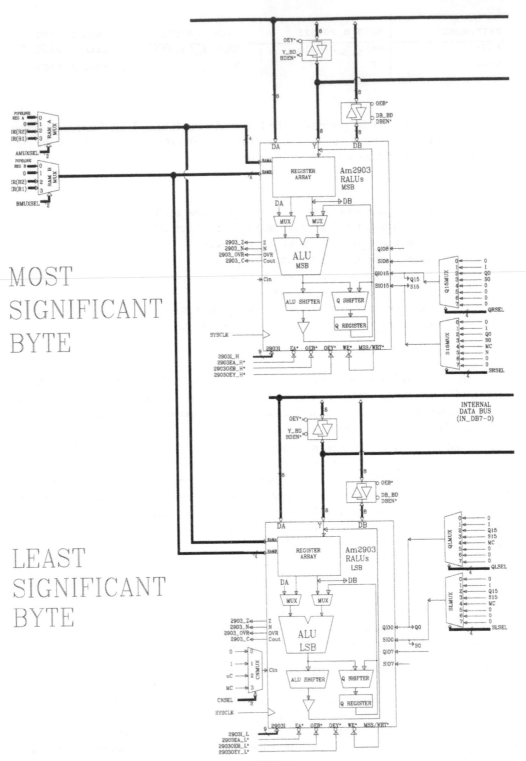

Figure 9-4: Complete byte manipulation modification.

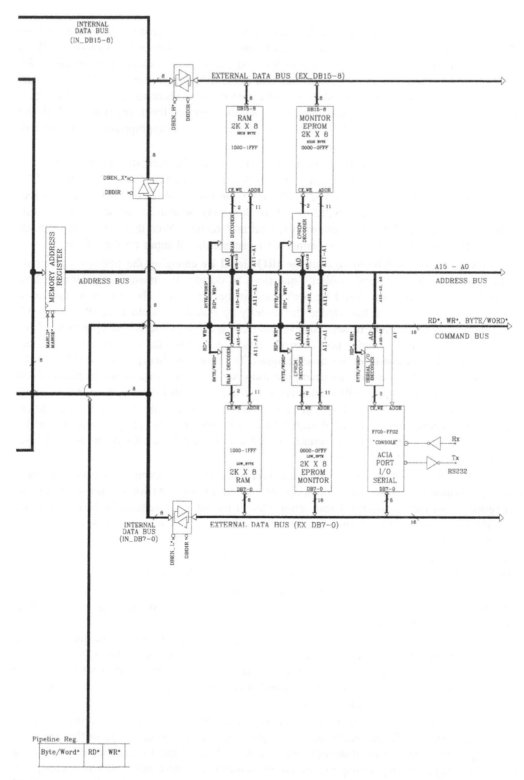

Figure 9-4: (Continued)

To create the interrupt requires that signals be added to the control bus connecting the CPU to the external architecture. Signals already present in the control bus include the Read and Write strobes that control data flow direction and timing for the external architecture. We will add a single input line to the control bus and name it *interrupt request* or IRQ. The purpose of this line is to allow controllers in the external architecture to request that the currently executing program be suspended and another second-level program, called the Interrupt Service Routine (ISR), be executed to perform the task appropriate to the cause of the interrupt.

We will first install the IRQ line in the hardware of the computer example; refer to Figure 8-1 showing the microprogrammed controller. The Condition Code Multiplexer, CC-MUX, in the center of the figure serves to multiplex the various flags from the first- and second-level status registers to the controller where they may be sensed one at a time. To this group we will add the IRQ line from the external architecture. Note that if we require more than one IRQ line, then each one is connected to an additional input on the CCMUX.

The next step will be to support the IRQ signal by changing the first-level program. At this point, we have several "policy" questions to decide. We must first determine how much of the requested task should be performed at the first-level. The answer to this question is based on how much knowledge the underlying machine should have of the second-level machine. The location in second-level address space and the actual operation of the I/O controller, i.e., is it a serial port or some other, are among the considerations that are only known to the second-level machine designer. The more we incorporate second-level design parameters in the first-level program, the more we limit the range of applications of our CPU. As CPU designers, we will implement a transfer of control operation so that program execution returns to the second-level machine at another point in the second-level program. This lets the second-level machine designer have complete freedom in specifying the architecture at that level. He can make suitable hardware/software trade-offs in the second level to support his intended application.

To do this, we need to provide a method whereby the interrupted second-level program can resume execution. This brings up the second question. At what point in the execution sequence of the second-level instruction should the interrupt request be honored? If we allow the interrupt to occur at any microinstruction boundary, then it should be clear that we will need to save more of the state of the first-level machine depending on where we are in an execution sequence. Temporary register contents and first-level flags are among the candidates since they would normally be used to support the second-level task.

Most CPU designers sense IRQ "between" second-level instructions to avoid this problem. Short execution sequences encourage this choice. In CISC designs, one uses long sequences to avoid second-level instruction fetching. Long sequences will increase the amount of time the CPU requires to respond to the interrupt. The total time for a CPU to respond is called "interrupt latency". If the application environment for the CPU requires a short interrupt latency, then long execution sequences must be interrupted in a CISC. Notice how the RISC design can be made to respond more quickly since it does not support any execution sequences longer than one cycle.

We will choose to implement the sensing of interrupts between second-level instructions in our example. In the event that we implement long sequences (e.g., in the course of providing instructions for the floating point data type) then we must reexamine this point and provide interrupt sensing during the execution sequence. Concomitant with this change in philosophy, we must arrange to save any registers, flags, or other internal facilities associated with the algorithms we have chosen to support at the first level. Sensing the interrupt request at the first level consists of pointing the CCMUX at the IRQ input and executing a conditional branch instruction in the microprogrammed controller. In Program Design Language (PDL) this might appear as follows:

```
IF (CCMUX(IRQ) == TRUE) THEN Goto IntServ ELSE Continue
```

This controller instruction is placed as the first instruction in the fetch sequence. A similar form given next may be placed as the last controller instruction in each of the execution sequences:

```
IF (CCMUX(IRQ) == TRUE) THEN Goto IntServ ELSE Goto Fetch
```

The two way branch may be more complex to implement in some microprogrammed controllers.

The interrupt implementation sequence beginning at the address labelled by "IntServ" is not an interrupt service routine in that it does not contain any code that is specific to the actual cause of the interrupt. The actual Interrupt Service Routine (ISR) is written in second-level language by the second-level machine designer or programmer. The task to be performed at the first level consists of saving a reasonable amount of the state of the second-level machine and initiating his ISR.

Since we have chosen to sense interrupts between second-level instructions, the state of the second-level machine that must be saved consists of the PC and possibly the second-level flags. While some early machines did not save the flags, this feature is generally needed since the first few instructions in the ISR will usually change them. If the flags are not automatically saved, we must provide a second-level instruction that saves and restores the flags. As we will see later in this section, the status register will include more information than just the arithmetic/logic flags. We will need to change some of this information in the course of implementing interrupts so that a status register saving mechanism under our control is necessary.

The remaining registers in the programming model are also candidates to be saved. If all are saved in a register-oriented machine such as ours, a large number of memory cycles will be needed. Each register saved increases the interrupt latency of our CPU design. The current thinking for general purpose register-based designs is to give the second-level programmer an efficient instruction that can save a selected group of registers. Since he knows the registers that his ISR needs to use, he can specify that only those be saved. The instruction is called variously SToreMultiple (STM) or MOVEMultiple (MOVEM) depending on the machine in question. At the cost of a single instruction fetch, the second-level programmer may move a list of registers specified in one of the operands of the instruction to main memory. A corresponding LoaDMultiple is also provided.

A PDL description of the state saving sequence is given next.

```
MAR := SP; SP := SP - 2;
MEM(MAR) := PC;
MAR := SP; SP := SP - 2;
MEM(MAR) := Second_Level_Status_Register;
```

In this example, the state of the system is saved on a pushdown stack in the external architecture. PC and Stack Pointer (SP) are each one word long and thus require 2 bytes of storage space on the stack.

It should be noted here that our requirement in this application is to save the state of the system. As was the case for the subroutine call sequence, the state may be saved in a number of places. We can use special registers in the CPU that are not visible to the second-level programmer or some of the registers in the programming model. Both methods are fast but restrict the number of pending, or "nested", interrupt operations that can occur at the same time. The programming model method was employed in the IBM /360 and its descendants. This is an effective method but nesting, as was true for subroutines, was a programming construct created by the second-level programmer called "save area chaining". The net effect was that the state was saved in the external architecture by the programmer, not by an automatic sequence as in our stack example. A stack may be implemented within the CPU; however, its restricted depth, necessitated by the high cost of fast parts, limits the depth of interrupt nesting. Other methods are possible and are used by some designers. Using some kind of stack in main memory is one of the slowest methods but it has the advantage of an "unlimited" nesting depth.

The last step in changing from the application task to the ISR is accomplished when the second-level PC is changed to the address of the first instruction in the ISR. There are two methods by which this is accomplished. The first method involves creating a number in the internal architecture and placing it in the PC. This requires that the second-level programmer begin his ISR at the address specified in the CPU's specifications. The second-level designer is therefore required to place memory at that location to support the program as well as limited in the space made available at that point by the CPU designer. In most cases where this method is used, the ISR writer places an unconditional branch to the actual entry point of his ISR located elsewhere in the external architecture. The second method, called a *vectored interrupt*, begins in the same way. The number created in the internal architecture is not placed in the PC, however, but in the MAR. The contents of the location to which it points is placed in the PC. While the length of an ISR is a function of its task, the number of bytes occupied by an address in most systems is fixed and known to both the CPU and second-level designer. The first method normally resulted in the ISR programmer providing a jump to his ISR anyway so the opcode fetch is avoided. The vectored interrupt is very commonly used in current designs since it affords the most flexible method by which the first- and second-level designers can attain their goal of a fast and efficient machine.

The two methods are shown in PDL next.

```
PC := Number; Absolute


MAR := Number; Vectored
PC := MEM(MAR);
```

The state saving and control transfer sequences that we have just shown are the entire execution sequence for the interrupt request. All of the actual interrupt service code is written for the second-level machine. While this results in a longer execution time for the ISR algorithm than if it had been written at the first level, the time has been traded for a larger application area for our CPU design. At each step we have tried to minimize the number of steps required so that the interrupt latency is not severely degraded.

Once the ISR is finished, a mechanism must be provided to allow the interrupted second-level program to resume. This requires no more hardware to be introduced since the return-to-task switching is synchronous with the ISR; i.e., the second-level ISR programmer knows when his ISR algorithm is completed. As CPU designers, we must provide an instruction to the second-level programmer to "undo" the interrupt task switch. In the course of implementing the task switch, we hope you noticed the similarity between the interrupt and the subroutine call. The subroutine call is initiated by an opcode during the normal fetch/decode/execute sequence. The interrupt sequence is initiated by sampling the IRQ line at the beginning of each fetch cycle. Both result in similar execution sequences. If we had not chosen to save the status register in the interrupt sequence, then the two would have been identical. To undo a subroutine task switch, we introduced a RETurn instruction. As a consequence, the execution sequence of the return recovered the old PC contents (i.e., the address of the next instruction in the calling task), and placed it in the PC. Our problem here is the same in addition to recovering the status register.

The new second-level instruction is given the name ReTurn from Interrupt (RTI) here and in some other CPUs. It requires no operands. A PDL description of its execution sequence that reverses the sequence given above for a stack based machine is given next.

```
SP  := SP + 2;        Back up to last used cell
MAR := SP;
Second_Level_Status_Register := MEM(MAR);
SP  := SP + 2;        Back up to where PC is stored
MAR := SP;
PC  := MEM(MAR);
```

The SP was left pointing at the next available cell when information was PUSHed. The current state of the SP as the result of a PUSH varies between designers and is usually left pointing at the last used cell or as is shown here. The SP must be changed first and be passed to the MAR. Notice the byte addressing. Consulting the RALU and its associated buses in the internal architecture should reveal whether this can be accomplished in the same clock cycle. In the computer example, see Figure 8-2, the SP may be modified by the ALU, i.e., incremented by passing it into the right side of the ALU and adding one via the Carry Input. It then can be conducted via the Y bus back to the Register Array and to the MAR. If the RALU's internal Y bus had not been connected to the internal data bus, these functions would need two cycles.

At this point, we have provided a mechanism whereby the second-level program can be interrupted by a hardware-induced subroutine call. A method to return to the interrupted program has also been provided. The implementation task is not yet complete since interrupts are always "enabled". Normally, the second-level program must initialize I/O controllers and other interrupting devices so that their mode of operation is consistent with the intent of the designers. Most I/O controllers are designed such that when reset they will not generate interrupts until properly initialized. We must provide a similar mechanism also. In addition, we need to provide a mechanism that allows some degree of control of whether an ISR may be interrupted. Such control is absolutely necessary if more than one IRQ line is included in the design.

Our specification for enabling interrupts reads as follows:

1. Interrupts must not be sensed until the second-level reset operation has been accomplished.
2. Interrupts must not be sensed inside of an ISR until allowed by the second-level programmer.
3. Some means to remember a pending interrupt request must be considered in the design.

A simple method to implement the first two specifications requires us to provide a single-bit storage element that is under control of both the first-level, and indirectly, the second-level programmers. A single RS flip-flop is provided for each IRQ line to serve as an interrupt mask register. The flip-flops will be included in the state-saving structure as part of the second-level status register so that when interrupt driven task switching occurs, the state of the interrupt mask register will be saved or restored as well. We will provide bits in the microinstruction to set and clear each bit in the mask register independently. The bits will be cleared in the course of executing the first-level reset sequence, thus inhibiting interrupt sensing. They will also be settable by the second-level programmer by way of an instruction, SETINT, that we will provide.

The output of each mask bit will, in our first attempt, be connected to separate inputs on the CCMUX. The interrupt sensing sequence will now become:

```
IF (Int_Mask_Bit_n == 0) THEN Goto Fetch ELSE Continue
IF (CCMUX(IRQ) == TRUE) THEN Goto IntServ ELSE Continue
Fetch: … ;Do remainder of Fetch sequence
```

A similar structure is possible if interrupt sensing is done in the last step of the execute sequence.

The number of clock cycles may be reduced by performing the logical combination implied by the two states shown above in a Boolean network before the IRQ line enters the CCMUX. The two signals may be ANDed, creating the masked IRQ signal which is sensed via the CCMUX in the Fetch sequence.

The second-level programmer may signal that interrupts are to be honored at any point in his program by invoking the SETINT second-level instruction. A corresponding CLRINT instruction is provided so that he may signal when it is unsafe to interrupt a given segment of second-level code. The action of these to instructions is to set or clear a particular interrupt mask bit. If more than one is available, the operand field of SETINT or CLRINT may be used to specify which one.

The first-level programmer will need to set and clear the interrupt mask bits also. We have already noted that the first-level Reset sequence must inhibit interrupts before the Fetch sequence begins. Once an interrupt is sensed and IntServ is entered in the microcode, we must also inhibit interrupts. There are two reasons for this. First, second-level instruction fetching will begin when the task switch is complete. The second-level ISR must determine the cause of the interrupt in order to turn off its requesting signal on IRQ. This may require polling one or more I/O controllers attached to IRQ. All during this time the first-level machine is fetching and implementing second-level instructions using the same sequences that were used before the task switch. To allow this to continue, we will inhibit the current IRQ by clearing its mask bit in IntServ after the state of the system is saved. When the ISR programmer restores the state of the system as a result of the RTI instruction, the interrupt-sensing state will be restored as well. In other words, restoring the status register sets the interrupt mask bit that we cleared in IntServ.

The second-level programmer may also reenable the interrupt mask bit during the ISR by simply writing a SETINT instruction at the proper place in his routine. The position should be after he has turned off the request at the I/O controller so that he won't enter an infinite loop processing the same request ad infinitum. This is a common error as the second-level programmer learns to write ISRs and its symptom is stack overflow in stack-based machines. The IntServ sequence is entered repeatedly without a corresponding RTI. Notice that giving the second-level programmer the SETINT instruction to use inside of an ISR allows him to decrease the interrupt latency of the second-level system.

In our examples, we have assumed that the I/O controller generating the interrupt request was capable of holding IRQ until some transaction occurred between it and the ISR. Some CPU designers assume this function in their own designs. In this case, the I/O controller generates a short pulse or logic transition that is trapped in a flip-flop in the CPU. It is then up to the CPU designer to provide a means to clear the request. In some microprocessors, a NonMaskable Interrupt (NMI) input is provided that is edge-sensitive. The fact that it is also not masked requires that the second-level system designer use I/O controllers that will not generate interrupts until enabled by the programmer. In most of these CPUs, the interrupt request flip-flop is cleared as a result of the RTI instruction so that this particular ISR is not itself interruptible. Other designers treat the interrupt request flip-flops (or register) as a polled I/O device or port. In this case, the second-level programmer must clear the interrupt request by using instructions already provided, i.e., AND and OR immediate. Do not confuse this *Interrupt Pending Register* with the *Interrupt Mask Register* discussed above.

A complete interrupt structure has been shown for the CPU at this time. We have alluded to the fact that adding more interrupt inputs consists simply of adding inputs to the CCMUX, lines of microcode to test each input, and second-level memory locations to contain the vectors to the ISRs. One ISR is provided by the second-level programmer for each IRQ signal. The job of the CPU designer is then complete. He merely supplies a task switch sequence for each IRQ line using one of the ISR vectors specified. All is not finished yet since second-level designers frequently employ more interrupting I/O controllers than there are IRQ lines. Most controllers are designed so that many can be "wire-ORed" to one IRQ input. The ISR is written to poll the I/O controllers attached to the active IRQ input. The order of polling determines the "priority" of service.

Normally, the first controller detected with an active interrupt request is serviced in the ISR. After the appropriate service is given, the ISR is exited by executing a ReTurn from Interrupt (RTI) instruction. If another I/O controller is also trying to drive IRQ, then its request is still active. This fact is detected when fetching the opcode of the next instruction in the interrupted second-level program. Most CPU designers will complete that instruction before honoring the next request to keep the interrupt structure from using all of the CPU cycles. The polling sequence results in a longer interrupt latency period. Several methods have been used to reduce the time before the highest priority ISR is dealing with the correct controller.

The first method discussed here reduces the time needed for the CPU to test the IRQ inputs. Instead of bringing each of the IRQ lines into the CCMUX, let us send them to a priority encoder. A truth table showing the logic for a four-input priority encoder is given at the top of the next page.

Each I/O controller is connected to one of the IRQn inputs to the encoder. The priority encoder outputs consist of a two bit address, S1 and S0, and the interrupt request signal IRQ. IRQ is connected to the CCMUX instead of the many IRQn. Signals S1 and S0 are ORed with two of the bits in the address input to the microprogram memory when IRQ is *True*. This construct results in a four-way branch structure seen in the ROM/Latch controller

IRQ0	IRQ1	IRQ2	S1	S0	IRQ
True	X	X	0	0	*True*
False	*True*	X	0	1	*True*
False	*False*	*True*	1	0	*True*
False	*False*	*False*	1	1	*False*

in Chapter 7. A single microword is needed to perform the dispatch to the correct IntServ routine for the corresponding IRQn input. It has the following form in PDL:

```
IF (CCMUX(IRQ) == TRUE) THEN Goto IntServ(S)
```

The symbol IntServ(S) stands for the next four microinstructions that follow this line. The first three microinstructions each unconditionally branch to the IntServ appropriate to the highest priority, active IRQn. The fourth microinstruction is the beginning of the normal Fetch sequence. Some economies are possible since subroutine calling is so efficiently performed in a microprogrammed controller. Each IntServ needs to save the state of the system, which can be handled in one subroutine. The difference between each IntServ routine lies in its last step, which is the value to use for the vector address. This difference can be accommodated in using a different immediate value for the address in the microword or by placing a table within the internal architecture accessed by the S parameters in the priority encoder output. The latter method reduces the microcode to that of a simple one IRQ design.

Other methods require cooperation from the controllers in the external architecture. The first method involves the generation of an Interrupt Acknowledge (IACK) signal from the processor. While there are many ways to generate the signal, let us assume a particularly simple method. Our processor must inform the external architecture that an IACK signal is currently available and it must present a code that represents the level of the interrupt being acknowledged. Remember that lower level interrupts may also be pending and the processor can now see only the highest priority active one. We will choose to place the interrupt level, i.e., the S code produced by the priority encoder on the least significant bits of the external address bus. The upper bits may be undefined or set to 0. One method by which the I/O controllers may cooperate is as follows. Each controller has an IACK input. A network that is a minterm of the CPU's IACK and the least significant address bits drives the controller's IACK input. The controller responds when this input is *True* by placing a pointer to its interrupt service routine on the external data bus. The CPU then completes a Read cycle just as it would have if it used the vectored interrupt address fetch described above. The following PDL describes the sequence.

```
MAR(1-0) := Priority_Encoder_S_Output(1-0); MAR(15,2) := 0; IACK := TRUE;
PC := I/O_Controller(MAR);
```

The notation I/O_Controller(MAR) is used to indicate that only the controllers can respond to the IACK signal since both Read and Write, the data bus direction control signals, are *False*. Notice that the data bus drivers at the interface between the CPU and external architecture must also respond to IACK as if it were a Read.

The net effect of this structure is that the I/O controllers supply the address of the correct ISR without further polling, thus reducing the interrupt latency. One question remains. How does the address of the correct ISR get into the I/O controller? Most I/O controllers present a simple array of registers to the CPU. Each register is mapped into a location in the memory space of the external architecture. We have indicated two of the registers in earlier examples. The data input and/or output registers are obviously present. These registers contain newly acquired data available to the second-level programmer or form a place to put data that the controller will send out of the computer. Two more, the *command* and *status registers*, serve to communicate the intentions of the programmer to the controller.

The command register is used to initialize the I/O controller, including giving permission to generate an interrupt request. The status register is used by the controller to tell the program that data is available or the controller is busy doing an earlier command. To this structure we can add a register that will contain the address of the ISR. When the controller is initialized by the second-level program and before it is given permission to interrupt, the ISR address register is filled with the address of the ISR entry point. After that, the controller can interrupt as needed.

The final improvements are usually made in the I/O controllers themselves. Most I/O controllers are complex devices in their own right. There usually are many possible causes of interrupts. In a simple serial port, one can name several, including data available (RDR), data sent (TDRE), and several error conditions. Within the controller, a prioritizing scheme can be used that is similar to the method shown above. In this case, an array of ISR entry points is kept in the controller instead of just one. The correct entry point is placed on the bus during the IACK cycle based on the priority established by the I/O controller's designer.

In the same vein, we find that there are usually more I/O controllers than there are IRQ lines, even with the prioritizing done at the processor. Each line can be expanded with the help of the I/O controller designer. Unbeknown to the CPU, the I/O controllers on a given IRQ line may carry on a conversation that results in only the one having the highest priority supplying its ISR address. While this sounds complex, the actual implementation is relatively easy. A "daisy chain" is created in the form of a pair of signals on each I/O controller: Chain_Enable_In and Chain_Enable_Out. If Chain_Enable_In is *True* and the controller has an interrupt request to make, then it will respond with its ISR vector on the next IACK cycle. It will also drive its Chain_Enable_Output *False*. On the other hand, if Chain_Enable_In is *False*, the I/O controller will not respond to the next IACK cycle. The controllers are wired in such a way that the Chain_Enable_Out of one controller is connected to the Chain_Enable_In of the next controller of lower priority. The "outermost" I/O controller's Chain_Enable_Input is connected to *True*, thus making it the highest priority controller.

The major attraction of this method is that prioritization is accomplished with two wires and the position of the controller in the daisy chain. At no time does the interrupt acknowledge operation exceed one Read cycle.

In conclusion, we should establish the speed domain that is occupied by interrupt driven data transfers. The introduction of the interrupt was intended to reduce the execution time cost of data transfers within the external architecture by reducing or eliminating the overhead incurred by polling. Even after the interrupt was introduced, a certain amount of polling was needed to determine which of the several I/O controllers attached to a given IRQ input was active. The CPU discriminated between different IRQ inputs very effectively by quickly vectoring to the correct ISR for that input with the assistance of hardware prioritization schemes. The problem of multiple I/O controllers on a single IRQ line was solved by getting help from the controller itself through IACK and daisy-chaining techniques. In the end though, a particular ISR still consists of many instructions to do the simple job of moving data from one place to another. All data transfers must pass through the processor in many CPU designs since the single bus architecture used outside the processor allows only one data transfer to occur at a time relative to the CPU. In summary, the interrupt technique occupies a middle ground among the various data movement methods. The next section will introduce two faster methods, while polling represents the slowest method.

9.2.4. Memory Access by Controllers Other Than the CPU

We will now attempt to remove all barriers associated with fast data transfer in the external architecture. The first method introduced will deal with the problem from within the CPU. Some processor designers have given their CPUs a set of block data transfer instructions that can move large blocks of data as fast as the external memory can cycle. The I/O controller must be able to absorb or supply the data at this rate since no "hand shaking" is provided, i.e., no polling occurs. A controller that buffers the data in a memory in its own architecture (e.g., some hard disk and video display controllers) is appropriate for this treatment. A short PDL sequence will summarize the execution phase of this type of instruction.

```
Reg_Array(Pointer1) := Beginning Address of Source Block;
Reg_Array(Pointer2) := Beginning Address of Destination Block;
Reg_Array(Counter) := Number of Words to Move;
LOOP:
MAR := Reg_Array(Pointer1); Reg_Array(Pointer1) := Reg_Array(Pointer1) + 1;
Reg_Array(Temp) := MEM(MAR);
MAR := Reg_Array(Pointer2); Reg_Array(Pointer2) := Reg_Array(Pointer2) + 1;
MEM(MAR) := Reg_Array(Temp);
Reg_Array(Counter) := Reg_Array(Counter) - 1;
IF (Reg_Array(Counter) > 0) THEN Goto LOOP ELSE Continue;
```

Two pointers to the source and destination blocks in the external architecture are defined to be in the register array in the RALU of the CPU. They may be registers that are visible in the programming model in which case the second-level programmer must initialize them as shown in the first two lines of the PDL description. A counter that contains the number of words to be moved during the loop is also maintained in the register array. This counter must also be initialized by the second-level programmer, either as an operand of the assembly language instruction or as being part of the programming model and therefore accessible to normal second-level instructions. The latter version was favored by the designers of the Z80 instruction LDDR where DE and HL contained the pointers and BC was the counter. The INTEL IAPX 86 and later designs contained the MOVS instruction with similar capabilities.

The method outlined above, where the loop is implemented at the first level, is one of the fastest ways to move data in the external architecture short of introducing a second control path. It has the shortcoming that the I/O controller is unable to provide data synchronization information but must handle the other half of the transaction at the CPU speed. The main speed improvement is accomplished by reducing the number of instructions fetched to execute the loop.

The method that gives this section its name results from recognizing that the CPU cannot provide data flow synchronization without some instruction fetching. We have agreed that building a knowledge of particular I/O controllers into the CPU is a good way to restrict the range of applications for our design. Our attack on the problem will be to introduce a new controller into the external architecture that is neither an I/O controller nor a CPU, but a device that can acquire the system buses and perform a fast, synchronized data transfer with as little interruption as possible. It is a state machine like the I/O controller and CPU but, unlike the CPU, it does not need to fetch instructions to determine the sequence it must perform; i.e., it plays only one tune.

Before we introduce the device, we need some new terms. The system buses consist of the data bus, address bus, and the control bus. The data bus can be driven by any element in the external architecture, including the CPU. It is bidirectional and up until now was controlled by the CPU. The address and control buses (particularly the Read and Write strobes) are generated by the CPU and provide signals and timing appropriate to the anticipated external architecture elements. These signals define the timing of the external bus cycle. The period of an external bus cycle has become a *de facto* unit of time in our discussions, i.e., the memory cycle time. This time is slower than the processor clock frequency, by design, since the external architecture trade-offs favored maximizing storage space at the expense of speed.

The external architecture control function performed by the CPU up to this point can be extracted and given the name *bus master*. Any device that can generate control signals and addresses is a bus master. Two types of masters are recognizable. The bus master from whom other potential masters must obtain permission to use the system buses is known as the *permanent bus master*, while the others are called *temporary bus masters*. Needless to say, a given bus group (data, address, and control) can have only one permanent bus master, while many temporary masters may exist. An element in the external architecture that may only drive the data bus and no other is called a *bus slave*. Until now, all elements in the external architecture have been bus slaves, i.e., the memory and I/O controllers.

Our plan is to introduce a controller that will acquire the system buses, in particular the address and control buses, and perform data transfers between elements in the external architecture while the CPU is "off-line". This controller, called in some circles a Direct Memory Access Controller (DMAC), is classed as a temporary bus master. In introducing the subject, we will augment the CPU so that it will be the permanent bus master. Later, we will study the permanent bus master algorithm in more detail and show how some systems separate this function from the CPU and implement an independent state machine whose sole task is to allocate the system buses.

There are two possible structures for the DMAC. The first, called the *two cycle controller,* uses the system buses only and must perform data transfers using its own architecture as a way station. Since there is only one system bus, the DMAC cannot simultaneously read data from one element in the architecture and write to another in the same memory cycle. In this type of controller, the data is read from the source device and latched in a data register inside the DMAC state machine's architecture. The next cycle is used to write the data from the DMAC's register into the destination in the external architecture. The speed of the controller is slower than the next structure, but the control signals is reduced.

The second structure is called a "one-cycle DMAC" since it can perform a data transfer between two points in the external architecture in a single memory cycle. To accomplish this, the DMAC must drive both the system buses and a separate control bus that is normally connected to the I/O controller. The second control bus must be able to specify which register is needed in the I/O controller as well as the direction of transfer. It is, in a sense, a small *control/address bus*. Data is transferred from source to destination via the system data bus. At the cost of an additional bus, the one-cycle DMAC represents the fastest possible DMAC structure.

Additional elements are needed in the architecture of the DMAC state machine. The algorithm that it performs, once it obtains the use of the system buses, is like that portrayed for the block data movement instructions discussed at the beginning of this section. The two-cycle controller must have two pointer registers, one for source and the other for the destination. The one-cycle DMAC requires only one pointer register since the second control bus is normally connected to the I/O controller. This ploy is appropriate since the device controller has a much smaller "address space" than the memory. The system buses are normally used to support the "memory" end of the data transfer, whether source or destination. Both types require a word count register to control the loop, shown above, that is the backbone of the algorithm.

The only new element in the DMAC algorithm is the method of manipulating bus mastership. In words,

1. The temporary bus master requests the use of the system buses from the permanent bus master via a signal called "Bus Request (BR)".

2. The permanent bus master completes the current bus operation and then ceases to drive the three buses. It asserts a reply to all temporary masters by way of the "Bus Grant (BG) line.

3. In some systems, the temporary master confirms the grant with another signal, "Bus Grant Acknowledge (BGACK)". In those systems, BGACK helps contending temporary masters in deciding who really owns the bus. Our simple introduction will assume only a single temporary master, our DMAC, and so we can dispense with BGACK.

4. The temporary master drives the system buses for one or two cycles, as appropriate, and then relinquishes them by tristating his system bus drivers and setting BR *False*.

5. The permanent bus master, sensing BR being set *False*, also sets BG *False*. Most temporary masters should not attempt another bus request until BG goes *False*.

If the permanent bus master is a CPU, as in our case, Rule 5 allows the CPU to have a chance at the system buses so that instruction cycling is not completely stopped. Notice that interrupts cannot be serviced as long as a temporary bus master has the buses since instruction fetching has stopped, at least in a von Neumann machine. We can avoid some of this penalty by introducing instruction pipelines and caches, which we will do in Section 9.2.5.

We will now adapt our CPU example to become a permanent bus master by giving it some additional hardware and sequences based on the rules above. The hardware additions are very simple. The bus request line, BR, enters the controller of the CPU and connects to the CCMUX just as an IRQ line did. The bus grant, BG, signal originates in a flip-flop that may be set or cleared by bits in the microword. Bus grant acknowledge, BGACK, if present would also connect to another input on the CCMUX. The algorithm implemented in the CPU follows the outline given above. Instead of sensing the line "between" second-level instructions, as was the case for interrupts, we will sense the BR signal at all convenient places in the execution sequence of instructions except during external bus cycles. Sensing is done by replacing all continue instructions in the execute sequences (outside of bus cycle sequences) with conditional subroutine call instructions that test BR as follows.

```
IF (BR == TRUE) THEN Call BRService ELSE Continue;
```

Normally, execution takes the Continue path as before until BR becomes *True*. At that time, a short subroutine, BRService, runs at the first level to tristate all bus drivers except the BG output, which it sets *True*:

```
BRService:
          RETurn; Buffers_Off; BG := TRUE;
```

Implementing BG in a flip-flop allows us to continue execution without having to worry about holding it. Furthermore, we can test BG at the beginning of each memory cycle via the CCMUX and halt execution at that point if BG is *True*. This allows us to continue long sequences that do not need memory access, while the temporary bus master uses the external architecture. Now that we have shown our strategy, we must go back and modify the BR detection code to make continuation possible once BR goes *True*. All Continue commands in execute sequences outside of memory cycles become:

```
          IF ((BR == TRUE) AND (BG == FALSE))
               THEN Call BRService ELSE Continue;
```

Once both BR and BG are *True* and we are executing outside of a memory sequence, the program flow will continue as before until our sequence needs a memory cycle. At that time, we will test the signals to determine the next move as follows.

```
Wait:
          IF ((BR == TRUE) AND (BG == TRUE))
               THEN Goto Wait ELSE Continue;
          IF ((BR == FALSE) AND (BG == TRUE))
               THEN Call BGService ELSE Continue;
```

First-level program execution will halt in a loop in the first instruction until BR is set *False* by the temporary bus master. At that time, we must set BG to *False* and enable the CPU's bus drivers before the bus cycle can continue. This step is signified by the Call to the subroutine BGService that accomplishes the enabling task. During the remainder of the bus cycle sequence that follows the above instructions, BR is not tested at all so as not to upset memory cycle timing specifications.

Notice that the state of the second-level machine need not be saved since no processing will be accomplished at that level. The relevant contents of the CPU, whether in the programming model or not, are not disturbed during the time the temporary master uses the bus. For estimating time requirements, it is reasonable to assume that in the worst case the CPU has begun a bus cycle, thus delaying the acquisition of the buses by one memory cycle time. The DMAC will require the use of the bus for one or two memory cycles per word, depending on its structure. Releasing the bus may, at worst, require one more memory cycle time depending on the means used to sample the BR signal in the CPU. In our case, the release takes, at worst, about two microcycles. The DMAC is fast but cannot, as described, move data with 100% efficiency because of bus arbitration.

What if we allowed the DMAC to acquire the system bus and then transfer many bytes of data before it relinquished it? Up until now, the CPU was guaranteed every other usable memory cycle. The overhead incurred by the bus arbitration is obviously hurting the data transfer rate. Let us distinguish two protocols: the first could be called "word" mode where, as just presented, the DMAC acquires the bus for each word to be moved. The second is called "burst" mode in that the bus is acquired and then a burst or block of data is moved before it is released. The second mode has much lower overhead and results in much faster transfer. However, during this time, the CPU is unable to continue execution beyond the current instruction. This greatly slows the application program running at the second level and completely stops the interrupt mechanism. While there are circumstances in which this mode is appropriate, they must be evaluated in terms of all other aspects of a system's design.

The DMAC, as a temporary bus master, may acquire the system's buses and transfer a block of data between elements in the external architecture. It should be emphasized that the second-level program requesting such service need only initialize the I/O controller and the DMAC to specify the number of words, the location of the block in memory, and the controller to use. After this, the second-level program and the CPU are no longer involved in the data transfer and need only be informed upon completion. The DMAC is generally given an interrupt-generating capability similar to those for I/O controllers discussed above. They can even be "strung" on the daisy chain and placed in the priority structure with other I/O controllers. The ISR for the DMAC can be used to communicate the completion of the transfer to the second-level program, i.e., the operating system or application as appropriate. It may also be used to reinitialize the DMAC for another transfer.

The last step above may also be accomplished by some DMACs themselves if the designer provides a second set of pointer/counter registers. The DMAC can "chain" two data structures together to *double buffer* data transfers between a particular I/O controller and two buffers in memory. Such a scheme is useful when the total amount of data being moved exceeds the capacity of memory in the external architecture. Data may enter from an analog-to-digital converter (A/D) and be buffered to disk. At any instant the A/D converter is filling one buffer, while the second-level program moves data out of the other buffer to disk.

Most current DMACs for microprocessors support multiple sets of pointer/counter registers. Each register/counter structure is called a "channel". Multiple channel DMACs allow data transfers between several devices and memory *simultaneously*. Obviously, only a single source/destination pair may be used in one memory cycle. As far as the CPU and the second-level program are concerned, there are several I/O controllers filling or emptying several buffers in main memory at the same time. The only time this activity is noticed by the CPU is when the DMAC interrupts it after completing the transfer of a block of data. Our A/D to disk transfer can then be accomplished using two channels of a DMAC, one for the A/D and the other for the disk. Each channel has a pair of pointer/counter structures to implement double buffering. Depending on the data transfer rate, the CPU appears to run the second-level program at a slower rate than normal but is otherwise unaware of the activity in the external architecture.

Our last consideration on this subject concerns the permanent bus master algorithm itself. It is obvious that once the CPU, as permanent bus master, reaches a point in which it needs to use the bus, it must stop if it has loaned the bus to a temporary master. We can forestall this time by using techniques for instruction pipelining and caching discussed in the next section. Another solution is possible however. First remove the permanent bus master algorithm from the CPU. This simplifies the CPU control algorithm, allowing us to implement more complex sequences. The permanent bus master, implemented as a dedicated state machine, may be optimized to handle bus protocol transactions as efficiently as possible since complications caused by instruction execution sequences are removed. The state machine may be very fast to accommodate more temporary masters.

The CPU must still stop when it needs the bus. To alleviate this problem, let us move the interface between the CPU and the system bus away from the CPU such that at least some memory is available to it. This scheme is implemented on most modern buses such that some memory is always available to the CPU and another block of memory is shared between the CPU and the system bus. Programs and data in the dedicated memory will allow the second-level program to continue execution until data is needed from the shared memory. The likelihood of a collision is greatly reduced relative to our earlier examples. Each CPU/ local memory structure forms a complete *single-board* computer. Each computer can communicate with others and global I/O controllers by way of the system buses that are governed by a dedicated permanent bus master.

9.2.5. Instruction and Data Pipelines and Caching

This section could well be titled "Extending the Illusion". The basic design principles of second-level machines have been based on the illusion that each memory element in the second-level architecture is connected by the operation specified by the programmer. For example, if the second-level programmer wrote z = x + y, then he has specified that the memory location "z" is connected to locations "x" and "y" by an arithmetic summation network for the proper data type. The designers of the CPU have created the illusion that such a network exists by performing a sequence of operations that: (1) brings "x" and "y" into the one summation network that does exist in an RALU in the CPU, (2) performs the summation, and (3) places the result in "z". While the alternative (i.e., a large collection of operation networks that can be sprinkled about the second-level architecture by the programmer to do the operations he desires) is faster at producing results from input data, "programming" time and hardware costs have traditionally been prohibitive. In the course of earlier chapters, we have shown how first-level machines may approach this goal since large powerful arithmetic networks are available at a reasonable cost. We retained the programming approach as a means to interconnect the elements, as opposed to the wiring process implied above, in order to control application development costs; i.e., programming is much less expensive than wiring as a method to adapt a design to a new problem.

Unfortunately, in implementing the second-level *computer example* in Chapter 8, we noticed that the internal architecture has periods of time when it is unused. While our system uses the same architecture for the Program Control Unit (PCU) as for instruction execution and thus makes better use of the RALU than a machine in which these have their own dedicated hardware, we still can identify several periods in which the RALU is idle. If we had extended the internal architecture to support other operations (e.g., a combinatorial multiplier, or other datatypes like a floating point unit), then these relatively expensive units would be idle most of the time. The two structures discussed in this section — pipelines and caches — are used to extend the illusion and make better use of existing resources or make less expensive resources appear to operate as fast as more expensive versions. An extended discussion of these subjects, along with references, is presented by Stone in *High-Performance Computer Architecture* (see Bibliography). Both networks allow the CPU designer to take advantage of the organization of second-level programs. Two features of the program organization are noted here. The first is that second-level programs are executed sequentially. An occasional branch is made for decision-making purposes or for modularity, i.e., "structured programming". The second feature is that the next instruction or data element needed is frequently near, in second-level address space, to the one currently used.

Pipelines are storage networks that allow instruction fetching or execution to be "staged" with respect to the underlying computation network. An example of a simple instruction pipeline, as applied to the computer example, can be made by using the instruction register. If we had exploited the presence of the Instruction Register (IR), thus using it as a pipeline, we would have written our microprogram in such a way that the fetch sequence for the next second-level instruction would begin immediately after the conclusion of the current fetch sequence. In the version of the microprogram shown in Chapter 8, the next fetch sequence starts after the current execution cycle ends. If the execution cycles were the same length as the fetch cycle, the speed of execution would be doubled.

As execution sequences get longer, there is more room to add computation modules to speed them up by reducing their length. Chapter 3 contained many networks that performed operations in a single clock cycle, thus reducing the execution sequence to a single state. An intermediate situation is possible that employs less hardware and still produces results at a rapid rate, possibly one for each state. As an example, let us use the Floating Point Adder/Subtracter structure shown in Figure 3-24. Instead of implementing this as a single combinatorial network as implied in the figure, we will break the network into sections along the lines of the modules within the figure. The first stage involves separating the floating point formatted numbers into exponents and fractions. This can be done through the routing of the data path. At the end of the clock period (i.e., one state), the data for operand 2 would be stored in a register or element in the data pipeline. During the next state, the hardware used to separate operand 2 would be used to separate operand 1 while the difference of the exponents of the two operands is calculated. The next state (or states if appropriate) would see operand 2 denormalized using a shifter under control of the difference of the exponents. The operations continue until the final results are calculated. Each stage uses a small amount of the hardware needed in the entire Floating Point Unit (FPU), e.g., a single register to hold the appropriate parts of the operand and an ALU and shifter. As soon as the first pair of operands have passed into the data separation stage, a second pair of operands may be entered. It should be clear that, if the problem involved two operand operations on a series of floating point numbers, then the pipelining of the data through the network would allow a result to be removed each clock cycle once the pipeline was filled. Notice that the total amount of hardware would be reduced relative to a complete FPU.

As can be seen from the previous examples, pipelines can make the use of the underlying hardware more efficient. Unfortunately, the fact that the pipeline exploits a perceived structure in the instruction or data flow also causes a serious shortcoming when that flow is not maintained. The flow is broken in an instruction pipeline when a branch is taken. The flags that are tested to determine if the control transfer is to be taken may not be produced until the data reaches the end of the pipeline. If the branch instruction immediately follows the data operation in the instruction pipeline while other instructions lie before the branch, these instructions should not be performed if the branch is taken. A large number of instruction fetches have been made and partially executed that should not have been if the branch is taken. The longer the pipeline, the larger the amount of work that must be undone. There are two solutions to this problem. The first requires that the control mechanism purge the current state of the pipeline. Preventing the partially digested data in the combinatorial networks from replacing previous results makes this solution challenging. The second solution requires the inclusion of enough NO Operation (NOOP) instructions after each branch instruction to fill the pipeline. This method requires no special hardware interference since there will be no partial results in the pipeline. In either case, the number of branch instructions in the instruction stream should be minimized if the full benefit of the pipeline is to be enjoyed.

We will now examine an alternative to the pipeline. Remember, a basic trade-off was made in specifying the external architecture. The external architecture was to contain maximum storage for programs and data at the minimum cost. The unfortunate side effect of this choice was that the speed of the external architecture was reduced, thus leading us to introducing means (like the pipeline) to bridge the gap. Let's buy some very fast, read/write memory and place it such that the CPU can receive its instruction stream only from the new memory. This memory is very expensive, so we are unable to afford much of it. As an example, if we are building a microprocessor, the memory should be placed on the same die with the rest of the CPU to derive the benefits we are after. This is a very expensive location for a large network. The instruction fetch sequence can now run without waiting on anything, just as the instruction pipeline did. If the data flow in the machine is not pipelined, we will then have no problem with branching. Unfortunately, the down side to the solution is that the size of our added memory is so small that it will not contain a large program. We must add a control function to manage these resources such that the CPU can execute instructions from the fast memory as much as possible.

A cache is a small temporary storage area. In our use, it is a very fast random access memory. A *cache controller* is a network that manages this small resource in such a way that the CPU thinks that it is very large; i.e., it appears to contain programs of "unlimited" size. A simple version of the cache management strategy goes like this. The CPU presents an address to the cache controller as part of an instruction fetch or data movement sequence. The cache controller determines if the cell at that address is present in the cache memory. If it is, the requested transfer is made. Notice that the design goal established that the transfer must run at the CPU system clock rate, the same as that of the RALU and other elements in the internal architecture. This means that the decision as to the presence of the desired cell must be made very rapidly. We will see how this is accomplished shortly; but first, we must see what happens if the selected cell is not present.

If the operation requested by the CPU involves the transfer of data or instructions to the CPU (i.e., a read), then the cache controller must perform a read operation on main memory in the external architecture. This operation is slower than when the data was present in the cache, so the CPU must wait on the transfer. Now, since programs are normally organized into local structures, it is highly likely that, after several of these main memory read operations are accomplished, data that is already in the cache will be requested again. With no further intelligence on the part of the cache controller, program organization will result in the cache being used for many memory references. There are two related means by which the cache controller can anticipate the needs of the CPU. It is highly likely that the next instruction to be requested is the one that follows the current instruction in main memory. This point was used in designing the instruction fetching strategy for the pipeline. The cache controller can be made to fetch the next several instructions without waiting on the CPU request. Since main memory is slower than the CPU, the method of doing this will require the rearrangement of the second-level architecture data path. One arrangement makes the external data bus wider by an integer multiple of the internal data bus. Each main memory fetch will bring n times the amount of information that the earlier arrangement did. This method may be used without the cache or pipeline; but anytime the flow is interrupted, or references are made to boundaries other than n times the data bus width, data routing multiplexers will be required as in the Byte/Word problem discussed earlier in this chapter. One can view a structure of n cells developing in the cache, every cell in the structure (or *line*, as it is known) is updated every time a reference is made to main memory for the contents of one cell.

A second, related method exploits the structure of the modern Dynamic RAM (DRAM) to create the appearance of a wide external data bus without paying for the wires. Many DRAMs are organized internally such that after a reference is made to a particular cell, the data in the next three cells is available in much shorter time than the first. The cache controller need only make one normal (slow) reference and then three very fast ones over the same data bus to fill one of the cache lines mentioned above. The net effect of either method is that the contents of n cells may be moved quickly, as a single addressable unit, to the cache and then the local nature of the program will keep the CPU executing its instructions from the cache instead of main memory.

The structure introduced above can be exploited in another way. We have neglected the size of the cache in our earlier discussion, but obviously a large program will not fit all at one time. The structures introduced can be exploited to allow noncontiguous pieces of the program to be in the cache at a given time. This means that branches between addresses separated by megabytes in main memory may be accommodated between cells in neighboring structures in the cache. The contents of the cache at any time appears to be several noncontiguous segments of the executing program.

Eventually, the cache gets full. The next reference to a cell not contained in the cache will cause the cache controller to overwrite data or instructions that are present. The cache controller must write the contents of the cache back to its proper location in main memory before using the structure for new material. In this case, the time required to get the CPU going again will be the time required to write the number of cells in the structure and read the same number of new cells from the selected location. Notice that the cache controller treats the structure as an indivisible data type of a size that is appropriate of the underlying external data bus. This eliminates the need for the routing structure used earlier. Even if the reference is made to a byte in the middle of the structure, the entire structure will be moved and the byte will be referenced from the cache. While there are many techniques that can be used to select the structure to be written back to main memory, the Least Recently Used (LRU) algorithm is most frequently employed. As its name implies, the data that have not been used for the longest time, probably will not be used in the next reference. If the oldest data are "swapped out", then they most likely will not be missed as soon as the data that were used in the last reference. The cache controller is responsible for keeping track of the how long a particular structure has been in the cache.

If the CPU operation requires the transfer of data to main memory, the data may be written to the cache at the CPU clock rate. This results in the data in the external architecture at the specified address having a value different from the data in the corresponding location in the cache. This loss of cache coherency will cause a problem eventually. If there is only one bus master, the CPU, then the swapping associated with maintaining space in the cache will eventually update main memory. A mechanism must be used at the end of the program to purge all of the cache so that the last data will be updated before termination. The first problem occurs when the program tries to communicate with the external world. The simplest example involves the CPU trying to output data to a port. The store operation attempted by the CPU results in the data landing in the cache and going no further. The *output device* never receives the information. In this case, the cache controller must be able to be told which region of the address space in the external architecture that must not be cached. Any transmission of data to this region must be passed on to the external architecture. A second circumstance occurs when there are other bus masters in the system. This highly likely situation requires a method to maintain coherency between the cache and memory; otherwise, the other controllers may well be using invalid data. A "write through" strategy is frequently employed where the data to be written is placed in the cache and also in main memory. The cache controller may perform the memory write after releasing the CPU to continue its work since the cell in the cache is correct. Other strategies have also been explored.

We will now consider the nature and implementation possibilities of a *cache controller*. The previous paragraphs have established the functions that a cache controller must perform and the speed domain in which it operates. The requirement that the controller must mediate any reference to main memory made by the CPU (i.e., that decisions about cache contents must be made within one CPU clock cycle) require that an autonomous control structure running as fast as any in the CPU is needed. A traditional state machine implementation is appropriate. Additional architecture is needed to assist the cache controller in determining the presence of a cell in the cache. Another memory is used to hold the addresses of the structures, or *cache lines*, that are present. This *address memory* contains the address tags, usually the most significant part of an address. Associated with each tag is a line of data in the cache itself. Every element of this *address memory* must be compared with the address presented by the CPU to determine if the desired address is present. To perform this comparison sequentially is unthinkable at the speeds anticipated, so a parallel comparison is made. One can imagine that each cell of the address or "tag" memory is combined with its own comparator so that each tag is simultaneously compared with the incoming address. The result of the comparison is used to inform the cache controller to transfer the data, because a hit occurred, or to run a main memory cycle, due to a miss. The least significant bits of the proffered address are used to select the cell within the line. A memory, formed of our tag memory and the data memory, is said to be associative. Notice that the size of the tag memory, or directory, is related to the size of the cache and the number of data elements in a line.

Unfortunately, a reasonably sized cache memory operating associatively is slower than a corresponding random access memory because of the added propagation time implied by the comparators. An alternate plan that uses RAM for the address memory as well as the data memory is exploited now. The tag and data RAMs are divided into sets. The least significant part of the address is used to select one element in each set of tags. Each set of tags has its own comparator that is used to determine if the most significant part of the address matches the contents of its tag. If the tag is present in the selected position in one of the sets, the data corresponding to that set/tag position is manipulated. The cache controller is informed if the tag is not present so that it can perform a main memory cycle. Such a memory is said to be *set-associative*. The number of sets, hence comparators, determines the number of "ways" of a set-associative memory. A four-way set-associative memory contains four tag sets and comparators.

The cache memory may be combined with a short pipeline to further improve performance. The pipeline takes advantage of the sequential nature of the program or data flow. Keeping the pipeline short reduces the effects of pipeline purges. In many machines, a second short pipeline is introduced in the instruction path to be filled with instructions along each possible branch path. The cache serves as a random access memory to the CPU that operates at internal architecture speeds. Some designers provide two cache structures — one for instructions and the other for data, so that the CPU can be designed as a Harvard architecture machine. This is a convenient structure for a microprocessor where a single bus can be brought off-chip at reduced cost while still maintaining a Harvard architecture via caches on-chip.

9.3. VARIATIONS

The principal subject of this book has been to discuss the implementation of algorithmic state machines, having one or two levels of control, using the microprogrammed method. As indicated in the first chapter, there are alternative methods. The synchronous method that predates the subject of our discussion was given the name "traditional". It uses minimized Boolean networks to determine the next state, as indicated in Chapter 4. The beauty of this method is that the next state can be determined using the minimum propagation delay relative to our memory-based method. The absolute speed king of synchronous methods is therefore the traditional one.

We have noted that the strength of the microprogrammed method lies in the ease of implementing long sequences since the architecture of the controller is not related to the details of the control algorithm, as in the traditional implementation. We have shown in this book how long sequences that are easy to implement can be used to build first- and second-level synchronous state machines that accomplish useful tasks. In this section, we would like to contrast systems built around the traditional controller to show how the designer can exploit the fastest, but "crankiest", of all controllers.

With the advent of the Programmable Logic Device (PLD), specifically Programmable Array Logic (PAL) and Programmable Logic Array (PLA) in the late 1970s, the size and ease of implementation of traditional state machines was improved. Software became available that allowed the designer to specify a machine by giving its Next State Table, its state determining equations or, ultimately, a Hardware Design Language (HDL) version of its ASM diagram. With efficient simulation tools, the traditional state machine became relatively easy to implement in the gate array and custom logic areas as well. We will start with this form of stand-alone state machine and progress to computers built using them as their controllers. While the traditional controller was used as the primary controller in early mainframes, they were driven out to a great extent by the microprogrammed controller. When microprocessors were developed, the earliest controllers were again traditional. Later, the microprogrammed controller was almost universally applied. With the return-to-basics movement as embodied in the Reduced Instruction Set Computer (RISC), the need for a very fast, compact controller to run short simple sequences has resulted in the return of the traditional controller.

RISC and CISC machines and their design philosophies will be contrasted. Both design philosophies hope to produce machines that calculate application results in the shortest possible time. It is instructive to view the contrasts between these philosophies since many of the considerations and inventions can serve to improve both types of machines. Two areas in which ASISPs have been integrated will be discussed as well. We will see how specialized hardware is added to the internal architectures of these processors to allow certain application specific tasks to be performed very rapidly. These techniques allow mass-produced processors to enter areas dominated by dedicated first-level machines.

This section will conclude with the introduction of a method to simplify the user interface to a first-level machine by creating the illusion of the presence of a second-level machine. The programmer writes code using instructions from a set that would be appropriate to a second-level machine. The simple, single-command, sequential instruction set reduces the parallel control nature of the first-level machine. This allows compilers and other high-level tools to be applied to the problem of constructing the control program. First-level machine code is produced by the software. Such a method is appropriate for the code that does not need to execute at high speed. High-speed code can still be embedded using normal microprogramming techniques.

9.3.1. Traditional Controller Implementation

As introduced in Chapter 4, the traditional controller achieves its speed by using minimized Boolean networks to determine the next state. This means that only those minterms needed by the design are implemented since the same limitation on the width of gates is present here as for Adders and other architectural networks. The memory element in the microprogrammed controller effectively contains all possible minterms of the "address" inputs. The selected minterms, those with "1s" in the memory locations corresponding to the minterms required by the sequence of states, represent a small portion of the possible terms. Providing the decoding structure for each minterm, whether used or not, results in a relatively slow network in comparison with the minimized version if both are implemented in the same technology. The difference in speed is not as great as that found between any first-level and any second-level implementation. The traditional controller is faster by a small amount than the microprogrammed implementation. Following behind at a considerable distance is the second-level, or computer-based, implementation.

The digital designer has employed the traditional state machine in many ways since the beginning of the history of the field. He has implemented computers, ranging from main frames to microcomputers, using the traditional controller. Many machines used around computers, such as serial I/O controllers and disk controllers, are usually implemented as mass-produced traditional machines. The DMAC mentioned earlier in this chapter is usually built this way.

Many state machines were present in the TTL Data Books in the form of counters of various types. The designer could use members of a logic family to create his own dedicated traditional machines, but the time investment was relatively high, as was the parts count. With the advent of Programmable Logic Devices (PLDs), the programming environment enjoyed by computer users was brought to bear on problems that could be built using traditional state machines. These solutions were easy to produce since a useful set of 30 to 100 states could be embedded in a single chip. The development software has kept pace also, allowing the programming environment to mimic an ASM diagram or HDL specification of the solution. The hours of progressing from the ASM diagram through the next state table, the logic equations, and finally through minimization and testing have now been reduced to minutes using design tools like Advanced Micro Device's PALASM and other tools. Simulation tools present in the software allows the designer to perform the same cycles of design, implementation, and testing that the programmer enjoys; hardware commitment need only be made when the logic of the design has been verified.

PLDs have also evolved to include flip-flops that require fewer minterms to implement arbitrary state variable combinations than the D or JK types. Counters are included to implement counted loops, thus reducing the number of states needed under certain circumstances. No subroutine calling capability is available however. The PLDs achieve their speed relative to the memory-based controllers by providing a programmable subset of minterms that the designer needs. The PLA-based controllers, called *programmable logic sequencers*, provide both programmable AND gates and programmable OR gates in a sum of products organization. Both the AND and OR gates have a limited number of inputs. The number of inputs on the AND gates is determined by the number of inputs and outputs to the integrated circuit housing the device. The number of OR gate inputs is usually less than the number of AND gates. The OR gates each serve as an input to a flip-flop that can be used to hold a state variable or a registered output.

The resulting part is frequently available in several technologies and represents a single-chip traditional controller solution of considerable power. The Complementary Metal Oxide Semiconductor (CMOS) versions are even erasable and reprogrammable. The current state of the art device consists of a part in a package that is comparable to the integrated microprogrammed controller solutions mentioned at the end of Chapter 4. The designer therefore has two equally easy to use solutions to his problems in the PLS or microprogrammed integrated controller.

What points affect his choice? The final selection is ultimately determined by the number of states required. The complexity of the next state network increases rapidly with the number of states in a traditional controller. The corresponding complexity in a microprogrammed controller is independent of the number of states, assuming the number of words needed by the solution is available in the memory chosen.

Interestingly enough, combinations of the two types are also frequently used in computer design. Controllers that appear microprogrammed on the surface have a lower level based on traditionally implemented sequences. In a situation where the absolute number of transistors must be reduced to the minimum, such as in a single-chip microprocessor, short traditionally implemented sequences can be "strung together" at a higher level of abstraction in a microprogram. We have exploited this concept continually in the course of this book in recognizing the various levels of control (or abstraction) that can be used to solve a problem. Here, the bottom level exploits the absolute minimum number of transistors that the traditional method employs.

9.3.2. Reduced Instruction Set Computer (RISC) Architecture

A particularly interesting combination of traditional controller implementation and computer design has enjoyed a growing popularity recently. The Reduced Instruction Set Computer (RISC) architecture seeks to address several recognizable bottlenecks that appear to be present in Complex Instruction Set Computers (CISCs), particularly those that are microprogrammed. Remember that many of the perceived shortcomings can be solved in the CISC design, but let us look at the problems and see how the RISC architecture attempts to solve them.

A RISC computer is intended to be fast, so fast that it must execute one instruction per "cycle" to achieve its speed goal. Ultimately, the meaning of "cycle" is at the center of the discussion. The desired goal is to interpret "cycle" as the period of the system clock. The system clock period is set, as for the CISC case, by the propagation time through the combinatorial networks in the internal architecture.

First, let us interpret the "cycle" time as the period of one external memory cycle; i.e., the time required to carry on a single transaction with main memory. This time is determined by the propagation delay from the MAR through the external architecture to the destination register inside the CPU.

Central to any RISC discussion is the question of the instruction set that the CPU will use. This affects both the programming model and address modes. In studies of production programs, various groups have noted the types and frequencies of occurrence of instructions. It was found that most programs, at the machine language level, used only those instructions that were included in the *Basic Instruction Set* introduced in Chapter 8. Unfortunately, the long sequence instructions present in many CISCs were used very infrequently.

Compiler writers have long contended that the simpler the instruction set, the more effectively a compiler can be written. Many compilers never take advantage of the composite instructions such as the Decrement and Branch, even when they are available on the host machine. The extended address modes, such as preindexed indirect that can improve subroutine parameter passing, are not used either. The long sequences, which are easy to implement using the microprogrammed controller, are not being used with enough frequency to justify the amount of silicon and propagation delays they incur.

The RISC designer merely says to discard all such sequences and produce a machine that executes the instruction set it is given as fast as possible, "no holds barred". Another factor that we noticed in the implementation of the Basic Instruction Set is that most of the Arithmetic/Logic/Shift instructions can be implemented in a single cycle on an RALU. These instructions are then candidates for the RISC. If the RISC designer requires a multiply instruction in his set, he must provide a combinatorial multiplier in the internal architecture connected such that it can do its job in a single system clock cycle. If he desires to support the floating point data type, for example, then he must provide a combinatorial network for any instruction on this data type in his instruction set.

Unfortunately, data movement between the external and internal architecture is not one of those things that can happen in a single clock cycle if the memory is slower than the CPU. The designer must break his single-cycle requirement for certain instructions and then provide a means by which it need not be used often. This latter requirement means that the number of registers in the internal architecture must be relatively large compared with the CISC. Some designs approach 400 such registers. The large number of registers means that the machine instruction format must be extended over that used by many CISCs. Our example in Chapter 8 found that a 16-bit word could adequately contain an opcode and two 4-bit register fields to point at two of the three operands needed by an instruction. While we could reduce the width of the opcode in a RISC, this does not increase the number of bits in the register fields as rapidly as we would desire.

Both CISC and RISC designers have found that increasing the width of the data bus improves performance of their machines. The CISC designer frequently could fetch two 16-bit instructions in a single bus cycle. As CPU speeds have outstripped main memory speed, this level of pipelining has been necessary. The RISC designer can use the wider word to contain wider register selection fields, allowing him to directly support more internal registers and thus reducing the number of times operands must be moved between the internal and external architectures.

Note should be made here of methods that allow the designer to use narrower instruction words. A group of methods based on "bank switching" have been adopted in very interesting ways. An instruction is supplied that allows the programmer to select a group of registers within the larger array of registers available in the internal architecture. The narrow register fields can then be used within each instruction to specify the desired source and destination operands. The most useful version of this method is used by the RISC designer to change "context" during a function call or interrupt. The bank is changed by an amount that gives the programmer access to some of the old set as well as a new set of registers. This makes function calling and interrupts particularly fast to implement since "context switching" requires only the changing of the bank specification register.

Another set of sequences that must be omitted are those for the address modes. The register/register address mode will be retained since it can be performed in the same clock cycle as the data combination in the RALU. This mode becomes the workhorse used by all instructions. Its relative, register indirect, will be available since it allows the programmer to point, using a register, at any location in the external architecture. The immediate address mode, in which data is located in the instruction, must be available to initialize both address and data registers. Needless to say, it will require two memory cycles to complete since the data value must be fetched after the operation code (opcode). Various formatting methods may be used to squeeze a short length value in the first word of the instruction at the expense of one of the register fields. Absolute addressing is even more expensive in time, but may be provided in certain designer's machines as a comfort to the assembly language programmer.

Notice that the sparseness of the instruction set and address modes may be compensated for by the production of the promised "efficient" compiler. A quality macroassembler, once provided with an extensive set of macros, may make the programmer think that he is using a CISC. If the rest of the RISC designer's goal is realized, then the fact that many algorithms that were at the first level in the CISC and are at the second level in the RISC need not concern the second-level RISC programmer. The task should execute as fast, or faster if one believes the advertising, on the RISC machine.

Now let us see what other steps are available to the RISC designer to reach his speed goal. We can see from above that he intends to fetch a complete instruction in every memory cycle except for the very few times when data must be moved between the CPU and the external architecture. The RISC machine will now be made to execute the previously fetched instruction while it is fetching the next one. At this point, it appears that an instruction is being executed every memory cycle. How is this accomplished and can we improve on it such that we can approach the execution of an instruction every clock cycle?

First, let us see how we can execute an instruction while simultaneously fetching the next one. In the computer example given in Chapter 8, we first executed the instruction fetch and decode sequence followed by the selected execution sequence. Our example was limited by an internal architecture that required at least two clock cycles to complete the fetch and decode because the MAR broke the propagation path from the PC to the external architecture. Further, the internal data bus had to carry the PC to the MAR as well as the external instruction to the IR. We could eliminate this conflict by making the PC part of a partially autonomous PCU whose work is coordinated with the activity in the *data execution unit*. We used a single controller in our Chapter 8 design for simplicity's sake, but that would just interfere in the RISC design.

Once the PCU becomes autonomous, it can manage a much more complex interaction with main memory to overcome the speed mismatch between the internal and external architecture. The first step employed by many designers is to introduce an instruction pipeline and various multiplexing schemes between the internal and external data buses. The external data bus threading main memory may be made very wide; i.e., some integer number of internal data bus widths. If instructions are fetched in sequence, then the main memory can run at a rate of 1/N times the speed of the internal bus (where N is the ratio of the bus widths) by multiplexing ("interleaving") the instructions onto the narrower internal bus. As long as instructions are fetched in order, the CPU can execute them at a rate of N times the main memory's access time.

The pipeline is normally run into the execution unit in such a way that the incoming data are "digested" partially at each step. Such assembly line methods can decrease the complexity of the execution unit's networks to further increase speed. In a highly pipelined system, such as that described, a new data value is "finished" every cycle of the system clock.

Unfortunately, one hallmark of a programmed system is the capability to make data-dependent decisions. When a branch is taken (i.e., the PC should not have been incremented but changed), the contents of the instruction and data pipelines are invalid. The instruction pipeline may be dumped and then slowly refilled at the speed of the memory system. The data pipeline is filled with partially processed data; i.e., the contents of registers may have been changed. Dumping the data pipeline will not restore the registers, thus more drastic means must be used. The method used by RISC designers marks a very interesting aspect of this "minimalist" design philosophy. No states are devoted to this clean-up in the controller. Remember, we are here to go fast not clean up messes. The RISC designer expects that problems like this are to be solved once and at as high a level as possible. This particular problem can be solved in software. The solution is to place NO OPeration (NOOP) instructions after each branch instruction to fill the instruction pipeline with "non-data modifying" instructions. Thus, the data pipeline will be "clean" if it must be dumped. It is the task of the assembler or high-level language compiler to provide the NOOP instructions in the correct quantity and at the proper place to ensure that the pipelines will perform properly under branching conditions. The hardware designer will not waste any of his precious states on housekeeping.

Notice that the housekeeping problem associated with pipelines is present in CISCs as well as RISCs. The CISC designer may well be tempted to solve the problem with his "unlimited" sequences and not involve the second-level programmer. The pipeline is a good idea; however, long pipelines do not perform well if branching, function calling, or interrupts occur frequently. The CACHE was discussed in an earlier section as a way to minimize the problem. This intermediary between the external and internal architecture helps both the CISC and RISC designer by allowing flexibility in the way data and instructions are retained in the CPU. The RISC designer normally exploits caches on both the instruction and data streams to mitigate the effects of a slow external architecture. Once the data stream is separated from the instruction stream, the Harvard architecture comes into full bloom. Such a structure is supported in most RISCs even to the point of bringing both buses off of the chip. Current CISC microprocessor designers may use the Harvard architecture internally by way of separate instruction and data caches but bring only a single bus off of the chip.

Once the RISC designer eliminates long sequences, he can take advantage of the inherently faster traditional controller in his design. The reduced transistor count allows him to concentrate more on the internal architecture to make the small number of states more effective.

The one shining point that makes a RISC machine even more effective as a computer is the fact that its interrupt response time, "latency", can be made very short since there are no long sequences to wait for completion. If the pipeline is kept short, the size of the "state of the system" is manageable. The designer may work particularly hard to keep as much of the state on the CPU chip as possible, thus reducing the number of memory cycles needed to respond to the interrupt. By reducing the size of the controller and increasing the register array, the RISC designer feels that he can solve this problem more efficiently than the CISC designer.

The wide instruction word has allowed the RISC designer to create some versions of instructions that appear to be CISC-like. They still require only a single clock cycle to perform, but they appear to have a "microprogram-like" quality. Some RISC designs provide both integer and floating point arithmetic units in their internal architecture. The internal data bus design is done in such a way that operations in each section can be performed simultaneously, as we have noted for microprogrammed machines above. The Motorola DSP96002 supports these data types as well as two separate data buses to the external architecture. The bus control unit may move data on both buses simultaneously and in parallel with the arithmetic units. Many current RISC machine also allow the program control unit to make decisions on data flowing through the unit as well. One particularly useful combination is similar to the counted loop structure in our microprogrammed controller. The RISC CPU may repeat a small block of instructions a specified number of times. Normally, the block need not be fetched from main memory after the first time through the loop as a consequence of the pipeline and cache architecture. Since each section mentioned in this paragraph is autonomous, it may accomplish its task, if appropriate, in parallel with the others. The programmer is charged with the task of determining "appropriateness".

As is always the case, the architecture can support certain groups of operations very efficiently and then performance is degraded for anything not anticipated by the designer. The Digital Signal Processor (DSP) and Video Display Processor (VDP) introduced in the next two sections are designs that have architectures optimized for the algorithms normally encountered in their respective areas.

In summarizing, we would like to make the point that both the RISC and CISC designer must ultimately solve the same types of problems. Their solutions are different but the final evaluation can come only at the time the two designs go "head to head" in the same application. As time passes, we will see more of these real comparisons instead of benchmarks which are artificial. The CISC designer is presenting a moving target as well. Any progress in the use of pipelines, caches, Harvard architecture, or other techniques has been incorporated in CISC designs also. The current weak point seems to be in compilers. There is more variation in the speed of execution of one high-level program as compiled by different compilers than between the same program on different architectures. If CISC designers and compiler writers would cooperate in such a way that the instruction sets and address modes needed by the compiler writers were provided by the CPU designers and vice versa, then we would see very fast CISC machines as well.

Since CISC machines with user programmable microcode memories are rare, one can not use the final solution to the problem caused by not having an application-specific instruction. This leaves the methods outlined for first-level machines in Chapter 7 as the ultimate high-speed solution for the application designer.

9.3.3. Digital Signal Processors (DSP)

The field of digital signal processing is principally concerned with computations based on the discrete Fourier transform and digital filtering. These computations have the Multiply/Accumulate (MAC) operation in common. Areas of application range from seismic and biomedical signal processing at the low-frequency end to radar and video image processing at the high-frequency end of the spectrum. Historically, as the MAC operation was performed faster, more high-frequency applications were added. Most of the examples given in Chapter 7 were derived from this area. Since many applications were served by combining a general purpose computer with a fast MAC, it was only natural that, as technology permitted, the two were more closely coupled in a single integrated architecture yielding the digital signal processor. Several companies make DSPs, including Texas Instruments (the TMS320), Motorola (the DSP56000 and DSP96000 series), and AT&T (the DSP32C). There is a common conceptual thread in all DSPs. They borrow from the integrated microcomputer design the idea of combining all the elements of a complete computer on a single chip; i.e., RAM, ROM, and I/O are integrated with the CPU. An expansion method is provided for those applications that require more RAM and ROM than can be integrated. Since high-speed serial communications over the telephone and other networks is an important application area, the serial ports have more extensive capabilities. The DSP frequently must operate in conjunction with another computer; therefore, it is normally provided with an 8-bit wide "host port" to enable high-speed, low overhead communication with its host. This port is also supported with extensive interrupt capabilities, both for the host and the DSP itself. In most cases, the host port may be used to download programs and data allowing the DSP to be used as a reprogrammable computation or I/O engine where the task and data are changed by the host program as needed.

We will now look at the core computer containing the CPU and on-chip RAM and ROM. All DSPs, as is now true of most general purpose microprocessors, support some form of Harvard architecture using this structure. Data and program transfers are handled separately. The cost of bringing this capability off-chip is frequently traded-off for other features. In some cases (e.g., the Motorola processors), the internal data paths are further split into two separate data memories. Transfers between the CPU and these memories may be accomplished in the same instruction cycle. We therefore have the capability to move the instruction and one or two operands in a single instruction cycle. The ALU is associated with several registers and one or more shifters in a structure reminiscent of the RALU described in Chapter 3. One of the paths into the ALU passes through a combinatorial multiplier, thus implementing the MAC structure. The data path width is single precision in the early integer units except around the output of the multiplier and in the entire ALU structure, where it is double precision. The TI integer devices handle 16-bit numbers, while the Motorola DSP56000 uses 24-bit numbers. The AT&T device is floating point. Later versions of DSP chips by all manufacturers have "hardware" floating point capability that allow single instruction cycle time execution of floating point operations. Please refer to the data books and user's manuals (listed in the Bibiliography) for the specific devices.

The program control capability of the DSP is similarly extended. The loop control mechanism is streamlined in such a way that instruction fetching is greatly reduced and gives the appearance that control operations are performed in the same instruction as architectural operations. We experienced this feature in our microprogrammed controllers in Chapter 7. Pipelining on all data paths is normally used with its corresponding strong and weak points. As in the RISC processor case, data may not be available for use by an instruction for several instruction cycles. The effect of this delay on a conditional branch is a normal part of writing a program for this type of processor.

The instruction set of the DSP is usually formed from a normal general purpose instruction set with additional instructions and address modes that are optimized for the principle task of the machine. The two tasks for which the DSP is optimized are the digital filter and the Fast Fourier Transform (FFT) algorithms.

The digital filter is directly calculated by evaluating the Arithmetic Sum of Products algorithm shown in Chapter 7. Two lists of numbers (vectors) are multiplied element-by-element to create a single scalar result. One vector is a list of constants that describe the spectrum of the desired filter; i.e., each type of filter, low-pass, high-pass, or band-pass, is described by a different constant vector. The details of the shape of the filter and the method of implementing the filter affects the number of elements in the list; e.g., the more rapidly the output amplitude decreases with frequency (the roll-off), the larger the number of elements in the vector. The other list contains samples of the function of time being filtered. Each element in the sample list represents a sample taken at a specific time. The sample list is normally ordered relative to the list of constants so that the 20th sample in the list, acquired 20 sample periods before the first sample, is multiplied by the 20th constant.

Two implementation methods are commonly used. One, the Finite Impulse Response (FIR) filter, does not feed back energy from the output. It can be thought of as a tapped delay line. The output is formed by scaling, or multiplying, the output of a tap by an element in the list of constants. A given tap is associated with a particular element in the list of constants. The second implementation method, the Infinite Impulse Response (IIR) filter, incorporates feedback which implies that energy is retained in the filter for an "infinite" length of time. The most important computational effect of this energy retention is that the number of elements in each vector is reduced relative to the FIR, but the number of bits needed to represent a variable is much greater. From a filter point of view, rapid transients will cause the output of an IIR filter to "ring" just as in the corresponding analog case. A fast transient response requirement in an application will normally imply a FIR filter.

The computation rate is determined by the sample rate and the number of elements in the constant vector needed to describe the filter. In a consumer high-fidelity audio application, for example, the sample rate is normally 44,100 samples per second. FIR filters are normally used as reconstruction filters during Digital-to-Analog (D/A) conversion; 40 taps are not uncommon and more are frequently used. Simple arithmetic shows that over 1.6 million multiply/accumulate operations are needed in each channel every second to implement the filter. This is the type of demand placed on a DSP. Note that similar filtering requirements are needed in video applications where a two-dimensional arrangement of data is used to represent the image.

Besides the fast multiply/accumulate operation provided by the DSP, we find very fast modulo-N arithmetic capabilities. This type of arithmetic is needed in pointing at samples in the lists described above. From a hardware point of view, the pointer into a list can be implemented as a counter. When the counter reaches the "end" of the list, its overflow "state" merely resets it to the beginning of the list. In the general purpose processor, this task is accomplished by comparing a register that serves as a pointer with a constant containing the termination value and reinitializing the register when the two are equal. This requires many instructions and their attendant execution time. Most DSPs support modular integer arithmetic operations that involve no additional instructions to manage. From a program execution point of view, the modular arithmetic is executed as rapidly as having a hardware register but with the flexibility of the software version. Note that many algorithms, especially the FFT whose description follows, were derived to contain lists and operation sequences that are 2^n long since the simplest reset signal for a binary counter can be derived without computation from its most significant bit carry out.

We will conclude this section with a short discussion of the FFT to show the architectural impact that it has had on the DSP. Spectrum analysis is the fundamental tool of signal processing. Virtually every scientific or engineering field uses this tool. Spectrum analysis results in converting the time-based information in a signal into the frequency domain. Our interest here is, in particular, implementations of the Discrete Fourier Transform (DFT), a spectrum determination tool that is appropriate to sampled systems. While other texts, e.g., Rabiner and Gold (see Bibliography) should be consulted for the theoretical basis of the FFT, we wish to examine the way the designers of DSPs have responded to the needs of the FFT algorithm. The FFT, in general, converts a complex vector describing a signal in time to a complex vector describing the signal in frequency space (and vice versa). The number of data points transformed should normally be taken from a signal that can be thought of as extending for all time (or the spectrum extends over all of frequency space). Since real signals are not constant in time, a great deal of work is done to get around this requirement; suffice it to say that this can be accomplished for many signals of interest.

Once a representative sample size (n) is determined, the DFT may be calculated by evaluating the equation

$$X(k) = \sum x(n) \, e^{-j(2p/N)nk} \quad k = 0,1,...,N\text{-}1.$$

This equation may be written in a more compact form as follows:

$$X(k) = \sum x(n) \, W^{nk}$$

These equations clearly show the scale of the computation problem for the DFT. Notice that the "W^{nk}" terms can be precalculated so that they form a list similar to the filter coefficients used in the previous example. The history of signal processing has been marked with the derivation of algorithms that reduced the number of multiplications and accumulations represented above. This version of the FFT is based on the discovery that dividing the sample into two parts and performing two DFTs results in fewer mathematical operations than working on the entire sample at once. Since this process can be continued until the number of points in each DFT is very small, say two, the mapping of the algorithm on the architecture greatly affects the computation speed for large N. This algorithm is known as the *decimation-in-time* version of the FFT which results in the number of multiplications tending to $N/2 \log_2 N$ for large N. Many of these multiplications can be done at the same time so that the algorithm lends itself to parallel processing.

The net effect of the improvements in evaluating the DFT is that the order of the data in the complex result vector is different from that in the input vector. As a historical point, it should be noted that a fast memory that could hold all of the samples for a given transform was very expensive. A major goal of the DFT algorithm is to reduce the amount of memory required for data during the computation. The result is that the DFT studied here will place the results in the same location as the input sample vector and will use virtually no temporary storage. Unfortunately, as mentioned, the output vector is ordered differently than the input vector. By a happy coincidence, it happens that if one thinks of the index into the input vector as being an unsigned binary number such that the largest number representable by this representation is N itself (an integer power of 2), then the location of the result corresponding to an input element is in a location specified by reversing the bits that represent the location of the input element. In other words, if the input vector is 16 samples long, then the index of the sixth element is 0101 when counting from 0. The location of the corresponding transformed element is at 1010 by reversing the order of the bits in the index. The need for bit reversal means that any DSP must have some method to compute indices that will allow the programmer to recover the resulting transformed data in the correct order without having to waste instruction cycles. The Motorola processors implement a reverse carry in the ALU, which along with the modulo-N capability, makes scanning through an output vector relatively simple.

In summary, the DSP is normally implemented as a general purpose, second-level processor with certain additional instructions and architecture added to make the computation of the arithmetic sum-of-products and the DFT as fast as possible.

9.3.4. Video Display Processors

One important application area for the methods presented in Chapter 7 is the generation of video display information. While no examples were given from the area in that chapter, we will outline some of the algorithms that are used and show an integrated second-level solution to the problem. The goal of any display system is to receive information that describes a picture and display it. The display, at the present time, is created as a two-dimensional array of picture elements or pixels. A monochrome picture, then, may be thought of as an array of 1s and 0s, where the 1 may represent white and the 0, black. Creating the visible entity from a computer array of this sort requires only placing the elements of the array on the display surface, a piece of paper or computer screen, while maintaining the proper spatial relationship between the elements. While, in the case of the computer screen, speed provides the need for hardware aids in this process, there are no data transformations required between the image in computer memory and the display surface. This version of a picture is known as a *bitmap*. A grey scale may be created on the display surface, if appropriate, by first encoding the value of grey for each pixel using a reasonable number of bits. The bitmap is a two-dimensional array of numbers each representing the grey value at one point. A 256-level grey scale can be represented using 1 byte of information for each pixel. In the case of a computer screen, digital data is converted to analog by a digital-to-analog converter and used to modulate the intensity of the electron beam as it sweeps across the screen. Color may be represented by setting up three two-dimensional arrays, each containing the "grey" or intensity information for a pixel of one of the colors used to build the picture. Normally, the display surface in a color screen is composed of an array of dots. Each pixel consists of three dots: one for red, one for green, and one for blue, each illuminated by its own electron gun. One of the three arrays is used to set the intensity of all of the dots of a given color; i.e., there is a "red" array, a "green" array, etc.

Any graphics system must perform the transmission of data from the array(s) in memory to the display device, e.g., a character printer, laser printer, or Cathode Ray Tube (CRT).

It must accommodate the characteristics of the device. A quick example will demonstrate some of these characteristics. Consider a simple dot-matrix character printer found with many personal computers. A head is moved from left to right across the paper while printing occurs to create a line of type. A page is built up of lines by spacing the paper upward at the end of each line. While some printers can print as the head is returned to the left edge of the paper to begin the next line (i.e., every other line is printed in reverse order), few printers can print from the bottom of the page to the top and none can feed the paper sideways. The simplest printer appears as a device that must receive characters in a certain order. If we let our two-dimensional display array contain the dot patterns that are needed for each character to be printed in our document, we can relate the rows of the array to the lines on the page while the array columns correspond to a character position in the line. To create a page of print, our computer program must supply elements from the array starting with element (0,0) and proceeding in order one row at a time. While our simple example has assumed that the printer is only printing a single row of dots at a time, modern printers of this type have from 9 to 24 pins arranged vertically in the print head to reduce the number of head movements. The program in question is normally implemented in a microcomputer in the printer itself where conversion from a character code (e.g., ASCII) to a bitmap is performed using a translate table as each character is printed. Two points are to be made: the order of placing the data on the print head is important and the speed of doing so is governed by the structure of the display device.

A CRT is used just like our simple example except that the scan rate (i.e., similar to the print head movement) is much faster. The electron beam is moved across the screen from 15 to 19 thousand times per second (kHz), while the vertical motion ranges from 50 to 80 Hz. If all of the scan lines needed for a picture are made in one vertical trace, the display is said to be non-interlaced. An interlaced picture is built by producing the even-numbered scan lines in one field and then returning the electron beam to the upper left corner to scan the odd-numbered line. The order of moving data from the memory arrays to the display is obviously affected by the choice. The basic display system must provide the means to move the data from memory to the display at the rates needed and also the electron beam trace synchronization signals. If the picture contains grey scales or color, the system must convert the digital data to analog form using a D/A for each "color". The speed required to paint a moderately high-resolution picture (e.g., 1024 X 768 pixels) is very high. If one assumes a 60-Hz rate to produce an entire picture, a new pixel must be displayed every 20 nanoseconds. This is a very high rate of data transfer that is not normally handled by the computer itself but by a specialized display system. The amount of data to represent a color image having 8 bits for each of the electron beams is on the order of 2 Megabytes. This amount of data must be moved to the display 60 times per second to preserve the picture on the screen for the user.

While the bitmap is relatively easy to display, it is very difficult to change. The actions needed to change the size of the image or rotate it relative to the frame result in a degraded picture at best. For this reason, information describing the picture is maintained in a form that can be dealt with mathematically. A popular form is that of the vector. A picture may be thought of as being composed of many lines, curves, etc. For many pictures, less information is needed to describe the image in this form. A vector is a line segment specified by the locations of its endpoints in a coordinate space. Rotation and scaling are then relatively straight-forward though computationally intensive tasks. Another process that needs to be performed when solid objects are displayed is "area filling" or the placing of a colored surface bounded by vectors. Here, a rapid means is needed to set the color of pixels within the bounded region.

To summarize the graphics problem, we see that the display surface (e.g., a CRT) requires that a bitmap be created containing a pixel-by-pixel description of the image. Specialized hardware must be provided to move the bitmap to the display surface, meeting the speed and order requirements established by the display itself. While bitmaps are difficult to manipulate, they will remain as the only way to represent some images, e.g., those captured from the real world. High-resolution color pictures require a large amount of memory. Storage of bitmaps will increasingly demand some means to "compress" the space required. The storage form needs to be expanded to the display form at the instant of display in a time that is too short to notice by the user. Signal processing is appropriate to accomplish data compression as well as image improvement. In many cases, the image may be described in terms of other, more economical and easy to manipulate forms (e.g., vectors). The image may be maintained and manipulated in display list (i.e., vector) form to make image rotation, scaling, and viewpoint changes easier. The display list form is converted to a bitmap when display is desired.

There are two classes of hardware present in the graphics system. The bitmap display system consists of a large array of very fast memory that, in a color system, supplies three streams of data to D/As to generate the electron beam modulation signal. Most video D/As provide a table-based mechanism to translate the bitmap pixel value on the way to the D/A. This allows the "mixing" of colors for standardization purposes to compensate for variations in CRTs and to vary grey-scales for image intensification purposes. A set of counters are normally provided to generate sweep synchronization signals for the CRT and addresses for the video memory containing the bitmap. The second class of hardware performs the computations necessary to convert a display list version of the image into the bitmap. This task is handled by the main computer in low-end systems, while machines built along principles discussed in Chapter 7 are used for high-end systems. The process of conversion as well as image manipulation is arithmetic intensive. For computation purposes, the display list should be represented using floating point numbers. Computation time is very high for this representation, so hardware assistance is needed as images get more complex. The speed requirement for this part of the task is established by the user, not the display hardware as was the case above. High-end graphics processors can take advantage of the Floating Point Unit described in Chapter 3.

An interesting integrated graphics processor has been introduced by Texas Instruments, Inc. which will be discussed here. The TMS340 Graphics System Processor (GSP) is a complete second-level machine that supports a general purpose instruction set extended by special instructions and hardware to support a serious graphics system. A short survey of the features of the processor, taken from its user's guide (see Bibliography) is given below:

1. It is a fully programmable 32-bit general purpose processor.

2. It has a 128 Megabyte address range and a 160 nanosecond instruction cycle time.

3. Its on-chip peripheral functions include:

    a. Programmable CRT control (horizontal sync, vertical sync, and blanking)

    b. Direct interfacing to conventional DRAMs and multiport video RAMs

    c. Automatic CRT refresh

    d. Direct communications with an external (host) processor

4. Its instruction set includes special graphics functions such as pixel processing, XY addressing and window clip/hit.

5. Programmable 1, 2, 3, 8, or 16-bit pixel sizes with 16 Boolean and 6 arithmetic pixel processing options are provided.

The first member of the family, the TMS34010, has a microprogrammed controller. Its 32-bit wide register/ALU architecture is extended with the inclusion of a barrel shifter that allows shifting or rotation of 32-bit operands over 1 to 32 bit positions in a single machine state. The counters for generating CRT sync signals and video RAM addresses are included on chip and are programmable to accommodate a variety of display requirements. The host interface gives the designer a simple but fast method to transfer commands and data to the graphics system. Specialized address modes that allow the definition and transfer of fields having a length between 1 and 32 bits are provided. Instructions are present that take advantage of the underlying architecture to implement part of a line drawing algorithm, fill arrays, and move pixels. A Floating Point Coprocessor (FPC), the TMS34082, is also available to improve floating point computation time. The FPC will perform floating point addition, subtraction, and multiplication in a single state as in the SN74ACT8847 presented in Chapter 3.

Video display applications also require a significant amount of signal processing. The video processor may be combined with a DSP to help in vector drawing or rendering of surfaces. The most common problem that spans both areas of application occurs when a line or edge must be drawn at an angle with respect to the scan lines on the CRT when the pixels are not "vanishingly" small. The edge of a line, or any region in which there is a great contrast between the colors or intensities of the adjacent areas, will be demarcated by a line that is "jagged" due to the shape of the pixels. A popular signal processing technique called "dithering" is used to smooth the edge by adding suitable amounts of noise to fool the eye. This computationally intensive task is appropriate for the video processor or a DSP added to the video subsystem.

9.3.5. Pseudo Second-Level Machine

An interesting variation of the first-level microprogrammed state machine is possible that allows for simpler user programming and retains the power of the many parallel data paths and combinatorial elements. This variation involves the creation of a "simulated" second-level machine on top of the first-level machine. While such a simulation was accomplished in Chapter 8 as in all standard microprogrammed computers by introducing the second-level instruction fetch cycle, this variation will achieve the same ends without paying the time penalty of second-level instruction fetching.

How is such a feat accomplished? It must be emphasized here that the solution is not without cost. Let us set the stage by recalling how one specifies a second-level machine. The three components of the specification are the instruction set, the address modes, and the programming model. These are the only items visible to the second-level programmer. We introduced the instruction fetch cycle to move instruction codes from the external architecture to our controller in the first-level machine in order for us to determine the steps needed to do the second-level programmer's bidding. Why not convert the second-level program directly into first-level sequences and bypass the instruction fetch cycle altogether?

Before we examine the method by which this feat is accomplished, let us examine the costs. A machine that is configured to support normal second-level operations uses compact program codes that are stored in a memory in the external architecture. These elements are shown in the von Neumann form below. Each program code is fetched into the controller and used to index the beginning of a sequence in the microprogram. This sequence runs to completion before the next second-level instruction is used. In a microprogrammed ISP implementation as is shown below, the direct control of all of the data paths and combinatorial elements in the machine is accomplished by each microinstruction. The relatively large number of elements to be controlled requires the use of many fields in the microword resulting in the first-level instruction containing many more bits than the second-level code. Computer design has concentrated its efforts toward maximizing the number of second-level instructions that can be held in a machine at one time by reducing the cost and size of memory in the external architecture. This has had the accompanying effect of slowing the speed of events in the external architecture far below that in the CPU.

Instruction Set Processor Implementation

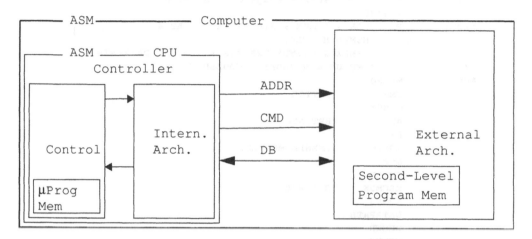

Several methods outlined in the earlier parts of this section were used to minimize the effect of this large speed difference on the performance of the machine. The end effect of the development, however, is that the second-level programmer can write programs of virtually unlimited length in terms of a small set of instructions whose realization sequences can be placed in the relatively small but wide microprogram memory. The current expression of this trade-off takes the form of using dynamic RAM containing more than 1 Megabyte per device at the cost of a few dollars to hold the second-level program while implementing the microprogram memory on the same die, with the CPU using the same expensive high-speed static circuit design as the rest of the CPU. In larger computers, the microprogram memory is normally implemented in separate devices which are as fast and costly as any other in the CPU. If we follow our proposal of directly replacing the second-level program, thus deleting the second-level program memory, we will be substituting the use of low-cost memory with that of a high-cost, limited resource. Normally, such a substitution is undesirable but there are certain mitigating circumstances that make this a reasonable step for some applications.

For dedicated systems where the major speed gain is achieved by architecture and not by extreme component speed, the actual cost of microprogram memory devices is not as high as required by normal computer designs. Large CMOS static RAMs are reasonably inexpensive. While not in the same league as the dynamic RAM-based bulk memories in the external architecture, they are considerably cheaper than the fastest memories used either inside microprocessors or mainframes. In these cases, a large microprogram is feasible. At the same time, the control program itself is not as large as those normally encountered in the general purpose computing world since the dedicated hardware should make the instructions more efficient.

With the foregoing limitations in mind, we will now examine methods to achieve a pseudo second-level machine. The first step is conceptually simple. Let us examine the second-level macro for the ADD instruction used in the basic computer example in Chapter 8.

```
ADD            MACRO         \1,\2
               DC.B          $A0
               DC.B          $10*\1+\2
          ENDM
```

The execution sequence for the Add Register/Register is shown in the following excerpted from the microprogram source program for the basic computer example.

```
     ; add register to register - Flags C, OVR, Z, N
           ORGA $140
;ADD1:    JMP ,FETCH & CCMUX & ALU RPT,ADD & AB &
;/        MSRLD ,MZNLD,MCOVLD,  &
;/        SHFT & CNMUX & REGMUX R2A,R1B & R2903 & DATAPATH &
;/        REGWR & MCMUX & NOYBDEN & NODBBDEN
ADD1:        MACRO
             JMP          ,FETCH
             CCMUX
             ALU          RPT,ADD
             AB
             MSRLD        ,MZNLD,MCOVLD,
             SHFT
             CNMUX
             REGMUX       R2A,R1B
             R2903
             DATAPATH
             REGWR
             MCMUX
             NOYBDEN
             NODBBDEN
             endsc
           ENDM
```

The instruction fetch cycle in the basic computer example moved the instruction code for the ADD from the external architecture into the microprogrammed controller in the CPU. The operation code, $A0, was translated by the MAP ROM to the address $140 which was used by the microprogram controller to point at the next microinstruction to be executed. We had previously placed the execution sequence for the ADD instruction, here a single statement shown above, at location $140 in the microprogram memory during initialization of the micromachine. Operands for the ADD instruction were passed through the limited space in the second-level machine instruction word. Notice that if the programmer desired to use one of the more complex address modes, then the second-level instruction would need to be extended to accommodate more address information. It is clear that the second-level programmer can communicate very little information to the first-level machine to assist it in executing his intentions. Even if a first-level machine contained many parallel data paths, the second-level programmer would be limited to a small number of sequences and paths that the first-level designer thought useful enough to include in the specification of the second-level machine.

Our variation, then, recognizes that the second-level programmer can be allowed direct access to the first-level machine by eliminating the instruction fetch cycle altogether and allowing him to write microcode. Remember, there are two shortcomings to this approach. One is the cost of first-level memory and the second is the shear complexity of the first-level machine when viewed without the simplifications given in the second-level specification. We will control the first shortcoming by restricting the application of this method to areas where the cost of the first-level memory can be contained, e.g., to dedicated first-level machines. The second will be dealt with by creating a pseudo or "paper" second-level machine that is even more tenuous than the one in Chapter 8.

The first step to the creation of a pseudo second-level machine consists of drawing up a specification just as was done for a conventional one. It consists of an instruction set, address modes, and programming model. For our purposes here, we will use the one given in Chapter 8 for the basic computer example. The next step consists of writing second-level macros that define the assembler instruction set to the second-level assembler *except* this time we will place the execution sequences from the microprogram in them as follows.

```
ADD          MACRO        \1, \2
             JMP          , FETCH
             CCMUX
             ALU          RPT, ADD
             AB
             MSRLD        , MZNLD, MCOVLD,
             SHFT
             CNMUX
             REGMUX
             R2903        \1, \2
             DATAPATH
             REGWR
             MCMUX
             NOYBDEN
             NODBBDEN
             endsc
             ENDM
```

The example for the ADD instruction should be virtually identical to that shown for the conventional second level shown earlier. The differences include:

1. The ADD macro does not generate a second-level opcode as it did earlier. There is no second-level opcode or operands in the machine representation of the instruction. Such operands occur only in the source form of the second-level instruction. Note: there is no second-level program memory either. This task has been taken over by the first-level, i.e., microprogram memory.

2. The second-level instruction no longer uses the operand fields of the instruction register to contain pointers to the internal architecture made visible in the programming model. The actual register specifications as given by the first and second variable symbols (i.e., \1 and \2 in the macro prototype statement) will be assembled directly into the microword register pointer fields. This step is accomplished by repointing the register multiplexers to the pipeline fields, the default REGMUX state, and using the variable symbols in the prototype statement as operands to initialize those fields (i.e., R2903 \1,\2).

Now, when programs written in terms of the second-level instructions are assembled using the new macro file the resulting object code is microcode, *not* second-level object code. It can be loaded directly into the microprogram memory in the controller and the system should operate as it did before except much faster since the time taken by the old instruction fetch sequence is now no longer required.

The structure of our second-level machine implementation shown above has undergone a transition. There is now no need for a second-level program memory; however, the external architecture still must contain data memory and I/O controllers. Generally, the sequential addressing philosophy is still useful in the external architecture since only one bus connects it with the CPU. The total address space in the external architecture certainly would be larger than that addressable by any field in the microword of the basic computer example.

Let's look at that example to see where these large addresses are obtained. Notice that any address mode that needed a pointer used a word that followed the first word of the second-level machine code. We have two methods by which we can support extended addressing. One is to place pointers in the data memory in the external architecture and the other is to designate a field in the microword that is wide enough to contain the desired pointer.

The first method is similar to that used in the conventional implementation in Chapter 8. Notice, by referring to the microprogram listing in Appendix H of Chapter 8, that the subroutine OPERA is similar to the instruction fetch sequence in that it moves a word from the second-level program memory into the CPU using the PC. The word is placed in the MAR where it serves as an address to data in the external architecture. The subroutine OPERR is similar, except the source of the address is already in one of the CPU registers. Our problem with using this method still involves generating a pointer that is large enough to let the pointer table be located anywhere in the external architecture.

The second method requires a reasonably wide field in the microword. A field that can be coupled to the internal data bus is also necessary in that the pointer must be moved ultimately to the Memory Address Register. Such a field may be shared with other functions if the user realizes that both functions may not be used simultaneously. This field is also useful in many other circumstances; e.g., for supplying constants during certain calculations. It is therefore worth adding to any first-level machine.

With the addition of the "constant" field to the microword, our first-level machine can easily support the complete set of address modes specified in Chapter 8. Let's consider what has been accomplished up to this point. We have created a machine that appears to the programmer to have a "simplified" specification similar to any computer. He can write programs in assembly language that disregard the details of the underlying architecture, just as he could with normal computers. For the bulk of the code that is not speed sensitive, this disregard suffices to accomplish the work the machine should do. In the parts of the program that are speed sensitive, the programmer can revert to the broad capabilities available in the underlying machine by, as it were, "opening the window" into the microprogram. Notice that, while we created our example macro using a normal second-level instruction, the programmer can create macros having varying degrees of access to the capabilities of the first-level machine. He is not permanently frozen into one programming model as he is with normal computers.

A second benefit appears once the user realizes that he can support many different second-level specifications on the same first-level machine. As we have seen before, the second-level machine is merely a figment of a particular computer designer's imagination. If the data paths in the first-level machine do not match the needs of the designer of the second-level instruction set, then he will need to use more states to create the sequences that he needs. This is reflected in the slower execution time of second-level programs. While this point applies to the conventional computer as well, it should be noted here that any second-level machine can be implemented on any first-level machine, albeit not with the same execution speed.

Let's apply what we have learned to the dedicated first-level machine that must support a microprogram that is longer than we desire to write using the assembly language syntax introduced in Chapter 5. Some people don't like to type as much as that. Our first step to reduce typing and, possibly, debugging would involve writing first-level macros that accomplished large pieces of our algorithm by exploiting our carefully designed architecture. We would be using all of the power of our problem-specific data paths and combinatorial elements. Since macros can contain macros, we can even create sequences of microinstructions each invoked by a one-word name, if appropriate. Furthermore, each macro can contain operands so we are able to change details of the sequence without having to retype the entire sequence; i.e., we are letting the macro expansion mechanism do the typing for us. Notice that there are error catching consequences of this method since the macro can be used to limit the number and types of operands that can appear in any given field.

Now we find that our microprogram has contracted in size since our instruction set contains just the instructions needed to solve our original problem. We have, however, not spent much time optimizing other tasks that, while not performed frequently in our algorithm, are necessary to it to complete the solution. There are always these "housekeeping" jobs that, for example, move data between the system and the outside world, prepare reports, or indicate the current state of the calculation to the user. Such jobs still require programming and, while not requiring the power of the underlying machine, must be written to run on it. It is here that a more conventional (i.e., simple) programming model can be imposed on the first-level machine with an assembly language that is easier to use and remember, albeit limited relative to the completely revealed first-level machine. The housekeeping tasks may then be quickly written and tested.

A last step may be appropriate in certain cases. Since a reasonably arbitrary programming model can be imposed on the first-level machine, it is not out of the question to provide a high-level language for the machine. While not generally applicable for the reasons given in earlier chapters (i.e., the machine is new or one-of-a-kind), the effort may be justified in some cases. All compilers are written with a programming model in mind. In many cases, the programming model is less complex than the machine that the compiler was intended to support. This point allows a given compiler to be moved, "ported", to many different computers. If we consider the overall organization of a compiler, as we did assemblers in Chapter 5, we find that the task of analyzing the source code is the same regardless of target machine. It is only when the procedural elements are translated into the machine code of the target machine in the compiler's last processing stage that any target machine dependence is encountered. In terms of the entire compiler, the code generating section in the last stage is very small. It consists primarily of an algorithm that takes internal representations of the original program's steps (i.e., tokens) and replaces them with small modules of the target machine's code. In most C compilers, the modules are written in the assembler language of the target machine. Compilation must therefore be followed by an assembly step. This allows the user to select a C compiler and, if the source code is available, modify the replacement modules to contain the assembly language for his target system. He may also select a C compiler and build a programming model on his target system that matches that required by the compiler. The assembler is produced as outlined in Chapter 5.

After all is said and done, the designer of a first-level machine can support his efforts with any degree of software tools that he feels is worth the effort. The software/hardware trade-off may be accomplished in many ways, including the creation of a completely "fictional" second-level machine.

9.4. FABRICATION CONSIDERATIONS AND CONCLUSION

The speed benefits described in earlier chapters cannot, unfortunately, be obtained until the digital circuit is fabricated. Some comments concerning fabrication are therefore appropriate at this point. The digital device itself is implemented using the same elements as any other circuit, i.e., transistors, capacitors, resistors, etc. Power supply specifications still exist, as well as loading and driving rules. A digital device must receive input from the physical world just as it must ultimately transmit the results of its computation to that world. At this point, a certain amount of analog circuit design must be performed in order to maintain the digital abstraction because digital integrated circuits will behave just like any other amplifier if various constraints are not met; i.e., outputs will not be predictable by digital logic.

We will make a few points here about fabrication, but will also advise the reader to consult other references for more information. Single copies of specialized networks may be fabricated using the wrapped wire technique. Here, the pins of an array of sockets are interconnected with prestripped AWG #30 wire to implement the circuit. This technique is useful at low clock speeds with the upper limit being in the 10 to 20 MHz region depending on wire layout details. Since the printed circuit board is so expensive, some speed may be gained by increasing the number of parallel (but slower processing elements) with the money saved by the fabrication technique. In an academic environment, this is an especially good method.

Lay out the sockets on a fiberglass-epoxy board in a rectangular array separating them by about 0.2 in., i.e., a compact array. The board should be perforated with a grid of holes having a 0.1-in. spacing. Place the devices such that signal wiring may be kept short. The signals involved change at a high rate of speed, especially the clocks. A power and ground grid should be built using heavier wire. Connect each power pin on a device to its four nearest neighbors. You are trying to approximate a sheet of copper with each grid. Build one grid for each type and polarity of power used. At each IC, place a 0.1-μFd ceramic or monolithic bypass capacitor between the power and ground pins using as short leads as possible. If there are multiple power and ground pins on a device, use a capacitor for each pair. Some research in the data book for the device should show which pins are associated. These bypass capacitors are the source of power when the outputs of the device change state. The power supply is too far away and too slow for the switching transient to be suppressed by it. Remember that the output structure of all currently used large devices short the power to ground during the few nanoseconds used for switching; thus a large amount of current is needed for a very short time, i.e., from the local bypass capacitor. Large electrolytic capacitors are placed where the power enters the board.

During the prototyping stage, it is useful to start testing the board before its construction is finished. Start construction with the clock oscillators and follow the flow of signals layer by layer, testing at each stage. The board can be used to help test itself. Newer parts contain registers at various locations normally occupied by the buffers that drive the pins. The idea here is to implement a test mechanism that allows the internal structure of the board to be interrogated without breaking signal paths. In the case of our microprogrammed system, the controller can be used to run test programs for the architecture as it is constructed and checked out. The main hazard of this technique is that some static sensitive ICs, especially CMOS circuits, are on the board during the construction phase. Observation of static protection methods, including personnel, tool, and circuit grounding, are appropriate.

The distribution of a clock signal about a large board should receive special mention in the implementation of microprogrammed state machines. Many parts (e.g., the pipeline, controller, RALU, and various registers) must received the active edge of the clock at the same instant. While our choice of the enabled, edge-triggered flip-flop reduces the timing problems caused by clock skew, gross arrival time errors or insufficient drive capability will still cause trouble. Such problems are minimized by buffering the clock oscillator. The drive requirements, found by adding the individual current requirements for each signal input, will normally be greater than the current available at the output of a normal buffer. Integrated circuits containing several copies of a buffer should be driven in parallel by the clock oscillator. Since all of the buffers used are implemented in the same IC, the delay across each buffer should be the same. The output of each buffer forms the beginning of one of several buffered clock lines. The clock lines, as well as any other high-speed lines, are frequently supplied with series resistors, in the range of 22 to 47Ω, to reduce ringing. This is especially necessary for lines driving CMOS inputs.

With the increase in access to computer-based design and layout tools, it is now possible to build printed circuit boards for prototype projects as well as production operations. The design tools contain many of the features discussed in Chapter 6 so the designer should have a solid understanding of the performance of his design by the time hardware commitment is made. Printed circuit board (PCB) prototyping is comparatively expensive and has a relatively long lead time, but current trends indicate that both are decreasing. The PCB is the only way to reach the full speed potential of the devices discussed in this book. Where design and layout facilities are available, this is the recommended method. Computer based tools, while not absolutely necessary, will save a great deal of time since a board will contain many narrow traces on several planes or layers. For the scale of the designs in this book, the PCB will contain six to eight layers, with two layers devoted to power planes. Bypass capacitors are placed at each IC as above. The number of signal layers allows the area of the board to be reduced to maintain a compact layout, thus reducing the lengths of signal propagation paths.

In the course of this book, we have attempted to present a methodology, with examples, that will allow the interested individual to attack his computationally intensive problems with more force. The special purpose or application-specific processor as described in Chapter 7 has the capability to deal with a diverse range of problems since it is composed of common, but powerful, building blocks and is programmable. The user selects the data paths and computation elements that fit his problem, not because 50 million copies must be sold. It is hoped that members of the scientific community, especially physics and astronomy, can benefit from the methods introduced so that they can create non-production versions of application specific machines in the manner that they have done so successfully over the years. Students of computer engineering should have experienced the scale and potential complexity of the problems that they will encounter in their professional work. They should also have acquired a familiarity with design techniques and practices that will allow them to tackle large system designs that are well beyond those they have encountered before. The examples of computer systems included were intended to show the application of our subject machines to ISPs. While modern CISCs and RISCs are more complex than the ones discussed, the student should recognize all of the components and be able to associate them with the tasks that were performed more simply here.

EXERCISES

1. If you have performed Exercise 9 in Chapter 8, apply the Am2903 Y Bus to MAR data path modification to your simulator. Examine your first-level source program and modify it to take advantage of the new path where appropriate. Get the first-level sequence that you wrote for the indexed-with-offset address mode in Exercise 6 in Chapter 8 and adapt it to take advantage of the modification. Add this address mode to the appropriate second-level instructions. Provide first- and second-level test programs that demonstrate the correctness of the implementation. You should specifically indicate the number of first-level machine states needed for each second-level instruction with this address mode.

2. Write the necessary DEF statements or macros that support the Byte/Word bit in the pipeline. Write a set of second-level arithmetic/logic/shift instructions that request byte-with-byte operations in the first-level machine. As outlined in this chapter, write first-level sequences that implement the instruction set for the byte data type, taking advantage of the hardware modifications described in Section 9.2.2. Be sure to preserve the correct second-level flags for use with the branch instructions. Is the upper or lower

RALU used to operate on the bytes moved from memory? How has this affected flag preservation?

3. Provide a pair of instructions that LOAD and STORE bytes for the case that sign-extends the byte into a word in the 16-bit RALU in the computer example. The LOAD is responsible for sign extension and should use the "logical" flag setting strategy. STORE does not affect the flags but should route the byte to the proper "half" of memory based on A0. Use the top-down approach at the second and then the first level to produce the proper microprogram source code. Support all the address modes shown in Chapter 8. Note that these sequences were written as part of Exercise 6 in Chapter 8 and Exercise 1 above.

4. Add three interrupt request lines to the computer example. Draw a logic diagram of the relevant section. Design the hardware and write the microprogram sequence that makes one line a Non-Maskable Interrupt (NMI). Provide a 2-bit mask register and masking mechanism that is used with the other two interrupt inputs, IRQ0 and IRQ1. Treat IRQ0 in such a way in the microprogram that it has the higher priority. Handle the mask register in the microcode such that, when an interrupt is detected on one IRQ line, further interrupts on that line will not be detected until after the second-level Return from Interrupt (RTI) instruction is executed. Save the minimum state of the system (i.e., the second-level flags [including the mask register] and the PC) on the stack. Use the vectored interrupt method to determine the location of the second-level ISR. Provide an RTI instruction to return from the second-level ISR. Note that the mask register is part of the system status. It must be moved to and from the stack as well as be manipulated by the first-level program. Assume an interrupt-sensing strategy in which the minimum state of the system is saved.

   a. What hardware refinements are needed if NMI is to be edge-sensitive?

   b. If IRQ0 and IRQ1 are to be level-sensitive, you will need to dictate a strategy for holding an interrupt pending until it is recognized. Which device should do this, the processor or interrupting device? If the interrupting device is responsible for latching the interrupt request, how is it informed to negate the request? Look at the appropriate data books for your favorite microprocessor and describe how this point is handled.

5. A simple DMAC was added to certain early microprogrammed computers by incorporating it into the CPU. Registers were designated in the RALU to be used as source and destination pointers and word counters. A DMAC request line was added to the CPU's command bus for each channel to allow the device to request service. The DMAC was normally implemented as a two-cycle controller within the CPU's microprogram to reduce the amount of extra control signals that needed to be added. In this version, the CPU did not actually release the system buses. The part of the microprogram that implemented the ISP "went to sleep" and the DMAC sequence ran instead.

   Add this type of DMAC to the computer example. Make appropriate logic diagrams showing the changes needed to the hardware. Discuss how and when DMAC request sensing is accomplished by the first-level machine. Show sections of the microprogram that demonstrate how sensing of the DMAC request is accomplished. Provide the microprogram source for the DMAC sequence itself. Use normal top-down design procedures so that the proper PDL, HDL, and logic diagrams accompany the discussion of the solution. Is there any need to save any of the machine state during a DMAC cycle?

6. In the discussion of Bus Masters in Section 9.2.4, the principal examples showed the CPU as the Permanent Bus Master (PBM). In many modern bus-oriented systems, one or more CPUs or single-board computers are treated as Temporary Bus Masters (TBM).

The permanent bus master protocol is implemented in a dedicated state machine controller. Adapt the hardware and microcode of the computer example to add the TMB capability. In overall appearance, this transaction should not be visible to the second-level programmer. He is merely running a program that needs data or programs in a part of his address space that is implemented as a shared resource (e.g., memory) on a bus presided over by a permanent bus master. The only effect visible at the second level is that the transaction with the PBM delays data access a little so the application runs slower. At the first-level, remember, the access is made within the TMB's address space and thus the bus cycle begins like any other. What must be done to accommodate the fact that the termination time of the cycle may be longer than others by an amount dependent on the activity on the bus. In other words, the bus cycle cannot complete until the PBM makes the bus available to the TMB. What signals are needed and how must they behave to implement this behavior? Are these useful in other cases; i.e., when the external architecture contains devices with varying access times? Identify these signals in some existing microprocessors.

7. Add the necessary bus buffers and other hardware and signals to the computer example to make it able to act as a permanent bus master (PBM). Make appropriate logic diagrams showing the changes needed to the hardware. Discuss how and when *bus request sensing* is accomplished by the first-level machine. Show sections of the microprogram that demonstrate how sensing of BR is actually accomplished. Provide the microprogram source for the BR/Bus Grant sequence itself.

8. Implement a pseudo second-level machine on the computer example. Show the hardware changes that may be made since there is now no second-level program memory. Add a field to the microword to contain a constant that may be used to help generate addresses in the second-level data memory. Demonstrate the appearance of the microcode for each member of the basic instruction set by implementing an example sequence from each group. How are the address modes implemented? Pay particular attention to passing the register and offset parameters in the macro. What happens to the PC?

9. Modify the computer example to support a Harvard architecture. Present a logic diagram showing the changes needed in the hardware to support the new architecture. For simplicity, retain the single RALU structure. Discuss how the microcode is modified to take advantage of the new hardware and the limitations imposed by the single internal architecture. Discuss how implementing the PCU as a separate ASM would improve the speed of the Harvard architecture.

10. Obtain the data book for one of the DSPs. Map the Sum-Of-Products (SOP) operation onto the architecture of the selected device. Examine the instruction set and write a short routine that performs the SOP. Take advantage of the instruction set and address modes by using the MAC instruction and any multiple operand moves that are available. Some DSPs can perform several operations at once besides the arithmetic and data movement. A repeat instruction that is performed without timing overhead is frequently provided. Use the support in the data book to get the best possible code and then estimate the computation speed based on a specified clock. Watch for, and discuss in your write-up, how the DSP designer provided data paths and other constructs that made his processor execute this program efficiently and rapidly.

BIBLIOGRAPHY

Advanced Micro Devices Inc., *Bipolar Microprocessor Logic and Interface Data Book*, Sunnyvale, CA: Advanced Micro Devices Inc., 1985.

Advanced Micro Devices Inc., *PAL Device Data Book Bipolar and CMOS*, Sunnyvale, CA: Advanced Micro Devices Inc., 1990.

Advanced Micro Devices Inc., *The Am2900 Family Data Book*, Sunnyvale, CA: Advanced Micro Devices Inc., 1979.

Advanced Micro Devices Inc., *32-Bit Microprogrammable Products Am29C300/29300 Data Book*, Sunnyvale, CA: Advanced Micro Devices Inc., 1988.

Altera Corporation, *User-Configurable Logic Data Book*, (EPS488 STAND-ALONE MICROSEQUENCER), Santa Clara, CA: Altera Corporation, 1988.

Anceau, F., *The Architecture of Microprocessors*, Reading, MA: Addison-Wesley, 1986.

Andrews, M., *Principles of Firmware Engineering in Microprogram Control*, Rockville, MD: Computer Science Press, 1980.

Ansari, A., *A 3x3 Multipurpose Bus-Connected Floating-Point Array Processor,* Masters Thesis, Gainesville, FL: University of Florida, 1990.

Baer, Jean-Loup, *Computer Systems Architecture*, Rockville, MD: Computer Science Press, 1980.

Bell, C. G., and Newell, A., *Computer Structures: Readings and Examples*, New York: McGraw-Hill 1971.

Borland International, *Turbo C Reference Guide*, Version 2, Scotts Valley, CA: Borland International 1988.

Borland International, *Turbo C User's Guide*, Version 2, Scotts Valley, CA: Borland International 1988.

Coelho, David R., *The VHDL Handbook*, Boston, MA: Kluwer Academic Publishers, 1989.

Cox, Brad J., *Object-Oriented Programming An Evolutionary Approach*, Reading, MA: Addison-Wesley Publishing Co., 1987.

Cypress Semiconductor Corporation, *CMOS BiCMOS Data Book*, Cypress Semiconductor Corporation, 1989.

Dasgupta, S., *Computer Architecture A Modern Synthesis*, New York: Wiley, 1989.

Digital Equipment Corp., *VAX MACRO and Instruction Set Reference Manual*, Maynard, MA: Digital Equipment Corporation, 1988.

Dowling, E., Griffin, M., Lynch, M., Smith, K., Taylor, F., "A Multipurpose VLSI Floating-Point Array Processor", 22nd Annual Asilomar Conference on Signals, Systems, and Computers, Pacific Grove, CA: October 1988.

Draft, G. D. and Toy, W. N., *Microprogrammed Control and Reliable Design of Small Computers*, Englewood Cliffs, NJ: Prentice-Hall, 1981.

Drafz, R., Smith, K., Dowling, E., Griffin, M., Lynch, M., and Taylor, F., Turn a PC Into a Supercomputer with Plug-In Boards, *Electronic Design*, 23 Nov. 1988, 89-93.

Eggebrecht, L.C., *Interfacing to the IBM Personal Computer*, 2nd Edition, Carmel, IN: Sams, 1990.

Foster, C. C., *Computer Architecture*, New York: Van Nostrand Reinhold, 1970.

Hill, F. J. and Peterson, G. R., *Digital Systems Hardware Organization and Design*, New York: Wiley, 1987.

Hewlett Packard, *System HILO® GHDL Reference Manual* USA: Hewlett Packard, 1989.

Husson, S. S., *Microprogramming: Principles and Practices*, Englewood Cliffs, NJ: Prentice-Hall, 1970.

IEEE Staff, *1076-1987 IEEE Standard VHDL Language Reference Manual (ANSI)*, Piscataway, NJ: The Institute of Electrical and Electronic Engineers, Inc., 1987.

IEEE Staff, *754-1985 Standard for Binary Floating Point Arithmetic (ANSI)*, Piscataway, NJ: The Institute of Electrical and Electronic Engineers, Inc., 1985.

Intel, *Microsystem Components Handbook*, Vol. 1 (8085), Santa Clara, CA: Intel Corp., 1984.

Kapps, C. and Stafford, R. L., *VAX Assembly Language and Architecture*, Boston, MA: Prindle, Weber & Schmidt, 1985.

Kernighan, Brian W. and Ritchie, Dennis M., *The C Programming Language*, Englewood Cliffs, NJ: Prentice Hall, Inc. 1978.

Langdon, G. Jr., *Computer Design*, San Jose, CA: Computeach Press Inc., 1982.

Lipsett, R., Schaefer, C. F. and Ussery, C. VHDL, *Hardware Description and Design*, Boston: Kluwer Academic Publishers, 1989.

Mano, M. M., *Computer System Architecture*, Englewood Cliffs, NJ: Prentice-Hall, 1982.

Mano, M. M., *Computer Engineering Hardware Design*, Englewood Cliffs, NJ: Prentice-Hall, 1988.

Mick J. and Brick J., *Bit-Slice Microprocessor Design*, New York: McGraw-Hill, 1980.

Microtec, *Meta Assembler Manual AMD2900 Microprocessor*, Sunnyvale, CA: Microtec, 1978.

Miller, Alan R., *Assembler Language for the IBM PC*, Berkeley, CA: Sybex, 1985.

Milutinovic, V. M., Editor *High-Level Language Computer Architecture*, Rockville, MD: Computer Science Press, 1988.

Motorola, *DSP56000/DSP56001 Digital Signal Processor User's Manual*, Rev. 1, Austin, TX: Motorola, Inc., 1989.

Motorola, *DSP96002 IEEE Floating Point Dual-Port Processor User's Manual*, Austin, TX: Motorola, Inc., 1989.

Motorola, *16-Bit Microprocessor User's Manual*, 3nd Edition, Englewood Cliffs, NJ: Prentice-Hall, 1982.

Motorola, *MC68030, Enhanced 32-Bit Microprocessor User's Manual*, 2nd Edition, Englewood Cliffs, NJ: Prentice-Hall, 1989.

Motorola, *MC68881/882 Floating-Point Coprocessor User's Manual*, 2nd Edition, Englewood Cliffs, NJ: Prentice-Hall, 1989.

Murray, W. D., *Computer and Digital System Architecture*, Englewood Cliffs, NJ: Prentice-Hall, 1990.

Pollard, L. H., *Computer Design and Architecture*, Englewood Cliffs, NJ: Prentice-Hall, 1990.

Rabiner, L. R. and Gold, B., *Theory and Application of Digital Signal Processing*, Englewood Cliffs, NJ: Prentice-Hall, 1975.

Rafiquzzaman M. and Chandra, R., *Modern Computer Architecture*, St. Paul, MI: West Publishing, 1988.

Richards, R. K., *Digital Design*, New York: John Wiley & Sons, 1971.

Schildt, Herbert, *Advanced Turbo C*, 2nd Edition, Berkeley, CA: Borland.Osborne/McGraw-Hill 1989.

Shiva, S. G., *Computer Design and Architecture*, Boston, MA: Little, Brown, and Co., 1985.

Siewiorek, K. P., Bell, C. G., and Newell, A., *Computer Structures: Principles and Examples*, New York: McGraw-Hill 1982.

Stone, H. S., Editor *Introduction to Computer Architecture*, Chicago, IL: Science Research Associates, Inc., 1980.

Stone, H. S., *High-Performance Computer Architecture*, Reading, MA: Addison-Wesley, 1990.

Tanenbaum, A. S., *Structured Computer Organization*, Englewood Cliffs, NJ: Prentice-Hall, 1981.

Tomek, I., *Introduction to Computer Organization*, Rockville, MD: Computer Science Press, 1981.

Texas Instruments Inc., *SN74AS888 SN74AS890 Bit-Slice Processor User's Guide*, Dallas, TX: Texas Instruments Inc., 1985.

Texas Instruments Inc., *SN74ACT8800 Family 32-Bit CMOS Processor Building Blocks Data Manual*, Dallas, TX: Texas Instruments Inc., 1988.

Texas Instruments Inc., *TMS34010 User's Guide*, Dallas, TX: Texas Instruments Inc., 1986.

van de Goor, A. J., *Computer Architecture and Design*, Reading, MA: Addison-Wesley, 1989.

White, D., *Bit-Slice Design: Controller and ALUs*, New York : Garland STPM Press, 1981.

Zilog, *1982/1983 Data Book*,(Z80),Campbell, CA: Zilog, Inc., 1982.

INDEX

Printed and bound by CPI Group (UK) Ltd, Croydon, CR0 4YY

17/10/2024

01775703-0001